PROCEEDINGS

of the

8th INTERNATIONAL CONFERENCE ON COORDINATION CHEMISTRY

Vienna, 7. — 11. September 1964

Edited by

V. GUTMANN

Technical University of Vienna

Springer-Verlag Wien GmbH

ISBN 978-3-7091-3652-2   ISBN 978-3-7091-3650-8 (eBook)
DOI 10.1007/978-3-7091-3650-8

Paper 6B1 has been substituted by the following paper:

THE ADDUCT OF NITROSYL FLUORIDE WITH PHENYLTETRAFLUORO-PHOSPHORANE - NEW FLUOROPHOSPHATES.

R. Schmutzler and G.S. Reddy

E.I. du Pont de Nemours and Co., Inc., Wilmington/Delaware, U.S.A.

Nitrosyl fluoride reacted with phenyltetrafluorophosphorane in FREON 113 [*]) (b.p. 46,6°) or, preferably, FREON 12 [*]) medium (b.p. -30°) between -80 to 0° to give a light orange, solid 1 : 1 adduct [I]. [I] is stable below 0°, but undergoes a sudden irreversible decomposition at ca. 10° with formation of undefined, purple products. NOF thus cannot be recovered directly from the adduct. The result of an F^{19} NMR study of [I] was not conclusive, due to its sensitivity both towards moisture and heat. Indirect evidence for the formulation of [I] as a nitrosyl salt, $NO^+C_6H_5PF_5^-$ was obtained from some of its reactions.

The formation of phenylphosphonic acid upon hydrolysis of [I] shows that the phenyl group remained unchanged during the reaction of $C_6H_5PF_4$ with NOF. Reaction of [I] with BF_3 in nitromethane gives $NOBF_4$ and $C_6H_5PF_4$, the latter was found to be unaffected by BF_3:

$$[I] + BF_3 \xrightarrow{< 0^0} NOBF_4 + C_6H_5PF_4$$

The following reactions, typical of nitrosyl compounds of type [I] were also observed:

$$[I] + ROH \xrightarrow{< 0^0} RONO$$

$$[I] + ArNH_2 \xrightarrow{< 0^0} ArF$$

$$[I] + R_2NH \xrightarrow{< 0^0} R_2N \cdot NO$$

The latter reaction is particularly interesting and was studied in most detail with respect to the formation both of the nitrosamine and a novel type of fluoro complex, $[C_6H_5PF_5]^-$, obtained as the dialkylammonium salt. The following stoichiometry was established for the reaction of [I] with secondary amines,

----
[*]) Registered trade mark, E.I. duPont de Nemours and Co., Inc.

$$NOF \cdot C_6H_5PF_4 + 2 R_2NH \longrightarrow R_2N \cdot NO + \overset{(+)}{R_2NH_2} \overset{(-)}{C_6H_5PF_5}$$

A number of the new fluorophosphates could be obtained as stable products, sufficiently volatile to be distillable or sublimable in a high-vacuum.

The following reaction provided a particularly easy route to salts containing the $[R(Ar)PF_5]^-$ anion, which were isolated in high yields, along with alkyl(aryl)-dialkylaminotrifluorophosphoranes,

$$1 \; R(Ar)PF_4 + 2 R_2'NH \xrightarrow{\sim 0^o} R(Ar)PF_3NR_2' + \overset{(+)}{R_2'NH_2} \overset{(-)}{R(Ar)PF_5}$$

A phenylpentafluorophosphate, together with a novel cationic species, also resulted from the isomerization of phenyldimethylaminotrifluoro-phosphorane. It occurs when the fluorophosphorane is allowed to stand at room temperature,

$$2 \; C_6H_5PF_3N(CH_3)_2 \longrightarrow \{C_6H_5PF[N(CH_3)_2]_2\} [C_6H_5PF_5]$$

The P^{31} NMR spectrum of the ion $[RPF_5]^-$ consists of a doublet of partially overlapping quintuplets. The observed pattern is consistent with an octahedral structure, four fluorine atoms occupying equatorial sites, and one being in axial position. Most indicative as to the nature of the fluoro anion is the P^{31} chemical shift, e.g. of $[C_6H_5PF_5]^-$: its very large positive value (+ 136 p.p.m.) suggests a highly shielded phosphorus atom, comparable to PF_6^- (δ_p = + 143, 7 p.p.m.). F^{19} NMR spectra are also consistent with the above formulation of the anions, $[RPF_5]^-$.

Both, P^{31} and F^{19} NMR spectra of alkyl(aryl)-dialkylaminotrifluoro-phosphoranes may be interpreted in terms of a trigonal bipyramidal model with the dialkylamino- and organic hydrocarbon group occupying equatorial sites.

The reaction of nitryl fluoride with phenyltetrafluorophosphorane is different from that of nitrosyl fluoride. Instead of formation of an adduct, nitration of the aromatic nucleus takes place. Hydrolysis of the product gives the known p-nitrophenylphosphonic acid, suggesting that nitration occurred largely in p-position to the $-PF_4$ group.

8 A 2 has been substituted by the following paper:

NQR SPECTRA AND STRUCTURE OF TETRACHLOROAURATES(III).

V. Caglioti, G. Sartori and C. Furlani
Istituto di Chimica Generale e Inorganica,
Università di Roma, Italy.

Nuclear quadrupole resonance spectroscopy is a very promising novel technique for the study of coordination compounds. Its applicability includes observation of quadrupolar nuclei of central metal atoms (e.g. ^{27}Al, ^{59}Co, ^{63}Cu, ^{65}Cu, ^{115}In, ^{193}Ir, ^{197}Au, etc.) as well as of ligand nuclei (^{14}N, ^{35,37}Cl, ^{79}Br, ^{81}Br, ^{127}I, ^{75}As, etc.), but the former possibility has been very little explored until now. Instrumental difficulties, namely poor versatility of NQR spectrometers and practical non availability of commercial apparatus have probably retarded the diffusion of NQR technique in the field of coordination chemistry, but vistas are now quite encouraging: e.g. unpaired electron spins do not prevent detection of NQR lines, and Mössbauer effect measurements can be complementary to NQR spectroscopy for nuclei which are quadrupolar only in excited states.

NQR data yield primarily informations on the p electron densities in the valence shells, from which the net resulting charge on the investigated atom can be easily inferred by means of some simple models, e.g. of the TOWNES-DAILEY theory. Usually the accuracy of experimental measurements is rather high, and is at present not matched by equally powerful and accurate inter-pretative theories. Therefore at this initial state of application in the field of coordination chemistry we are devoting more attention to the experimental results of NQR spectroscopy, their immediate interpretation and their close connection with X-ray structural data.

NQR of ligand halogen nuclei in halo-complexes has been reported first by ITO, NAKAMURA and KUBO for several hexacoordinated complexes of transition elements (1), which possess almost invariably regular octahedral structure and exhibit therefore often only one halogen resonance line. Square planar halo-complexes have been much less thoroughly investigated, except Me_2PdX_4 (2) and Me_2PtX_4 (2,3), which have again only one resonance fre-quency, and have therefore presumably regular square planar structure.

We have investigated salts of the $[AuCl_4]^-$ anion which is roughly but not exactly square planar [also $[AuBr_4]^-$ is distorted square (4)] giving therefore a more complex NQR spectrum (see also reference (5)).

Especially in the sodium salt very close agreement is obtained between NQR (two resonance lines) and X-ray data (two non-equivalent types of chlorines in the [AuCl$_4$] unit).

The chlorine coupling constant is rather high, and a simple interpretation following the TOWNES-DAILEY scheme indicates a charge of about -0,4 per chlorine. Although we cannot distinguish between effects of s- or d-hybridization, of ionic bonding and π-bonding, the observed high values are strongly suggestive of a high degree of covalency of the coordinative bonds which is in line with the very high optical electro-negativity reported for Au (III).

NaAuCl$_4$.2 H$_2$O, rhombic prisms, space group Pmna, contains planar [AuCl$_4$] units, with three long (2,27 Å) and one short (2,22 Å) Au-Cl bonds. Long-bonded chlorines are relatively close to Na$^+$ ions (3,0 - 3,2 Å), whereas the short-bonded chlorine is not less than 4,4 Å from the nearest sodium. We have therefore two types of non-equivalent chlorines in each [AuCl$_4$]$^-$ unit (6).

Ammonium chloroaurate is not isomorphous with the sodium salt; preliminary X-ray data seem to indicate monoclinic structure with presumably lower microsymmetry of the [AuCl$_4$]$^-$ unit.

NaAuCl$_4$.2 H$_2$O as microcrystalline powder has two ^{35}Cl resonance lines (27,36 and 28,33 Mc/sec at 25°). Both frequencies have a slightly negative temperature coefficient; e.g. values of 27,81 and 28,76 Mc/sec are observed at -140°.

Below -140°, a sudden reversible shift to 28,8 and 29,3 Mc/sec is observed which suggests that a solid state transition is taking place. The average of the frequencies observed, at room temperature is consistent, according to the TOWNES-DAILEY model and assuming 15% s-hybridization, with a charge distribution of +0,61 e on Au and -0,40 e on Cl.

Potassium chloroaurate has again two frequencies (26,75 and 27,36 Mc/sec at 18°), whereas ammonium chloroaurate has four frequencies in the ranges 26,62 - 26,98 - 27,43 - 27,62 Mc/sec, which indicate presence of four sets of chemically non-equivalent chlorine atoms in the crystal.

References

1. NAKAMURA, D., KUBO, M., ITO, K., and cow., J.Am.Chem.Soc., **82**, 5783 (1960) and **84**, 163 (1962); Bull.Chem.Soc.Japan, **35**, 518 (1962) and **36**, 1056 (1963); Inorg.Chem., **2**, 61 and 690 (1963).
2. ITO, K., NAKAMURA, D., KUNITO, Y., ITO, K., and KUBO, M.,

J.Am.Chem.Soc., <u>83</u>, 4526 (1961).

3. MARRAM, E.P., McNIFF, E.J., and RAGLE, J.L., J.Phys.Chem., <u>67</u>, 1719 (1963).

4. COX, E.G., and WEBSTER, K.C., J.Chem.Soc. (London), <u>1936</u>, 1635.

5. LEGEL, S.L., and BARNES, R.G., Catalog of nuclear quadrupole interactions and resonance frequencies in solids - U.S.A.E.C., I, 520 (1962).

6. BONAMICO, M., DESSY, G., and VACIAGO, A., Unpublished results.

8th INTERNATIONAL CONFERENCE ON COORDINATION CHEMISTRY

Organised by

VEREIN ÖSTERREICHISCHER CHEMIKER

Sponsors:

INTERNATIONAL UNION OF PURE AND APPLIED CHEMISTRY

DER BUNDESMINISTER FÜR UNTERRICHT
DER BÜRGERMEISTER DER STADT WIEN

President of the Conference:

H. NOWOTNY, Vienna

Executive Committee:

V. GUTMANN, Vienna (Chairman)
A. MASCHKA, Vienna (Secretary)
E. HAYEK, Innsbruck
H. MALISSA, Vienna
O. POLANSKY, Vienna
U. WANNAGAT, Graz

This Conference has been supported partially by the

United States Government

through its European Research Office, Rheingau Allee 2, Frankfurt (M), Germany, and through the

Bundesministerium für Unterricht

Wien I, Minoritenplatz 5

PLENARY LECTURES

(to be published in the Journal
of Pure and Applied Chemistry)

H. HARTMANN, Frankfurt/Main (Germany): "Neue Ansätze in der Theorie
der Komplexionen und ihre physikalische Begründung".

J. LEWIS, Manchester (UK): "Metal-Metal-Interactions in Coordination
Compounds".

E. L. MUETTERTIES, Wilmington/Delaware (USA): "Tetrahedral and Octahedral Cationic Chelates of the Main Group Elements".

F. G. A. STONE, Bristol (UK): "The Role of Organometallic Compounds in
the Development of Coordination Chemistry".

J. A. SYRKIN, Moscow (USSR): "The Nature of Bonding of Water Molecules
in Crystallohydrates and Clathrates".

A. A. VLČEK, Prague (CSSR): "Intermediates of Electroreduction of Transition
Metal Complexes".

PREFACE

International Conferences on Coordination Chemistry have been so successful, that the Organisers of 8 I.C.C.C. have experienced great difficulty in the attempt to maintain the previous high standard. The main problem was the increased interest in presenting papers. While, apart from plenary lectures, 68 papers (all by invitation) were read at 6 I.C.C.C. in Detroit in 1961, 285 papers were presented at 7 I.C.C.C. in Stockholm in 1962. An even higher number was to be expected for 8 I.C.C.C. unless restrictions were made. It was finally decided to follow the pattern of 6 I.C.C.C. on a somewhat larger scale and to admit approximately 150 papers for oral presentation, allowing each speaker 25 minutes, including time for discussion. Some of the more distinguished Coordination Chemists have been asked to act as Session chairman and not to present papers. In addition selection of papers cannot be completely free from arbitrariness, so that we may have missed some excellent papers. On the other hand we hope to have gained proper presentations of the papers admitted by providing sufficient time for each speaker.

The following main subjects were selected:

(A) The Nature of the Chemical Bond in Coordination Compounds,

(B) Novel Coordination Compounds,

(C) Methods and Results in the Coordination Chemistry in Solution.

It is true that this classification is not completely clear-cut, but it served as a frame work for the selection of papers. The first number given to a paper indicates the Session number and the following capital letter refers to the Section (main field, see above), while the last number indicates the order of presentation within the Session.

This book could not have been produced had I not enjoyed the perfect understanding and co-operation by all contributors. The authors submitted their manuscripts in April 1964, allowing sufficient time to have the book published in July, well in advance of the Conference. Thus it will be possible to provide each participant with a copy several weeks before the Conference and I have every hope that participants will read the abstracts before the Conference, so that discussions may be considerably "catalysed".

I have to apologize for misprints or errors which may still be found in this volume, since authors had no chance to read the proofs. In this way certain misprints in the manuscripts could not be eliminated. Such short-comings may be compensated, however, by the early publication of the Proceedings of 8 I.C.C.C, wich contain long abstracts of 164 papers.

Many thanks are due to the President of the Conference, Professor H.NOWOTNY and to all members of the Executive Committee for their valuable advice and their continuous help. In particular I have to express my gratitude to Professor A.MASCHKA and Dr.O.POLANSKY, who have done an excellent job in assisting me both in editorial work and in proof-reading.

Vienna, 20th May 1964   VIKTOR GUTMANN

SECTION A

NATURE OF THE CHEMICAL BOND IN COORDINATION COMPOUNDS

SECTION B

NOVEL COORDINATION COMPOUNDS

SECTION C

METHODS AND RESULTS IN THE COORDINATION CHEMISTRY
IN SOLUTION

APPLICATION OF THE MÖSSBAUER-EFFECT IN COORDINATION CHEMISTRY.

Ekkehard Fluck, Werner Kerler and Wolfgang Neuwirth
Anorganisch-Chemisches Institut and II. Physikalisches Institut,
Universität Heidelberg, Germany.

While nuclear magnetic resonance spectroscopy is quite common today in most chemical laboratories, Mössbauer spectroscopy is not developed beyond the early stages as far as its application in chemistry is concerned; for reviews see e.g. (1-3). Both methods are based on physical phenomena of the atomic nuclei and can serve to elucidate chemical bond characteristics of the atoms. Especially questions concerning oxidation number, coordination number and the symmetry of the environment of the atom as well as chemical bond types may be answered.

The effect discovered by MÖSSBAUER (4-6) concerns the resonance fluorescence of so-called recoil-free γ-radiation of atomic nuclei. The same principles which are valid for fluorescence phenomena of the atomic shell apply also to nuclear resonance fluorescence; a nucleus is raised to its first excitation level by absorption of a γ-quantum of appropriate energy and then returns to its ground state emitting a γ-quantum. The γ-quanta required for the excitation are emitted by excited nuclei being produced by decay of a suitable isotope, e.g. by decay of ^{57}Co for the investigation of iron and iron compounds, as shown in Fig. 1. Because of their considerably higher energy compared to light quanta, γ-quanta generally impart a considerable recoil to the nucleus during emission, at the same time losing part of their excitation energy. The emitted γ-quantum therefore has insufficient energy to excite a nucleus again. In addition to this an analogous loss by recoil occurs during absorption. In the arrangement according to MÖSSBAUER the loss resulting from recoil is avoided. If the radiating atom is solidly built into a crystal lattice for some isotopes a sufficiently large probability exists, that the crystal as a whole will absorb the recoil momentum and no energy transfer to the crystal occurs. Therefore the condition for resonance is fulfilled or quanta emitted and absorbed by such atoms. If, for some reason, the first energy level of the atomic nuclei of the absorber is shifted upwards or downwards the radiation source and absorber have to be moved relative to each other in order to obtain resonance (Doppler shift).

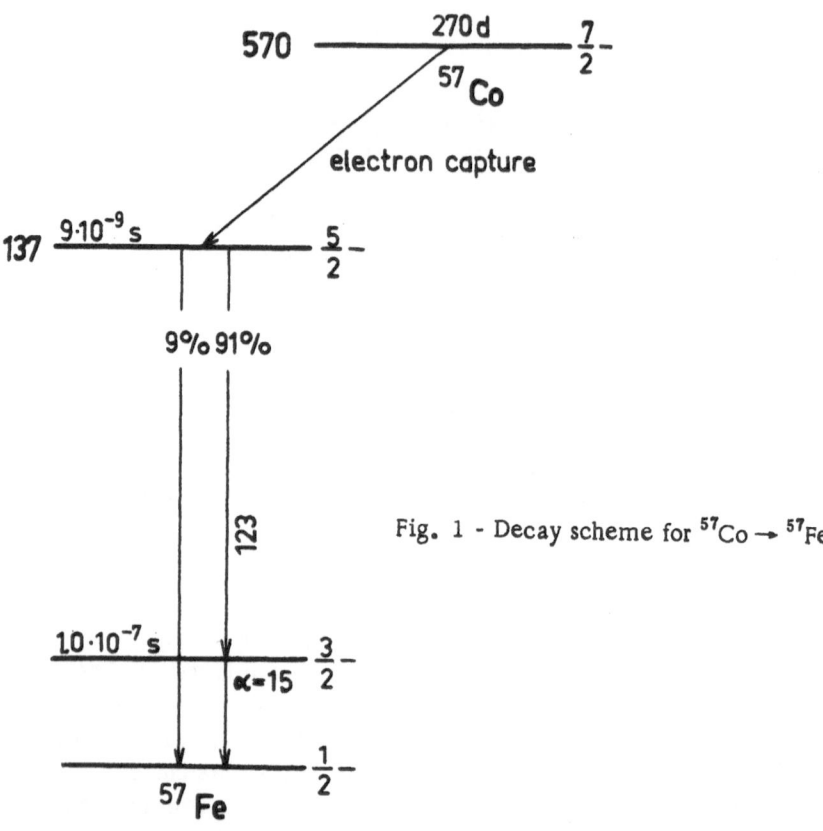

Fig. 1 - Decay scheme for ^{57}Co → ^{57}Fe

If the γ-radiation emitted by the radiation source strikes an absorber (the substance to be investigated), the atoms of which are present in the same chemical compound and in the same crystal lattice as are the atoms of the source, resonance occurs when source and absorber have the relative velocity zero. This is no longer the case when the nuclei in the absorber belong to atoms which exist in a state of chemical combination different from that of the atoms of the source. In order to fulfil the condition for resonance it is then necessary, as a rule, to move the absorber relative to the radiation source with a certain velocity and thus to change the energy of the quanta received in such a way as to correspond to the excitation energy of the nuclei in the absorber substance. This velocity relative to the zero velocity of an arbitrary radiation source is called the line shift δ. It is made up of the isomeric shift and the temperature shift.

The isomeric shift is a linear function of the s-electron density at the nucleus (p-, d- and f-electrons influence the electron density at the nucleus

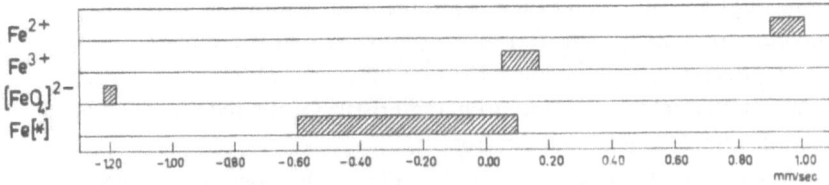

Fig. 2 - Values for the isomeric shifts in iron, iron alloys
and iron compounds
Radiation source: [57]Co in Pt at 25°
Abscissa: Velocity of the source relative to the
absorber [mm/sec]
Fe[*] = iron in complexes and metals

by screening the s-electrons). The isomeric shift decreases linearly with
increasing s-electron density, i.e. an increasing s-electron density causes a
shift of the resonance line toward negative velocity values. According to this
the isomeric shifts of compounds with different oxidation states of the element
in question fall into regions characteristic for these states, as is shown for iron
in Fig. 2. The contribution of the temperature shift to the total line shift is
generally small in relation to the isomeric shift. The temperature shift reflects
the properties of the vibrational spectrum of the crystals.

Often the velocity spectrum of a sample consists of two lines even if all
atoms of the absorber are in the same state of chemical bonding and in
corresponding lattice positions. The quadrupole splitting of the resonance line
in the case of iron is caused by the interaction of the electric field gradient
around the nucleus with the electric quadrupole moment of the excited [57]Fe-
nucleus (I=3/2). The field gradient around the nucleus depends on the electronic configuration of the nucleus and on its environment.

Octahedral coordination compounds of iron in which the iron has the
oxidation number +2 have the 3d-orbitals completely filled. If all ligands
are the same, as is for instance the case in $[Fe(CN)_6]^{4-}$, no splitting of the
resonance line is observed because of the spherical symmetrical charge
distribution around the nucleus. With different ligands, however, as in
$[Fe(CN)_5NH_3]^{3-}$, a pronounced splitting occurs which is not dependent on
temperature. Generally it is true that no splitting occurs when at least two
three- or four-fold symmetry axes are existent (7, 8).

Coordination compounds containing iron of the oxidation number +3
which lack one electron for completing the 3d-orbitals show splitting, even
when all ligands are of the same type as in $[Fe(CN)_6]^{3-}$. This splitting is
strongly dependent on temperature since the electron terms are occupied in

different ways as a function of temperature. The quadrupole splittings in iron coordination compounds fall into a region of 0 to 2,6 mm/sec.

With prussiates, quadrupole splitting is observed which is always independent of temperature. Its size is characteristic for the nature of the bond between the single ligand and the iron.

In hexacyanoferrate(II), $[Fe(CN)_6]^{4-}$, the cyano groups are bonded to the central iron atom via σ-bonds. In addition, each bond has some π-bonding, thus avoiding a high negative charge on the iron ("backdonation"). This is true for most of the stable iron coordination compounds. The charge is transferred from the central atom to the ligands via the π-bonds. This corresponds to Paulings charge neutrality principle, according to which the central atoms should never have an electric charge greater than +1 or -1. The similar bond types in all stable coordination compounds of iron cause their isomeric shifts to fall into a limited region. This region also includes metallic iron, thus substantiating the reality of Paulings principle.

If in $[Fe(CN)_6]^{4-}$ a cyanide-ligand is substituted by another ligand forming a stronger π-bond with iron, as is the case with NO^+, the electron density of the 3d-orbitals, from which the electrons used for the π-bonding are taken, is diminshed. Therefore the nitroprussiate ion, $[Fe(CN)_5NO]^{2-}$ shows a negative shift relative to $[Fe(CN)_6]^{4-}$. If the cyano group, on the other hand, is substituted by a ligand such as NH_3 or NO_2^- which is not able to form a π-bond the reversed effect occurs. A shift to higher δ-values relative to $[Fe(CN)_6]^{4-}$ is then observed. A weak π-bond is to be expected in the case of the sulfito-compound $[Fe(CN)_5SO_3]^{5-}$, since only the diffuse d-orbitals are available for the formation of a π-bond. Thus with decreasing strength of the π-bond of the single ligands

$$NO^+ > CN^- > SO_3^{--} > NO_2^- = NH_3$$

the observed isomeric shifts of the prussiates move towards more positive values of δ. Nitrito- and ammin-prussiates show the same isomeric shift.

References

1. FLUCK, E., KERLER, W., and NEUWIRTH, W., Angew.Chem. 75, 461 (1963); Internat.Edition in English Vol.2, 277 (1963).
2. GOLDANSKY, V.I., Atomic Energy Review, Vol.1, No.4, Vienna 1963.
3. BRADY, P.R., WIGLEY, P.R.F., and DUNCAN, J.F., Rev.pure appl. Chem. 12, 165 (1962).
4. MÖSSBAUER, R.L., Z.Physik 151, 124 (1958).
5. MÖSSBAUER, R.L., Naturwissenschaften 45, 538 (1958).
6. MÖSSBAUER, R.L., Z.Naturforschg. 14a, 211 (1959).

7. KERLER, W., NEUWIRTH, W., and FLUCK, E., Z.Physik. **175**, 200 (1963).
8. KERLER, W., NEUWIRTH, W., FLUCK, E., KUHN, P., and ZIMMER-MANN, B., Z.Physik **173**, 321 (1963).

1 A 2

INVESTIGATIONS ON THE ELECTRONIC STRUCTURE OF TRANSITION METAL COMPLEXES USING THE MÖSSBAUER EFFECT.

J. Danon

Centro Brasileiro de Pesquisas Físicas, Rio de Janeiro, Brasil.

Several features of the Mössbauer effect (resonante absorption of recoil-free gamma radiation) which arise from electron-nucleus interactions are directly connected with the electronic structure of the molecule where the resonant nucleus is located.

One is the isomer shift (I.S.), which is due to the interaction of the nuclear charge distribution with s-electrons. The I.S. gives a relative measure of the total s-electron density at the nucleus. For a free iron ion (d^n) the total s-electron density is due to the filled 1s + 2s + 3s shells. On going from d^n to d^{n-1} configurations the total s-electron density increases due to the decrease in shielding of the 3s electrons by the 3d shell (1). In a bonded iron ion the total s-electron density also depends on the partial occupation of the 4s shell by electrons from the ligands and on the alterations of the shielding of s-electrons due to the σ and π-bonding involving d electrons (2).

The close relation between the total s-electron density and the degree of covalent bonding is well characterized through the parallelism between I.S. and nephelauxetic effect (N.E.) (3). The partial occupation of the 4s orbitals and the bonding involving d electrons are included respectively in the central field covalency and in the restricted symmetry covalency, which determine the N.E. (4).

In order to translate the relative values of I.S. in terms of total s-electron density a calibration has been made by assuming a negligible occupation of the 4s orbitals in iron complexes with F, H_2O (1). However, even for the more ionic Fe(III) complexes the N.E. is large, indicating a relatively strong perturbation of the central ion by the ligands.

Similar observations were made on the basis of measurements of effective charges η of iron in complexes by X-ray spectra (5). In Table I we compare the values of η with those calculated from the principle of electronegativity equalization according to the method of FERREIRA (6).

TABLE I

Complex	Charge on Fe	η
[Fe(II)F$_6$]	1,4	1,9
[Fe(III)F$_6$]	0,9	1,2

In agreement with the determinations of η, the theoretical calculations give an effective charge for Fe(III) smaller than that of Fe(II), which is quite satisfactory considering the present limitations of the method (7).

These results show that it is necessary to recalibrate the scale of I.S. in terms of total s-electron density proposed in ref. (1). Preliminary results were obtained by calculating the 4s orbital occupation using the principle of electronegativity equalization and from the relation between N.E. and I.S. for some Fe(III) complexes. A scale was obtained with the values +0,10 cm/sec for the relative I.S. of Fe^{3+} ion (instead of +0,055 cm/sec in the previous calibration) and +0,14 cm/sec for the Fe^{2+} ion. Table II compares the values of 4s occupation given by the two calibrations with that calculated by molecular orbitals with the Wolfsberg-Helmholtz approximation for the tetrahedral [FeCl$_4$]$^-$ complex (8).

TABLE II

Complex	Isomer Shift	4s (1)	4s (this work)	4s (8)
[FeCl$_4$]$^-$	+0,040	0,05	0,33	0,36

These considerations cannot be applied to spin-paired complexes of Fe(II) and Fe(III) with vacant π-orbital ligands. For such cases the metal-to-ligand π-bonding delocalizes the d-electrons with the consequent decreases of the shielding of s-electrons and strong raise of the s-density at the nucleus. This becomes apparent when the comparison is made between the s-electron densities of complexes with iso-electronic ligands:

$$[Fe(II)(CN)_6] \quad < \quad [Fe(II)(CN)_5CO] \quad < \quad [Fe(II)(CN)_5NO]$$

This particular order shows that the s-electron density increases with increasing ligand electronegativity or the increasing tendency of the ligands to withdraw electrons from the neighborhood of the iron nucleus (3).

The contribution of π-bonding to s-electron density appears so strong that it breaks the parallelism between I.S. and N.E.: the I.S. with CN complexes indicates a higher s-electron density than with Br or S, whereas the N.E. is higher with complexes of the latter ligands.

The high s-electron density observed with $[Fe(CN)_5 NO]^{2-}$ (I.S. = -0,012 cm/sec) indicates a strong π-bond in the molecule (2). The unusual order of these bonds in this complex was confirmed by infrared spectral measurements (9).

The other feature of the Mössbauer effect which can be related to the molecular structure is the nuclear quadrupole interaction, which is due to the coupling of the quadrupole moment Q of the nucleus with an electric field gradient q. The latter arises from the asymmetry of external charges and its magnitude and its asymmetry parameter can be related with the electronic configuration of the central ion.

In a diamagnetic complex as the $[Fe(CN)_5 NO]^{2-}$ the electric field gradient at the iron nucleus is originated from the asymmetric expansion of the filled $3d_\epsilon$ shell due to π-bonding toward the ligands. On the basis of the angular variation of the Mössbauer spectra with single crystals of the complex we found that the electric field gradient is positive and axially symmetric. The only possible expansion of the filled $3d_\epsilon$ shell which accounts for these results is that in which the pairs of electrons in d_{xz} and d_{yz} orbitals are delocalized in strong π-orbitals. This result is in agreement with the molecular orbital bonding scheme proposed for transition metal-nitrosyl complexes (10).

References

1. WALKER, L.R., WERTHEIM, G.K., and JACCARINO, V., Phys. Rev. Letters, 6, 98 (1961).
2. DANON, J., J.Chem.Phys., 38, 266 (1963).
3. DANON, J., III. International Conference on Mössbauer Effect, Cornell University, Ithaca, Cornell (1963).
4. JØRGENSEN, C.K., The Nephelauxetic Series, in Progress in Inorganic Chemistry, 4, 73, Interscience Publishers, N.Y. (1962).
5. GOLDANSKY, V.I., Atom.Energy Rev., 1, 3 (1963).
6. FERREIRA, R., Trans.Far.Soc., 59, 1064 (1963).
7. FERREIRA, R., private communication.
8. ZASLOW, B., and RUNDLE, R.E., J.Phys.Chem., 61, 490 (1957).
9. TÓSI, L., and DANON, J., Inorg.Chem., 3, 150 (1964).

10. GRAY, H.B., BERNAL, I., and BILLIG, E., J.Am.Chem.Soc., **84**, 3404 (1962).

A 3

MÖSSBAUER SPECTRA OF IRON COMPLEXES WITH THIOSEMICARBAZONE OF DIACETYLOXIME AND THEIR INTERPRETATION.

A.V. Ablov, I.B. Bersuker and V.I. Gol'dansky
Academy of Science, Moscow, UdSSR.

The use of Mössbauer effect permits to investigate the displacements of the electronic cloud arising due to the chemical changes in the molecule. Mössbauer spectra were obtained for iron compounds given below. The compound {I} represents an addition product of iron(II) chloride with thio-semicarbazone of diacetyloxime ($DTOH_2$). On loss of protons from oximide groups inner complex salt $Fe(DTOH)_2$ {II} is forming. The compound $Fe(DTOMe)_2$ {III} represents also an inner complex salt, but in this case the protons from thiosemicarbazone grouping are lossed. Although in the mentioned compounds the nearest environment of the central atom remains unchanged, however their Mössbauer spectra are distinguishable (Table I). It means that the re-distribution of the electronic density in the molecules connected with the loss of far from iron atom located protons reflects on the electric field in close vicinity of the central atoms nuclei. All the three compounds are dia-magnetic, it means that they are low-spin complexes. In this case more appropriate description is possible only on the base of MO LCAO method. The expressions for wave functions of the bonding orbitals are:

$$\Psi_1 (a_g) = \alpha_1 s + \beta_1 (\sigma_1 + \sigma_4) + \gamma_1 (\sigma_2 + \sigma_5) + \delta_1 (\sigma_3 + \sigma_6)$$

$$\Psi_2 (b_{1u}) = \alpha_2 p_z + \beta_2 (\sigma_1 - \sigma_4)$$

$$\Psi_3 (b_{2u}) = \alpha_3 p_x + \beta_3 (\sigma_2 - \sigma_5)$$

$$\Psi_4 (b_{3u}) = \alpha_4 p_y + \beta_4 (\sigma_3 - \sigma_6)$$

$$\Psi_5 (b_{3g}) = \alpha_5 d_{z^2} + \beta_5 (\sigma_1 + \sigma_4) - \gamma_5 (\sigma_2 + \sigma_5) - \delta_5 (\sigma_3 - \sigma_6)$$

$$\Psi_6 (b_{2g}) = \alpha_6 d_{x^2-y^2} + \beta_6 (\sigma_2 + \sigma_5) - \gamma_6 (\sigma_3 + \sigma_6)$$

Where s, p_x, p_y, p_z, d_{z^2} and $d_{x^2-y^2}$ are wave functions of the iron atom and σ_1, σ_2, ...σ_6 represent the ligand σ-functions. Utilizing these wave functions one may obtain the gradient magnitude of the electric field in the vicinity of iron nucleus (q). Hence, firstly one may ascertain the gradient a sign and, secondly, to evaluate the inductive effect (assuming its direction to be known). Thereby we make some usual assumptions and, in particular, we suppose that one may neglect with ligands σ -function value near the iron nucleus. It has been established that $q_I < 0$, $q_{II} < 0$ and $q_{III} > 0$, e.g. on passing from compound I to II and from II to III the electric field gradient changes the sign. This conclusion doesn't result direct from the form of the spectra and is obtained as a direct consequence of the quite substantiated assumption about the direction of the inductive effects. There were valued also the changes of the α_i^2-constants on passing from one to another compound:

$$\alpha_{3I}^2 - \alpha_{2I}^2 = 0,092; \quad \alpha_{3II}^2 - \alpha_{3I}^2 = 0,39 \text{ and } \alpha_{2III}^2 - \alpha_{2I}^2 = 0,216$$

$$FeCl_2 \cdot 2DTOH_2 \qquad \{I\}$$

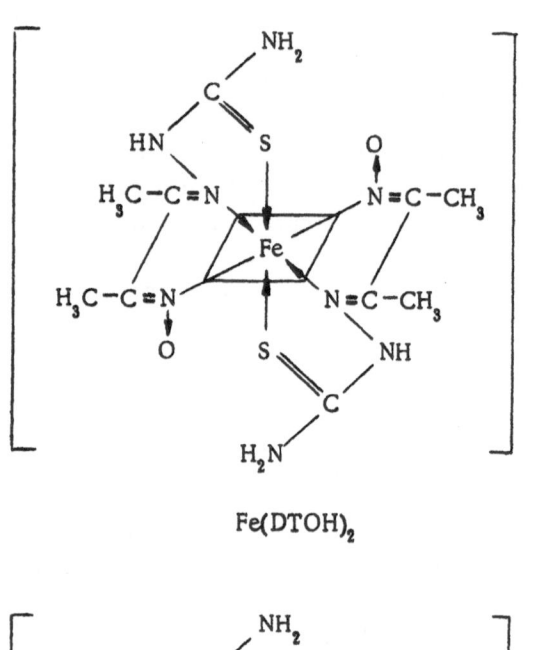

Fe(DTOH)$_2$ {II}

Fe(DTOMe)$_2$ {III}

Fig. 1

TABLE I

ISOMER SHIFT δ AND QUADRUPOLE SPLITTING Δ
FOR IRON(II) COMPLEXES WITH DTOH$_2$ AND DTOHMe [*)]

No	Compound	δ mm/sec		Δ mm/sec	
		78°K	300°K	78°K	300°K
I	[Fe(DTOH$_2$)$_2$]Cl$_2$	0,47	0,42	0,65	0,66
II	[Fe(DTOH)$_2$]	0,29	0,22	2,02	2,02
III	[Fe(DTOMe)$_2$]	0,50	0,45	0,88	0,88

[*)] 1 mm/sec = $4,8.10^{-8}$ ev; accuracy in determination
of δ and Δ ± 0,01 mm/sec

1 A 4

THE MÖSSBAUER EFFECT AND CHEMICAL THEORY.

J.F. Duncan and R.M. Golding
Victoria University of Wellington and Department of Scientific
and Industrial Research, Wellington, New Zealand.

The Mössbauer effect is rapidly assuming an important place as a new technique for investigating chemical problems. It may be used to obtain evidence about the crystallographic and defect structures of solids and their surface properties. But for chemists perhaps more important are the inferences which may be made about the electronic configurations, and the internal magnetic and electric fields in atoms.

In this abstract we discuss only those aspects of our work which fall under the last head. The relation between the Mössbauer parameters ΔE_Q (quadrupole split) and δ (the chemical shift) on the one hand and current chemical theory (especially ligand field theory) is now so well understood that it is possible to predict the way they will change as chemical features are altered, and how they are related to the results of other physical measurements. We here present a brief description of the theoretical features and the way they explain certain experimental results reported previously.

The internal magnetic field of an atom arises through the Fermi contact term - the interaction of the s-electrons with the nucleus. In many iron compounds, the internal field may be about 10^6 gauss, but this does not normally influence the Mössbauer spectrum. Thus, if the lattice electron relaxation time is short compared with the Larmor frequency of the nucleus in the magnetic field, the averaged field to which the nucleus reacts is zero, so that only electric-field interactions (ΔE_Q) are observed.

The values of ΔE_Q expected for octahedral iron complexes may be estimated as follows. The significant ground terms for a d^5 configuration are $^6A_1$ and $^2T_2$ for high and low spin respectively. For d^6, the corresponding terms are $^5T_2$ and $^1A_1$. From a spherical ground state (A_1) a zero electric field gradient and ΔE_Q will be obtained, but finite values are given by the T_2 terms. In addition ΔE_Q will be greater when there are a larger number of unpaired electrons. This is in fact observed, as may be represented as follows,

$$
\begin{array}{lll}
d^5 & Fe^{3+} & Fe^{III} \\
\hline
d^6 & Fe^{II} & Fe^{2+}
\end{array} \longrightarrow \Delta E_Q
$$

where the arabic numerals (2+, 3+) refer to the high spin case, and the roman numerals (II, III) to the low spin case.

We may also estimate the difference between the energy separations of the excited and the ground states of the emitter and absorber (δ) from

$$\delta = A\{|\psi_a(0)|^2 - |\psi_s(0)|^2\}$$

where A depends solely on the physical properties of the nucleus, and $\psi_a(0)$ and $\psi_s(0)$ are respectively the total s electron density at the nucleus of the absorber and source. $\psi_a(0)$ arises from spin polarisation of the core electrons by the unpaired d-electrons, the magnitude of which is proportional to

$$\sum_{\text{s shells}} \{|\psi_\uparrow(0)|^2 - |\psi_\downarrow(0)|^2\}$$

where $\downarrow$ and $\uparrow$ represent the two spin states of the s electrons. This is greatest when the number of unpaired electrons is largest. Thus δ decreases in the order $^6A_1 > {}^1A_1$ and $^5T_2 > {}^2T_2$. The s-electron density also markedly depends on the environment of the Mössbauer nucleus and we expect the effect of the d-induced polarisation of the s-shell will be least for a symmetrical ground state. Experimentally this is observed, as may be represented as follows,

$$
\begin{array}{lll}
d^5 & Fe^{III} & Fe^{3+} \\
\hline
d^6 & Fe^{II} & Fe^{2+}
\end{array} \longrightarrow \delta
$$

   Whilst these conclusions are only strictly valid for a completely
symmetrical octahedral arrangement of orbitals and ligands they afford a
satisfying justification of the main features of the Mössbauer results for iron
compounds. These may be discussed under the following heads 1 and 2.
   1. Correlation diagram.
   When ΔE_Q is plotted as a function of δ, characteristic areas are observed
to which compounds of the four types approximate even when the symmetry
is seriously distorted (1). In Fig. 1 the heavily shaded areas show those deduced
for symmetrical octahedral systems from the above considerations. The lightly
shaded areas give those obtained from the results of about twenty different
compounds of all types and indicate the positions where compounds of different
type are usually found. A similar diagram has been obtained by us for Sn^{119}.

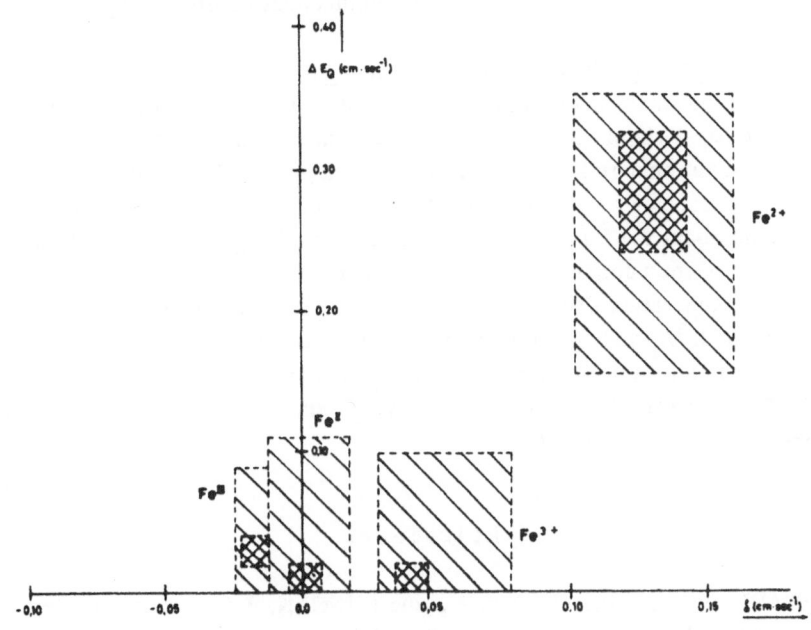

Figure

Correlation diagram for Fe^{57}, showing areas where compounds of different
type usually appear. The heavily shaded areas show those obtained for
compounds with a symmetrical (octahedral) arrangement of ligands.

2. Absorption spectra.

For a series of low spin compounds of given type (e.g. $[Fe(II)(CN)_5 X]^{3-}$) ΔE_Q and δ will be directly related to the ligand field strength of X, which is conveniently measured by observing the absorption peaks in the visible region. For $X = CN^-$, NO_2, NH_3, H_2O, and NO^+ a very good straight line has been obtained between both ΔE_Q and δ for a d-d transition.

3. Magnetic susceptibilities.

A linear relation over a wide range of values has been obtained by us for Fe^{3+} compounds, when ΔE_Q is plotted against the magnetic susceptibility. This suggests the presence of low lying electronic excited states.

4. Nuclear Magnetic Resonance.

A reasonable linear relation has been observed between δ and the NMR proton resonance in a series of compounds of the carbonyl type (3). Although this cannot be directly predicted from the above considerations we intuitively expect such a relation to be obtained for compounds of similar type and stereochemistry.

5. Temperature dependence of ΔE_Q and δ.

Since the wave function at the nucleus does not vary significantly with temperature, δ should be temperature independent. Experimentally a small, but significant change in δ with temperature is found. The values obtained for many compounds are almost identical ($2 - 6 . 10^{-5}$ cm sec^{-1} degree^{-1} for iron), and about the same as that expected from changes in the oscillation amplitude of the nucleus with respect to the electron cloud.

The temperature dependence of ΔE_Q depends on the type of compound present. With $^6A_1$ and $^1A_1$ ground states, there are no excited levels to provide a means by which temperature dependence could be observed. But with $^5T_2$ a ^5E excited state may be low lying, and consequently a temperature dependent ΔE_Q could result.

References

1. BRADY, P.R., WIGLEY, P.R.F., and DUNCAN, J.F., Australian Rev. of Pure and Appl. Chem., 10, 165 (1962).
2. BRADY, P.R., DUNCAN, J.F., and MOK, K.F., forthcoming publication.
3. HERBER, R.H., KING, R.B., and WERTHEIM, G.K., Inorganic Chemistry, 3, 101 (1964).

RAMAN EFFECT AND COMPETITIVE FORMATION OF TETRAHEDRAL MIXED HALIDE COMPLEXES BY Ga³⁺ IN SOLUTION.

L.A. Woodward

Inorganic Chemistry Laboratory, University of Oxford, U.K.

Raman spectroscopic evidence has shown (1) that in presence of excess halogen acid HX (where X is Cl, Br or I) Ga³⁺ forms $[GaX_4]^-$ as the sole detectable complex. The present paper describes some preliminary studies of solutions containing two halogen acids, in which mixed halide complexes are present in labile equilibrium.

The spectra obtained are exactly analogous to those of binary mixtures of stannic halides (2), where analogous labile mixed molecules are formed. Each of the five species present is unmistakably characterised by an intense polarized Raman line of different frequency. Table I shows the close analogy between the chloro-bromo complexes of Ga³⁺ and the stannic chloro-bromides.

TABLE I - WAVE NUMBERS (cm⁻¹) OF CHARACTERISTIC LINES

Species type	MBr_4	MBr_3Cl	MBr_2Cl_2	$MBrCl_3$	MCl_4
ν (M = Sn)	220	235	250	270	367
ν (M = Ga³⁺)	210	226	243	264	346

Frequencies for the iodide-bromide and iodide-chloride cases are given in Table II.

TABLE II - WAVE NUMBERS (cm⁻¹) OF CHARACTERISTIC LINES

Anion type	$[GaY_4]^-$	$[GaY_3X]^-$	$[GaY_2X_2]^-$	$[GaYX_3]^-$	$[GaX_4]^-$
X = Br⁻; Y = I⁻	145	156	170	187	210
X = Cl⁻; Y = I⁻	145	167	193	224	346

For each characteristic Raman line the intensity is proportional to the amount of the particular species present; but only for the extreme species $[GaX_4]^-$ and $[GaY_4]^-$ (which can be obtained alone in solution in known concentration) is a direct determination of specific intensity (i.e. per unite concentration) possible. The mixed species are labile, and so their specific intensities cannot be thus determined. Nevertheless, use can be made of the fact that, as the relative amounts of X and Y in the system are varied, the position of the intensity maximum for any mixed species must be the position of its maximum concentration.

Statistical considerations show that (provided sufficient of X and Y are present) the fractions of the central ion M in the form of the different complexes are: for $[MY_4]^-$, $(1 - N_X)^4/F$; for $[MY_3X]^-$, $4(1 - N_X)^3 wN_X/F$; for $[MY_2X_2]^-$, $6(1 - N_X)^2(wN_X)^2/F$; for $[MYX_3]^-$, $4(1 - N_X)(wN_X)^3/F$; and for $[MX_4]^-$, $(wN_X)^4/F$. In these expressions N_X is the fraction of X in the total halide, and w is a weighting factor signifying that the attachment of an X to a site on M is w times more probable than the attachment of a Y. The normalizing denominator F has the value $[1 + (w - 1)N_X]$ (4). Of course the actual fractions of M present in the form of the different complexes will only be given by the above expressions provided that the amounts of X and Y in the system are sufficient for their realization; if this is not so, the significance of w is lost.

The fractions of the mixed complexes $[MY_3X]^-$, $[MY_2X_2]^-$ and $[MYX_3]^-$ have maxima at the respective N_X values $N_{max.1} = 1/(1 + 3w)$, $N_{max.2} = 1/(1 + w)$ and $N_{max.3} = 3/(3 + w)$. Hence observations of the N_{max} values (which do not require any knowledge of specific intensities) can be used to obtain three independent values for w. The concordance of these will indicate the degree of success of the simple one-parameter statistical treatment.

So far only preliminary results have been obtained. They are only approximate, being based on visual estimates of intensity trends of the characteristic lines in the Raman spectra of series of solutions photographed under the same conditions. More precise studies by means of photoelectrically recorded spectra are being undertaken. The system most extensively explored so far is that involving competition between bromide ions X^- and iodide ions Y^-. The solutions were all 1,5 M in Ga and the total halide concentration was 10 M throughout. Eight different values of N_X, ranging from 0 to 1 were studied. The intensity changes of the characteristic Raman lines with changes of N_X showed clearly that w > 1, i.e. that the probability of complex formation is greater for Br^- than for I^-. This qualitative conclusion follows, for instance, from the observation that in going from $N_X = 0$ (only $[GaI_4]^-$ present) to $N_X = 0,1$ the line of $[GaI_2Br_2]^-$ appears, whereas in going from $N_X = 1$ (only $[GaBr_4]^-$ present) to $N_X = 0,9$ the same line remains too weak to be detected. Clearly a given amount of Br^- is more effective in displacing I^- from $[GaI_4]^-$ than is the same amount of I^- in displacing Br^- from $[GaBr_4]^-$.

More particularly, the intensity maxima were observed to occur at approximately $N_{max.1} = 0,2$, $N_{max.2} = 0,4$ and $N_{max.3} = 0,65$ to $0,7$.

Thus all are definitely shifted towards lower N_X relative to the positions they would have had (0,25, 0,5 and 0,75) if w had been unity. This is in accord with the conclusion (above) that $w > 1$. From each of the N_{max} values we can deduce a value for w, using the statistical expressions already given. We obtain respectively $w = 1,3_3$, $w = 1,5$ and $w = 1,6$ to 1,3. The rough agreement indicates that the simple statistical treatment is at least roughly applicable.

Corresponding results so far obtained for the bromide-chloride and iodide-chloride systems, though less extensive, show definitely that the probability of attachment is greater for Cl^- than for Br^-, and of course greater for Cl^- than for I^-. The w-value favouring Cl^- relative to Br^- appears to be close to that for Br^- relative to I^-, i.e. about 1,3 to 1,5. The value for Cl^- relative to I^- is expected to be the highest, since it should be the product of the other two. In agreement with this expectation, the preliminary observations indicate a value greater than 2.

The order $Cl^- > Br^- > I^-$ is in harmony with the inclusion of Ga in group 1 of AHRLAND's classification (3), for which this is the order of stability of halide complexes.

The author thanks Mr. R.C. WAGGETT and Mr. D.C. HARRIS for assistance with the experimental work.

References

1. WOODWARD, L.A., and NORD, A.A., J.Chem.Soc. (London), 1956, 3721; 1955, 2655; WOODWARD, L.A., and SINGER, G.H., ibid., 1958, 716.
2. DELWAULLE, M.-L., and FRANCOIS, F., Compt.rend., 219, 64 (1944).
3. AHRLAND, S., Acta Chem.Scand., 10, 723 (1956).

2 A 2

DETERMINATION OF CONSTITUTION, STRUCTURE, AND BONDING OF COMPLEX IONS IN SOLUTION BY RAMAN SPECTROSCOPY.

Robert A. Plane
Cornell University; Ithaca, New York, U.S.A.

Of the various methods employed for characterizing complex ions in solution, among the most direct is that of Raman spectroscopy. This method determines the vibrational spectrum of the complex species with a minimum of interference from the solvent and from other species present in the solution. By means of such studies in our laboratories, we have quantitatively character-ized the complexes formed in the following aqueous systems: the cyanides of

Cu(I), Ag(I), Zn(II), Cd(II), Hg(II), Fe(II), Co(III) (1,2); the bromides of
Zn(II), Cd(II), and Ga(III) (3,4). We have investigated the nitrates, sulfates, and
perchlorates of a wide variety of cations and found Raman evidence for complexes with nitrate of Ca(II), Ce(III), Th(IV), Cu(II), Ag(I), Zn(II), Hg(II),
Al(III), In(III) (5-7); evidence of inner-sphere sulfate complex formation was
shown only by In(III) (5); no evidence was found for perchlorate complexes
even from a careful study of Raman intensities and polarization ratios (in
addition to the frequency measurements) (7); a Raman line characterizing the
hydrated cation was found for Mg(II), Cu(II), Zn(II), Hg(II), Ga(III), and
In(III) (7).

Measurement of Raman intensities yields, not only information regarding
concentration of species, but also information regarding the nature of chemical
bonding within a species. The Raman intensity of a normal vibrational mode,
Q, of a molecule is given by the expression

$$I = \frac{KM (\nu_0 - \nu)^4}{\nu [1-\exp(-h\nu /kT)]} \cdot 45 \left(\frac{\partial \bar{\alpha}}{\partial Q}\right)^2 \left(\frac{6}{6 - 7\rho}\right)$$

where K is a constant, M the molar concentration of the molecule, ν_0 the
excitation frequency, ν the Raman frequency shift, $\bar{\alpha}$ the mean molecular
polarizability, and ρ the degree of depolarization. The quantity $\partial \bar{\alpha}/\partial Q$ is
zero for all vibrational modes other than totally symmetric ones. When the
transformation from internal coordinates to normal coordinates is known, it is
possible to express $\partial \bar{\alpha}/\partial Q$ in terms of derivatives with respects to bond lengths,
angles, etc. For example, in the case of the A_1 mode of a tetrahedral XY_4
molecule, $\partial \bar{\alpha}/\partial Q = 2 M_y^{-1/2}\bar{\alpha}'_{xy}$, where M_y is the mass of the Y atom and
$\bar{\alpha}'_{xy}$ is the derivative of the mean polarizability with respect to the extension
of the xy bond. It has been suggested that the quantity $\bar{\alpha}'$ is a bond property
resulting from the fact that when a bond stretches the resulting change in
molecular polarizability is localized in that bond (8); so that the value of $\bar{\alpha}'$
gives information concerning the electrons in the bond. Indeed, this quantity
has been experimentally shown to be directly related to the degree (order) of
covalent bonding (8-11). A direct comparison can be made only between isoelectronic molecules, because it has been shown that the derived bond
polarizability is directly dependent on the atomic numbers of the two atoms
forming the bond (9). However, the power of this method for determining
covalent bond order can be utilized for certain select systems.

Raman frequencies, intensities, and depolarization ratios (ρ) were
measured for a series of tetrahedral chloride and bromide complexes. These

complexes were chosen for study since they are isoelectronic with some of the group IV tetrahalides studied by WOODWARD and LONG (9). By comparison, covalent bond order should be obtained. In Table I, experimental values are given as measured with a Carey 81 Raman spectrophotometer by methods previously described (11). From these, the derived bond polarizabilities were calculated, which led to the covalent bond orders listed. The latter values were obtained by direct comparison with the isoelectronic group IV tetra-halides, or in cases where the corresponding group IV compound had not been measured, by using the atomic number rule which correctly described all those previously measured (9).

TABLE I

Complex	A_1 [cm^{-1}]	Relative Intensity[a]		$\bar{\alpha}'$, A^2	Order
$[ZnCl_4]^{2-}$	285	0,709	(ρ 0,19)	0,87	0,18
$[CdCl_4]^{2-}$	259	0,900	(ρ 0,04)	1,04	0,16
$[HgCl_4]^{2-}$	273	4,04	(ρ 0,13)	2,10	0,22
$[GaCl_4]^-$	346	1,04	(ρ 0,21)	1,12	0,23
$[ZnBr_4]^{2-}$	172	1,58	(ρ 0,06)	1,80	0,27
$[CdBr_4]^{2-}$	166	3,08	(ρ 0,08)	2,46	0,29
$[HgBr_4]^{2-}$	168	15,2	(ρ 0,18)	5,11	0,44
$[GaBr_4]^-$	210	4,39	(ρ 0,10)	3,08	0,47

[a] Relative to ClO_4^- ($\partial \bar{\alpha}/\partial r = 1,73$ A^2); all concentrations reduced to one molar; corrections made only for instrument sensitivity.

Complexes containing the polyatomic ligand NH_3 are summarized in Table II.

TABLE II

Complex	A_1 [cm^{-1}]	$\partial \bar{\alpha}/\partial Q$	Order	Force Const. [dyn/cm]
$[Zn(NH_3)_4]^{2+}$	427	0,45	0,48	$1,8 \cdot 10^5$
$[Cd(NH_3)_4]^{2+}$	350	0,39	0,29	$1,2 \cdot 10^5$
$[Hg(NH_3)_4]^{2+}$	410	0,96	0,47	$1,7 \cdot 10^5$

For such species, $\overline{\alpha}'$ cannot be exactly obtained; however, within the series of like complexes $\partial \overline{\alpha}/\partial Q$ can be directly compared. The values are seen to pass through a minimum. Relative bond orders were determined by comparison with values in Table I. In order to substantiate the minimum at Cd, simple valency force constants were calculated by treating NH_3 as a point mass of 17 amu.

In all complexes thus far discussed, Raman intensities have indicated bond orders less than unity. In at least one case, acetyl-acetonate complexes, we have found Raman evidence for π bonding (7). The frequencies and derived polarizabilities in Table III refer to symmetrical stretching of the MO_6 octahedron for both tris-oxalato (included for comparison) and tris-acetylacetonato complexes. The large derived polarizabilities for the acetylacetonates can either indicate that the bound oxygen atoms have effective masses many times that in the oxylates, or much more likely, that the bonds in the acetyl-acetonates significant π character. We have previously shown (11) that π bonding of O atoms in ClO_4^-, SO_4^{2-}, and NO_3^- leads to large values of derived polarizability.

TABLE III

	Tris-Oxalato		Tris-Acetylacetonato	
	A_1 [cm^{-1}]	$\partial \overline{\alpha}/\partial Q$	A_1 [cm^{-1}]	$\partial \overline{\alpha}/\partial Q$
Al	585	0,29	465	1,33
Ga	573	0,44	460	1,37
In	-	-	444	1,47

It is a pleasure to acknowledge financial assistance from the Air Force Office of Scientific Research, the Office of Saline Waters, and the National Science Foundation; and most especially to note the scientific contributions of G.W.CHANTRY, R.E.HESTER, G.S.KUDRAK, J.NIXON, K.E.TAYLOR, and W.YELLIN.

References

1. CHANTRY, G.W., and PLANE, R.A., J.Chem.Phys., 33, 736 (1960).
2. CHANTRY, G.W., and PLANE, R.A., J.Chem.Phys., 35, 1027 (1961).
3. YELLIN, W., and PLANE, R.A., J.Am.Chem.Soc., 83, 2448 (1961).
4. NIXON, J., and PLANE, R.A., J.Am.Chem.Soc., 84, 4445 (1962).
5. HESTER, R.E., PLANE, R.A., and WALRAFEN, G.E., J.Chem.Phys., 38, 249 (1963).

6. HESTER, R.E., PLANE, R.A., J.Chem.Phys., **40**, 411 (1964).

7. HESTER, R.E., and PLANE, R.A., Inorg.Chem., in press.

8. WOLKENSTEIN, M., Compt.rend.Acad.Sci.U.S.S.R., **32**, 185 (1941).

9. WOODWARD, L.A., and LONG, D.A., Trans.Faraday Soc., **45**, 1131 (1949).

10. YOSHINO, T., and BERNSTEIN, H.J., Molecular Spectroscopy (Proceedings of the Institute of Petroleum and Hydrocarbon Research Group Conference, London), Pergamon Press, New York, 1959.

11. CHANTRY, G.W., and PLANE, R.A., J.Chem.Phys., **32**, 319 (1960).

2 A 3

INFRARED SPECTRA AND METAL-LIGAND FORCE CONSTANTS IN COORDINATION COMPOUNDS.

Ichiro Nakagawa, Takehiko Shimanouchi and Jiro Hiraishi
Department of Chemistry, Faculty of Science,
The University of Tokyo, Bunkyo-ku, Tokyo, Japan.

Infrared spectra of metal complexes have already been intensively studied and the effects of a metal ion on the vibration frequencies in the ligand have been investigated (1). However, not so many investigations have been reported for the metal-ligand stretching and deformation vibrations. Recently we have measured the low frequency infrared spectra of ammine, halogeno, nitro, cyano, aquo and isothiocyanato complex ions. The assignments of the bands observed have been made on the basis of a normal coordinate analysis and the force constants associated with the metal-ligand bonds have been obtained. The results may be useful for the understanding of the nature of the metal-ligand bonds and give further information on the already recognized tendencies of electrons (2). The present paper is a summary of the results of our study.

I. Ammine complex ions. The octahedral complex ions, $[M(NH_3)_6]^{3+,2+}$, and the square planar complex ions, $[M(NH_3)_4]^{2+}$, have been studied. The infrared active vibrations belong to F_{1u} species for the octahedral complex ions of the molecular symmetry O_h and belong to A_{2u} and E_u species for the square planar ions (D_{4h}). Table 1 shows the observed and calculated frequencies of these infrared active vibrations. Table 2 lists the values of the force constants calculated based on the modified Urey-Bradley potential (5). K, H and F are the force constants for the bond stretching, the angle deformation and the repulsion between non-bonded atoms. $F_{dia}(MN)$, $F_{dia}(rock)$ and $F_{dia}(sym)$ represent the diagonal elements of the F-matrix for the M-N stretching, NH_3 rocking and NH_3 symmetric deformation modes, which were obtained by the procedure described in WILSON's GF-matrix method (6). These values represent the strengths of the forces opposing the corresponding vibration modes. The values for the metal-nitrogen stretching force constants vary in the order Pt > Pd > Co(III) > Cr > Cu > Ni > Co(II), showing that the degree of the covalent character decreases in the same order. It is to be noted that the values of the force constants, H(HNM), F_{dia}(rock) and F_{dia}(sym), vary in a similar manner. However, the vibration frequencies in the ligand are more or less affected by the outer ion through hydrogen bonding (7) and therefore these force constants do not supply such direct information about the nature of coordination bond as does the metal-nitrogen stretching force constant.

The halogenopentamminecobalt(III) and trans-dihalogenotetrammine-cobalt(III) complex ions have been studied. The observed frequencies of the infrared bands can be reasonably explained by normal coordinate analysis. The force constants determined in $[Co(NH_3)_6]^{3+}$ ion were transferred to $[CoX_2(NH_3)_4]^+$ and $[CoX(NH_3)_5]^{2+}$ ions. New force constants K(CoX) and H(NCoX) were determined from the observed frequencies of the A_{2u} species for trans-$[CoX_2(NH_3)_4]^+$ ion. (Table 3.)

II. <u>Halogeno complex ions</u>. Table 4 lists the observed frequencies and the calculated values of the metal-halogen stretching force constants. The bond between Pt(IV) and Cl is stronger than that between Pt(II) and Cl. This is also the case for Pd(IV) and Pd(II). This result is in agreement with that of the pure quadrupole resonance study (8).

Table 1. Observed and calculated frequencies of metal ammine complex ions (cm^{-1})

			$\nu(NH)_a$	$\nu(NH)_s$	$\delta(NH_3)_d$	$\delta(NH_3)_s$	$\delta(NH_3)_r$	$\nu(MN)$	$\delta(NMN)$
$[Pt(NH_3)_4]^{2+}$	calc	E_u	3231	3155	1619	1344	846	509	295
Cl_2	obs		3236	3156	1563	1325	842	510	297
$[Pd(NH_3)_4]^{2+}$	calc	E_u	3240	3164	1613	1304	774	493	296
Cl_2	obs		3268	3142	1601	1285	797	491	295
$[Co^{III}(NH_3)_6]^{3+}$	calc	F_{1u}	3240	3164	1615	1323	830	501	328
Cl_3	obs		3240	3170	1600	1325	820	503	325
$[Co^{III}(ND_3)_6]^{3+}$	calc	F_{1u}	2396	2265	1165	1009	661	454	291
Cl_3	obs (3)		2450		1155	1016	665		310
$[Cr(NH_3)_6]^{3+}$	calc	F_{1u}	3268	3198	1612	1292	759	474	267
Cl_3	obs		3260	3205 (3140)	1600	1310	745	470	(285)(4)
$[Cu(NH_3)_6]^{2+}$	calc	E_u	3310	3231	1610	1251	708	419	249
SO_4	obs			3270	1610	{1270 1240	713	420	250
$[Ni(NH_3)_6]^{2+}$	calc	F_{1u}	3393	3310	1606	1197	672	335	214
Cl_2	obs			3370 br	1610	1175	678	330	215
$[Co^{II}(NH_3)_6]^{2+}$	calc	F_{1u}	3338	3258	1605	1171	625	323	184
Cl_2	obs (3)		3330	3250	1605	1160	634	318	

"ν" stretching, "δ" deformation, "a" antisymmetric, and "s" symmetric. $(NH_3)_d$, $(NH_3)_s$ and $(NH_3)_r$ are NH_3 degenerate deformation, symmetric deformation and rocking vibrations.

The vibrations of A_{2u} species of the square planar complex ions are not given, because they are the out-of-plane modes and were not used for the calculation of the force constants.

Table 2. Force constants of metal ammine complex ions

	K(MN) md/A	F_{dia}(MN) md/A	K(NH) md/A	H(HNM) md/A	F_{dia}(rock) md·A	F_{dia}(sym) md·A	H(NMN) md/A	F(N···N) md/A
$[Pt(NH_3)_4]^{2+}$	1,92	2,21	5,58	0,180	0,441	0,452	0,400	0,040
$[Pd(NH_3)_4]^{2+}$	1,71	1,97	5,61	0,150	0,354	0,427	0,397	0,050
$[Co^{III}(NH_3)_6]^{3+}$	1,05	1,39	5,61	0,176	0,386	0,443	0,402	0,050
$[Cr(NH_3)_6]^{3+}$	0,94	1,25	5,71	0,128	0,329	0,423	0,233	0,030
$[Cu(NH_3)_4]^{2+}$	0,84	1,13	5,86	0,114	0,296	0,398	0,255	0,030
$[Ni(NH_3)_6]^{2+}$	0,34	0,65	6,16	0,105	0,268	0,366	0,150	0,050
$[Co^{II}(NH_3)_6]^{2+}$	0,33	0,65	6,00	0,073	0,239	0,351	0,095	0,030

The values of H(HNH), F(H···H) and F(H···M) are 0,546, 0,060 and 0,100 md/A for all. The values of H(HNH) and F(H···H) were taken from $NH_3(ND_3)$ molecule and $NH_4^+(ND_4^+)$ ion.

Table 3. Force constants of $[CoX_2(NH_3)_4]^+$ and $[CoX(NH_3)_5]^{2+}$ ions (md/A)

	K(MX)	F_{dia}(MX)	H(NMX)	F(N···X)
trans-$[CoCl_2(NH_3)_4]^+$, $[CoCl(NH_3)_5]^{2+}$	1,00	1,10	0,095	0,050
trans-$[CoBr_2(NH_3)_4]^+$, $[CoBr(NH_3)_5]^{2+}$	1,15	1,26	0,079	0,050
$[CoF(NH_3)_5]^{2+}$	1,00	1,08	0,120	0,050
$[CoI(NH_3)_5]^{2+}$	0,80	0,92	0,052	0,050

Table 4. The observed frequencies and the calculated values of the metal-halogen stretching force constants of the halogeno complex ions

	$[Pt(IV)Cl_6]^{2-}$	$[Pd(IV)Cl_6]^{2-}$	$[Pt(IV)Br_6]^{2-}$	$[Ir(IV)Cl_6]^{2-}$
ν(MX), δ(XMX) cm^{-1}	343, 182	340 (355, 332), 175	244, 90	332,
K(MX) md/A	1,86	1,51	1,54	1,60
	$[Os(IV)Cl_6]^{2-}$	$[Rh(III)Cl_6]^{3-}$	$[Pt(II)Cl_4]^{2-}$	$[Pd(II)Cl_4]^{2-}$
ν(MX) cm^{-1}	323	322	324	334
K(MX) md/A	1,50	1,20	1,50	1,30

III. Nitro complex ions. X-ray analysis of the hexanitrocobalt complex salts, $M_3[Co(NO_2)_6]$, shows that the crystal structure of the K, Rb and Cs salts is cubic and the molecular symmetry of complex ion is T_h (9). No structural analysis has been made for the Na salt. The infrared spectrum of the Na salt is quite different from those of the K, Rb and Cs salts as shown in Fig. 1. Since this difference is too large to be explained by the simple outer ion effect, we may expect the difference in the molecular structure to be caused by complex formation. The vibration selection rule shows that the spectrum of the Na salt is consistent with the deformed structure of the S symmetry in which the NO_2 plane rotates about the Co-N axis.

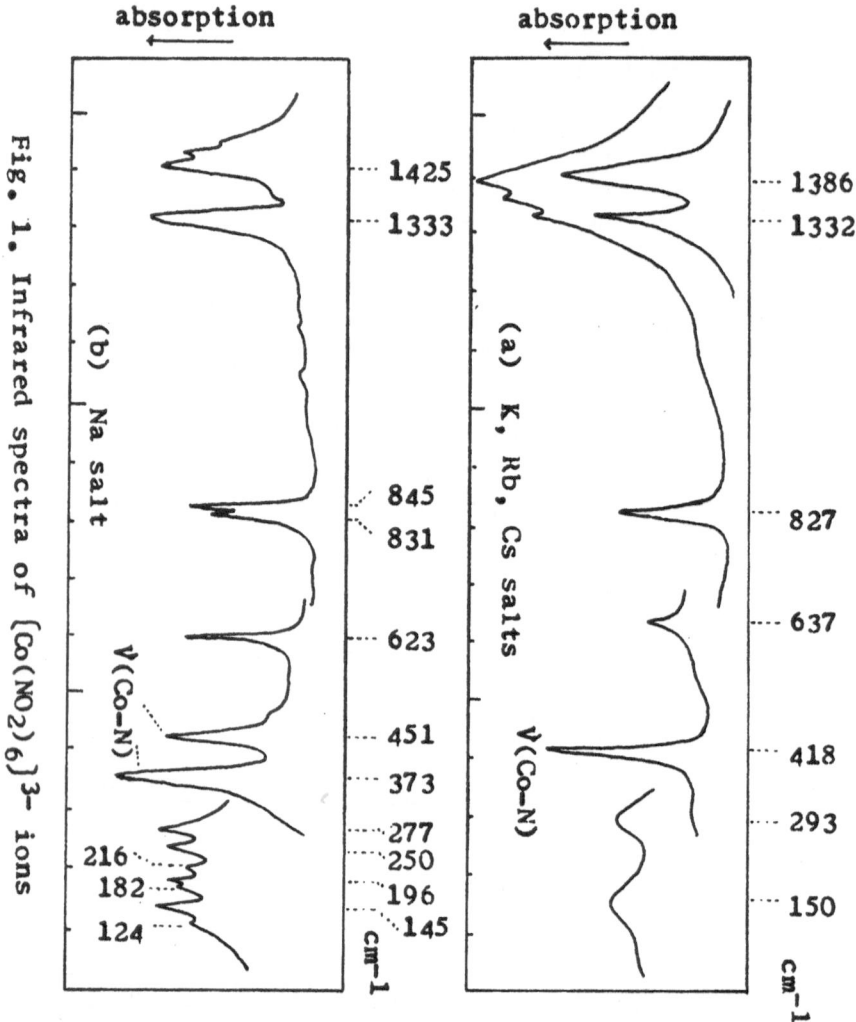

Fig. 1. Infrared spectra of $[Co(NO_2)_6]^{3-}$ ions

The force constants were determined by using the observed frequencies of the K salt (T_h symmetry). The values of K(Co-N) and F_{dia}(Co-N) are calculated to be 1,16 md/A and 1,59 md/A, respectively. The normal vibration calculation, obtained by using these force constants, shows that the two strong bands at 452 and 373 cm^{-1} of the Na salt corresponds to the two Co-N stretching frequencies of the E_u and A_u species, respectively, for the S_6 structure with a twisting angle of 30°.

The observed frequencies of the nitroamminecobalt(III) complex ions can be explained by the force constants transferred from $[Co(NH_3)_6]^{3+}$ and $[Co(NO_2)_6]^{3-}$.

IV. Other complex ions. In the cyano complex ions, $[M(CN)_6]^{3-,4-}$, there exists such a relationship between $K_1(M-C)$ and $K_2(C\equiv N)$ that as the K_1 value increases the K_2 value decreases. The K_1 decreases according to the sequence Os > Ru > Fe(II) > Co(III) > Fe(III) , whereas the K_2 value increases in the same order. The lowering of $K_2(C\equiv N)$ and the rising of $K_1(M-C)$ may be connected with the delocalization of π electrons.

The K(Cr-NCS) value determined from the observed frequencies of $[Cr(NCS)_6]^{3-}$ and $[Cr(NCS)_4(NH_3)_2]^-$ ions is 1, 40 ~ 1, 56 md/A.

Summary. Sufficient data are not avilable for comparing the force constants of complexes containing the same metal ion. However, with the information obtained in this study the force constants of coordination bonds formed by various ligands indicate the following sequence:
$CN^- > NCS^- > NO_2^- > NH_3 > Cl^- > H_2O.$

References

1. COTTON, F.A., "The Infrared Spectra of Transition Metal Complexes" in LEWIS, J., and WILKINS, R.G., "Modern Coordination Chemistry", Interscience, 1960; NAKAMOTO, K., "Infrared Spectra of Inorganic and Coordination Compounds", Wiley, 1963.
2. ORGEL, L.E., "An Introduction to Transition-metal Chemistry", Mathuen, 1960.
3. MIZUSHIMA, S., NAKAGAWA, I., and QUAGLIANO, J.V., J.Chem. Phys., 23, 1367 (1955); BERTIN, E.P., NAKAGAWA, I., et al, J.Am. Chem.Soc., 80, 525 (1958).
4. BLYHOLDER, G., and VERGEZ, S., J.Phys.Chem., 67, 2149 (1963).
5. SHIMANOUCHI, T., Pure & Applied Chem., 7, 131 (1963).
6. WILSON, Jr., E.B., J.Chem.Phys., 7, 1047 (1939); 9, 76 (1941).
7. FUJITA, J., NAKAMOTO, K., and KOBAYASHI, M., J.Am.Chem.Soc., 78, 3295 (1956).
8. ITO, K., et al, J.Am.Chem.Soc., 83, 4526 (1961).
9. DRIEL, M., and VERWEEL, H.J., Z.Krist., A 95, 308 (1936).

3 A 1

THERMOCHEMISTRY OF SOME TRANSITION-METAL TETRAHALOCOMPLEXES.

P. Paoletti and A. Sabatini
Istituto di Chimica Generale dell'Universita,
Via J.Nardi 39, Firenze, Italy.

Until recently only few direct measurements of the strength of the coordinate bond habe been made. In this note are illustrated the results of calorimetric measurements from which may be derived the relative bond-energies of the metal-chlorine and metal-bromine bonds in the tetrahedral complexes $[MX_4]^{2-}$; M = Mn, Fe, Co, Ni, Cu, Zn; X = Cl, Br. The compounds studied are: Rb_2CoCl_4, Cs_2CoCl_4, Rb_2ZnX_4, Cs_2ZnX_4, $[Et_4N]_2[MX_4]$, $[Me_4N]_2[MCl_4]$.

The thermochemical cycle used is illustrated below:

$$2M(I)X \quad + \quad M(II)X_2 \xrightarrow{\quad \Delta H_r \quad} M(I)_2[M(II)X_4] \quad \text{(solid state)}$$

$$\downarrow 2\Delta H_1[M(I)X] \qquad \downarrow \Delta H_1[M(II)X_2] \qquad \downarrow \Delta H_1[M(I)_2\, M(II)X_4]$$

$$2M^+ + 2X^- + M^{2+} + 2X^- \xleftarrow{\qquad\qquad} 2M^+ + [M(II)X_4]^{2-} \quad \text{(gaseous state)}$$
$$4\overline{D}\,[M\text{-}X]$$

ΔH_r was obtained calorimetrically from measurements of the heats of solution in water of the species implied by the reaction in the solid state. $4\overline{D}\,[M\text{-}X]$ represents the enthalpy of dissociation of the gaseous complex ion. The lattice enthalpies ΔH_1 of the simple chlorides were calculated using the Born-Haber cycle.

Assuming that the complexes $[Me_4N]_2[MnCl_4]$ and $[Me_4N]_2[ZnCl_4]$ have the same lattice enthalpy, one obtains:

$$4D\,[Mn\text{-}Cl] - 4D\,[Zn\text{-}Cl] \quad =$$

$$= \Delta H_r\,[Me_4N]_2\,[ZnCl_4] - \Delta H_r\,[Me_4N]_2[MnCl_4] + \Delta H_1\,[MnCl_2] - \Delta H_1\,[ZnCl_2]$$

by using the two thermochemical cycles described. Thus one obtains values for the strength of the metal-chlorine bond relative to the strength of the zinc-chlorine bond: these values are given in the second column of Table 1. Similarly for the series of complexes $[Et_4N]_2[MCl_4]$; the values given in the third column are in excellent agreement with those of the $[Me_4N]_2[MCl_4]$ series, showing that the lattice enthalpy does not change within a series or that the variations are exactly parallel in the two series. Following the same method, the values in the fourth column of Table I were obtained for the bromo-complexes $[Et_4N]_2[MBr_4]$.

In Fig. 1 the relative metal-chlorine and metal-bromine bond strengths are plotted against the atomic number of the metal. The filled-in circles and squares represent the values corrected for the crystal field stabilisation energy.

In order to calculate the absolute values of all the metal-chlorine and metal-bromine bonds it is necessary now to know the value for the zinc

TABLE I - VALUES OF THE DISSOCIATION ENERGIES (kcal)

Central Ion M	$[4\bar{D}[M\text{-}X] - 4\bar{D}[Zn\text{-}X]]$			$4\bar{D}[M\text{-}X]$ and thermochemical radii (Å)	
	X = Cl		X = Br		
	M(I)=Me$_4$N	M(I)=Et$_4$N	M(I)=Et$_4$N	X = Cl	X = Br
Mn(II)	-54,2	-54,2	-56,7		
Fe(II)	-33,1	-33,1	-34,7		
Co(II)	-13,3	-13,3	-14,9	602,4 (3,14)	
Ni(II)	-8,4	-8,1	-6,3		
Cu(II)	5,7	4,9	6,5		
Zn(II)	0	0	0	607,6 (3,02)	592,4 (3,16)

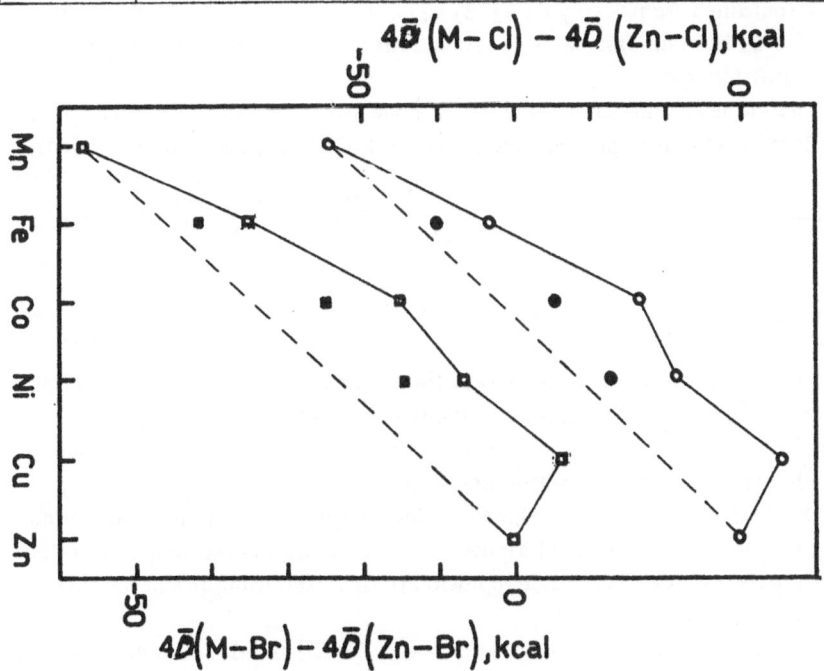

Fig. 1

complex, and for this the lattice energy of the complex must be calculated. This may be done with the aid of the KAPUNSTINSKII-YATSIMIRSKII (1) formula. By substituting thermochemical values for the complexes Rb_2ZnCl_4 and Cs_2ZnCl_4 into this equation and using the thermochemical cycles described one obtains the equations:

$$\Delta H_r [Rb_2ZnCl_4] + \frac{1728,2}{1,49+x} \left(1 - \frac{0,345}{1,49+x}\right) + 15 + 4\overline{D}\,[Zn-Cl] =$$

$$= 2\Delta H_1 [RbCl] + \Delta H_1 [ZnCl_2]$$

$$\Delta H_r [Cs_2ZnCl_4] + \frac{1723,2}{1,65+x} \left(1 - \frac{0,345}{1,65+x}\right) + 15 + 4\overline{D}\,[Zn-Cl] =$$

$$= 2\Delta H_1 [CsCl] + \Delta H_1 [ZnCl_2]$$

where x is the thermochemical radius of the ion $[ZnCl_4]^{2-}$. The values given in Table I are obtained by solvind these equations simultaneously. The values of $4\overline{D}\,[Co-Cl]$ and $4\overline{D}\,[Zn-Br]$ are obtained by the same process.

The ratio of the thermochemical radius to the radius of the circumscribed sphere (obtained from X-ray data) (2) is: $[CoCl_4]^{2-}$ 0,78; $[ZnCl_4]^{2-}$ 0,76; $[ZnBr_4]^{2-}$ 0,73. This last value is rather lower than is usually found for tetrahedral ions (3).

The values of the Zn-Cl and Zn-Br bond strengths have also been calculated using an improved version of GARRICK's (4) electrostatic model:

$$-4D\,[Zn-X] = -\frac{nze^2}{r} + \frac{1}{2} \sum_k \sum_{k'} \frac{e^2}{s_{kk'}} - \frac{nzpe}{r^2} + \sum_k \sum_{k'} \frac{pe}{s^2} \cos\Theta_{kk'} +$$

$$+ \frac{1}{2} \sum_k \sum_{k'} \frac{p^2}{s^3_{kk'}} (1 + \cos^2\Theta_{kk'}) + \frac{np^2}{2\alpha} + \frac{n\lambda}{8r^8} + \frac{1}{2} \sum_k \sum_{k'} \frac{\lambda'}{ms^m_{kk'}}$$

where the terms represent respectively the energy of a) coulombic attraction between Zn^{2+} and X^-; b) coulombic repulsion between X^- and X^-; c) interaction between Zn^{2+} and the dipole induced on each X^- by its surrounding total electric field; d) interaction between the electric field generated by three X^- ions and the dipole induced on the fourth X^- ion by the surrounding total electric field; e) induced dipole-induced dipole interaction; f) creation of the dipoles; g) Van der Waals repulsion (calculated using the LENNARD-JONES (5) potential).

The values obtained in this way are $4\overline{D}\,[Zn-Cl] = 591,8$ and $4\overline{D}\,[Zn-Br] = 570,5$ kcal./mole in good agreement with the "experimental" values. Consideration of the polarisability of the ions increases the bond energy over that

calculated for non-polarisable ions by 30,6 and 42,9 kcal/mole, i.e. ca. 5%
of the total bond energy.

References

1. YATSIMIRSKII, Russ.J.Inorg.Chem., 6, 265 (1961).
2. MOROSIN and LINGAFELTER, Acta Cryst., 12, 611, 744 (1959).
3. YATSIMIRSKII, Thermochemie von Komplexverbindungen (german trans.,
   Akademie-Verlag, Berlin, 1955).
4. GARRICK, Phyl.Mag. IX, 131 (1930).
5. LENNARD-JONES and TAYLOR, Roy.Soc.Proc., A 109, 476 (1925);
   LENNARD-JONES, ibid, p.584.

3 A 2

MICROCALORIMETRIC DETERMINATION OF THE ENTHALPY OF FORMATION OF THE COMPLEX IONS OF TRIVALENT PLUTONIUM, AMERICIUM AND LANTHANIUM WITH EDTA.

J. Fuger[*] and B.B. Cunningham
Department of Chemistry and Lawrence Radiation Laboratory,
University of California, Berkeley, Calif., U.S.A.

As to date, quantitative data on actinide complex species are very scarce
and, when existing often limited to stability constants determinations. No
direct calorimetric measurement of the enthalpy of complexing have sofar
been published for those elements. The present communication deals with
such measurements for the complexing of Pu^{3+} and Am^{3+} with EDTA. We also
report data on La^{3+} complexing for comparison purpose. From the enthalpy
change and knowing the stability constants of the chelates, the entropy
variation on complexing can easily be calculated.

We have compared the heat of solution of crystalline anhydrous trichlorides in dilute solutions of EDTA partially neutralized with KOH at a
carefully chosen pH value, with the heats of solution of the same chlorides
in solutions of similar ionic strength, but containing no EDTA.

From the acid dissociation constants of EDTA (1,2) it can be calculated
that between pH 7,4 and 6,6 an EDTA solution is represented to an extent
greater than 99,8% by the forms HY^{3-} and H_2Y^{2-}. If the EDTA is in sufficient
excess with respect to the cation studied, the pH of the solution, on complexing, will remain between the above limits and the experimental reaction
will be described unambigously by the equation:

[*] Present address: Université de Liège, Laboratoire de Chimie Nucleaire,
    2, rue A.Stévart, Liège (Belgium).

$$MeCl_{3(c)} + 2 HY^{3-}_{aq} \longrightarrow MeY^-_{aq} + 3 Cl^-_{aq} + H_2Y^{2-}_{aq}; \quad \Delta H_1$$

Using the heats of solution of the trichlorides (3-5) in noncomplexing media of similar ionic strength, and the heats of third and fourth dissociation of EDTA measured by ANDEREGG (6), the enthalpy of complexing can be calculated.

Among the arguments in favor of the present method, in our particular case, the most important are that it is easier to handle small amounts of crystals (even though they are hygroscopic), than prepare, handle and keep small volumes of accurately known of those highly radioactive species; secondly, due to its own alpha activity, plutonium cannot be kept in a solely 3+ state for periods exceeding a few hours; moreover the oxygen formed by radiolysis of water and accumulated in such solutions would immediately oxidize the plutonium in the chelate at the time of reaction. Visible and Near Infra Red spectra of the chelates have been used to identify the species.

Some features of the microcalorimeter used for our measurements have been described previously (7) but the instrument has since been modified to increase its stability and converted to a recording instrument. It has a heat capacity of approximately 9 calories, a temperature sensitivity of approximately $1.10^{-5\,o}$ and a thermal leakage modulus of about 5.10^{-3} min^{-1}. It was thermostated at 25°. Samples of 2 to 8 milligrams of trichlorides have been used, leading to experimental reaction heats of a few tenths of a cal. The thermodynamic quantities for the formation of the chelates at 25° and $\mu = 0,1$ are summarized in Table I.

TABLE I

THERMODYNAMIC QUANTITIES FOR THE FORMATION OF THE
CHELATES WITH EDTA AT 25° AND $\mu = 0,1$.

Species	ΔH^o (kcal. mole^{-1})	pK_c	ΔG^o (kcal. mole^{-1})	ΔS^o (cal. mole^{-1}. degree^{-1})
PuY$^-$	$-4,23 \pm 0,25$	$18,07$[a]	$-24,67$	$67,8 \pm 2$
AmY$^-$	$-4,67 \pm 0,25$	$18,16$[b]	$-24,79$	$67,4 \pm 2$
LaY$^-$	$-1,15 \pm 0,20$	$15,88$[c]	$-21,66$	$67,4 \pm 2$

(a) value from reference (8), corrected from 20° to 25°.
(b) value from reference (9).

(c) value from reference (10), corrected from 20^0 to 25^0 and recalculated using CABELL's (2) dissociation constants for EDTA instead of SCHWARZENBACH's (1) in order to normalize pK_C as well as possible. We have noted previously (11) that the two sets of EDTA dissociation constants lead to a discrepancy of about 0,4 pK_C unit in the pH range considered.

As the first observation we note that the values in Table I indicate, as expected, that the entropic term is the predominant contributing factor to the stability of those chelates but very little more can be concluded from the entropy change because of the limited number of chelates investigated in the present study.

If we compare our results with those of MACKEY and coworkers (12) who give data for the complete lanthanide series, we observe that for lanthanum, a descrepancy of about 1,8 kcal. mole^{-1} exists. However, differences in the auxiliary data used can account for deviation up to 0,5 kcal. mole^{-1}. On the other hand, if we compare the actinides studied with the lanthanides having approximately the same ionic radii (Pr and Nd), we note that the enthalpy change for the actinides is more negative by an amount greater than 1 kcal. mole^{-1}.

Even if the disagreement between the data of MACKEY and coworkers and ours for La is reflected by a systematic discrepancy throughout the lanthanide series, it would make the difference between the actinides studied and the lanthanides, bigger. We observe also that the stability constants of the Pu^{3+} and Am^{3+} chelates are on the order of 10 times bigger than those of the lanthanides (10) of similar ionic radii (Pr^{3+}, $pK_C = 16,80$; Nd^{3+}, $pK_C = 17,01$, the same correction being applied as to the lanthanum pK_C in Table I) and indeed the elution behaviour (11) of mixtures of actinides and lanthanides in presence of EDTA leads to the same conclusion. Therefore the increased stability of the actinide complexes is reflected mainly by a bigger enthalpic term.

In conclusion, we believe that the differences between the complexes of actinides studied and those of the corresponding lanthanides are due to other contributing factors in the interaction between central ion and ligand rather than to structural differences. Thermodynamic data on complexes of other trivalent actinides as well as study of homologous chelating agents would of course be necessary to understand better the phenomena.

This work was done under the auspices of the United States Atomic Energy Commission. L'Institut Interuniversitaire des Sciences Nucléaires, Belgium, provided part of the financial support.

References

1. SCHWARZENBACH, G., and ACKERMANN, H., Helv.Chim.Acta, 30, 1798 (1947).

2. CABELL, M.J., AERE C/R 813 (1951), Analyst, 77, 859 (1952).

3. FUGER, J., and CUNNINGHAM, B.B., J.Inorg.Nucl.Chem., 25, 1423 (1963).

4. WESTRUM, Jr., E.F., and ROBINSON, H.P., The Transuranium Elements (edited by G.T.Seaborg, J.J.Katz and W.M.Manning), Plutonium Project Records NNES, Div. IV vol 14 B, p.922, McGraw Hill, New York (1949).

5. SPEDDING, F.H., and FLYNN, J.P., J.Am.Chem.Soc., 76, 1477 (1954).

6. ANDEREGG, G., Helv.Chim.Acta, 46, 1833 (1963).

7. WESTRUM, Jr., E.F., and EYRING, L., J.Am.Chem.Soc., 74, 2045 (1952).

8. FOREMANN, J.K., and SMITH, T.D., J.Chem.Soc. (London), 1957, 1752.

9. FUGER, J., J.Inorg.Nucl.Chem., 5, 332 (1958).

10. SCHWARZENBACH, G., GUT, R., and ANDEREGG, G., Helv.Chim. Acta, 37, 936 (1954).

11. FUGER, J., J.Inorg.Nucl.Chem., 18, 263 (1961).

12. MACKEY, J.L., POWELL, J.L., and SPEDDING, F.H., J.Am.Chem. Soc., 84, 2047 (1962).

A 3

THERMOCHEMISTRY OF SOME ADDUCTS OF GROUP III ELEMENTS.

N.N. Greenwood and P.G. Perkins
University of Newcastle upon Tyne, U.K.

Thermochemical data for the following series of acceptors is presented:

A) BPh_3, $AlPh_3$, $GaPh_3$, $InPh_3$; (ligand = pyridine)

B) $GaCl_3$, $PhGaCl_2$, Ph_2GaCl, $GaPh_3$; (ligand = pyridine)

C) $GaCl_3$, $GaBr_3$, (ligands: Et_2O, Me_2S, Et_3N)

It is well-known that the halides of the Group III elements are good electron acceptors and previous work (1,2) has explored the thermochemistry of adduct formation in order to obtain information about the strength of the donor-acceptor bond. The results of this work have been interpreted in terms of the steric, inductive, and π-bonding effects of the groups attached to the donor and acceptor atoms. The dominant effect, at least in the halide acceptors, appears to be the last. When aryl groups are attached to the acceptor atom π-bonding must stem from filled π molecular orbitals on the phenyl groups rather than from non-bonding electrons as in halides. There is evidence for such behaviour in silicon chemistry (3) but no quantitative data exists.

The heats of formation of the crystalline adducts of pyridine with the triphenyls of B, Al, Ga and In were obtained by direct reaction calorimetry thus

$$MPh_3^{(c)} + (n+1) Py^{(l)} = MPh_3 \cdot Py \text{ (in n moles of pyridine)}; \; -\Delta H_1$$

$$MPh_3 \cdot Py^{(c)} + n \, Py^{(l)} = MPh_3 \cdot Ph \text{ (in n moles of pyridine)}; \; -\Delta H_2$$

Thus $MPh_3^{(c)} + Py^{(l)} = MPh_3 \cdot Py^{(c)}; \; -\Delta H_f = -H_1 + \Delta H_2$

The corresponding heat of formation of the adduct in the gas phase, a more fundamental quantity, is obtained by additional measurement of the latent heats of sublimation of the solids. Because of the very low vapour pressures of the compounds at ordinary temperatures these measurements are difficult to make. The heats of formation of the complexes in the condensed phase are (in kcal. $mole^{-1}$)

$$BPh_3 \cdot Py, \; 18,2; \qquad AlPh_3 \cdot Py, \; 22,3;$$

$$GaPh_3 \cdot Py, \; 19,4; \qquad InPh_3 \cdot Py, \; 13,8.$$

Data for the gas phase reaction will also be presented. The results suggest that, in the condensed phase, the order of acceptor strength is $AlPh_3 > GaPh_3 > BPh_3 > InPh_3$. The same order may be inferred for the tri-methyl compounds of the elements towards the same ligand in the gas phase though complete data for proper comparison are lacking. It is noteworthy that the actual heats of formation are rather similar.

The position of boron in the series is significant. In the case of BMe_3 the reason is probably steric hindrance from the methyl groups and no doubt this also diminishes the acceptor strength of the B atom in BPh_3 since the phenyl rings are twisted out of the plane (4). An additional factor however is the π-bonding between the phenyl group and the empty p_z orbital of the boron atom. This is consistent with the UV spectrum of BPh_3 in which the first transition band associated with the benzene ring (260 mμ; weak) exhibits a bathochromic shift and a pronounced increase in intensity (5). The π-bonding may be expected to be less evident in $AlPh_3$ and $GaPh_3$ due to the less favourable overlap of the orbitals involved.

In all cases the triphenyls are weaker acceptors than the trichlorides of the elements, and, since calculation has shown that there is considerable

π-bonding in the monomeric chlorides of B, Al (6) and Ga, then the difference in acceptor strength appears to depend largely on inductive and steric factors.

Thermochemical data for the adducts of the new dimeric compounds Ph_2GaCl and $PhGaCl_2$ with pyridine are also being studied and it seems that these compounds, in the monomeric state, fall between $GaPh_3$ and $GaCl_3$ in acceptor strength, probably $GaCl_3 > PhGaCl_2 > Ph_2GaCl > GaPh_3$.

In a parallel study the relative acceptor strengths of $GaCl_3$, $GaBr_3$ and GaI_3 towards a variety of ligands has been measured calorimetrically. The new adducts of $GaBr_3$ and GaI_3 with ether, dimethylsulphide and triethyl-amine have been prepared and characterized by phase studies and their heats of formation will be reported and compared with those of the corresponding adducts of gallium trichloride.

The experimental work reported in this paper has been done in association with M.E. TWENTYMAN and T.S. SRIVASTAVA.

References

1. STONE, Chem. Reviews, <u>58</u>, 101 (1958).
2. GREENWOOD and WADE, Friedel-Crafts and Related Reactions, Ed. G. Olah, Interscience, New York, 569 (1963).
3. CHATT and WILLIAMS, J. Chem. Soc. (London), <u>1954</u>, 4403.
4. SHARP and SHEPPARD, J. Chem. Soc. (London), <u>1957</u>, 674.
5. NIKITMA, VAVER, FEDATOV and MIKHAILOV, Optics and Spectroscopy, <u>7</u>, 389 (1959).
6. COTTON and LETO, J. Chem. Phys., <u>30</u>, 993 (1959).

A 4

SOME CALORIMETRIC MEASUREMENTS IN METAL COMPLEX REACTIONS.

G. Anderegg

Laboratory of Inorganic Chemistry, ETH, Zürich, Switzerland.

Previous equilibrium studies (1) in aqueous solution of the reaction between Ca^{2+} or Cd^{2+} and the EDTA homologues:

$$^-OOC-CH_2 \diagdown \qquad \diagup CH_2-COO^-$$
$$N-(CH_2)_n-N$$
$$^-OOC-CH_2 \diagup \qquad \diagdown CH_2-COO^-$$

(n = 2 für EDTA; n = 3 für TMTA; n = 4 für TETA; n = 6 für HDTA; and n = 8 für ODTA, generally abbreviated L) have revealed that increase of the number n of methylene groups causes the stability constant of the 1:1 complex ML to drop off rapidly, especially in the range n = 2 to n = 4. As generally expected, complexes with 5-membered rings were found to be much more stable (2) than those containing 6- or 7-membered rings. The measurements have now been extended to many other divalent metals, and to lanthanum.

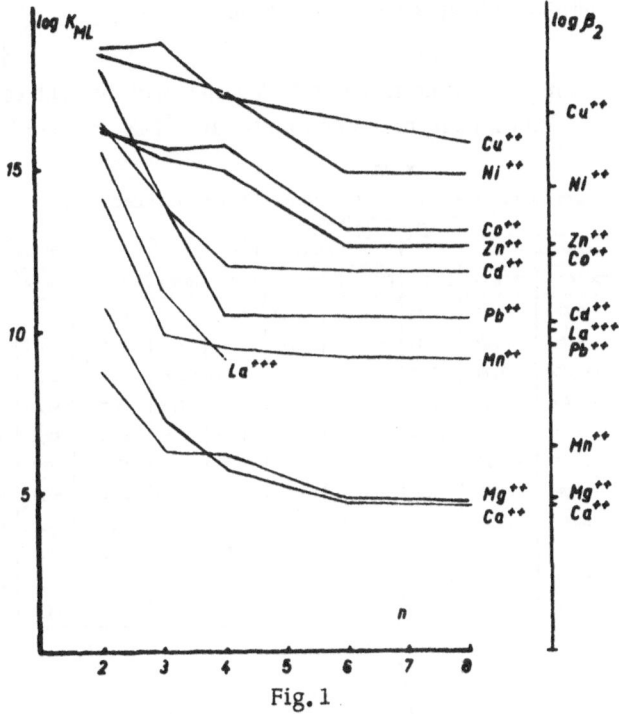

Fig. 1

Stability constants of the EDTA homologous against
numbers of -CH_2- of the alkylene chain.
Stability constants log β_2 of the iminodiacetate complexes.

The results show (Fig. 1) that the drop in stability when n > 2 is usually not as pronounced as had been believed. In some cases pronounced change in stability occurs only after introducing the 5th and the 6th methylene group into the ligand molecule, which corresponds to an 8- respectively a 9-membered chelate ring.

For quantitative discussion we compare, as in (2), chelation with L and that with two independent iminodiacetate ions Im, and define the 'chelate effect' by

$$\text{Chel} = \log K_{ML} - \log \beta_2 \qquad \{1\}$$

$$(K_{ML} = [ML]/[M][L]; \quad \beta_2 = [MIm_2]/[M][Im]^2)$$

Our results show that the chelate effect depends not only on the chain length, but also on the kind of complexed metal ion. To find the underlying principles the enthalpy changes for the reactions:

$$M + L \longrightarrow ML \qquad \{2\}$$

were measured. They are listed in Table I, together with the values of ΔG obtained from the known equilibrium constants, and of $\Delta S = (\Delta G - \Delta H)/T$.

TABLE 1

THERMODYNAMIC DATA FOR COMPLEX FORMATION REACTIONS WITH EDTA HOMOLOGUES AT
20° AND $\mu = 0,1$ (KNO$_3$).

Metal ion	ΔG in kcal Mol^{-1}				ΔH in kcal Mol^{-1}				ΔS in e.u.			
	EDTA	TMTA (3)	TETA (3)	HDTA	EDTA (4)	TMTA	TETA	HDTA	EDTA (4)	TMTA	TETA	HDTA
Mg^{2+}	-11,65	- 8,33	- 8,34	- 6,44	3,49	9,09	8,50	-	51,0	59	54,0	-
Ca^{2+}	-14,35	- 9,76	- 7,59	- 6,17	- 6,55	- 1,74	0,88	-	26,6	27,35	29,7	-
Mn^{2+}	-18,51	-13,40	-12,78	-12,07	- 4,56	- 0,72	3,41	0,87	46,6	52,9	55,2	44,2
Co^{2+}	-21,87	-20,80	-21,0	-17,5	- 4,2	- 2,6	- 1,56	- 4,56	60,3	62,2	66,3	44,1
Ni^{2+}	-24,97	-24,34	-23,28	-18,53	- 7,55	- 6,76	- 6,95	- 8,5	59,4	60,3	55,5	34,2
Cu^{2+}	-25,21	-25,37	-23,25	-	- 8,15	- 7,74	- 6,52	-	58,2	60,1	57,0	-
Zn^{2+}	-22,13	-20,38	-20,13	-17,0	- 4,85	- 2,27	- 3,48	- 4,00	59,0	61,77	56,8	44,4
Cd^{2+}	-22,07	-18,64	-16,12	-15,96	- 9,05	- 5,44	- 2,88	- 4,26	44,4	45,0	45,2	39,9
Pb^{2+}	-24,19	-18,37	-14,12	-13,93	-13,20	- 6,4	- 4,85	- 7,53	37,5	40,8	31,6	21,8
Hg^{2+}	-29,23	-26,71	-28,13	-28,94	-18,9	-18,9	-19,1	-20,97	35,5	26,6	30,8	27,25
La^{3+}	-20,79	-15,06	-12,24	-	- 2,8	3,76	0,1	-	61,4	64,2	42,0	-

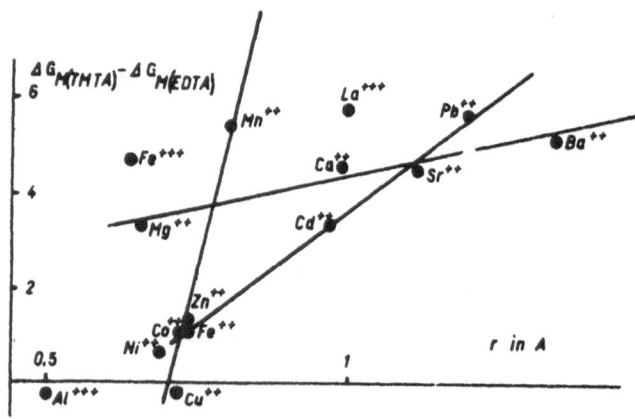

Fig. 2

Differences of ΔG values between TMTA and EDTA as functions of the ionic radii of the central ions.

In fig. 2 the differences of the ΔG values between EDTA and TMTA complexes are plotted against the ionic radius of the central ion, including values for three trivalent metal ions. In a given group of cations (transition metals; earth alkali metals; Zn, Cd, Pb) is appears that an approximately linear relationship exists between the size of the ion and the difference of the ΔG values, and that TMTA prefers small metal ions. The increase of ΔG going from EDTA to TMTA is paralleled by an increase in ΔH, which is especially large for Ca^{2+}, Pb^{2+}, and La^{3+}. However, it is most significant that there is no corresponding change in ΔS. There can therefore be no decrease in the number of ligand atoms tied to the central atom. Such a decrease would mean the release of fewer H_2O molecules upon chelation, corresponding to a decrease in ΔS, whereas experimentally ΔS increases, if anything (excepting Hg).

For further discussion we consider that the entropy change $\Delta S \equiv \Delta S_{ML}$ for reaction {2} is the difference between the standard entropy (in solution) of the complex S_{ML}, and those of metal ion S_M and ligand S_L:

$$\Delta S_{ML} = S_{ML} - S_M - S_L \qquad \{3\}$$

For the replacement reaction between two ligands L and L':

$$ML + L' \longrightarrow ML' + L \qquad \{4\}$$

the entropy change is:

$$\Delta S_{ML'} - \Delta S_{ML} = S_{ML'} - S_{ML} - (S_{L'} - S_L) \qquad \{5\}$$

Putting L' = EDTA and L = TMTA and considering (Table I) that $\Delta S_{ML'} - \Delta S_{ML} \sim 0$ it follows:

$$S_{ML'} - S_{ML} = S_{L'} - S_L \qquad \{6\}$$

This shows that the increases of S_L and of S_{ML} are approximately equal when n is changed from 2 to 3. On the other hand, for TETA (n = 4) some increase and some decrease of the ΔS values are found. Finally for the higher homologue HDTA (n = 6) the ΔS values decrease (Table I). Using COBBLE's (5) differential relationship between the entropies of homologuous charged organic ligands we estimate that S_L increases by about 4-5 e.u. when one methylene group is added. It therefore appears from the results and (3) that when n is changed from 2 to 3 that S_{ML} increases by about 5 e.u. An increase of S_{ML} of about 8-10 e.u. is found changing n from 2 to 4. Further increase of n leads to small increases of S_{ML} only, because otherwise ΔS would continue to remain constant, contrary to Table I. It appears however, not unreasonable that the steps from n = 2 to n = 3 and n = 3 to n = 4 lead to a significant entropy increase of the complex, while further increase of n has much less influence; for n = 2 the chelate rings are known to be rigid; while for n = 3

and 4 a reasonable amount of motion may be possible and further increase
of the ring size may not add significantly more possibilities for ring motion.
Unfortunately, changes in ΔS due to changes in the number of atoms coordinated are also possible and cannot be excluded at this time. Calorimetric
experiments to elucidate these points are in progress. Furthermore, the conclusions based on the ΔS variation must be taken with some caution, because
this change is small compared to the large ΔS values themselves, obtained
when two ions of high charge and therefore large hydration react and many
water molecules are released.

The parallel increase of the ligational enthalpy and free enthalpy going
from EDTA to higher homologues must be caused by steric hindrance. Such
steric hindrance can be seen by comparing on the one hand the stability
differences of complexes of 1:3 diaminopropane and of ethylenediamine with
those of TMTA and EDTA. For example, the 1:1 complexes of 1:3 diaminopropane with Cd^{2+}, Ni^{2+}, and Cu^{3+} (6) are all equally less stable by a factor
10 than those of ethylenediamine whereas there are large variations (Fig. 1)
between the stabilities of the corresponding TMTA and EDTA complexes. The
change in stability is particularly large for Mn^{2+}, Cd^{2+}, Pb^{2+}, and La^{3+}.

References

1. SCHWARZENBACH, G., SENN, H., and ANDEREGG, G., Helv.Chim.
   Acta, 40, 1886 (1957).
2. SCHWARZENBACH, G., Helv.Chim.Acta, 35, 2344 (1952).
3. Unpublished measurements of ANDEREGG, G., and L'EPLATTENIER, F.
4. ANDEREGG, G., Helv.Chim.Acta, 46, 1833 (1963).
5. COBBLE, J.W., J.chem.Phys., 21, 1451 (1953).
6. Stability constants, Special Publication No. 7, The Chemical Society,
   London, 1958.

A 1

THE AFFINITY OF CUPRIC(PYRIDINE) COMPLEX
TOWARDS A FIFTH LIGAND.

H.C. Volger and W. Brackman
Koninklijke/Shell-Laboratorium, Amsterdam,
Shell Research N.V., Netherlands.

The further coordination of copper pyridine complexes with other ligands
in methanol has been studied by investigation the absorption spectra and by
studying equilibria of cupric nitrate and pyridine in methanol in the presence
of solid $Cu(Pyr)_4(NO_3)_2$. If addition of a ligand X to such a system leads to

the formation of a new complex, it will do so at the expense of the solid phase. The formation constant of the new complex may then be estimated from the change in the total quantity of dissolved copper as a function of the amount of X added to the system.

Combination of the two methods leads to the following results. A maximum of four molecules of pyridine can be coordinated in a cupric complex. No evidence has been found for the existence of $Cu(Pyr)_5$ complex at pyridine concentrations up to 2 M. At pyridine concentrations between 0.15 and 2 M $[Cu(Pyr)_3]^{2+}$ is present in solution, besides $[Cu(Pyr)_4]^{2+}$ owing to a much smaller stability constant for the $[Cu(Pyr)_4]^{2+}$ complex compared to the $[Cu(Pyr)_3]^{2+}$ complex. If sodium methoxide is added to a $[Cu(Pyr)_4]^{2+}$ solution the colour of the system changes from dark blue to green. The extinction at 610 mμ increases only slightly; the major effect is the appearance of a new absorption band at 361 mμ. The extinction at this wavelength increases linearly with the amount of added base up to a $NaOCH_3$/Cu(II) ratio of 1. At higher ratios no further changes occur.

It has been found that triethylamine added to the $[Cu(Pyr)_4]^{2+}$ solution has the same effect as $NaOCH_3$ in spite of the fact that the amine is a relatively weak base. Apparently, the affinity of $[Cu(Pyr)_4]^{2+}$ for methoxide ions is so great that CH_3O^- ions are effectively withdrawn from the equilibrium

$$(C_2H_5)_3N + CH_3OH \;\rightleftharpoons\; (C_2H_5)_3\overset{\oplus}{N}H + CH_3O^{\ominus}$$

by the cupric pyridine complex.

Acetate ions were found to have a similar affinity towards $[Cu(Pyr)_4]^{2+}$. They produce a shift of the absorption maximum to 662 mμ and an increase in extinction, which again is a linear function of the amount of added ion, up to a 1:1 ratio of acetate to copper. From our experiments the formation constants of $[Cu(Pyr)_4(OCH_3)]^+$ and $[Cu(Pyr)_4(OAc)]^+$ were estimated to be greater than 10^5 and 10^4, respectively.

The general structure $[Cu(Pyr)_4L]^+$ for the new complexes is based upon the following arguments. First, according to BJERRUM (1), the average affinity of cupric ions towards pyridine is large compared to acetate ions. Hence a ligand exchange reaction, which leads to the $[Cu(Pyr)_3(OAc)]^+$ complex does not seem likely. Secondly, at a pyridine concentration at which the solution contains mainly the $[Cu(Pyr)_4]^{2+}$ complex, acetate also forms a 1:1 complex (presumably $[Cu(Pyr)_3(OAc)]^+$) the spectrum of which is, however, entirely different from that of the assumed $[Cu(Pyr)_4(OAc)]^+$ complex.

Similar arguments lead to a structure $[Cu(Pyr)_4OCH_3]^+$ for the complex formed from $[Cu(Pyr)_4]^{2+}$ and CH_3O^- ions.

The high affinity for small or "pointed" negative ions of the $[Cu(Pyr)_4]^{2+}$ ion is attributed to a combination of steric and electrostatic factors.

For the cupric pyridine complexes scale models suggest that the first four pyridines sterically obstruct the coordination of a fifth pyridine ligand. The $[Cu(Pyr)_4]^{2+}$ complex has two relatively narrow holes at the 5th and 6th positions, which are only accessible to small ligands. The fact that these holes are largely surrounded by coordinated pyridine molecules probably creates an additional electrostatic effect leading to a strong preference of the holes for negatively charged small (or pointed) ligands like CH_3O^- or $CH_3.COO^-$.

We have not found any evidence for coordination of $[Cu^{2+}(Pyr)_4(OCH_3^-)]$ or $[Cu^{2+}(Pyr)_4(CH_3COO^-)]$ with a second negative ion. This is attributed to a closing of the second "hole" due to a slight deformation of the $Cu(pyridine)_4$ moiety by the first coordinated ion. Thus a five-fold coordination of Cu^{2+} would arise similar to the trigonal-bipyramidal complexes $Cu^{2+}(phen-anthroline)_2X$ and $[Cu^{2+}(bipyridine)_2X]$ described by HARRIS and coworkers (2).

References

1. BJERRUM, J., Chem.Revs., 46, 394 (1950).
2. HARRIS, C.N., LOCKYER, T.N., and WATERMAN, H., Nature, 192, 424 (1961); BARCLAY, G.A., and KENNARD, C.H.L., Nature, 192, 425 (1961).

4 A 2

ELECTRON PARAMAGNETIC RESONANCE OF CERTAIN COMPLEX COMBINATIONS OF Cu(II).

Petru Spacu, Marieta Brezeanu, Constanta Gheorghiu
and Valeriu Voicu
Department of Chemistry, University of Bucuresti, Roumania.

By producing EPR spectra of certain complexes of Cu(II), there has been discovered, on the one hand, the behaviour of the free electron of the central ion in this combinations and, on the other hand, as far as possible, its group of crystal symmetry.

The existence of the signal and its narrowing, in the case of certain combinations can be explained through the localization of the unpaired electron in one of the free orbitals of the central ion, that is, a limited exchange interaction between complex ions; in the case of other combinations EPR is proved to be absent, which can be explained through the delocalization of the electron and its redistribution in the ion as a result of a powerful interaction with the π orbitals of the ligands.

The deviation of the value of the g factor from the corresponding value of the free electron (2,0023), is a measure for the deviation from pure ionic bond, and may be due either to the crystal field created by the ligands or to the interaction of the electron with the π electron system of the ligands.

EPR has been measured at 9270 Mc/s, at room temperature on granulation powders $\leq 0,2$ mm.

There have also been studied different types of complex combinations of Cu(II), as shown in the following tables.

TABLE I

Complex	g^+ Factor 0,003			$\Delta H^1/2$ (gauss)
	g_1	g_2	g_3	
$[Cu\ py_2(NCS)_2]$	2,051	2,063	2,268	-
$[Cu\ py_4](C_6H_5COO)_2$		2,086		117,24
$[Cu(ampy)_2(NCS,Y)_2]$		-		-
$[Cu(ampy)_4]SO_4$		-		-
$[Cu\ dpy_2]SO_4$	2,065	2,078	2,186	-
$[Cu\ pip\ X_2]$		-		-
$[Cu\ pip(NCS)_2]$	2,0519	2,0598	2,2537	288,80
$[Cu(phtal)_2X_2]$		-		-
$[Cu(phtal)_2(NCS)_2]$	2,0567	2,0801	2,1903	196,24
$[Cu\ an_2(NCS)_2]$	2,056	2,082	2,260	-
$[Cu\ an_4]SO_4$	2,057	2,082	2,260	-
$[Cu\ tld\ (NCS)_2]$		2,0030		112,31
$[Cu\ bzd\ (NCS)_2]$	2,045	2,063	2,226	-
$[Cu\ bzd\ J_2]$	2,060	2,069	2,259	-
$[Cu\ bzd\ Cl_2]$	2,036	2,066	2,197	232,2
$[Cu(\beta\text{-napht})_2SO_4]$	2,0637	2,0880	2,2752	298,25
$[Cu(\beta\text{-napht})_2MoO_4]$	2,0650	2,0903	2,2758	296,81

TABLE (contd.)

$[Cu(\beta\text{-napht})_2 WO_4]$	2,0643	2,0883	2,2710	291,79
$[Cu(\beta\text{-napht})_2 CrO_4]$	2,0645	2,0867	2,2747	296,19
$[Cu(\beta\text{-napht})_2 Cr_2O_7]$	2,0637	2,0867	2,2693	290,69
$[Cu(\beta\text{-napht})_2(VO_3)_2]$	2,0647	2,0901	2,2730	293,68
$[Cu(\beta\text{-napht})_2(NCS)_2]$	2,0695	2,1069	2,1936	180,91

py = C_5H_5N; ampy = [structure] $-NH_2$; dpy = [structure] ; pip = [structure] ;

X = SO_4^{2-}, MoO_4^{2-}, WO_4^{2-}, CrO_4^{2-}, $Cr_2O_7^{2-}$, $(VO_3^-)_2$; phtal = [structure] ;

an = $C_6H_5NH_2$; tld = $(NH_2-$ [structure with CH_3] $)_2$; bzd = $(C_6H_4NH_2)_2$;

β-napht = [structure] $-NH_2$.

The majority of these combinations have in general similar values for g_1, whereas those for g_2 and g_3 are more different. This leads to the attribution of a square planar configuration to the respective combinations.

In a series of compounds, the anisotropy of the g factor disappears; among these is, notably, the compound of tolidine for which $g = 2,0030 \pm 0,003$ which may be used as "marker".

In combination of the type $[Cu\ phtal_2(NCS)_2]$ {I} $[Cu\ pip(NCS)_2]$ {II} and $[Cu(\beta\text{-naphtyl})_2(NCS)_2]$ {III} a pronounced narrowing of the signal is observed in comparison with the compounds of the same type but containing other anions. That narrowing is from the infinite to 196 {I} , respectively to 288 {II} and from 294 to 180 {III} . This behaviour can be explained through the localization of the electron in the practical free orbitals of the central ion.

In combination $[Cu\ pip(NCS)_2]$ and $[Cu\ bzd(NCS)_2]$ a hyperfine structure has been observed.

TABLE II

Complex	g^+ Factor 0,003			$\Delta H^1/2$ (gauss)
	g_1	g_2	g_3	
[Cu en$_2$]NO$_3$.(H$_2$O)$_2$	2,046	2,098	2,172	-
[Cu pald Cl$_2$]	2,042	2,058	2,210	239,6
[Cu pald$_2$](OH)$_2$		2,077		103,9
[Cu pald$_2$]SO$_4$		2,053		134
[Cu pald$_2$](SCN)$_2$		2,073		104

en = C$_2$H$_4$(NH$_2$)$_2$; pald = ClC$_6$H$_4$ $-NH-\underset{\underset{NH}{\|}}{C}-NH-\underset{\underset{NH}{\|}}{C}-NHC_3H_7$.

Compounds with the form [Cu pald$_2$]X$_2$ reveal analogous behaviour with regard to g and $\Delta H^1/2$ values; they differ from the [Cu pald Cl$_2$] compound which reveals anisotropy for the g factor.

TABLE III

[CuCl$_4$]H$_2$.en	g	$\Delta H^1/2$
[CuCl$_4$]H$_2$.2 Py	2,135	185,6
[CuCl$_4$]H$_2$.2 bzd	2,137	184,2
[CuCl$_4$]H$_2$.2 tld	2,068	53,7
[CuBr$_3$]H.py	2,087	264,5
[CuBr$_3$]H.pald	2,079	230,3

As it may be observed, there is in all compounds of this type a disappearence of the anisotropy of the g factor.

4 A 3

ESR SPECTRA OF CUPRIC COMPLEXES WITH SUBSTITUTED PYRIDINES.

A. v. Zelewsky and W. Schneider
Laboratory of Inorganic Chemistry, ETH, Zürich, Switzerland.

It has been shown, that the ESR spectra of transition metal complexes that exhibit a resolved ligand hyperfinestructure (hfs), give particularly useful information to perform semiempirical LCAO-MO calculations of such complexes (1, 2). The ESR data give a good measure of the degree of delocalisation of the unpaired electrons in the complex. MAKI and McGARVEY (3) have given a theory of the ESR spectra of copper(II) complexes, and they have shown, that the theory gives reasonable values for the bonding parameters of these complexes.

We have investigated the ESR spectra of a series of copper(II) complexes with substituted pyridines as ligands to see whether the substitution of the pyridine ligand has an effect on the bonding parameters of the complex. Furthermore we have compared the complex with unsubstituted pyridine in two different host lattices. To avoid any steric effects, only complexes with pyridine derivatives that are substituted in 4-position were studied.

All measurements were performed on magnetically diluted single crystals. As diamagnetic host lattices the corresponding cadmium(II)-complexes and in one case also the corresponding platinum(II) complex were used. The analysed diamagnetic compounds had the following compositions:

(i) $Pt(PyH)_4(NO_3)_2$

(ii) $Cd(PyH)_4(Ts)_2$ The symbols following Py in the

(iii) $Cd(PyCH_3)_4(Ts)_2$ bracket are the substituents in
                                        the 4-position.

(iv) $Cd(PyN(CH_3)_2)_4(Ts)_2$ Ts stands for the anion of para-

(v) $Cd(PyCN)_6(ClO_4)_2$ toluenesulfonic acid.

Single crystals were grown from aqueous solution (i), methanol (ii, iii, iv) or methyltetrahydrofuran (v). Copper(II) was introduced as a substantial impurity by growing single crystals from solutions containing a few percent of the corresponding copper complex. The estimated concentration of copper in the single crystals was about one percent.

We assumed, that the complexes possess a point group symmetry D_{4h}. In this symmetry (neglecting quadrupole interactions and the direct action of the magnetic field on the nucleus) the spin Hamiltonian takes the simple form of:

$$H_s = \beta \left[g_{\parallel} H_z S_z + g_{\perp}(H_x S_x + H_y S_y) \right] + A_{cu} S_z I_z + B_{cu} (S_x I_x + S_y I_y)$$

In D_{4h} symmetry the interaction of the unpaired electron with the nitrogen nuclei can be described by two further terms with coefficients A_N and B_N. The assumption of D_{4h} symmetry was justified, because neither the g tensors nor the copper hfs tensors showed any deviation from axial symmetry inside the experimental error.

Since the crystal structure of none of these compounds is known, the direction of the four-fold symmetry axis had to be found from the ESR spectra themselves.

Fig. 1 shows a typical ESR spectrum of one of these compounds (v). The angle between magnetic field and the four-fold axis of the complex is 42°. The splitting into the four main groups is due to the spin of the copper nucleus $\left(I = \frac{3}{2}\right)$. Each of these groups is split in 9 lines with relative intensities 1 : 4 : 10 : 16 : 19 : 16 : 10 : 4 : 1, which is a clear indication that the four nitrogen nuclei (I = 1) are in equivalent positions. The narrow peak on the right side of the spectrum is due to the free radical dpph, which was used to calibrate the field. As the angle between field and four-fold axis is increased, the interval between the four groups of lines decreases and they begin to

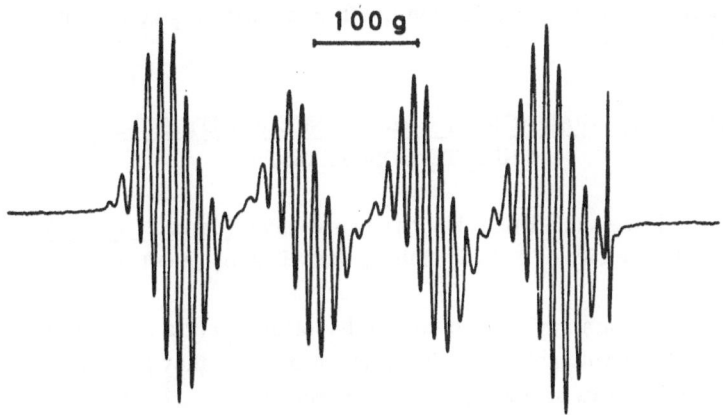

100 g

Fig. 1

merge; when the field is in the plane of the four nitrogens, only 14 lines are left in a single group. When the field is nearly parallel to the four fold axis, the spectrum becomes more complicated, because the small difference in the magnetic moments of the two copper isotopes (Cu^{63} and Cu^{65}) causes additional splitting in the outer wings of the four groups. Second-order effects in the copper hfs were found for all compounds. These effects have been accounted for in deducing the parameters from experimental data as listed in Table I.

TABLE I

PARAMETERS OF THE SPIN HAMILTONIAN FOR THE FIVE MEASURED COMPOUNDS.

Compound	$g_{\parallel}$	$g_{\perp}$	$A_{cu} \cdot 10^4 \, cm^{-1}$	$B_{cu} \cdot 10^4 \, cm^{-1}$	$A_N \cdot 10^4 \, cm^{-1}$	$B_N \cdot 10^4 \, cm^{-1}$
(i)	2,236	2,050	192,1	26	12,6	15,7
(ii)	2,290	2,050	161,5	21	11,2	13,1
(iii)	2,283	2,055	170,5	23	12,5	13,5
(iv)	2,283	2,055	166,4	25	12,5	15,1
(v)	2,274	2,066	173,4	22	11,2	13,5

(i)   The crystal is monoclinic, and has two paramagnetic complexes in the unit cell. The angle between the two four-fold axes is 78° 15'.

(ii) and (iii) Monoclinic, two complexes per unit cell, $\sphericalangle = 36^{\circ}$ and 72° respectively.

(iv) and (v) One complex per unit cell.

Calculation of the delocalization coefficients within a LCAO-MO scheme are in progress.

5 A 1

ELECTRON PARAMAGNETIC RESONANCE STUDY OF Ti(III) CHELATE COMPLEXES.

Shizuo F u j i w a r a , Kozo N a g a s h i m a and Maurice C o d e l l
University of Tokyo, Bunkyo-ku, Tokyo, Japan.

The electron paramagnetic resonance hyperfine spectra of the transition elements are of considerable scientific interest because they permit the study of the electronic states of the unpaired electrons in terms of the interaction with the nucleus.

Chelating reagents have been used in chelatometric titrations and as masking agents for colorimetric determinations and for separations by pre-cipitation or solvent extraction. This investigation has provided the first indication that the acids are an integral part of the complexes formed with Ti(III). It was found that Ti(III), EDTA, and oxalic acid form a very stable complex of equimolar composition which shows an absorbance maximum at 720 mµ. Tartaric acid appears to form a similar type of complex which is much less stable and also shows an absorbance maximum at 720 mµ. With this complex, however, the optical absorbance and EPR signal intensity increase with the addition of larger amounts of tartaric acid.

The three species listed in the following table are proposed to represent the complexes which exist in the solutions we have studied.

TABLE I

Ti(III) EDTA SPECIES IN SOLUTION

Formula	Formation in	Absorbance maxima	Remarks
$[Ti(III) EDTAH_2]^{2-}L_2$	1 acid solution	550 mμ	No EPR signal
$[Ti(III) EDTAH]^{3-}$	11 neutral solution	550 mμ	EPR signal g = 1,960
$[Ti(III) EDTA]^{4-}$	111 basic solution	580 to 600 mμ	EPR signal g = 1,956

$[EDTAH_2]^{2-}$, $[EDTAH]^{3-}$, and $[EDTA]^{4-}$ represent ions where two, three and four hydrogen ions are removed from the EDTA molecule containing four acid groups. L represents ligands other than EDTA such as H_2O, Cl^-, and NH_3. The K_a values can be calculated from the following equation.

$$[Ti(III) EDTAH]^{3-} \rightleftharpoons [Ti(III) EDTA]^{4-} + H^+$$

$$K_a = \frac{{}^a[Ti(III) EDTA]^{4-} \cdot {}^a H^+}{{}^a[Ti(III) EDTAH]^{3-}}$$

Table II gives relative areas of reconstructed EPR curves and calculated K_a constants for the system Ti(III) EDTA.

TABLE II

pH	Relative areas of separated and integrated curves	K_a (m/l), calc.
6,9	1:1,8	$2,1.10^{-7}$
7,7	1:6,8	$1,4.10^{-7}$
8,2	1:24	$1,5.10^{-7}$
8,6	1:30	$0,7.10^{-7}$

The temperature and salt concentration were not maintained constant for the data listed in table II. The calculated K_a values are not in excellent agreement; the effect of temperature and foreign salts such as ammonium chloride and sodium chloride, which usually effects the stability of chelates, must be taken into consideration.

For the first time, simultaneous signals from two complexes of Ti(III) with complexing reagents, with and without the addition of organic acids, have been obtained. These complexes can be observed because they have slightly different g values. It was found that the equilibrium mixtures of these complexes could be varied at will, which enabled the determination of factors which change the concentration of individual compounds. Corroboration of the results obtained by EPR was made by a series of spectrophotometric studies. These studies also show the existence of complexes which do not produce EPR signals, and by comparison of spectrophotometric and EPR spectra, a logical explanation of the chemistry of several classes of chelate compounds has been given.

During the couse of studying the EPR spectra of chelate compounds, it was noticed that Ti(III) with triethanolamine gives a signal with two peaks. This signal was obtained with $TiCl_3$. The signal could not be obtained with pure $TiCl_3$ solutions and an investigation showed that the presence of zinc in the solution was necessary in order to obtain the signal. It was found that signals could be obtained in the absence of triethanolamine by the interaction of the proper proportions of Ti(III), tartaric acid, and zinc, and that a fair degree of similarity was exhibited by the spectra of the two systems. The existence of another unusual complex formed by the interaction of Ti(III) and triethanolamine within the narrow pH range of 1,8 to 3,6 was demonstrated by the broad EPR signal exhibited by this complex. The complex appears to be fairly unstable since it cannot exist in the presence of tartaric acid.

Nitrilo triacetic acid (NTA) and hydroxyethylamino diacetic acid (HEID) have been shown to form complexes of 1 part Ti(III) to 1 part chelating reagents, and 1 part Ti(III) to 2 parts chelating reagents. Some of the complexes formed give EPR signals. The compositions of these complexes were spectrophotometrically established. Under certain conditions it was found that a signal of two peaks could be obtained for complexes formed by the interaction of Ti(III) and HEID and that this affords a means of calculating equilibrium constants for the mixture.

It has been demonstrated that a complex of great stability is formed by the interaction of Ti(III) and diethylenetriaminepentaacetic acid (DTPA), which gives large signals over the pH range of less than 1 to 13,5. A signal with two peaks is obtained by a mixture of Ti(III) and ethylenediamine-di-

-o-hydroxyphenylacetic acid (EDDHA). The difference in g values of these peaks is greater than those obtained with other chelating reagents tested.

5 A 2

ELECTRON TRANSFER IN TRIS-2,2'-DIPYRIDYL-COMPLEXES AS STUDIED BY ELECTRON PARAMAGNETIC RESONANCE.

Edgar König
Mellon Institute, Pittsburgh 13, Pennsylvania, U.S.A.

In order to determine experimentally the character of metal-ligand bonds in tris-2,2'-dipyridyl-complexes of transition metals in low oxidation states the electron paramagnetic resonance of the ions $[Cr(I)dip_3]^+$, $[V(O)dip_3]$ and $[Ti(-I)dip_3]^-$ was studied. The electronic configuration of these compounds is $(3d)^5$. Their paramagnetism is due to one unpaired electron as demonstrated by the magnetic moments of 2,05, 1,80 and 1,72 B.M. respectively (1).

The solution spectrum of the $[Cr(I)dip_3]^+$ ion shows two weak satellite lines, which were assigned to $m_I = +3/2$ and $m_I = -3/2$ due to interaction with the nuclei of the isotope Cr^{53}, the lines with $m_I = \pm 1/2$ being hidden by the strong central line due to Cr^{52}. The isotope V^{51} accounts for the eight line spectrum observed in solutions of $[V(O)dip_3]$. In both spectra every line is further split by interaction with the N^{14} nuclei of the 2,2'-dipyridyl ligands into 13 hyperfine lines. But, no such hyperfine structure is observed in the spectrum of the $[Ti(-I)dip_3]^-$ ion. The positions of the lines in the EPR spectra of these ions are given by the spin Hamiltonian $H = g \beta S.H + a S.I + \sum_N a^N S.I^N$,

where the last term describes the interaction with the nitrogen nuclei of the ligands. The g factors and isotropic hyperfine constants obtained are listed in Table I.

The spin density at the V(O) and Cr^+ nuclei may be given in terms of the parameter χ defined by ABRAGAM et al. (2):

$$\chi = \frac{4\pi}{S} \left\langle \sum_k \delta(r_K)s_{kz} \right\rangle_{S_z=S} = -\frac{3}{4} \frac{|A|}{g_I \beta \beta_N}$$

The values of χ obtained, in atomic units, are -2,51 for $[V(O)dip_3]$ and -3,05 for $[Cr(I)dip_3]^+$. The effective field at the nucleus, $H_e = (8\pi/3) g \beta S |\varphi(O)|^2$, arising from the contact term

$$H_c = -\frac{8\pi}{3} g g_I \beta \beta_N \delta(r) S.I$$

is found to be -106 kgauss and -126 kgauss respectively. These values are similar to those of other ions of the first transition series (2,3). Therefore, it seems possible to explain the hyperfine splitting (due to the odd isotopes of the central ions) by exchange polarization of inner core s electrons only. A more complicated model would have to be used however, to account for the observed ligand HFS. Since the overlap integral between the 3d orbital of the unpaired electron and a nitrogen 2s orbital is essentially zero, the N^{14}-HFS could be understood assuming exchange polarization of ligand π electrons and an additional unpairing of nitrogen s electrons similar to that in organic radicals. Splittings originating from this mechanism are, however, expected to be small.

Therefore, it seems reasonable to discuss a 4s electron contribution to the HFS. The closeness of the observed g factors to the free electron g factor of 2,0023 shows that the ground state is orbitally non-degenerate. In D_3 symmetry, the only orbital singlet $t_{2g}^0 = \frac{1}{\sqrt{3}} (d_{xy} + d_{xz} + d_{yz})$ mixes by configuration interaction with the 4s orbital giving for the molecular orbital of the unpaired electron

$$a_1 = \sin\theta \, [\beta \, t_{2g}^0 - \beta' \Pi_0^+] - \cos\theta \, [\alpha \, (4s) - \alpha' \Sigma_0^+] \qquad ,$$

where

$$\text{tg } 2\theta = - \frac{2 \langle t_{2g}^0 | V_{trig} | 4s \rangle}{E \{a_1(4s)\} - E\{t_{2g}^0\}}$$

and Π_0^+ and Σ_0^+ are the appropriate π- and σ-bonding linear combinations of ligand orbitals of symmetry A_1. Estimating the hyperfine splitting due to exchange polarization from an extrapolation of the results of SCF calculations (different orbitals for different spin) for the $(3d)^5$ configuration one obtains -134 gauss for V(O) and -26,4 gauss for Cr^+ (3). This gives a contribution of the 4s electrons to the HFS of +50,5 and +4,6 gauss respectively. The density of the unpaired electron $\rho_{4s,M}$ in the 4s orbital on the metal and $\rho_{2s,N}$ in the 2s orbital on a nitrogen atom of the ligand are obtained from the relation $a = a_0 \cdot \rho_i$; a_0, the splitting constant due to an electron occupying the orbit i in a free atom, has been calculated to be $a_{0,V(O)}^{4s} = 923,1$, $a_{0,Cr^+}^{4s} = 422,5$ and $a_{0,N}^{2s} = 546,1$ gauss. The results are

	$\rho_{4s,M}$	$\rho_{2s,N}$
[V(O)dip$_3$]	0,0547	0,0042
[Cr(I)dip$_3$]$^+$	0,0109	0,0056

The numerical values obtained for the electron density indicate strong σ-bonding between metal 4s and ligand Σ_0^+ orbitals which increases in the order (Ti$^-$), V(O), Cr$^+$. The coefficients α and α' resulting from the densities are understood, if one assumes that the ligand orbitals involved are higher in energy than the metal 4s orbital. The amount of configuration mixing between the metal t_{2g}^0 and 4s orbital is given by $\cos\theta = 0,258$ and $0,265$ respectively.

Two other possibilities which might have been invoked to explain the observed hyperfine splitting are here ruled out: (i) Since the sign of a_{Me} could not be determined in this investigation, it could be assumed that it is, in contrast to the usual case, positive for V(O) and negative for Cr$^+$. The contribution of the 4s electron to the HFS would be $+217,5$ and $+48,2$ gauss respectively, which results in the mixing of $23,6\%$ and $11,4\%$ 4s-character into the t_{2g}^0 orbitals. As this is even more than that observed in sandwich molecules in the same oxidation state, it therefore does not seem reasonable. (ii) Assuming a_{Me} to have the same sign as in (i), an interpretation of the observed HFS as originating from 4s electrons only (4), also seems unreasonable in view of the recent calculations of exchange polarization for a series of transition metals (3).

TABLE I

SUMMARY OF EPR RESULTS FOR TRIS-2,2'-DIPYRIDYL-
COMPLEXES OF Cr(I), V(O) and Ti(-I).

Compound	Solvent/ Concentration (mole liter^{-1})	g	$|A_{Me}|$ (gauss)	$|A_N^{14}|$ (gauss)
[Cr(I)dip$_3$]ClO$_4$	CH$_3$OH/ $5,8.10^{-4}$	$g_{Cr}^{52} = 1,9971$ $\pm 0,0002$ $g_{Cr}^{53} = 1,9973$ $\pm 0,0002$	$a_{Cr}^{53} = 21,8$ $\pm 0,5$	$3,05$ $\pm 0,05$
[V(O)dip$_3$]	THF/ $5,3.10^{-4}$	$1,9831$ $\pm 0,0002$	$a_V^{51} = 83,5$ $\pm 1,0$	$2,3$ $\pm 0,1$
Li[Ti(-I)dip$_3$]	THF/ $2,6.10^{-4}$	$2,0074$ $\pm 0,0002$		

References

1. PERTHEL, R., Z.Phys.Chem. (Leipzig), 211, 74 (1959).
2. ABRAGAM, A., HOROWITZ, J., and PRYCE, M.H.L., Proc.Roy.Soc., A 230, 169 (1955).

3. WATSON, R.E., and FREEMAN, A.J., Phys.Rev., 123, 2027 (1961).
4. WEBER, G., Z.Phys.Chem. (Leipzig), 218, 204, 217 (1961).

A 3

ISOTROPIC NMR SHIFTS IN PARAMAGNETIC COBALT AND NICKEL PHOSPHINE COMPLEXES.

William DeW. Horrocks, Jr. and Gerd N. LaMar
Department of Chemistry, Princeton University, Princeton, N.J., U.S.A.

The conditions necessary for observation of proton magnetic resonance spectra in paramagnetic systems are well established (1). Either the electronic spin-lattice relaxation time, T_1, or a characteristic electronic exchange time, T_e, must be short compared with the isotropic hyperfine contact interaction constant, A_i, in order for resonances to be observed. Proton resonances in paramagnetic systems are often shifted hundreds of cps from their values in the diamagnetic substances. These isotropic resonance shifts may arise from two causes, the hyperfine contact and pseudocontact interactions. The contact shift arises from the existence of unpaired spin-density at the resonating nucleus and is described by {1} (2) for systems obeying the Curie law,

$$\left(\frac{\Delta H}{H}\right)_i = -A_i \frac{\gamma_e}{\gamma_H} \frac{g \beta (S+1)}{6 kT} \qquad \{1\}$$

where $\left(\frac{\Delta H}{H}\right)_i$ is the contact shift for the ith proton, A_i is the hyperfine contact interaction constant, γ_e and γ_H are the gyromagnetic ratios for the electron and proton respectively, g is the spectroscopic splitting factor, and β is the Bohr magneton and the other symbols have their usual significance. For unpaired electron spin-density delocalized in the π-orbitals of aromatic systems, the spin-density on the ith carbon atom, ρ_i, is related to the hyperfine contact interaction constant by relation {2} (3)

$$A_i \cong Q_{CH} \rho_i \qquad \{2\}$$

where Q_{CH} is a negative constant approximately equal to -22,5 gauss, the value obtained for the benzene negative ion (4). An analogous expression for the hyperfine interaction of a methyl group may be written, but where the proportionality factor Q_{CH_3} is positive. The admirable work of PHILLIPS, BENSON and their associates on the Ni(II)aminotroponeimineates (5) is satis-

factorily interpreted in terms of contact shifts due to unpaired spin-density
in the π-systems of these unsaturated ligands.

We have measured the PMR spectra of some bis(triarylphosphine) complexes of the Co(II) and Ni(II) halides in CDCl₃ solution. These compounds
are paramagnetic with approximately tetrahedral coordination about the metal
atom (6-8). Isotropic shifts for the various protons in these complexes in
solution at 25°, referred to the diamagnetic ligand resonances, are given in
Table I, and are shown schematically in Fig. 1.

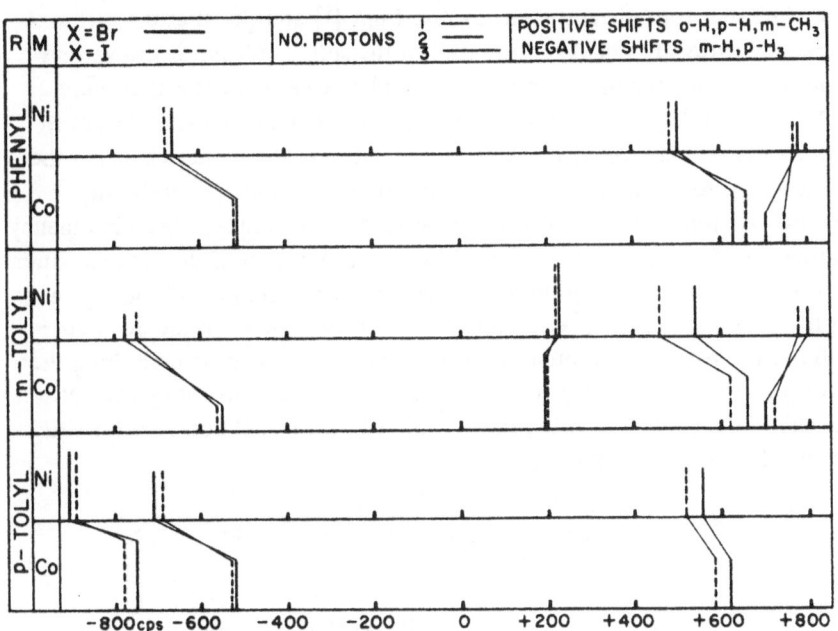

Fig. 1. - Schematic Representation of the
Observed Isotropic NMR Shifts in the Cobalt and
Nickel Complexes.

The alternate up-field, down-field shifts of adjacent protons indicates
that some spin-density is delocalized into the π-orbitals of the phenyl rings.
Unpaired electron spin-density is transferred from the metal d orbitals to the
vacant 3d orbitals of the P atoms (dπ-dπ bonding). Positive spin-density on
the P is transmitted by a dπ-pπ interaction with the phenyl rings so as to place
positive spin-density at the ortho and para carbons. A negative spin-density
will result at the meta position through spin-correlation. {1} and {2} indicate
that up-field shifts are to be expected for ortho and para hydrogens and meta
methyl groups and down-field shifts for the meta hydrogen and para methyl

resonances from this mechanism. The isotropic shifts given in Table I are in accord with these expectations.

However, if we consider the ratios of the observed isotropic shifts for the para/ortho and meta/ortho proton resonances shown in the first four columns of Table II, an interesting fact is noted. While the corresponding ratios (p/o and m/o) agree fairly well within the sets of cobalt and nickel compounds respectively, there is a considerable discrepancy between the corresponding ratios in the cobalt and nickel complexes (compare Table II: col. 1 with col. 3; col. 2 with col. 4). If only a π-type contact interaction is responsible for the observed isotropic shifts, it is clear from {1} and {2} that the ratios of the observed shifts are equal to the ratios of the spin-densities at the various carbons of the phenyl ring. Spectroscopic evidence (9) indicates that there is little or no conjugation through the P atom between phenyl rings bound to a P which has no lone pairs. It seems highly unlikely then, that changing the metal from Co to Ni would cause the major redistribution of spin-density which the difference in the ratios in Table II seems to imply. This discrepancy in isotropic shift ratios can be removed by a consideration of the pseudocontact interaction. Pseudocontact shifts in PMR spectra arise from an anisotropy in the g tensor of the paramagnetic complex and they depend on the geometric position of the resonating proton in the molecule. An equation describing the pseudocontact shift in axially symmetric molecules was derived by McCONNELL and ROBERTSON (10). Following those authors La MAR (11) botained {3} valid for molecules of C_{2v} symmetry,

$$\left(\frac{\Delta H}{H}\right)_i = -\frac{\beta^2 S(S+1)}{27\,kT}\,(g_1 + g_2 + g_3)\,\left\{(g_1 - 1/2\,g_2 - 1/2\,g_3)\right.$$

$$\left[\frac{3\cos^2 \chi_i - 1}{r_i^3}\right] + 3/2\,(g_2 - g_3)\left[\frac{\sin^2 \chi_i\,(\cos^2 \Omega_i - \sin^2 \Omega_i)}{r_i^3}\right]\right\}$$

$$\{3\}$$

where g_1 is the g-factor along the principal (C_2) axis, g_2 and g_3 are the g-factors in the planes including the two P atoms and the two halogen atoms respectively, r_i is the distance from the metal to the ith proton, χ_i is the angle that a radius vector from the metal to the ith proton makes with the principal axis, and Ω_i is the angle that its projection in the g_2, g_3 plane makes with the g_2 axis.

We are able to explain the difference in isotropic shift ratios between the cobalt and nickel compounds on the basis of a pseudocontact interaction if it is assumed that the shifts in the nickel compounds observed by us and by LaLANCETTE and EATON (12) are due only to a π-type contact interaction. Further, we postulate that the observed isotropic shifts in the Co compounds arise from both contact and pseudocontact interactions. Supporting this hypothises is the fact that the spin-density ratios in the nickel phosphine complexes are similar to those of the N-phenyl derivatives of the nickel(II)aminotropone-imineates where the pseudocontact contribution is believed to be negligible (1).

The g-value anisotropies for the complexes are unknown, but the geometric factors which give the ratios of the pseudocontact shifts of the various protons can be calculated assuming reasonable structural parameters (13). (An IBM 7094 computer was used to calculate the average values of the geometric factors for all internal rotations about the metal-phosphorous and phosphorous-carbon bonds.) The ratios of these calculated factors for the $\left\langle \dfrac{3 \cos^2 X_i - 1}{r_i^3} \right\rangle_{av.}$ term for the various protons fall in the sequence -ortho/meta/para/para-methyl/meta-methyl = 1,000/0,247/0,357/0,098/0,050.

We were able to obtain contact shifts for the cobalt compounds by subtracting a pseudocontact contribution from the observed shifts. If X_0 is the value of the pseudocontact shift of the ortho proton resonance in a cobalt complex, the pseudocontact shift for the other protons can be obtained from it using the calculated geometric ratios, i.e., the pseudocontact shift for the meta proton is $0,247 X_0$. It was possible to find a unique value of the pseudocontact shift, X_0, which, with corresponding pseudocontact shifts for the other protons, when subtracted from the observed cobalt resonance shifts, gave corrected cobalt contact shifts whose ratios matched the observed nickel ratios satisfactorily. The values of X_0 and the corrected shifts are given in Table I. The pseudocontact shift ratios calculated for the second geometric factor,

$$\left\langle \frac{\sin^2 X_i (\cos^2 \Omega_i - \sin^2 \Omega_i)}{r_i^3} \right\rangle_{av.}$$

are remarkably similar to the first and a somewhat less satisfactory agreement of contact shift ratios can be obtained assuming that this term is dominant. A contribution to the pseudocontact shift from both terms is certainly possible.

A calculation of Q_{CH_3}, assuming identical spin density at the para position in the phenyl and p-tolyl compounds, yields $Q_{CH_3} \cong 26,3$ gauss using the observed nickel data and $Q_{CH_3} \cong 23,6$ gauss from the uncorrected cobalt shifts. The pseudocontact correction brings Q_{CH_3} up to 27,5 gauss for the cobalt compounds. A value somewhere around 27 gauss might be expected (14).

Our observation (15) of isotropic shifts of the butyl proton resonances in the ionic compound $[(C_4H_9)_4N]^+[(C_6H_5)_3P\,CoI_3]^-$ and the corresponding nickel complex affords direct evidence for a pseudocontact interaction in these related systems. In these compounds, which exist as ion pairs in chloroform solution, there is no way for spin-density to arrive on the tetrabutyl ammonium cation.

TABLE I

Compound	Observed Isotropic Shifts at 25° (cps)				Isotropic Shifts of Cobalt Compounds Corrected for Pseudocontact Interaction (cps)				
	ortho	meta	para	-CH₃	ortho	meta	para	-CH₃	X₀ᵃ
$[(C_6H_5)_3P]_2CoBr_2$	630	-510	705	-	420	-568	630	-	210
$[(C_6H_5)_3P]_2CoI_2$	662	-519	746	-	417	-587	659	-	245
$[(C_6H_5)_3P]_2NiBr_2$	505	-660	780	-	505	-660	780	-	0
$[(C_6H_5)_3P]_2NiI_2$	487	-678	771	-	487	-678	-771	-	0
$[(p-CH_3C_6H_4)_3P]_2CoBr_2$	609	-523	-	-759	436	-570	-	-776	173
$[(p-CH_3C_6H_4)_3P]_2CoI_2$	638	-528	-	-779	432	-584	-	-799	206
$[(p-CH_3C_6H_4)_3P]_2NiBr_2$	548	-712	-	-909	548	-712	-	-909	0
$[(p-CH_3C_6H_4)_3P]_2NiI_2$	512	-693	-	-896	512	-693	-	-896	0
$[(m-CH_3C_6H_4)_3P]CoBr_2$	615	-547	700	196	420	-600	629	186	195
$[(m-CH_3C_6H_4)_3P]CoI_2$	620	-560	720	197	380	-625	634	185	240
$[(m-CH_3C_6H_4)_3P]NiBr_2$	540	-775	801	240	540	-775	801	240	0
$[(m-CH_3C_6H_4)_3P]NiI_2$	460	-750	776	222	460	-750	776	222	0

a. X_0 is the pseudocontact shift of the ortho hydrogen (cps).

TABLE II

RATIOS OF ISOTROPIC SHIFTS (OBSERVED AND CORRECTED)

Compound/M	Co observed		Ni observed		Co Corrected for pseudocontact shift	
ratio	p/o	m/o	p/o	m/o	p/o	m/o
$[(C_6H_5)_3P]_2MBr_2$	1.119	-0.809	1.545	-1.306	1.503	-1.348
$[(C_6H_5)_3P]_2MI_2$	1.126	-0.784	1.583	-1.392	1.578	-1.406
ratio	p-CH₃/o	m/o	p-CH₃/o	m/o	p-CH₃/o	m/o
$[(p-CH_3C_6H_4)_3P]_2MBr_2$	-1.246	-0.859	-1.660	-1.302	-1.778	-1.307
$[(p-CH_3C_6H_4)_3P]_2MI_2$	-1.221	-0.828	-1.751	-1.354	-1.851	-1.352
ratio	p/o	m/o	p/o	m/o	p/o	m/o
$[(m-CH_3C_6H_4)_3]_2MBr_2$	1.139	-0.889	1.483	-1.436	1.497	-1.429
$[(m-CH_3C_6H_4)_3]_2MI_2$	1.162	-0.903	1.688	-1.630	1.669	-1.645

This research was supported in part by the Air Force Office of Scientific Research (AF-AFOSR-242-63) and by the Petroleum Research Fund (PRF-1179-A1).

References

1. EATON, D.R., JOSEY, A.D., PHILLIPS, W.D., and BENSON, R.E., J.Chem.Phys., 37, 347 (1962), and references cited therein.
2. McCONNELL, H.M., and CHESTNUT, D.B., J.Chem.Phys., 28, 107 (1958).
3. McCONNELL, H.M., J.Chem.Phys., 24, 632 (1956).
4. WEESSMAN, S.I., TUTTLE, T.R., and deBOER, E., J.Phys.Chem., 61, 28 (1957).
5. EATON, D.R., JOSEY, A.D., PHILLIPS, W.D., and BENSON, R.E., J.Chem.Phys., 39, 3513 (1963); EATON, D.R., PHILLIPS, W.D., and CALDWELL, D.J., J.Am.Chem.Soc., 85, 397 (1963); EATON, D.R., JOSEY, A.D., PHILLIPS, W.D., and BENSON, R.E., Discuss. Faraday Soc., 34, 77 (1962); and references cited therein.
6. COTTON, F.A., FAUT, O.D., GOODGAME, D.M.L., and HOLM, R.H., J.Am.Chem.Soc., 83, 1780 (1961).
7. COTTON, F.A., FAUT, O.D., and GOODGAME, D.M.L., J.Am. Chem.Soc., 83, 344 (1961).
8. VENANZI, L., J.Chem.Soc. (London), 1958, 719.
9. JAFFÉ, H.H., J.Chem.Phys., 22, 1430 (1954).
10. McCONNELL, H.M., and ROBERTSON, R.E., J.Chem.Phys., 29, 1361 (1958).
11. LaMAR, G.N., THESIS, Ph.D., Princeton University, 1964.
12. LaLANCETTE, E.A., and EATON, D.R., to be published.
13. GARTON, G., HENN, D.E., POWELL, H.M., and VENANZI, L.M., J.Chem.Soc. (London), 1963, 3625.
14. BOLTON, J.R., CARRINGTON, A., and McLACHLAN, A.D., Mol. Phys. 5, 31 (1962).
15. LaMAR, G.N., HORROCKS, Jr., W.D., and ALLEN, L.C., to be published.

6 A 1

PURE QUADRUPOLE RESONANCE STUDIES ON HEXAHALORHENATE(IV) AND HEXACHLOROTUNGSTATE(IV).

Daiyu Nakamura, Ryuichi Ikeda and Masaji Kubo
Department of Chemistry, Nagoya University, Chikusa, Nagoya, Japan.

We have already studied the pure quadrupole resonance of halogens in potassium hexahaloplatinate(IV) (1), hexachloroiridate(IV), and hexachloroosmate(IV) (2). It was shown that metal-ligand bonds in the latter two complexes, which are paramagnetic, involve π-bond character, because the $5d_\epsilon$ orbitals of iridium and osmium ions in the cubic field have one and two holes, respectively. This paper presents the results of further studies on paramagnetic potassium hexahalorhenate(IV) and hexachlorotungstate(IV), rhenium and tungsten ions in which have three and two electrons in their degenerate $5d_\epsilon$ orbitals leaving three and four vacancies, respectively.

Potassium hexachlororhenate(IV) shows a single resonance line for each chlorine isotope at room and dry ice temperatures in agreement with the cubic potassium hexachloroplatinate(IV) structure of this compound in the crystalline state, but two lines at liquid nitrogen temperature indicating the existence of crystallographically nonequivalent chlorine atoms in crystals. Potassium hexabromorhenate(IV) gives rise to a single Br^{79} resonance line at room temperature, two lines at dry ice temperature, and three lines at liquid nitrogen temperature. Phase transition takes place at about -4^0, -16^0, and -27^0 (Fig. 1).

Potassium hexaiodorhenate(IV) yields v_1 and v_2 resonances, which are triplets at all temperatures studied, the asymmetry parameter being almost equal to zero ($\eta \le 0.07$). For potassium hexachlorotungstate(IV), only a weak single line was observed at $10-35^0$.

With an assumption that the asymmetry parameter, η, is zero for hexachloro and hexabromo complexes, the nuclear quadrupole coupling constant, eQq, was evaluated and the ionic character, i, of metal-ligand bonds was calculated by a modified Townes-Dailey formula (2),

$$eQq = [(1 - s)(1 - i - \pi) - \pi/2](eQq)_{atom}$$

where $(eQq)_{atom}$ stands for the atomic quadrupole coupling constant. The extent of s-character in the sp hybridized bonding orbitals of halogens was assumed to be s = 15%. From electron spin resonance experiments carried out by GRIFFITHS, OWEN, et al. (3-5), the π-bond character of metal-ligand bonds in hexachloroiridates(IV) is estimated to be $\pi = 5.3\%$. From the observed g-values in the electron spin resonance of hexachloro- and hexabromoiridate(IV) ions, GRIFFITHS and OWEN (4) have shown that Ir-Cl and Ir-Br bonds in these complexes have nearly the same π-bond character. Since similar experiments have not yet been performed on other complexes under investigation, it was assumed that the d_π-p_π bond character, π, of metal-halogen bonds is proportional to the number of electronic vacancies in the d_ϵ orbitals of the central metal ion. The calculated ionic character of metal-ligand bonds in complexes $K_2[MCl_6]$ falls in a narrow range (44% for platinum, 47% for iridium, 47% for osmium, 44% for rhenium, and 44% for tungsten complexes). The ionic character of metal-halogen bonds decreases with increasing atomic number of halogens (44%, 39%, and 32% for potassium hexahalorhenates(IV)).

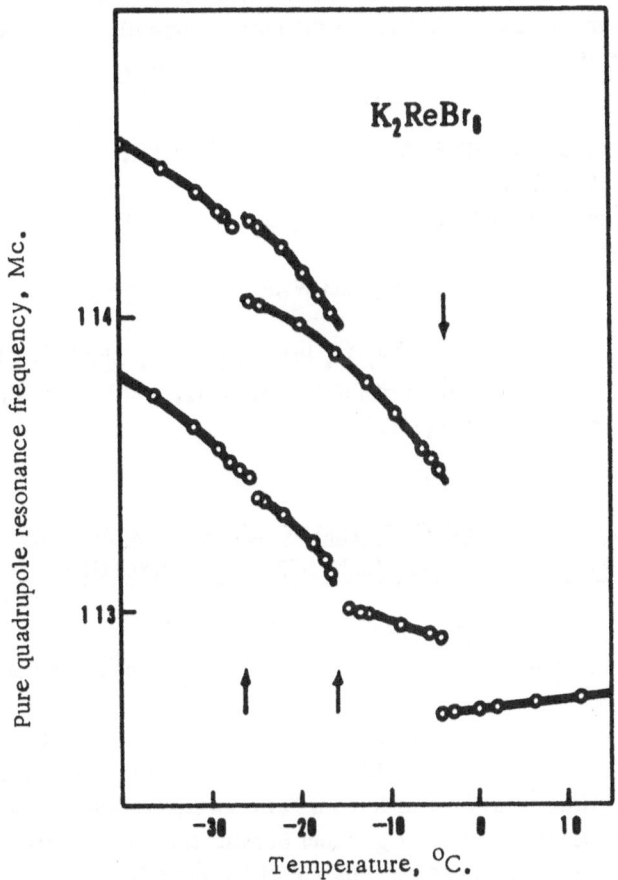

Fig. 1. Pure quadrupole resonance frequencies of Br^{79} in potassium hexabromorhenate(IV) showing the existence of three transition points at about -4°, -16°, and -27°C.

The temperature coefficient (at room temperature) of resonance frequencies of potassium hexachloro complexes increases progressively with decreasing atomic number of the central metal atom, i.e., with increasing electron deficiency of the central metal ions from Pt(IV) to W(IV) (-1, 0,

-0,54, -0,22, +0,13, and +0,44 kc/deg). It is quite remarkable that the temperature coefficient of resonance frequencies of potassium hexachloro-rhenate(IV), hexachlorotungstate(IV), and hexabromorhenate(IV) (+2,8 kc/deg) is positive, whereas in the majority of cases, the temperature coefficient of pure quadrupole resonance frequencies is negative. The progressive increase of the temperature coefficients suggests that some close relationship exists between the temperature coefficient of the pure quadrupole resonance frequencies and the extent of π-bond character of metal-ligand bonds.

References

1. NAKAMURA, D., KURITA, Y., ITO, K., and KUBO, M., J.Am.Chem. Soc., 82, 5783 (1960).
2. ITO, Ko., NAKAMURA, D., ITO, K., and KUBO, M., Inorg.Chem., 2, 690 (1963).
3. GRIFFITHS, J.H.E., OWEN, J., and WARD, I.M., Proc.Roy.Soc., A219, 526 (1953).
4. GRIFFITHS, J.H.E., and OWEN, J., Proc.Roy.Soc., A226, 96 (1954).
5. CIPOLLINI, E., OWEN, J., THORNLEY, J.H.M., and WINDSOR, C., Proc.Phys.Soc., 79, 1083 (1962).

6 A 2

PMR-SPEKTROSKOPISCHE UNTERSUCHUNGEN AN AROMAT- -METALL(O)-TRICARBONYLEN DER VI. NEBENGRUPPE.

C.G. Kreiter und H.P. Fritz
Institut für anorganische Chemie der Universität München, Germany.

Die Ringprotonen monosubstituierter Benzole bilden ein A_2B_2C-Spinsystem (1), dessen komplizierte PMR-Spektren eine vollständige Analyse erschweren. Bei Disubstitution resultiert im Falle zweier verschiedener Substituenten in p-Stellung oder zweier gleicher Substituenten in o-Stellung ein A_2B_2 oder A_2X_2-Spinsystem, dessen Analyse direkt ohne Näherungsverfahren fast sämtliche PMR-Parameter liefert (2). Die einzig noch verbleibende Schwierigkeit ist die Zuordnung der beiden erhaltenen chemischen Verschiebungen zu der richtigen Protonensorte. Diese kann entweder aus analogen, bereits richtig gedeuteten Spektren erfolgen, oder durch die Einführung von Substituenten, welche in geringem Ausmaß in eine Spin-Spin-Wechselwirkung mit den Ringprotonen treten. Günstig ist eine Methylgruppe, deren Protonen mit den zu ihr o-ständigen Ringprotonen stärker als mit den m-ständigen koppeln (0,7 bzw. 0,3 Hz). Die PMR-Spektren von p-substituierten Toluolen sind einer Analyse gut zugänglich, da die Ringprotonen ein einfaches A_2B_2-System bilden und die Signale der zur Methylgruppe o-ständigen Protonen eine deutliche Verbreiterung durch die Spin-Spin-Kopplung mit den Methylprotonen zeigen. Bei der Einführung dieser Aromaten in Übergangsmetallkomplexe sollte auch sofort eine richtige Zuordnung und Analyse möglich sein.

Aromat-metall(O)-tricarbonyle sind allgemein durch die Reaktion der Aromaten mit den entsprechenden Hexacarbonylen in geeigneten Lösungsmitteln, bei erhöhter Temperatur im offenen System in guten Ausbeuten zugänglich (3-5).

$$M(CO)_6 + Ar \longrightarrow ArM(CO)_3 + 3\ CO. \qquad (M = Cr,\ Mo,\ W)$$

Die Veränderungen, die die PMR-Spektren p-substituierter Toluole durch diese Komplexbildung erfahren, betreffen im wesentlichen nur die Signale der Ringprotonen. Die Signale anderer, ebenfalls protonenhaltiger Gruppen, die am Ring gebunden sind, erscheinen an nahezu der gleichen Stelle wie im freien Aromaten. Die Signale der Ringprotonen werden jedoch durch die Komplexbildung um etwa 1,5 bis 2,0 ppm nach höheren Feldern verschoben.

Dies läßt sich mit einem einfachen Modell erklären:

Im Benzol-chrom(O)-tricarbonyl ist der Molekülbereich, den das Zentralmetall mit dem Benzolring bildet, als rotationssymmetrisches Elektronensystem beschreibbar, auf welches die Beziehung (6)

$$\sigma_{ar}(G) = \frac{1}{3r^3}(3\cos^2\Theta - 1)(\chi_L - \chi_T)$$

angewendet werden kann. Die Länge von r gibt die Entfernung eines Protons vom Schwerpunkt der G Elektronen an, Θ ist der Winkel, den r mit der Rotationsachse bildet, χ_L und χ_T sind die longitudinale und transversale Suszeptibilität. Nimmt man an, daß durch die Komplexbildung die Differenz der beiden Suszeptibilitäten von der gleichen Größenordnung bleibt, wie im freien Benzol, so führt der durch Dipol- (7-9) und IR (10)-Messungen bewiesene Elektronenzug des Zentralmetalls zu einer Entfernung des elektronischen Schwerpunkts aus der Ringebene zum Zentralmetall hin. Das führt zu einer Vergrößerung von r und einer Verkleinerung des Winkels Θ. Beide Veränderungen haben eine starke zahlenmäßige Verkleinerung der negativen Abschirmung zur Folge, die einer Verschiebung der Signale nach höheren Feldern entspricht. Das von uns angegebene Modell liefert jedoch nur die Richtung des Effekts, da das Fehlen der Meßgrößen χ_L, χ_T, r und Θ eine zahlenmäßige Abschätzung unmöglich macht, und die angegebene Formel nur für den idealen Fall exakt gilt, daß sich die G Elektronen punktförmig am elektronischen Schwerpunkt befinden.

Die von uns zur Komplexbildung verwendeten p-substituierten Toluole trugen drei Sorten von Substituenten:

a) Elektronenschiebende Gruppen: OH, NH_2, OCH_3, $N(CH_3)_2$, $Si(CH_3)_3$
b) Elektronenziehende Gruppen mit freien Elektronenpaaren: F, Cl, Br
c) Elektronenziehende Gruppen: $COOCH_3$, CHO

Die beobachteten Verschiebungen der Ringprotonensignale sind stark von der Art der Substituenten abhängig. Die Signale von den zu den stärker elektronenziehenden Substituenten o-ständigen Protonen erfahren eine geringere Verschiebung nach höheren Feldern als die m-ständigen. Das führt mit der Tatsache, daß die ersteren immer bei niederen Feldern erscheinen als die letzteren, zu einer Vergrößerung der chemischen Verschiebung zwischen beiden Protonensorten um einen geringen Betrag.

Die Verschiebung des gesamten Liniensystems nach höheren Feldern ist von der Summe der elektronischen Wirkung der Substituenten abhängig. Überwiegt der elektronenabgebende Einfluß, so ist die Verschiebung geringer als im umgekehrten Fall. Diese eigentlich überraschende Erscheinung legt die Annahme nahe, daß die Abschirmung der Ringprotonen im Komplex von der Art der Substituenten unabhängiger ist als in den freien Aromaten. So erscheinen z.B. die Ringprotonensignale von Benzol und p-Xylol bei $\tau = 2,73$ und 3,06, die der entsprechenden Chrom-tricarbonylkomplexe bei $\tau = 4,68$ und 4,74 gegen intern. TMS. Die elektronenabgebende oder elektronenziehende Wirkung von Substituenten wird weitgehend vom Zentralmetall aufgefangen. Dieses Ergebnis ist mit Dipolmessungen (7-9) gut in Einklang zu bringen. Komplexe mit elektronenziehenden Substituenten am Benzolring zeigen ein kleineres Dipolmoment als solche mit elektronenabgebenden. Das entspricht einer geringeren bzw. größeren π-Elektronenabgabe an das Zentralmetall.

Von den vier Kopplungskonstanten, die zwischen zwei identischen Spinpaaren denkbar sind, erfährt die zwischen o-ständigen Protonen die auffälligste Änderung. In den Chromkomplexen erniedrigen sich die Werte von etwa 8,8 Hz bis 7,8 Hz auf 6,8 bis 6,4 Hz. Auch die Kopplungskonstanten zeigen eine ähnliche Tendenz wie die chemischen Verschiebungen: die Komplexbildung bewirkt eine stärkere Angleichung der Werte untereinander.

In den gleichen Aromaten erniedrigen sich die o-Kopplungskonstanten bei der Komplexbildung an Cr um 1,4 Hz, an Mo um 1,2 Hz und an W um 1,5 Hz. Die Gesamtänderung der Kopplungskonstanten ist größer, als der Beteiligung der π-Elektronen daran zukommt (11). Die Komplexbindung ändert mithin auch das σ-Bindungsgerüst.

Zu dem unsteten Verhalten der Änderung der o-Kopplungskonstante, wenn man von Cr über Mo zum W übergeht, kommt noch die chemische Verschiebung der Ringprotonen. Z.B. bei den p-Xylolkomplexen τ = 4,74, 4,37, 4,55. Die PMR-Daten geben das aus der homologen Reihe fallende Verhalten der Molybdänverbindung besser wieder als IR- und Dipolmessungen. Die Labilität der Molybdänkomplexe wird durch die geringste Änderung der o-Kopplungskonstante und der chemischen Verschiebungen relativ zum freien Aromat gekennzeichnet.

Literatur

1. CORIO, P.L., und DAILEY, B.P., J.Am.Chem.Soc., 78, 3043 (1956); BOTHNER-BY, A.A., und GLICK, R.E., J.Am.Chem.Soc., 78, 1071 (1956).
2. McCONNELL, H.M., McLEAN, A.D., und REILLY, C.A., J.Chem. Phys., 23, 1152 (1955); POPLE, J.A., SCHNEIDER, W.G., und BERNSTEIN, H.J., Can.J.Chem., 35, 1060 (1957).
3. FISCHER, E.O., ÖFELE, K., ESSLER, H., FRÖHLICH, W., MORTENSEN, J.P., und SEMMLINGER, W., Chem.Ber., 91, 2763 (1958).
4. NICHOLLS, B., und WHITING, M.C., Proc.Chem.Soc., 1958, 152.
5. NATTA, G., ERCOLI, R., und CALDERAZZO, F., La Chim.e 1'Ind., 40, 287 (1958).
6. McCONNELL, H.M., J.Chem.Phys., 27, 226 (1957).
7. FISCHER, E.O., und SCHREINER, S., Chem.Ber., 92, 938 (1959).
8. RANDALL, E.W., und SUTTON, L.E., Proceed. 1959, 93.
9. STROHMEIER, W., und HELLMANN, H., Z.Elektrochem., 67, 190 (1963).
10. FISCHER, R.D., Chem.Ber., 93, 165 (1960); HUMPHREY, R.E., J.Am. Chem.Soc., 82, 93 (1960).
11. McCONNELL, H.M., J.Chem.Phys., 30, 126 (1959).

6 A 3

NMR STUDIES OF MICROSCOPIC SITES OF PROTONATION AND METAL-BINDING WITH MULTIDENTATE COMPLEXING AGENTS.

Charles N. Reilley
Department of Chemistry, University of North Carolina,
Chapel Hill, North Carolina, U.S.A.

I. Protonation Sites.

NMR techniques have been used to indicate the time-average location of protons on multidentate complexing agents. For example, a proton spends an equivalent time on each of the nitrogens in symmetrical compounds such as ethylenediamine, EDTA, EGTA, etc. This conclusion is reached from the fact that the chemical shifts of $-CH_2-$ protons adjacent to the N-ligand atom are equal. In a species such as DTPA (Y^{5-}), where the "middle" acetate $-CH_2-$ protons are distinguishable from the other four $-CH_2-$ acetate protons, the relative extent of chemical shifts downfield suggests that the first proton added (forming HY^{4-}) spends more of its time on the central N-atom. After a second equivalent of acid has been added (forming H_2Y^{3-}), the middle acetate $-CH_2-$ proton peak shifts upfield, very close to the field at which it occurred for the completely non-protonated basic form (Y^{5-}). The down-field shift of all the remaining acetate $-CH_2-$ protons then shows that the H_2Y^{3-} species is chiefly protonated on the two end nitrogen atoms. Similar results have been obtained with the polyamines. Quantitative evaluation of the percent protonation of the various ligand atoms has been made through computer computations using substituent constants obtained from model compounds.

II. Formation Constants.

Metal-complex formation constants may be determined through NMR by observing the pH-dependency of certain proton chemical shifts in solution com-posed of the organic ligand and a non-paramagnetic metal ion.

III. Kinetics.

A. Intra-Chelate Exchange.

EDTA contains both oxygen and nitrogen ligand atoms, and, according to the lability of the various metal-ligand bonds, several situations may arise:

1. The lifetimes of both the metal-oxygen and metal-nitrogen bonds are short.

2. The lifetime of the metal-oxygen bond is short while that of the metal-nitrogen bond is long.

3. The lifetime of the metal-oxygen bond is long while that of the metal-nitrogen bond is short. In view of the structure of metal-EDTA complexes, it is unlikely that the metal-nitrogen bond could be broken without prior rupture of the metal-oxygen bonds on its two associated acetate groups. Investigations thus far have tended to confirm this conclusion in that they have not provided an example of this type.

4. The lifetimes of both the metal-oxygen and metal-nitrogen bonds are long.

In the first situation, the proton NMR spectra of complexes of diama-gnetic metal ions which have negligible abundances of isotopes with spin one-half are simple and exhibit two sharp peaks corresponding to the ethylenic and to the acetate protons. The complex should exist in various stages of unwrapping of the EDTA molecule, with the more complete stages of un-wrapping being less prevalent. Although the rates of unwrapping and wrapping may be fast, the lifetime of a given EDTA molecule on a given metal ion may be long. Thus, any possible multiplets caused by proton-proton splitting are collapsed because of internal exchange achieved by rapid umbrella inversion of the nitrogen. If the metal ion has an appreciable abundance of isotopes with spin of one-half, metal-proton splitting occurs and persists, despite the short lifetime of the metal-nitrogen bonds and rapid inversion of the nitrogen, because the nitrogen-metal bond always reforms to the same metal ion, and, hence, all nuclei are still in the same spin states and the coupling is not relaxed. An example of this type of system is lead-EDTA [lead 207, $I = 1/2$, occurs in a natural abundance of 21 per cent].

In the second situation, in which the metal-nitrogen bond has a long lifetime, while the metal-oxygen bond lifetime is short, the acetate and ethylenic protons exhibit different NMR patterns. These two ethylenic con-formations can rapidly interconvert and, thus, average out any differences of the two protons on a given ethylenic carbon. However, the two protons on a given acetate are not equivalent. With a fixed metal-nitrogen bond, the acetate groups are still free to rotate about the acetate-carbon to nitrogen bond and may be depicted by three rotational staggered configurations. These three positions are not equivalent. An example of a metal-EDTA complex where the long lifetime of the metal-nitrogen bond causes an AB splitting pattern of the acetate is cadmium-EDTA.

For the fourth situation, in which both the nitrogen-to-metal bonds each have relatively long lifetimes, rather complex spectra are expected because the acetate groups themselves are not equivalent. In the case of an octahedral complex where all ligand atoms are bound, dl forms are present. In a given optical form, two of the acetate groups are in the ethylenic nitrogen-to-metal plane and are identical, and the other two acetate groups are out of this plane and identical. The two methylene protons on a given acetate group are in different magnetic environments and yield an AB pattern. Therefore, the two types of acetates exhibit two different AB patterns. Also the rigidity of the structure causes the ethylenic protons to be no longer equivalent.

B. Interchelate Exchange.

If a metal M rapidly attacks a complex M*Y and displaces the metal M*, this will be evident in the broadening or the loss of both of the two types of splitting patterns mentioned above. If the attacking metal ion simply forms an unstable MM*Y species without complete exchange, by bonding to acetate functions (leaving the original metal-nitrogen intact and not decreasing its lifetime), the AB splitting of the acetate $-CH_2-$ protons may be broadened without seriously affecting the splitting caused by the metal M*. Likewise, if a ligand Y rapidly attacks a complex MY* and displaces the ligand Y*, broadening of both types is seen. It is important to note that M and M* and Y and Y* may be the same metal ion or chelon.

In favorable cases, the magnitude of the rates of exchange can be obtained. The exchange rate can often be suitably controlled by conditions such as pH, relative concentration of ligand and metal ions, and temperature, hence permitting production of the sharp and broadened splitting patterns required for evaluation of the rates.

From an analytical point of view, it is obvious that NMR measurements permit quantitative estimates of certain metal ion mixtures (i.e., Cd-Pb mixtures) and of isotope content (i.e., ^{207}Pb in lead). This procedure also forms the basis for slower isotope exchange studies (i.e., $^{207}PbY + Pb^{++} \longrightarrow$ $PbY + ^{207}Pb^{++}$). The kinetic study of chelate exchange reactions ($^{207}PbY^* + PbY \rightleftarrows PbY^* + ^{207}PbY$) by normal mixing methods is also possible by employing deuterated chelates, Y*; the NMR technique has the advantage of continuous measurement of the course of reaction, thus avoiding the usual time-consuming separations and their possible effect on the amount of exchange and the use of radioactive isotopes. One of the chief disadvantages of the NMR technique is the concentration level which must be employed.

C. Use of Paramagnetic Ions.

If small incremental quantities of copper(II) ion are added to a 0,5 M solution of ethylamine, at approximately 10^{-4} M copper ion only the $-CH_2-$ peak will be broadened and at higher concentrations will eventually disappear, showing that those proton spin states have been rapidly relaxed by the paramagnetism of the nearby copper ion. If EDTA is now added, the peak reappears, showing the copper ion to be tied up in a slow-exchanging complex. If a bidentate ligand with distinguishable and widely separated coordination sites is used in this experiment, one site may be preferentially relaxed, showing the preferred coordination site.

A 1

DAS VALENZWINKELPROBLEM IN DER THEORIE DER SPINVALENZ.

K.H. Hansen

Institut für Physikalische Chemie, Universität Frankfurt (M), Germany.
(to be presented in English)

In der Theorie der Valenzstrukturen stellt man die üblichen s,p,d-Atom-

eigenfunktionen für die Bildung von hybridisierten Bindungseigenfunktionen
zur Verfügung. Die Forderung der Gleichwertigkeit von n Mischfunktionen
führt dann zu den bekannten Aussagen über die Symmetrie der Anordnung von
n "Liganden", z.B.:

n = 2 :	sp	linear,	p^2	angular
n = 3 :	sp^2	trigonal eben,	p^3	trigonale Pyramide
n = 4 :	sp^3	Tetraeder,	sp^2d	tetragonal eben
n = 6 :	sp^3d^2	Oktaeder.		

Die Theorie der Valenzstrukturen ist sinnvoll, wenn die Bindungsenergie
des Moleküls groß gegen die Abstände der aus einer Atomkonfiguration her-
vorgehenden Terme ist. Baut man dagegen das Molekül aus fertigen Atomen
auf, wie es in der Theorie der Spinvalenz üblich ist, so kann man mit Hilfe
gruppentheoretischer Methoden ebenfalls zu Aussagen über die Struktur der
Moleküle gelangen. Bei Zentralatomen, deren Konfigurationen nur aus s-
und p-Elektronen aufgebaut sind, bekommt man dieselben Ergebnisse wie in
der Theorie der Valenzstrukturen. Bei Konfigurationen mit d-Elektronen ergeben sich jedoch Abweichungen. Die Abweichungen liegen alle in Richtung
auf höhere Symmetrien. Die Konfiguration sp^2d gestattet so z.B. in der
Theorie der Spinvalenz bereits eine tetraedrische Anordnung der "Liganden",
während sie in der Theorie der Valenzstrukturen erst die niedrigere Symmetrie
des Quadrats erlaubt.

Die Ergebnisse für die Koordinationszahlen n = 2, 3, 4, 5, 6 werden
mitgeteilt. Die Abweichungen von den Aussagen der Theorie der Valenzstrukturen werden erklärt. Die gruppentheoretische Ableitung der Ergebnisse
wird angedeutet.

Insgesamt ergibt sich die Einsicht, daß den Aussagen der Theorie der
Valenzstrukturen keine absolute Richtigkeit zukommt, daß man vielmehr,
je nach dem, ob das betrachtete Molekül in den Geltungsbereich der Theorie
der Valenzstrukturen oder den der Theorie der Spinvalenz gehört, zu verschiedenen Aussagen über die Molekülstruktur kommen kann.

7 A 2

OPTICAL ELECTRONEGATIVITIES.

Chr.Klixbüll Jørgensen
Cyanamid European Research Institute, Cologny (Geneva), Switzerland.

Recent quantum-mechanical investigations have clarified certain properties of the exact wavefunctions (which never are explicitly known in many-electron systems). Thus, all the observable quantities pertaining to a given stationary state can be described by the first-order density matrices (or rather operators), i.e. the density of electrons having each of the two possible spin directions in our three-dimensional space, and the analogous second-order density operators describing the relative distribution of two electrons considered at the time, in a six-dimensional space (1-3). Hence, if we concentrate our interest on the groundstate alone, or a single excited state, we have only a basis for talking about individual orbitals in an indirect way (introducing natural spin orbitals) except in the case of systems with positive total spin S where the density of uncompensated spin is observable in principle. On the other hand, if we consider transitions from the groundstate to the various excited energy levels, we get another, but not necessarily consistent, picture of the individual orbitals. Such transitions can be divided to a first approximation into zero-electron jumps (belonging to the same M.O. configuration, the two energy levels consisting of degenerate states, in average having the same electron density in our space but different second-order density operators), one-electron jumps (the electron density varying in a characteristic way which can be interpreted as a transition from one orbital a to another b) and more-electron jumps, rarely observed as optical absorption bands.

The detailed structure of the second-order density contributes to the total energy in three distinct ways. Numerically most important are the A-integrals corresponding to the difference between the ionization energy and the electron affinity of a given set of degenerate orbitals. Next comes the spin-pairing energy which can be written for one partly filled shell

$$D[\langle S(S+1) \rangle - S(S+1)] \qquad \{1\}$$

where the spin-pairing parameter D, like A, is inversely proportional to the average radius of the partly filled shell. The explicit formula for the average value $\langle S(S+1) \rangle$ for a given configuration is known (2). Finally, if different energy levels of the same configuration present the same S, their energies are separated by relatively small parameters, the coefficients of which depend on the symmetry type (L or Γ_n) of the terms.

In complexes, the spin-forbidden intra-subshell transitions are zero-electron jumps as defined above, representing spin-pairing energy and symmetry-type-dependent interelectronic repulsion, and also the transitions between different subshells are both discussed in ligand field theory. However, for chemical purposes it is perhaps more interesting to study electron transfer spectra. These, more intense, absorption bands usually correspond to one-electron jumps from filled orbitals mainly concentrated on the ligand atoms X to the empty or partly filled shell predominantly localized on the central atom M. It is possible to correct (2) the observed wavenumbers with the appropriate expression {1} for S of the central atom and to define optical electronegativities x_{opt} from the corrected wavenumbers

$$\sigma_{corr} = [x_{opt}(X) - x_{opt}(M)] \cdot kK \qquad \{2\}$$

where the energy 30 kK is equivalent to 3.7 eV or 85 kcal/mole. The Table gives the optical electronegativities defined from {2} where electronegativities for the four halogens fix the numerical scale (4). This scale has been extended to a variety of central atoms and ligands (5, 6). In the case of lanthanides (9) and 5f group elements (10), the Table gives the x_{opt} values without correction for effects of interelectronic repulsion and relativistic effects as well as the corrected values. In this particular category, the uncorrected values are of considerable chemical interest, since they also represent the oxidizing characters of the central atoms. Recently, certain electron transfer bands occurring in solids containing Ag(I) and Tl(I) and 5d group hexahalides have also been discussed (11).

For a given oxidation number z, the x_{opt} values are linear functions of the number q of electrons in the partly filled shell:

$$
\begin{array}{lll}
3d & \text{(tetrahedral } MX_4\text{)} & z = +2: \ 0.5 \ + 0.2 \ q \\
   & \text{(} MA_5X \text{)} & +3: \ 1.2 \ + 0.2 \ q \\
4d & & +3: \ 1.1 \ + 0.2 \ q \\
   & & +4: \ 1.6 \ + 0.2 \ q \\
5d & & +4: \ 1.4 \ + 0.2 \ q \qquad (3) \\
   & & +6: \ 2.0 \ + 0.3 \ q \\
4f & & +3: \ 0.5 \ + 0.1 \ q \\
   & & +4: \ 2.05 + 0.2 \ q \\
5f & \text{(} MX_6 \text{)} & +4: \ 1.0 \ + 0.25 \ q
\end{array}
$$

It is unexpected that the slope of these linear functions (4) so frequently is so close to 0.2. Actually, for the gaseous ions, the electron affinities have larger coefficients to q:

$$
\begin{array}{llll}
3d \quad z = +2 \rightarrow +1: 0.30 & \qquad & 4d \quad z = +2 \rightarrow +1: 0.37 \\
 +3 \rightarrow +2: 0.45 & & +3 \rightarrow +2: 0.51
\end{array}
$$

$$(4)$$

whereas the oxidation potentials of 3d group $[M(H_2O)_6]^{++}$ in aqueous solution indicate (12) the coefficient 0,26 to q.

It is a fascinating problem why the orbitals derived from optical excitations; the strongly occupied natural spin orbitals; and the occupied Hartree-Fock orbitals all are so conspicuously similar. This problem should not be masked by deductive, but highly approximate, attempts of calculations, but should rather encourage the chemical inductive approach. The Wolfsberg-Helmholz model has recently been very popular (13, 14) and from this point of view, it is obvious that a close connection exists between the optical electronegativities and the nephelauxetic effect (15, 16). In general, one may classify the complexes (17) according to the constituent chromophores MX_q, usually of rather high symmetry permitting sets of necessarily degenerate orbitals, and extensive progress oan still be made in the description applying molecular orbital theory though a priori calculations at present are impossible.

	Lower sub-shell	Upper sub-shell
$3d^3$ Cr(III)	1,8	1,0-1,3
$3d^5$ Fe(III)	2,1-2,5	-
$3d^6$ Co(III)	(2,4)	1,6-1,9
$3d^7$ Co(II)	(1,9)	1,7
$3d^8$ Ni(II)	(2,1)	1,9
$3d^9$ Cu(II)	(2,3)	1,9-2,1
$4d^0$ Mo(VI)	2,1	-
$4d^3$ Tc(IV)	2,2	1,4
$4d^4$ Ru(IV)	2,4	1,6
$4d^5$ Rh(IV)	2,6	-
$4d^6$ Pd(IV)	(2,7)	1,9
$4d^3$ Mo(III)	1,7	1,1
$4d^5$ Ru(III)	2,1	1,1-1,5
$4d^6$ Rh(III)	(2,3)	1,3-1,7
$4d^8$ Pd(II)	(2,4)	1,7
$5d^0$ W(VI)	2,0	-
$5d^2$ Os(VI)	2,6	-
$5d^3$ Ir(VI)	2,9	-
$5d^4$ Pt(VI)	3,2	-
$5d^3$ Re(IV)	2,0	1,1
$5d^4$ Os(IV)	2,2	1,3

	Direct	Corrected
$4f^0$ Ce(IV)	2,1	2,05
$4f^2$ Nd(IV)	3,0	2,5
$4f^8$ Dy(IV)	3,0	3,6
$4f^5$ Sm(III)	1,45	0,95
$4f^6$ Eu(III)	1,75	1,1
$4f^{12}$ Tm(III)	1,3	1,6
$4f^{13}$ Yb(III)	1,6	1,75
$5f^2$ U(IV)	1,8	1,5
$5f^3$ Np(IV)	1,9	1,75
$5f^4$ Pu(IV)	2,1	2,05

	π	σ
$[CN]^-$	-	2,8
$[C_5H_5]^-$	2,3?	-
NH_3	-	3,3
diethylenetriamine	-	3,2
o-phenanthroline	2,6	-
H_2O	3,5	-
$[H_2PO_2]^-$	3,3	-
$[SeO_4]^{--}$	3,3	-
$[SO_4]^{--}$	3,2	-

$5d^5$	Ir(IV)	2,35	1,5	CH_3OH	3,1	-
$5d^6$	Pt(IV)	(2,6)	1,6-1,7	acetylacetonate	2,7	-
$5d^5$	Os(III)	1,9	1,1	F^-	3,9	4,4
$5d^6$	Ir(III)	(2,25)	1,3-1,5	phosphines R_3P	-	2,6
$5d^8$	Au(III)	(2,8)	1,8	$(C_2H_5)_2S$	2,9	-
				$[(C_2H_5O)_2PS_2]^-$	2,7	-
				$[(C_2H_5)_2NCS_2]^-$	2,6?	-
				Cl^-	3,0	3,5
				arsines R_3As	-	2,5
				$[(C_2H_5O)_2PSe_2]^-$	2,6	-
				Br^-	2,8	3,3
				I^-	2,5	3,0

References

1. LÖWDIN, P.O., J.Phys.Chem. 61, 55 (1957).
2. JØRGENSEN, C.K., Orbitals in Atoms and Molecules, Academic Press, London 1962.
3. KUTZELNIGG, W.Z., Naturforsch. 18a, 1058 (1963).
4. JØRGENSEN, C.K., Mol.Phys. 6, 43 (1963).
5. JØRGENSEN, C.K., Acta Chem.Scand. 16, 2406 (1962).
6. JØRGENSEN, C.K., "Inorganic Chromophores" in the 60-year birthday volume for G.Schwarzenbach (ed.W.Schneider). Birkhäuser Verlag, Basel 1964.
7. BARNES, J.C., and DAY, P., private communication.
8. SYKES, K.W., private communication.
9. JØRGENSEN, C.K., Mol.Phys. 5, 271 (1962).
10. RYAN, J.L., and JØRGENSEN, C.K., Mol.Phys. 7, 17 (1963).
11. JØRGENSEN, C.K., Acta Chem.Scand. 17, 1034 (1963).
12. JØRGENSEN, C.K., Acta Chem.Scand. 10, 1505 (1956).
13. KIDA, S., FUJITA, J., NAKAMOTO, K., and TSUCHIDA, R., Bull. Chem.Soc. Japan 31, 79 (1958).
14. LOHR, L.L., and LIPSCOMB, W.N., J.Chem.Phys. 38, 1607 (1963).
15. JØRGENSEN, C.K., Progress Inorg.Chem. 4, 73 (1962).
16. JØRGENSEN, C.K., PAPPALARDO, R., and SCHMIDTKE, H.-H., J.Chem.Phys. 39, 1422 (1963).
17. JØRGENSEN, C.K., Inorganic Complexes. Academic Press, London 1963.

SPECTRA OF SOME Rh(III)- AND Ir(III)-COMPLEX COMPOUNDS.

Hans-Herbert Schmidtke

Cyanamid European Research Institute, Cologny (Geneva), Switzerland.

During the last two years a series of Rh(III)- and Ir(III)-compounds have been prepared by us and spectroscopically investigated. Some of them have been known for a long time, but the spectra in the visible and UV range were not recorded; others are entirely new. For all of them the spectrum together with modern theoretical procedures can give some insight in the chemical bonding, charge distribution and certain other properties. Furthermore it is possible from visible and UV spectroscopy to assign the steric isomers in the case of low symmetry complexes. This has its theoretical foundations in the results of YAMATERA (1) and McCLURE (2), who investigated the energy level splittings of non cubic complex ions by means of crystal field and MO theory. It is known that bands in UV and visible spectra can be divided in ligand field bands, charge transfer bands (ctf) from the ligand to the central ion or in opposite direction, and in bands belonging essentially to the perturbed electronic system of the ligands. All these transitions are found in the spectra of the compounds investigated here, however, as can be seen from the table the classification is no longer so clearcut for low symmetry complexes.

The table is ordered within a certain symmetry type according to the spectrochemical series of the ligands (3). The molar extinction coefficients, assignments, crystal field parameters Δ, RACAH parameters B and the nephelauxetic ratio β (4) are given in so far as they could be calculated from the position of the ligand field bands. For non cubic cases Δ and B have of course only approximate values.

The spectra of thiocyanato and thiourea complexes are dominated by bands belonging to internal transitions of the ligands which prove to be very different from those of the free ions or molecules. The low Δ value of $[Rh(SCN)_6]^{3-}$ shows that the thiocyanate groups are linked by their sulfur atoms to the central metal. The reflection spectrum of $K_3[RhF_6]$ represents a "ligand field spectrum" in its purest form. The excited $t_{2g}^5 e_g$ configuration of octahedrally coordinated Rh(III) gives rise to the states $^3T_{1g}$, $^3T_{2g}$, $^1T_{1g}$ and $^1T_{2g}$. Two of the four transitions from the groundstate $^1A_{1g}$ are spin allowed, leading to bands with appreciable extinction coefficients while the others are spin forbidden and should only show up weakly in the spectrum as is actually found in the $[RhF_6]^{3-}$ spectrum. For the amines $[Rh(NH_3)_6]^{3+}$, $[Rh\,pn_3]^{3+}$, $[Rh\,den_2]^{3+}$ the iodides have been also prepared in order to look for charge transfer bands due to ion pair formation which were reported by LINHARD (5) and WEIGEL (5, 6) in cobaltamine complexes. No such bands however could be found in the case of Rh(III)-complexes. Compounds of the type $[Ir(NH_3)_5X]^{n+}$ follow closely the spectrochemical series having the

$^1T_{1g}$ transition at the longest wave length for $X = I$ and at the shortest for $X = NH_3$. The spectra of DELÉPINE's (7) $[Ir\,py_4\,Cl_2]^+$ salts provide evidence for an assignment to the cis- and trans-isomers. Theory (1, 2) also supplies some proof for the assignment of the two spectroscopically different $[Rh\,py_3\,Cl_3]$ salts to their geometric isomers. In the case of $[Rh\,den\,Cl_3]$ we assign the one component found to the (1,2,6) isomer on the basis of the spectrochemical series which predicts the $^1T_{1g}$ transition for the den complex at higher wave numbers than for the pyridine complex.

Table. (pn 1,2 propanediamine; den diethylenetriamine; py pyridine).

Complex	Absorption maximum [cm^{-1}]	Extinction coefficients	Assignment	Δ [cm^{-1}]	B [cm^{-1}]	β
$K_3[Rh(SCN)_6]$	19 600	275	$^1T_{1g}$	~20 300		
	34 700	40 500	} internal			
	49 000	30 500	} thiocyanate			
$[Rh(thiourea)_6]Cl_3$	22 200	450	$^1T_{1g}$	~22 900		
	34 200	24 500	} internal			
	41 200	47 100	} thiourea			
$K_3[RhF_6]$	15 500		} $^3T_{1g}\ ^3T_{2g}$	22 300	460	0,64
	16 500					
	21 300		$^1T_{1g}$			
	27 800		$^1T_{2g}$			
$[Rh(NH_3)_6]Cl_3$	32 800	134	$^1T_{1g}$	34 000	430	0,60
	39 200	101	$^1T_{2g}$			
$[Ir(NH_3)_6]Cl_3$ **	39 800	92	$^1T_{1g}$	41 200	470	0,71
	46 800	160	$^1T_{2g}$			
$[Rh\,pn_3](ClO_4)_3$	33 100	236	$^1T_{1g}$	34 300	401	0,56
	39 100	192	$^1T_{2g}$			
$[Rh\,den_2]Cl_3$	33 300	274	$^1T_{1g}$	34 600	420	0,58
	39 600	186	$^1T_{2g}$			

Table. (contd.)

K₃[Rh(CN)₆]	44 500	527	$^1T_{1g}$	~45 500		
[Ir(NH₃)₅I]Cl₂ **	29 700	372	$^1T_{1g}$	~31 500		
	42 700	4600	ctf.I_π→ Ir			
	46 500	6 700	ctf.I_σ→ Ir			
[Ir(NH₃)₅Br]Br₂ **	33 000	100	$^1T_{1g}$	~34 800		
	43 500	800	ctf.Br→ Ir ?			
[Ir(NH₃)₅OH]²⁺ **	37 000	100	$^1T_{1g}$	38 100	330?	
	42 000	302	$^1T_{2g}$			
[Ir(NH₃)₅(H₂O)]Cl₃ **	38 800	90	$^1T_{1g}$	40 300	510	0,77
	46 500	204	$^1T_{2g}$			
tr[Rh py₄ Br₂]Br *	22 600	133	$^1E_g(D_{4h})$	23 700	630?	
	31 300	2 280	singlet or ctf.			
tr[Ir py₄ Cl₂]Cl *	24 400	26	$(^3T_{1g}, {}^3T_{2g})$			
	35 100	7 500	ctf.Ir→py			
cis[Ir py₄ Cl₂]Cl *	25 600	55	$(^3T_{1g}, {}^3T_{2g})$			
	35 100	8 200	ctf.Ir→py			
(1,2,6)[Rh py₃ Br₃] *	21 700	186	$^1T_{1g}$	22 500	300	
	26 100	1 150	$^1T_{2g}$			
	30 800	2 600	ctf.Br→Rh ?			
(1,2,6)[Rh py₃ Cl₃] *	23 400	73	$^1T_{1g}$	24 100	240?	
	27 000	~74	$^1T_{2g}$?			
	42 600	~25 000	ctf.Cl→Rh			
(1,2,3)[Rh py₃ Cl₃] *	24 500	140	$^1T_{1g}$	~25 300		
	41 700	-	ctf.Cl→Rh			
(1,2,6)[Rh den Br₃] *	22 800	104	$^1T_{1g}$			
	35 300	2 350	ctf ?			
	41 700	24 000	ctf.Br→Rh			
(1,2,6)[Rh den Cl₃] *	24 000	86	$^1T_{1g}$	25 100	590?	0,82
	32 200	330	$^1T_{2g}$?			
	47 200	34 000	ctf.Cl→Rh			

* internal pyridine transitions and
** singlet-triplet transitions are not listed.

References

1. YAMATERA, H., Naturwiss. 44, 375 (1957), Bull.chem.Soc.Japan 31, 95 (1958).
2. McCLURE, D.S., Advances in the chemistry of the coordination compounds, Proc.6.I.C.C.C., Detroit 1961, p.498.

3. TSUCHIDA, R., Bull.chem.Soc.Japan 13, 388, 436, 471 (1938).

4. SCHÄFFER, C.E., and JØRGENSEN, C.K., J.Inorg.Nucl.Chem. 8, 143 (1958).

5. LINHARD, M., Z.Elektrochem. 50, 224 (1944).

6. LINHARD, M., and WEIGEL, M., Z.anorg.allg.Chem. 266, 73 (1951).

7. DELÉPINE, M., and LARÈZE, F., C.r.256, 3912 (1963), 257, 3772 (1963).

7 A 4

DYNAMICAL JAHN-TELLER EFFECTS IN d^4 SYSTEMS - AQUEOUS MANGANESE(III) AND OTHER COMPLEXES OF Mn(III) AND Cr(III).

John P. Fackler, Jr., David G. Holah and I.D. Chawla
Case Institute of Technology, Cleveland 6, Ohio, U.S.A.

The known spectral properties of transition metal complexes in which the metal ion has the $3d^4$ configuration are quite limited. Several recent books (1, 2) emphasize this fact by their rather meager presentation of data. Of these discussions, BALLHAUSEN (1) presents the most nearly complete consideration of the data.

$CsMn(SO_4)_2 \cdot 12 H_2O$ crystals were studied originally by HARTMANN and SCHLÄFER (3, 4). These authors found a band at $21\,000$ cm^{-1} which is assigned to the $^5T_{2g} \longleftarrow {}^5E_g$ absorption in O_h symmetry (4). Aqueous manganese(III) prepared by addition of $KMnO_4$ to a ~ 1 M solution of $Mn(ClO_4)_2 \cdot 6 H_2O$ in ~ 3 M $HClO_4$ also shows a band near $21\,000$ cm^{-1} (5). However, the intensity of this band in aqueous solution is approximately 15 times stronger than the $^5T_{2g} \longleftarrow {}^5E_g$ band in $CsMn(SO_4)_2 \cdot 12 H_2O$. Addition of chloride to this solution causes the appearance of a band near $12\,000$ cm^{-1}. In fact, the spectrum becomes similar to that previously reported (6) for manganese(III) in 10 N HCl. The absence of a low energy band in Mn^{3+}(aq) and the concomitant appearance of such a band when chloride is added suggests that solutions containing coordinating anions such as chloride or oxalate (7) may contain tetrahedral manganese(III) species.

Addition of fluoride to Mn^{3+}(aq) produces a drastic lowering of intensity accompanied by a splitting of the $21\,000$ cm^{-1} band. No new band at lower energies is produced. These results, along with the spectral data of RUNCIMAN and SYME (8), agree well with the hypothesis that these $3d^4$ complexes are not centro-symmetric at ambient temperatures but undergo dynamical oscillations from one potential minimum to another (9). However, addition of fluoride tends to "freeze out" these oscillations, producing species of the type MnF^{2+}, MnF_2^+ which do not tend to minima exchange in solution (10).

Similarly (see the table), low temperatures cause the bands in $CrCl_2 . 4 H_2O$ to be resolved into the three spin-allowed components.

Preparations and properties of other $3d^4$ complexes including those of $CrI_2(CH_3CN)_2$, $Cr(DPM)_2$, $Cr(AA)_2$, $Mn(DPM)_3$, $Mn(DIBM)_3$, $Mn(AA)_3$, $Mn(F_3AA)_3$ and $Mn(F_6AA)_3$, where AA = acetylacetone, DIBM = diisobutyl-methane and DPM = dipivaloylmethane, will be discussed if time permits.

This work was supported by the National Science Foundation.

TABLE - LIGAND FIELD SPECTRA OF SOME $3d^4$ COMPLEXES

Species	Solvent	Wave number . 10^{-3} cm^{-1}
Mn^{3+}, ClO_4^-	1-3 M $HClO_4$	21,0 ($\sim$87)[a]; 14,5 (25)
	3 M $HClO_4$, added 0,2 M NaCl	22,2 (80); 11,7 (36,5)
	3 M $HClO_4$, added 0,2 M NaF	23,2 (24); 18,6 (19); 13,1 (9)
$Mn(DPM)_3$	Cyclohexane	17,9 ($\sim$200); 8,7 (110)
$Mn(F_6AA)_3$	Cyclohexane	18,7 ($\sim$240); 11,6 (89)
$Mn(AA)_3$	Cyclohexane	17,9 ($\sim$200); 9,52 (110)
$[MnCl_6]^{3-}$	pressed salt[b]	22,4; 17,5
$CrCl_2 . 4 H_2O$	crystal[c] 77°K ambient	18,9, 15,2, 13,1 14,3 (broad)
$CrI_2(CH_3CN)_2$	solid, reflectance	30,3; 25,0; 15,9; 9,7

a. Molar extinctions in parentheses
b. Ref. (11)
c. Ref. (8)

References

1. BALLHAUSEN, C.J., "Introduction to Ligand Field Theory", McGraw-Hill Book Co., New York (1962).
2. a) JØRGENSEN, C.K., "Absorption Spectra and Chemical Bonding in Complexes", Addison-Wesley Book Co., Reading, Mass. (1962).
   b) DUNN, T.M., in "Modern Coordination Chemistry", J.Lewis and R.Wilkins, editors, Interscience, New York, 1960.
   c) JÄFFE, H., and ORCHIN, M., "Theory and Applications of Ultraviolet

Spectroscopy", John Wiley and Sons, Inc., New York, 1962.

3. HARTMANN, H., and SCHLÄFER, H.L., Z.Naturf., 6a, 754 (1951).

4. HOLMES, O., and McCLURE, D.S., J.Chem.Phys., 26, 1686 (1957).

5. FACKLER, J.P., Jr., and CHAWLA, I.D., Inorg.Chem., to be published.
   ROSSEINSKY, D.R., J.Chem.Soc. (London), 1963, 1181;
   DIEBLER, H., and SUTIN, N., J.Phys.Chem., 68, 174 (1964).

6. IBERS, J.A., and DAVIDSON, N., J.Am.Chem.Soc., 72, 4744 (1950).

7. PIPER, T.S., and CARLIN, R.L., J.Chem.Phys., 35, 1809 (1961);
   Inorg.Chem., 2, 260 (1963);
   FURLANI, C., and CIANI, A., Ann.Chem. (Den.), 48, 286 (1958).

8. RUNCIMAN, W.A., and SYME, R.W.G., Phil.Mag., 8, 605 (1963).

9. LIEHR, A.D., "Progress in Inorganic Chemistry", Vol. 3, F.A.Cotton,
   editor, Interscience, pp. 218-314 (1962).

10. Minima exchange is possible if energy level splittings are not too large.

11. HATFIELD, W.E., FAY, R.C., PFLUGER, C.E., and PIPER, T.S.,
    J.Am.Chem.Soc., 85, 265 (1963).

8 A 1

OBSERVABILITY OF SPLITTINGS OF EXCITED ENERGY LEVELS CAUSED BY SPIN-ORBIT COUPLING OR LOW SYMMETRY COMPONENTS OF THE CORE FIELD.

Claus Schäffer

The H.C.Ørsted Institute, University of Copenhagen, Denmark.

It has been discussed by various authors how the observed absorption bands of octahedral complexes can be represented analytically as approximately gaussian error curves. Some authors have preferred the wavelength as the independent variable (1), others the wavenumber in which case the half width of the curves sometimes were introduced as two parameters (2), one for each side of the gaussian maximum. Because of the better fit in the region of most interest (relatively near the maximum), and because of the simplicity in having only a single value of the half width we shall apply the wavelength representation here. However, the conclusions that we shall reach only depend on the approximate shape of our curves and will be valid also for other gaussian-like representations.

We shall here first consider the diagonal spin-orbit splitting of a composite energy level. We shall assume

a) that an experimental absorption band can be considered as the sum of the spin-orbit components;

b) that the half-width of the spin-orbit components will be the same;

c) that their height will be proportional to their degeneracy number.

ad (a) This assumption is probably bad for small splittings when the energy difference is comparable to vibrational energies (3). However, any vibronic interactions between levels carrying the character of broadness are likely to have a levelling out effect rather than one of producing narrow sub-components.

ad (b) This is the most important assumption and it is hard to see this assumption breaking down before effects of non-diagonal elements with spin-orbit levels carrying the character of narrowness become important.

ad (c) This is of little importance for our arguments but as long as we are considering diagonal spin-orbital effects only it is a consequence of assumption (b).

As an illustration we shall take a $^4T_2$ level, whose spin-orbit components are "J" $= \frac{1}{2} (\Gamma_7)$, "J" $= \frac{3}{2} (\Gamma_8)$ and "J" $= \frac{5}{2} (\Gamma_6 + \Gamma_8)$, Γ_6 and Γ_8 in parenthesis being degenerate for pure d-functions (4). This sum curve can be approximated quite well by one gaussian whose half-width in this particular case is about 50% larger than those of the components. This half-width corresponds to $1\frac{1}{4}\zeta_{nd}$. Now the experimental half-widths for 5d transition metal ions vary between 3000 - 5000 cm^{-1} thus being able to completely hide diagonal splittings of the energy levels corresponding to $\zeta_{5d} = 2500 - 4000$ cm^{-1}. We conclude that such diagonal splittings will be observable only in rare cases where a combination of anomalously narrow bands and high spin-orbit coupling constants occurs.

If, however, energy levels carrying the character of narrowness get close to levels showing up as broad bands, a mixing may occur.

It is now the question whether it is possible in a case where an overlapping band (even after the mixing) has a pronounced narrowness to make an estimation of its intensity through a gaussian analysis, and thus to make possible an estimation of the matrix elements of spin-orbit coupling.

We shall tentatively mention an example of how such an analysis may lead to the estimation of the one electron spin-orbit coupling parameter ζ_{3d} in the case of CrF$_3^-$. Fig. 2 shows the observed first spin-allowed absorption band corresponding to $^4A_{2g} \longrightarrow {}^4T_{2g}$. The composite structure is most certainly caused by interactions with the narrow spin doublets ^2E_g and $^2T_{1g}$, the coefficients to ζ_{3d} of the matrix elements being known from GRIFFITH for d-functions.

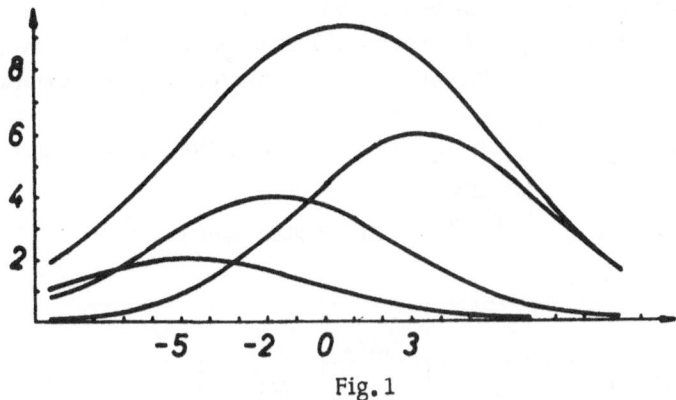

Fig. 1

Gaussian sums representing diagonal elements of spin-orbit coupling for a cubic $^4T_2$ level. Abscissa unit $\frac{1}{12}\zeta_{n1}$. Multiplet width $\frac{8}{12}\zeta_{n1}$. Half width of components $\frac{10}{12}\zeta_{n1}$. Half width of sum curve $\frac{15}{12}\zeta_{n1}$.

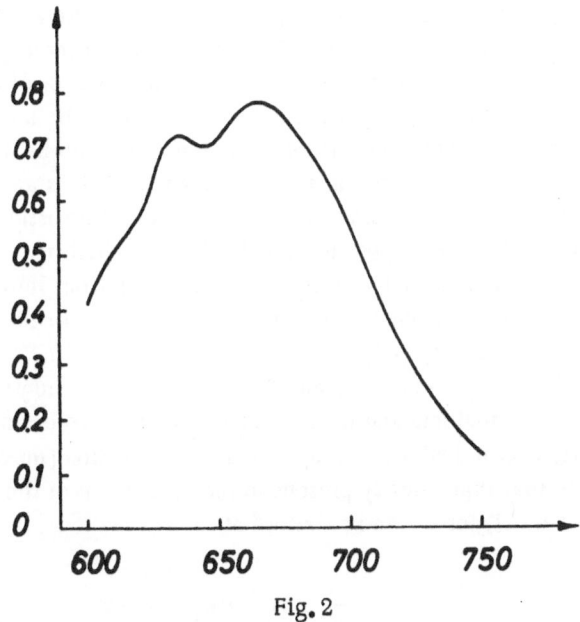

Fig. 2

Experimental absorption curve for CrF_6^{3-}
Abscissa: wavelength (nm)
Ordinate: absorbancy

We have used a regression analysis digital computer program (5) to pre-liminary investigate such a gaussian analysis. This program works iteratively from an input of initial parameters and minimizes the mean square deviation through successive linear approximations.

The experimental curve can be resolved into the following three gaussians (of the form $D_{max} \cdot 2^{-\left(\frac{\lambda - \lambda_0}{1/2 \, \delta}\right)^2}$) with assigned names

	D_{Max}	λ_0 (nm)	δ (nm)
$^4T_{2g}$ $(\Gamma_6 + \Gamma_7 + \Gamma_8 + \Gamma_8)$	0,764	665,3	100,2
^2E_g (Γ_8)	0,116	631,2	15,6
$^2T_{1g}$ $(\Gamma_6 + \Gamma_8)$	0,177	604,4	35,0

The standard deviation on the parameters representing the position of the maxima is small (< 2 nm) in all cases, that of the height and half-width of $^4T_{2g}$ is also small, 1-2%, whereas it is larger on ^2E_g, about 20%, and on the half-width of $^2T_{1g}$ it amounts to about 50%. By resolution of the experimental curve into four gaussians the variance can be reduced by a factor of 5, thus bringing the deviations from the experimental curve down to the uncertainty of the absorption measurements. This resolution, however, can be done in an infinite number of ways without increasing this variance. This is an example of the general property of the gaussians of being little characteristid. It may not sound encouraging for our purpose but for the fact that both with the resolution into three gaussians and with two different resolutions into four gaussians (these being distinguished by highly different choice of initial parameters) the parameters of $^2E(\Gamma_8)$ vary only about 10%.

Writing $\Psi_2 = \Psi_2' + \alpha \Psi_4'$, where Ψ_2' and Ψ_4' are the mixing functions corresponding to pure spin doublets and quartets and Ψ_2 is the mixed function which is essentially a spin doublet, the approximate expression (introducing far less uncertainty that that already present in the estimation of the intensity of the spin doublet level) for α can be written as

$$\frac{1}{3} \Sigma \alpha^2 = \frac{\frac{1}{\nu_2} \cdot D_2 \cdot \Delta \nu_2}{\frac{1}{\nu_4} \cdot D_4 \cdot \Delta \nu_4} \qquad \text{and}$$

$$k\,\zeta_{3d} = \frac{(\nu_2 - \nu_4)\,\alpha}{1 + \alpha^2}$$

where D and $\Delta\nu$ represent the optical density of the maximum and the half-width of the bands, and k the angular non-diagonal matrix elements for d-functions given by GRIFFITH (4). The sub-indices refer to the levels essentially spin-doublets and quartets, respectively, and the summation should be carried out over the two Γ_8 components of the quartet level, the factor $\frac{1}{3}$ occurring because only $\frac{1}{3}$ of the quartet level is acting in each case.

The result for ζ_{3d} is 185 cm^{-1} which in view of the standard deviation on the parameters of ^2E should be good to about 30%.

Other implications from the properties of gaussians on the observability of split components caused by low symmetry will be discussed. Also the significance of such considerations in interpreting polarized spectra and circular dichroism curves will be stressed.

Acknowledgement. The author is much indebted to Dr. ERIK JØRGENSEN in particular for his philosophy of experiment and its numerical analysis (5).

References

1. MEAD, A., Trans. Faraday Soc., 30, 1055 (1934).
2. JØRGENSEN, C.K., Acta Chem. Scand., 8, 1495 (1954).
3. SCHÄFFER, C.E., Symposium on the Structure and Properties of Coordination Compounds. Bratislava, September 1964.
4. GRIFFITH, J.S., The Theory of Transition Metal Ions. Cambridge University Press. 1961.
5. JØRGENSEN, E., Framework program for non-linear regression analysis with tests. Danish Institute of Computing Machinery. November 1963.

8 A 2

LUMINESCENCE SPECTRA OF TRANSITION-METAL COMPLEXES.

V. Caglioti, G. Sartori and C. Furlani
Istituto Chimico, Università di Roma, Italy.

Photoexcited luminescence spectra of transition metal complexes have been reported until now in the literature in several instances, some of which (e.g. Cr^{3+} in ruby (1) and in other complexes (2)) have acquired major importance in the elucidation of the symmetry type and of the electronic structure of coordination compounds; other literature reports include

Mn^{2+} (3) and Mn^{3+} (4).

We shall report here the results of a systematic investigation of $d^{2\cdots\cdots 8}$ electronic spectra in photoexcited luminescence; although a study of absorption spectra leads in principle to the knowledge of the same system of d^n-energy levels as the study of emission spectra, the latter technique can be sometimes superior to the former one in detecting spin-forbidden transitions (since longer-lived excited states are more apt to fluoresce), despite instrumental draw-backs (such as the dependence of emission intensities on both source intensity and frequency), and the need for further improvements in the instrumentation. Also, comparison of energy levels inferred from absorption and from lumine-scence spectra, which differ slightly because of different vibronic levels implied, supplies more informations about vibronic coupling, Franck-Condon mechanisms, Morse curves, and possibility of nonradiative transitions, than could be deduced from absorption spectra above, and suggests criticism of the significance of the ligand field parameters as observable electronic quantities. Eventually useful informations are expected from measurements of luminescent state decay and mean lifetime.

We have measured excitation and emission spectra, excited by Xe arc lamps, of many coordination compounds with $3d^n$ configurations (n = 2.....8), both high-spin and low-spin, mostly in rigid glass media at low temperatures, and occasionally also at room temperature (either in the solid state or in solution), whenever luminescent emission was still detectable. In summary, the most relevant results of these measurements include:

$\underline{d^2}$-Vanadium (III) complexes exhibit two emission bands, at slightly lower wavenumbers than the first two spin-allowed absorption bands (e.g. 15,8 and 21,8 kK in $[V(urea)_6]^{3+}$ against 16,2 and 24,2 kK in absorption). Since the $^1A_{1g}$ (γ_5^2) singlet is expected very close below b^3T_{1g}, the higher emission frequency could be due to phosphorescence of the latter singlet rather than to fluorescence from the spin-allowed excited triplet $^3T_{1g}$.

$\underline{d^3}$ - Besides confirming the position and shifts of the historically important "ruby lines" due to $(^2E_g$ $^2T_{1g})$ and to $^2T_{2g}$ (γ_5^3) (and occasionally also of the fluorescent level $^4T_{2g}$), we have had indications of other higher doublet levels in several Cr^{3+} complexes, and detected corresponding emission bands also in other examples of d^3-configurations, namely in complexes of V^{2+}, Mn^{4+} and Mo^{3+}.

$\underline{d^5}$ Emission due to the $^4T_1$ (^4G) level in tetrahedral Mn^{2+}-complexes occurs about 19 kK (absorption frequency is 21,7 kK), that is with a smaller red shift than reported for octahedral Mn^{2+} complexes (3), probably because the $^4T_1$ (^4G) state has more pronounced γ_5^4 γ_3^1 character in octahedral and more γ_5^3 γ_3^2 character in tetrahedral complexes of Mn^{2+} (5).

d^6 low-spin - Cobalt (III) complexes show commonly two spin-allowed emissive transitions, corresponding to the first two absorption bands $^1A_{1g} \rightarrow {}^1T_{1g}, {}^1T_{2g}$ (O_h), with red shifts, since both transitions are of the $\gamma_5^5 \gamma_3^1 \rightarrow \gamma_5^6$ type; emission from excited triplet levels is encountered much less frequently. In $[Fe(CN)_6]^{4-}$ however, two emission bands are observed, whose frequencies fit very well the expected positions of $^3T_{1g}$ and $^3T_{2g}$ of $\gamma_5^5 \gamma_3^1$.

d^6 high-spin - Among the triplet states observed (6) and assigned (6) in the absorption spectra of octahedral Fe^{2+} complexes, at least two levels (possibly the lowest $^3T_{1g}$ and $^3T_{2g}$ of $\gamma_5^4 \gamma_3^2$) give rise to emission bands in the region 20 - 26 kK, with negligible red shifts since no rearrangement of sub-shell occupancy is required.

d^7 - Several luminescence bands due to doublet levels lying at or about the region of spin allowed absorption (ca. 18 - 22 kK) in the visible spectrum of octahedral Co^{2+} complexes, have been detected.

d^8 - In octahedral Ni^{2+} complexes, excited triplet levels appear to be little effective in emission, whereas almost complete emission spectra from the spin-forbidden levels (a ^1E_g, $^1A_{1g}$, a $^1T_{2g}$, $^1T_{1g}$ and b ^1E_g or b $^1T_{2g}$) have been observed; the red shift is particularly small in $^1A_{1g} \rightarrow {}^3A_{2g}$, which occurs within $\gamma_5^6 \gamma_3^2$.

References

1. SUGANO, S., and TANABE, Y., Disc.Faraday Soc., 26, 43 (1958).
2. FORSTER, L., and de ARMOND, K., J.Chem.Phys., 34, 2193 (1961); PORTER, G.P., and SCHLÄFER, H.C., Z.Phys.Chem.N.F., 37, 109 (1963), and several others.
3. ORGEL, L.E., J.Chem.Phys., 23, 1958 (1955).
4. KRÖGER, F.A., J.Chem.Phys., 20, 345 (1952).
5. NEGOIN, D., and FURLANI, C., Atti Accad.Naz.Lincei, 35, 58 (1963).
6. FURLANI, C., and SARTORI, G., Gazz.Chim.Ital., 87, 371, 380 (1957).

8 A 3

SOME ASPECTS OF THE THEORY OF ATOM-BRIDGE BONDS.

B. Jeżowska-Trzebiatowska and W. Wojciechowski
Department of Inorganic Chemistry, University, Wrocław, Poland.

An atom may form a bridge-bond between two other atoms, when it considerably differs in electronegativity. There are two extreme examples of such type of bond: X-H-X hydrogen bond and X-O-X (1) oxygen bond.

Hydrogen bond may arise when at least 2 s-orbitals of hydrogen are excited. Oxygen bond is formed when metal d-orbitals are available, or else due to excitation of d-orbitals of the non-metal elements such as Si, P etc (3,4).

Hydrogen bond has two wells of the potential, corresponding to two different positions of proton between two X atoms. Two electron pairs are found in the range of proton influence, causing the increase of hydrogen radius to about 0,7 Å. The radius of bridge oxygen is about twice longer (1,42 Å). Hydrogen and oxygen bonds are in a sense reciprocally reversed and may be presented:

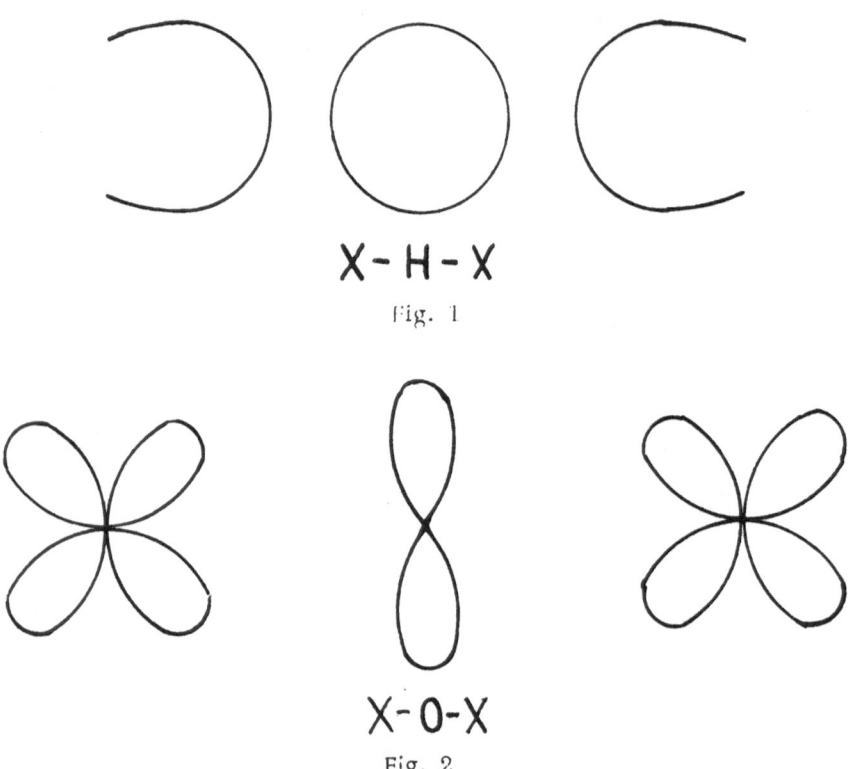

X- H - X

Fig. 1

X- O-X

Fig. 2

An oxygen bond is particularly stable when its angle amouts to 180°. Thus, a deviation of that angle could be a measure of bond strength (5). However the hydrogen bond strength is actually independent of the angle. The reason is, that both the 1-s-orbital and the excited 2 s-orbital as distinguished from d-orbitals, have a spherical symmetry.

An oxygen bond is stronger than a hydrogen bond and causes essential

changes in magnetic and spectroscopic properties of the bound atoms (2, 6).
In case of paramagnetic ions of the transition elements the coupling of the
electron spins occurs, and the result is a diamagnetic compound. Molecular
orbitals are then formed in the X-O-X nucleus in a following sequence (2, 6):

$$E_g^b \quad E_u \quad A_{1g} \quad A_{2u} \quad E_g^a \qquad \{1\}$$

Compounds are diamagnetic when the nucleus contains 4, 8, 10, 12 and
16 electrons. However, with 6 and 14 electrons a paramagnetic compound is
obtained. Examples of diamagnetic complexes are:

10 electrons	Cr-O-Cr	(7, 8),	Re-O-Re	(9)	$\{2\}$
12 electrons	Ru-O-Ru	(10) ,	Os-O-Os	(11)	$\{3\}$

The postulated species by HARE, BERNAL and GRAY (12) with Mo-O-Mo
nucleus, containing 6 electrons, may serve as an example of a paramagnetic
complex.

Oxygen plays an important role in hydrolysis processes of both the transition
and the even series elements. The formation of polycompounds such as poly-
phosphates, silicates etc. is due to contribution of the excited state d-orbitals
in the molecular orbitals arising.

Bi-nuclear complexes with $Me\!\!\begin{smallmatrix}O\\ \diagup \diagdown \\ \diagdown \diagup \\ O\end{smallmatrix}\!\!Me$ nucleus may be formed also

as for instance in the Mo(V) compounds: $[Mo_2O_4(C_2O_4)_2(H_2O)_2]^{2-}$ (13)
$[Mo_2O_4(OH)_2(H_2O)_4]$ 4, 5 H_2O (14). The molecular orbital method was used in
calculation of energy levels. The sequence of M.O. and their electrons
explain magnetic and optical properties of these compounds. Between the two
metal atoms a bond arises. The two electrons occupying a non degenerated
A_g^a orbital, cause diamagnetism. A characteristic feature of Me-O-Me and

$Me\!\!\begin{smallmatrix}O\\ \diagup \diagdown \\ \diagdown \diagup \\ O\end{smallmatrix}\!\!Me$ nucleus is an alteration of their properties, due to protons being

captured by oxygen-bridge atoms (7, 8). The compounds with Me-OH-Me or

$Me\!\!\begin{smallmatrix}OH\\ \diagup \diagdown \\ \diagdown \diagup \\ OH\end{smallmatrix}\!\!Me$ nucleus become weak paramagnetic (8, 14) or diamagnetic

(15-17).

No paramagnetic resonance absorption is observed in compounds with

Me–O–Me or $Me \big\langle^{O}_{O}\big\rangle Me$ nuclei; however in the hydrogenized $Me\big\langle^{OH}_{OH}\big\rangle Me$ and Me–OH–Me bridges the paramagnetic absorption does occur.

Another type of oxygen-bridged d-complexes are peroxycomplexes with

$Me\big\langle^{O}_{O}\big\rangle Me$ nucleus / e.g. cobalt and rhenium peroxycomplexes

$[Co_2O_2(NH_3)_{10}]^{5+}$, $[Re_2O_2Cl_{10}]^{4-}$.

Their orbital system may be represented as below:

bond between oxygen atoms $\qquad A_{1g}^b \qquad A_{1u}^a$

O_2 group π bonds with metals $\qquad E_g^b \qquad E_g \qquad E_u^a$

This system of MO supplies interesting information about properties and structure of the peroxycomplexes of that type. The same d_{xz}, d_{yz} and d_{z^2} orbitals of metal atoms which take part in sandwich-complexes, contribute also in the bonds of metal atoms with the formal O_2^{2-} group. These complexes therefore may be recognized as π complexes. It agrees with VANNEBERG and BROSSET's opinion (18) concerning the peroxycomplex of cobalt, which is based on the X-ray and IR investigations.

Peroxycomplexes with a d^3-electronic structure (rhenium complexes) are paramagnetic (19), because most probably they have the following MO closed:

bond between oxygen atoms $\qquad (A_{1g}^b)^2 \qquad (A_{1u}^a)^0$

O_2 group π bonds with metals $\qquad (E_g^b)^4 \qquad (E_g)^4 \qquad (E_u^a)^2$

Bond rupture between oxygen atoms due to the attachment of H^+, causes the closing of A_{1u} orbital and therefore the availability of E_u^a orbital and the appearance of diamagnetism.

The model of oxygen bond, based on MO theory presented here, requires still a lot of investigations. It seems to us, however, that it does contain a basical truth concerning the specific role of oxygen in formation and development of chemical compounds.

References

1. JEŻOWSKA-TRZEBIATOWSKA, B., Symposium "Theory and Structure of Complex Compounds", Wrocław 1962. Pergamon Press, Oxford, London, New York, Paris, WNT-Warszawa 1964, 1.
2. JEŻOWSKA-TRZEBIATOWSKA, B., and WOJCIECHOWSKI, W., J.Str.

Chim., $\underline{4}$, 764 (1963).

3. GILLESPIE, R.J., J.Am.Chem.Soc., $\underline{82}$, 5978 (1960).

4. GILLESPIE, R.J., and ROBINSON, E.A., Spectroch.Acta, $\underline{19}$, 741 (1963).

5. JEŻOWSKA-TRZEBIATOWSKA, B., Essay in Coordination Chemistry, Zürich 1964.

6. JEŻOWSKA-TRZEBIATOWSKA, B., and WOJCIECHOWSKI, W., J.Inorg.Nucl.Chem., $\underline{25}$, 1477 (1963).

7. Ref. 1, p.375.

8. WOJCIECHOWSKI, W., and JEŻOWSKA-TRZEBIATOWSKA, B., Bull. Acad.Polon.Sci., $\underline{11}$, 79 (1963).

9. JEŻOWSKA-TRZEBIATOWSKA, B., and WAJDA, S., Bull.Acad.Polon. Sci., $\underline{2}$, 249 (1954).

10. MELLOR, D.P., J.Roy.Soc. (N.S.Wales), $\underline{77}$, 145 (1943).

11. SZUSTROWICZ, E.M., J.Str.Chim., $\underline{4}$, 244 (1963).

12. GRAY, H., BERNAL, I., and HARE, C., VII International Conference on Coordination Chemistry, Stockholm 1962.

13. FRENCH, C.M., and GARSIDE, J.H., J.Chem.Soc. (London), $\underline{1962}$, 2006.

14. JEŻOWSKA-TRZEBIATOWSKA, B., RUDOLF, M., and WOJCIECHOWSKI, W., not published.

15. JAKÓB, W., OGORZALEK, M., and SIKORSKI, H., Roczniki Chemii, $\underline{35}$, 3 (1961).

16. MULAY, L.N., and SELWOOD, P.W., J.Am.Chem.Soc., $\underline{77}$, 2693 (1955).

17. SOCCONI, L., and CINI, R., J.Am.Chem.Soc., $\underline{76}$, 4239 (1954).

18. VANNERBERG, N.G., and BROSSET, C., Acta Cryst., $\underline{16}$, 247 (1963).

19. JEŻOWSKA-TRZEBIATOWSKA, B., and PRZYWARSKA, H., Bull.Acad. Polon.Sci., $\underline{3}$, 429 (1955).

9 A 1

POLARIZED CRYSTAL SPECTRA OF SOME NICKEL(II)CYANIDE COMPLEXES.

C.J. Ballhausen
Chemical Laboratory IV, Dept.of Physical Chemistry,
H.C. Ørsted Institute, Copenhagen, Denmark.

In order to settle some questions as to the bonding in square-planar nickel complexes the following compounds were investigated:

$$Na_2Ni(CN)_4 \cdot 3 H_2O \quad \text{and} \quad K_2Ni(CN)_4 \cdot 3 H_2O$$

which are isomorphous having triclinic symmetry,

$CaNi(CN)_4 \cdot 5 H_2O$ which is orthorhombic and

$SrNi(CN)_4 \cdot 5 H_2O$ and $BaNi(CN)_4 \cdot 4 H_2O$ both being monoclinic. The polarized spectra were measured with the light parallel and perpendicular to the c-axis; this axis being nearly, but not quite perpendicular to the planes of the nickel cyanide complexes. The spectra were measured both at room temperature and at liquid nitrogen temperature.

The general features are as follows: The crystals are highly dichroic, absorption setting in with the light polarized parallel at about 500 mµ. In perpendicular polarization the crystals are optically clear much further out towards the blue, and the observed absorption bands are both sharper and more welldefined than are the parallel ones. The overall spectrum is as follows: Broad parallel bands are seen at ~ 23000 cm^{-1} and at 27000 cm^{-1}, and a sharp perpendicular band is seen at 27000 cm^{-1}. The molar extinction coefficients of these bands are below 100. These numbers differ somewhat from complex to complex, but the general pattern seems to be the same.

In addition to these features we see, for the Barium salt alone a sharp peak in perpendicular polarization at 22000 cm^{-1}. The intensity of this peak goes down steeply as a function of temperature.

At the time of writing the interpretation of these facts are by no means clear. A vibronic intensity giving mechanism is hardly feasible, since the only bands which could be forbidden are parallel transitions. No small static perturbation to the D_{4h} molecular symmetry can give the observed experimental data. All the bands described are too intense to be spin-forbidden bands. What is left is considerations of factor group mixtures of high excited states into the predominantly even to even transitions or a complete change of molecular structure in the excited states. This latter explanation is suggested in view of the fact that no fluorescence is observed in the complexes. The interpretation of the polarizations of the electronic absorption bands is therefore proposed which makes use of the bonding diagram for $[Ni(CN)_4]^{2-}$ calculated by GRAY and BALLHAUSEN (1), but assuming a non-linear configuration of the Ni-C-N bond in the low excited states.

The author wants to thank Professor CURT R.HARE and mag.scient. NIELS BJERRUM for doing the experimental work described here. The research reported has been made possible through the support of the U.S. Department of Army under Contract Number DA-91-591-EUC-3153.

Reference

1. GRAY, H.B., and BALLHAUSEN, C.J., J.Am.Chem.Soc., 85, 260 (1963).

SPECTRA AND NATURE OF BONDING IN
METAL-α-DIIMINE AND RELATED COMPLEXES.

P. Krumholz

Research Laboratory of Orquima S.A., São Paulo, Brasil.

The strongly colored complexes of the d^6 ions Fe^{2+}, Ru^{2+} and Os^{2+} with organic ligands containing the α-diimine grouping $N=C-C=N$, are similar in many respects to organic dyes. The visible and near UV spectrum of tris--α-diimine iron(II) complexes consists of two composite bands with $\lambda_{max} = 500-600$ mμ (A) and $350 - 400$ mμ (B), respectively. The form of the A-band is nearly identical for a wide variety of diimine ligands (band width, $\delta- = 900 \pm 100$ cm^{-1}, shoulder at $1200 - 1500$ cm^{-1}). The broad B-band is less characteristic and of low intensity in complexes of aliphatic diimines.

The color of these iron(II) complexes can be related with the presence of the iron(II)-α-diimine chromophore, $\{1\}$ (1,2)

The influence of "auxochromic" groups on λ_{max} of the A-band is shown in Table I for complexes of 2-pyridinalimines and of 2-pyridyl ketoimines, $C_5H_4N.CR=NR'$.

TABLE I

ABSORPTION MAXIMA (λ_{max} in mμ) OF
$[Fe(C_5H_4N.CR=NR')_3]^{2+}$ COMPLEXES IN H_2O.

R' \ R	H	CH$_3$	C$_6$H$_5$	R' \ R	H	CH$_3$	C$_6$H$_5$
H		572	590	CH$_2$.C$_6$H$_5$	562		
CH$_3$	551	558	565	C$_2$H$_4$.C$_6$H$_5$ **	564	568	
C$_2$H$_5$	556	565	572	C$_6$H$_5$	574	568	579
C$_3$H$_7$	560	566	573	C$_6$H$_4$NH$_2$(m)	574	568	579

Table. (contd.)

C_4H_9	560	566	573	$C_6H_4NH_3^+$(m)		565	579
C_5H_{11}	562	567	574	$C_6H_4NH_2$(p)	583	578	590
C_6H_{11}*	565			$C_6H_4NH_3^+$(p)		565	578
$C_2H_4N(Et)_2$	563	567		C_6H_4COOet(p)		563	

* = cyclohexyl;   ** = 2-phenylethyl.

The bathochromic effect of phenyl substituents depends on the angle between the benzene and the chelate rings, suggesting that resonance effects are involved. This is supported, on the whole, by the spectral behaviour of the NH_2, NH_3^+ and COOR substituted derivatives. However, the effect of p-NH_2 substituents does not show the expected dependence on the aforesaid twisting angle. The hypsochromic shift on passing from NH to NCH_3, is also observed in some cyanine dyes.

The spectra of the iron(II) complexes of tridentate, "terpyridine-like" ligands containing one of the groupings $N=C-C=N-C-C=N$ (a), $N=C-C=N-C-C-N$ (b) or $C=N-C=N-C=C-N$ (c), are all alike but differ chracteristically from the spectra of the bidentate diimine complexes. δ - of the A-band is now only 550 ± 100 cm^{-1}, band B is occasionally missing and a new band appears at +2500-4500 cm^{-1} from the A-band. These spectral properties can be again correlated with the presence of a common chromophoric group {2}. The conjugated double bond in the second chelate loop of IIc causes a large ($\simeq 2000$ cm^{-1}) red-shift of the A-band. This bond belongs thus rightly to the chromophore.

Fusing of aromatic rings on chromophores {1} and {2} results in a blue shift. Similar effects are observed and theoretically expected (3) for derivatives of fulvene {3}.

On the whole, the behaviour of chromophores {1} and {2} provides support for the belief that metal orbitals of π-symmetry enter into the conjugation present in the ligands (1,2). A naive V.B. description involving metal-ligand double (d-π) bonding, may account for some ground state properties and, in terms of crossed resonance, for the blue-shift on ring fusion (2). With respect to the spectral effect of aromatic substituents, one is restricted to the rather trivial statement that they do extend the conjugated system.

According to WILLIAMS (4) and JØRGENSEN (5) the color of iron(II) diimine complexes is caused by electron transfer from a filled t_{2g} orbital, essentially localised on the metal, to empty π*-orbitals, essentially localised on the ligands. In a naive interpretation of this picture, one would expect electron-releasing substituents to produce a blue shift (and vice versa). Data in Table I fail to agree with this expectation. Likewise, no relation seems to exist between the wavenumbers of the electron transfer (A) bands, ν_{vis}^{2+}, and of the first strong UV bands (internal ligand transitions), ν_{uv}^{2+}, listed in Table II. Such correlations are more properly to be expected in the series of the related iron(III) diimine complexes, the color of which is likely caused by electron transfer from filled ligand π-orbitals, to the vacant t_{2g} orbital of the d^5 metal ion. In fact, $\Delta = \nu_{uv}^{2+} - \nu_{vis}^{3+}$ is found to vary but little over a nearly 1:2 range of ν_{vis}^{3+}. Coincidently, $\Sigma = \nu_{vis}^{2+} + \nu_{vis}^{3+}$ happens to differ by only about 1200 ± 500 cm^{-1} from ν_{uv}^{2+}.

TABLE II

WAVENUMBERS ($\nu . 10^{-3}$, cm^{-1}) OF VISIBLE AND UV ABSORPTION
BANDS OF IRON(II) AND IRON(III)* DIIMINE COMPLEXES.

Ligand	ν_{vis}^{2+}	ν_{vis}^{3+}	ν_{uv}^{2+}	Δ	Σ
o-phenanthroline	19,6	16,6	34,5	17,9	36,2
2,2'-bipyridine	19,1	16,2	33,7	17,5	35,3
2-pyridinalmethylimine	18,1	17,9	35,2	17,3	36
glyoxalbismethylimine	18	27	44	17	45
2,2',2"-terpyridine	18,1	14,2	31,4	17,2	32,3

* measured in 10 M H_2SO_4.

The spectra of the complexes of symmetrical and highly assymmetrical bidentate diimine ligands are nearly identical. Low over-all symmetry seems thus not to disturb significantly the local (cubic) microsymmetry. The appearance of a new transition in the complexes of tridentate ligands suggests, however, that perturbation caused by spatial conditions and preferential d-π bonding along the z-axis (6), is large enough to cause a splitting of the t_{2g} level.

References

1. KRUMHOLZ, P., J.Am.Chem.Soc., 75, 2163 (1953).
2. BUSCH, D.H., and BAILAR, Jr., J.C., J.Am.Chem.Soc., 78, 1137 (1956), refer to this group as iron(II)-methine chromophore.
3. PULLMAN, B., and PULLMAN, A., "Les Theories Electroniques de la Chimie Organique", Masson et Cie, pp. 495-497, Paris (1952).

4. WILLIAMS, R.J.P., J.Chem.Soc. (London), 1955, 137.

5. JØRGENSEN, C.K., Acta Chem.Scand., 11, 166 (1957).

6. FIGGINS, P.E., and BUSCH, D.H., J.Phys.Chem., 65, 2236 (1961).

9 A 3

CHEMICAL STRUCTURE DETERMINATION OF LANTANIDES' COORDINATION COMPOUNDS ON THE BASIS OF THEIR ABSORPTION SPECTRA.

K.B. Yatsimirsky, N.K. Davidenko, N.A. Kostromina, T.V. Ternovaya
Ukrainian Academy of Science, Kiev, USSR.

Heretofore the majority of investigators considered the chemical bond in coordination compounds of the rare earths elements to be of ionic nature. Within the last years, however, many arguments have been submitted in favour of a partial covalency of the formed bonds. One of the reasons points out at the impossibility to deduce thermodynamical characteristics of the complexes on the mere basis of the ions charges and radii. An ample information on the geometry and chemical structure of the rare earth coordination compounds may be obtained on the basis of their absorption spectra.

Absorption spectra of these compounds consist of a large number of narrow bands, originated from f-f transitions. Differing from the spectra of gaseous ions, those bands in the spectra of coordination compounds are being displaced ("nephelauxetic effect") and splitted up to components ("Crystal field effect"). Measuring their displacement we can evaluate the deviation from the ionic type of chemical bonding, while the magnitude and character of splitting enable us to judge upon the strength of the ligand field and the geometry of coordination compounds.

Contrary to the coordination compounds with d-transition elements the nephelauxetic effect in the series of lanthanides is not constant, but decreases with the rise of atomic number. The red shift may be explained by the action of two factors: delocalization of σ-antibonding orbitals and diminution of the effective charge owing to partial transition of the ligand electrons to 6s or to other penetrating orbitals of the central ion. It is possible that both factors contribute to the shift in the beginning of the series, but later, in proportion to f-orbitals filling with electrons, only the second factor is acting (central-field covalency).

The absorption and reflection spectra of praseodymium and neodymium compounds were studied. That enabled us to build up the "nephelauxetic series" of ligands for these elements:

$$F^- < H_2O < Tar^{2-} < Acac^- < Benz^- < NH_3 < EDTA^{4-} < Dipy < Phen <$$
$$< Cl^- < Br^- < I^- < O^{2-}$$

The magnitude of the red shift in the spectra of praseodymium and neodymium compounds correlates with the values of the nephelauxetic effects (β) of d-transition element complexes with the same ligands.

The red shift of $^4J_{9/2} \longrightarrow {}^2P_{1/2}$ transition band maximum was investigated in the case of coordination compounds of neodymium with some complexones and oxycarbonic acids in aqueous solution. The red shifts of the maximum were determined in comparison with its position in the neodymium aqua-ion. The shifts's magnitude depends on the nature of bonding of the central ion with ligands and represents an additive quantity.

Shift magnitudes ($\Delta\nu$) determined for the compounds of known composition, were compared. The increments, contributing to the shift in the presence of neodymium bond with nitrogen and oxygen atoms of the carboxy-, oxy- and hydroxy-groups were established. These increments may be used to investigate composition and structure of the neodymium compounds. Table I shows the red shift of the $^4J_{9/2} \longrightarrow {}^2P_{1/2}$ transition band maximum in neodymium complexes.

The splitting of the ground level of Nd^{3+} ($^4J_{9/2}$) and of the excited level of Eu^{3+} ($^5D_2$) were investigated in complexes of those rare earths with EDTA, NTA, DTPA and gluconic acid in aqueous solution. In aqua-ion as well as in all investigated Nd^{3+}-complexes a splitting of the ground level to 5 sublevels was observed, which corresponds to non-cubic fields of symmetry.

In order to establish the type of non-cubic symmetry for those events the splitting of the level $^5D_2$ in analogous complexes of Eu^{3+} was studied. It was found that the aqua-ion and complexes with EDTA belong to tetragonal symmetry while other complexes refer to tri- or hexagonal symmetry.

Taking into account those data, particularly the structure of ligands, we propose the probable structure formulas of the complexes: $[MX_2]^{3-}$ and MZ^{2-} (symmetry C_{3v}), $[MY(OH)]^{2-}$ (symmetry C_{4v}).

The spectrochemical series for ligands was established as follows:

$$H_2O < Gluc^-, Ac^-, Lac^- < 2 Gluc^- < 2 Gluc^{2-} < EDTA (YOH^{5-}) <$$
$$< 2 NTA (X^{3-}) < DTPA (Z^{5-})$$

There was established a correlation between spectral and thermodynamical data.

TABLE I - THE RED SHIFT OF THE $^4J_{9/2} \longrightarrow {}^2P_{1/2}$ TRANSITION BAND MAXIMUM IN THE NEODYMIUM COMPLEXES

Compounds	$\Delta \nu$	Atoms or groups bonding to Nd			
		$>N-$	$-COO^-$	$\geqq CO^-$	HO^-
$[NdLac]^{2+}$, $[NdGluc]^{2+}$, $[NdHMal]^{2+}$, $[NdHTar]^{2+}$	25	-	1	-	-
$[NdLac_2]^+$, $[NdGluc_2]^+$, $[NdHTar]^+$, $[NdTri]^+$	50	-	2	-	-
$[NdMal_2]^-$, $[NdTar_2]^-$	100	-	4	-	-
$[NdGluc]^+$	66	-	1	1	-
$[NdGluc_2]^-$	126	-	2	2	-
$[NdMalMal]^{2-}$, $[NdTarTar]^{2-}$	121	-	3	1	-
$[NdMal_2]^{3-}$, $[NdTar_2]^{3-}$	142	-	2	2	-
NdX^0	77	1	2	-	-
$[NdX_2]^{3-}$	141	2	4	-	-
$[NdY(OH)]^{2-}$	165	2	3	-	1
$[NdY]^-$	115	2	3	-	-
$[NdZ]^{2-}$	131	3	3	-	-

Symbols: Acac = acetylacetone, Benz = benzoyl-acetone, Dipy = dipyridil, Phen = phenantroline, Ac^-, Lac^-, $Gluc^-$, Mal^{2-}, Tar^{2-}, Tri^{2-} = anions of acetic, lactic, gluconic, tartaric, trioxyglutaric acids. The touch above the anion means that one proton of the oxygroup of the oxycarbonic acid is dissociated. X^{3-} or NTA = anion of nitrylotriacetic acid, Y^{4-} or EDTA = anion of ethylendiamintetraacetic acid, Z^{5-} or DTPA = anion of diethylentriaminpentaacetic acid.

References

1. JØRGENSEN, C.K., "Absorption spectra and chemical bonding in complexes" Pergamon Press, Oxford, London, New York, Paris (1962).

2. JØRGENSEN, C.K., J.Chem.Phys., _39_, 1422 (1963).

STRUCTURE OF A SYNTHETIC MOLECULAR OXYGEN CARRIER.

James A. Ibers and Sam J. LaPlaca
Chemistry Department, Brookhaven National Laboratory,
Upton, New York 11973, U.S.A.

VASKA (1) has characterized a simple synthetic molecular oxygen carrier $O_2IrCl(CO)[P(C_6H_5)_3]_2$ prepared by the addition of molecular oxygen to $IrCl(CO)[P(C_6H_5)_3]_2$ in benzene solution. This synthetic oxygen carrier, which is a reasonably stable crystalline solid, differs from most reported in the past in that there is but one molecule of oxygen per metal atom, as there is in oxyhemoglobin. The mode of attachment of oxygen to iridium in this compound is thus of interest both in itself and especially in relation to theories of the mode of attachment of molecular oxygen to the iron atom in oxyhemoglobin.

Excellent crystals of $O_2IrCl(CO)[P(C_6H_5)_3]_2$ were very kindly supplied by VASKA. The material crystallizes with two molecules in a triclinic cell of dimensions a = 19,02, b = 9,83, c = 9,93 Å, α = 94,0, β = 64,9, γ = 93,2°. A very sensitive piezoelectric test (2) was negative and the space group $P\bar{1}$ assumed. Data were collected at room temperature with Zr filtered MoKα radiation on a Nonius integrating Weissenberg camera. The intensities of approximately 1100 reflections were estimated visually and reduced in the normal manner to structure amplitudes. The crystal was carefully goniometered and its dimensions measured and an accurate correction for absorption applied.

The structure was solved by the usual combination of three-dimensional Patterson, least-squares, and Fourier methods. As in the case of $HRh(CO)[P(C_6H_5)_3]_3$ (3), the phenyl groups were constrained to their well known geometry during refinement, both because this makes physical sense and because it reduces the number of variables. The structure refined readily, but it was not possible to distinguish on the electron density maps the CO from the Cl, although the two oxygen atoms were clearly visible and resolved. Thus Fig. 1 shows a section through the three-dimensional electron density map in the plane of the CO, Cl, and O_2 groups. Clearly, the CO and Cl replace one another at random. This presumably is the result of molecular oxygen attacking the parent square-planar compound from either side. If then the triphenylphosphine ligands are equivalent in solution, and are of sufficient bulk to control the packing, such disorder is not unexpected. The final refinement converged to a very low conventional R factor of 7,1%.

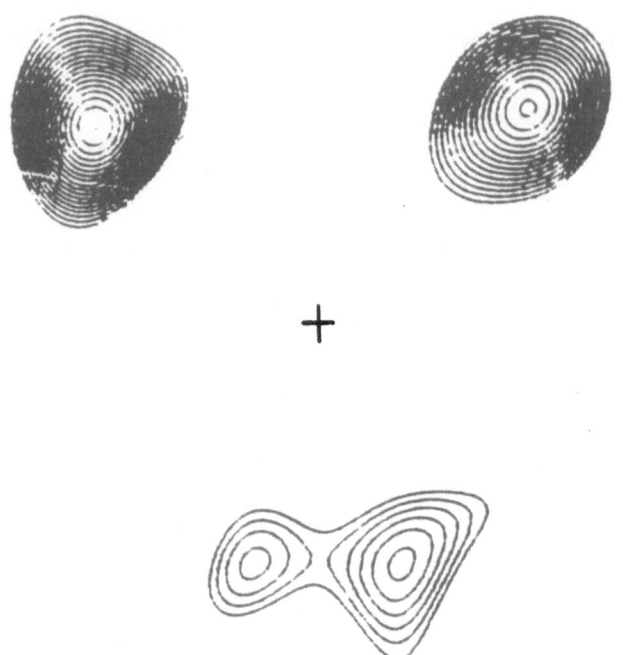

Fig. 1 - Electron density in the plane of the CO, Cl, Ir, and O_2. The contribution of Ir has been subtracted for clarity, but the Ir position is marked with a cross.

The molecular conformation may be described as a planar arrangement of Cl, CO, Ir, and O_2, with P above and below this plane and P-Ir-P perpendicular to the plane. Figure 2 shows this conformation in perspective. (We have designated the 1/2 Cl-1/2 CO positions by "X".) The conformation may either be described as that of a trigonal bipyramid if one wishes to consider O_2 as a single ligand, or as distorted octahedron, if one considers the two oxygen atoms separately. The important points are that the oxygen atoms, within the limits of error of this study, are equidistant from the Ir atom (mean Ir-O is 2,06 ± 0,02 Å) and the O-O bond length is 1,30 ± 0,03 Å. This bond length is characteristic of O_2^- (1,28 Å) but not of O_2^{2-} (1,49 Å). The Ir-X distances do not differ significantly and average 2,40 ± 0,02 Å and the Ir-P distances are equal (2,37 ± 0,01 Å).

The arrangement of oxygen relative to iridium is similar to the arrange-ment proposed by GRIFFITH (4) for the bonding of molecular oxygen to the iron in oxyhemoglobin in that the oxygens are equidistant from the iridium.

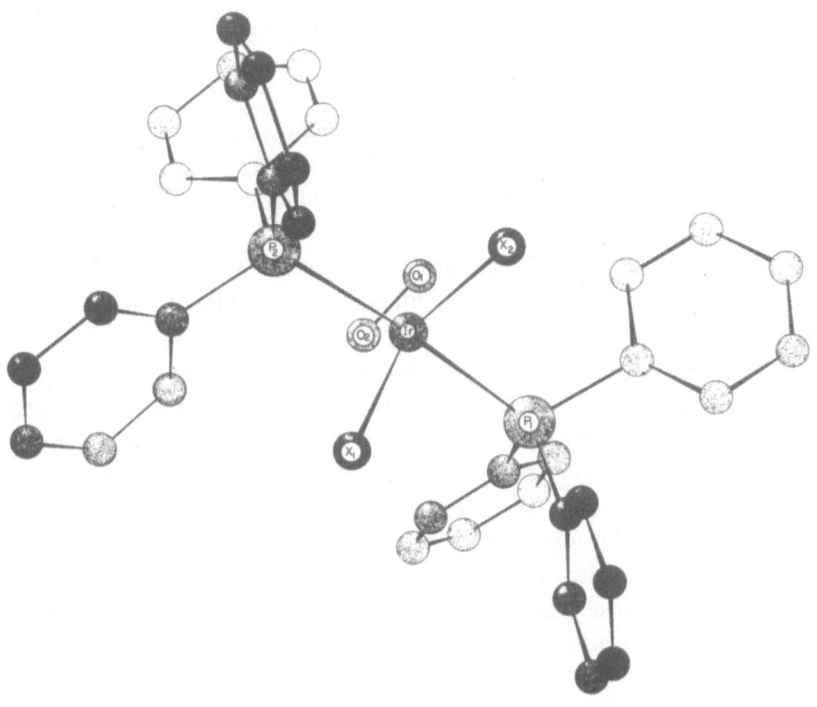

Fig. 2 - Perspective drawing of the molecular structure of
$O_2IrCl(CO)[P(C_6H_5)_3]_2$. The half-Cl-half-CO positions are
denoted by "X".

In place of molecular oxygen one can substitute SO_2 (5). This compound
is under active study and it is hoped that its molecular configuration can be
reported at the meeting.

This research was performed under the auspices of the U.S. Atomic
Energy Commission.

References

1. VASKA, L., Science, 140, 809 (1963).
2. We are indebted to HOLTZBERG, F., for this measurement.
3. LaPLACA, S.J., and IBERS, J.A., J.Am.Chem.Soc., 85, 3501 (1963).
4. GRIFFITH, J.S., Proc.Roy.Soc., A 235, 23 (1956).
5. VASKA, L., unpublished work.

O A 2

ÜBER DIE KATALYTISCHE WIRKSAMKEIT VON ÜBERGANGSMETALL-π-ALLYLVERBINDUNGEN.

G. Wilke, P. Heimbach, B. Bogdanović,
W. Oberkirch und K. Tanaka
Max-Planck-Institut für Kohlenforschung, Mühlheim-Ruhr, Germany.

Bei der Umsetzung von Übergangsmetallverbindungen mit metallorganischen Verbindungen, wie z.B. Aluminiumalkylen, entstehen im allgemeinen instabile Übergangsmetallalkyle bzw. -aryle, deren Metall-Kohlenstoff-Bindungen im Sinne einer Homolyse zerfallen. Neben der Bildung von Kohlenstoffradikalen werden die Metallatome reduziert, was bis zur Abscheidung eines Metallspiegels führen kann. Wird diese Reduktion in Gegenwart von Elektronendonatoren vorgenommen, so lassen sich vielfach die entstehenden Metallatome in Form von π-Komplexen abfangen. Auf diesem Wege können unter schonenden Bedingungen auch sehr labile Komplexe hergestellt werden. Die geschilderte Methode wurde z.B. bei der Züchtung von Katalysatoren für die Cyclooligomerisation von Butadien zu Cyclooctadien-(1,5) oder Cyclododecatrien-(1,5,9) sowie für die Mischoligomerisation von Butadien und Äthylen zu Cyclodecadien-(1,5) angewandt.

Die Cyclooligomerisationen verlaufen - wie nachgewiesen werden konnte - über die Zwischenstufen von π-Allyl-Metallverbindungen, die sowohl im Falle der Cyclooctadien- als auch bei der Cyclododecatriensynthese in kristallisierter Form isoliert werden konnten. In diesem Zusammenhang war es von besonderem Interesse, reine π-Allyl-Metallverbindungen zunächst des Nickels und dann auch von andern Übergangsmetallen zu synthetisieren.

Die Synthese gelingt allgemein durch Umsetzung von Allyl-Grignard-Verbindungen mit Übergangsmetallverbindungen. Prototyp der dabei entstehenden Komplexe ist das Bis-π-allyl-nickel, eine leicht flüchtige, bei tiefen Temperaturen kristallisierte Verbindung. Analog erhält man das Bis-π-methallyl-nickel, dessen Einkristall-Röntgenstrukturanalyse zeigte, daß es sich um eine "Sandwich"-Verbindung mit antiständigen Methylgruppen handelt. Das Protonenresonanzspektrum stimmt mit dieser Struktur überein.

Weitere, auf eine größere Zahl von Übergangsmetallen ausgedehnte Untersuchungen lassen erkennen, daß offensichtlich die meisten Elemente der Nebengruppen reine Allylverbindungen zu bilden vermögen, in denen zwei, drei und vier Allylgruppen an das Zentralmetall gebunden sind. Die Stabilität dieser Allyl-Metall-Verbindungen weist erhebliche Unterschiede auf. Sie nimmt jedoch im allgemeinen innerhalb einer Gruppe des periodischen Systems von den niederen zu den höheren Elementen hin zu. Von mehr als 10 verschiedenen Metallen konnten bisher entsprechende Verbindungen in kristallisierter Form isoliert werden.

Von besonderem Interesse ist die von diesen Verbindungen ausgehende katalytische Aktivität, die - wie gezeigt werden konnte - immer dann besonders hoch ist, wenn die Allyl-Metallverbindungen wenig stabil sind. Auf der Basis von Allyl-Metallverbindungen lassen sich hochwirksame Katalysatoren züchten, mit deren Hilfe in homogener Phase Reaktionen selektiv ausgelöst werden können. Je nach Wahl des Metalls läßt sich z.B. Butadien mit den entsprechenden reinen Allylverbindungen selektiv in Cyclododecatrien, 4-Methylheptatrien-(1,3,6) oder aber 1,2-Polybutadien verwandeln. Substituiert man einzelne Allylgruppen durch bestimmte andere Liganden, so entstehen Katalysatoren, die bei gleichem Metall z.B. statt 1,2-Polybutadien Cyclododecatrien oder statt 4-Methylheptatrien-(1,3,6) cis-1,4-Polybutadien entstehen lassen. Ebenfalls mit einer reinen Allyl-Metallverbindung als Katalysator geling es in Lösung, Äthylen zu hochmolekularem Polyäthylen zu polymerisieren, ohne daß der Katalysator - wie im Falle der Ziegler-Katalysatoren - Halogenatome oder metallorganische Verbindungen der Metalle der I. bis III. Hauptgruppe des periodischen Systems enthält. Etwas abgewandelte Übergangsmetall-π-allylverbindungen vermitteln eine wahlweise lenkbare Dimerisation von Olefinen, wie insbesondere von Propylen.

10 A 3

HOMOGENEOUS ACTIVATION OF MOLECULAR HYDROGEN BY FOUR-, FIVE-, AND SIX-COORDINATE TRANSITION METAL COMPLEXES.

Lauri Vaska
Mellon Institute, Pittsburgh, Pa., U.S.A.

Certain neutral complexes of bivalent ruthenium (1) and osmium (1,2), and tervalent and univalent iridium (3,4) have been discovered to activate molecular hydrogen in solution at normal conditions. These reactions are of considerable interest because they permit comparison of hydrogen-activating characteristics of metal complexes with different coordination numbers and formal oxidation states, two properties which are believed to represent important factors in determining catalytic activation of hydrogen.

Some experimental observations on hydrogen-deuterium exchange are summarized below. The reactions were carried out in toluene solution (10^{-2} M); D_2 (one at.) was used in large excess, and the products were isolated and characterized as pure deuteride complexes.

$$IrHCl_2(PPh_3)_3 + D_2 \longrightarrow IrD_2Cl(PPh_3)_3 + HCl \quad \{1\}$$

$$IrH(CO)(PPh_3)_3 + D_2 \longrightarrow IrD(CO)(PPh_3)_3 + HD \quad \{2\}$$

$$MHX(CO)(PPh_3)_3 + D_2 \longrightarrow MDX(CO)(PPh_3)_3 + HD \quad \{3\}$$

$$(M = Os, Ru; X = Cl, Br)$$

There are apparently two types of mechanisms represented. The six-coordinate and tervalent iridium complex in $\{1\}$ yields a product which is analogous with the one we have previously found to result from a four-coordinate and univalent Ir compound $\{4\}$ (5, 6).

$$IrX(CO)(PPh_3)_2 + H_2 \rightleftharpoons IrH_2X(CO)(PPh_3)_2 \quad \{4\}$$

$$(X = Cl, Br, I, etc.)$$

Reaction $\{1\}$ is therefore likely to commence with a dissociative step $\{5\}$, followed by addition of D_2, as in $\{4\}$, to a four-coordinate complex.

$$IrHCl_2(PPh_3)_3 \rightleftharpoons IrCl(PPh_3)_3 + HCl \quad \{5\}$$

The results of reactions $\{2\}$ and $\{3\}$, on the other hand, suggest that the H-D exchange in the five- and six-coordinate complexes may involve, respectively, seven- and eight-coordinate intermediates, and the mechanism may thus be expressed as in $\{6\}$ (for reaction $\{2\}$).

$$IrH(CO)(PPh_3)_3 + H_2 \rightleftharpoons IrH_3(CO)(PPh_3)_3 \quad \{6\}$$

Some of these compounds have been found to catalyze hydrogenation of ethylene and other substrates. Since the molecular structures of these catalysts are known (4, 6, 7), discussion of reaction mechanisms can now start with an evidence previously not available for hydrogen-activating systems (8).

References

1. VASKA, L., and DiLUZIO, J.W., J.Am.Chem.Soc., 83, 1262 (1961).
2. VASKA, L., J.Am.Chem.Soc., 86, in press.
3. VASKA, L., J.Am.Chem.Soc., 83, 756 (1961); VASKA, L., and DiLUZIO, J.W., ibid., 84, 4989 (1962).
4. BATH, S.S., and VASKA, L., J.Am.Chem.Soc., 85, 3500 (1963).
5. VASKA, L., and DiLUZIO, J.W., J.Am.Chem.Soc., 84, 679 (1962).
6. To be published.

7. LaPLACA, S.J., and IBERS, J.A., J.Am.Chem.Soc., 85, 3501 (1963); ORIOLI, P.L., and VASKA, L., Proc.Chem.Soc., 1962, 333.

8. See, for example, SLOAN, M.F., MATLACK, A.S., and BRESLOW, D.S., J.Am.Chem.Soc., 85, 4014 (1963); this article gives references to other recent as well as earlier papers in this field.

11 A 1

A NEW METHOD FOR THE DETERMINATION OF CIRCULAR DICHROISM.

Peter F. Arvedson and Edwin M. Larsen
University of Wisconsin, Madison, Wisconsin, U.S.A.

Recent papers on the relationship between optical rotatory dispersion and circular dichroism of optically active systems have stressed the superiority of the latter in interpreting the data. Unfortunately, circular dichroism measurements have been either difficult or expensive to obtain. Recently available instruments are costly and the reliability of the data these instruments afford certainly has not reached its maximum. Universal polarimeters have been described, but they require either major changes in existing instruments or special construction of new instruments. This paper describes a method of modifying the RUDOLPH Spectropolarimeter, Model 200 AS, to measure the ellipticity of elliptically polarized light and thus indirectly measure circular dichroism. This modification requires no major changes and virtually no expense.

Figure 1 is a schematic diagram of the optic train. Monochromatic plane polarized light enters the optically active solution and emerges (in general) as elliptically polarized light (E_1), whose azimuth (orientation of major axis with respect to the basic axes of the system) can readily be determined. The ellipticity, e, of this elliptically polarized light, which is proportional to the difference in extinction coefficients of the left- and right-handed circularly polarized components of the original plane polarized light, is to be determined in this experiment. This elliptically polarized light then strikes a thin piece of mica, situated perpendicular to the path of propagation, or arbitrary thickness and orientation. (Note: it is not necessary that this mica be a quarter-wave plate at the frequency of the light being used as in other ellipsometers). The vibration is resolved into two components, one along the fast axis of the mica and the other along the slow axis. Upon emerging from the mica the vibrations recombine to produce elliptically polarized light (E_2) which is (in general) different from the polarized light (E_1) which entered the mica. This light strikes the analyzing prism, which is used in the conventional

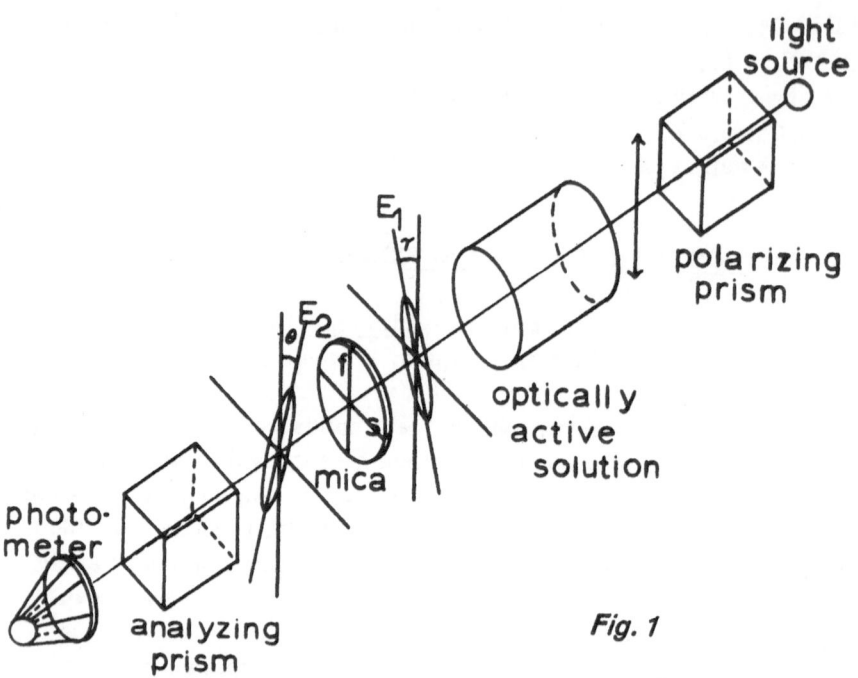

Fig. 1

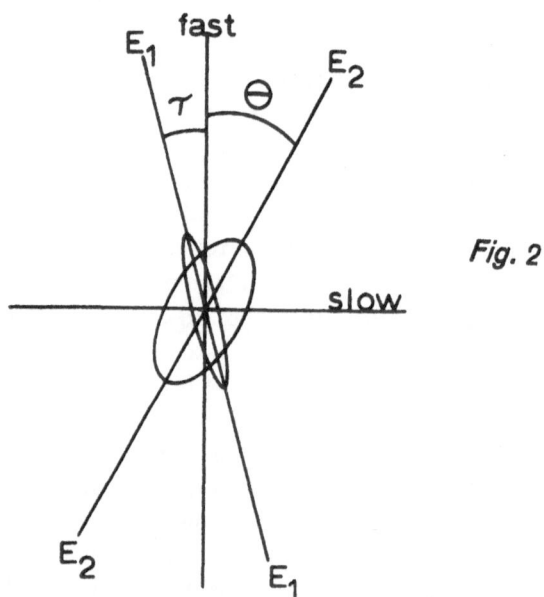

Fig. 2

manner to determine the azimuth of the elliptical vibration.

The effect of the mica on elliptically polarized light can be represented as in Figure 2. For convenience the axes of the mica are used as the basic reference axes. The original elliptically polarized light (E_1) has ellipticity e and its major axis forms the angle τ with the fast axis of the mica. The velocity of the vibration along the slow axis is less than the velocity along the fast axis by a factor δ, called the retardation. The major axis of the resultant elliptically polarized light (E_2) forms the angle Θ with the fast axis of the mica. It will be shown that this angle is related to the ellipticity e of (E_1) by the following expression:

$$\tan 2\,\Theta = \tan 2\,\tau \;.\; \cos\delta \;-\; \frac{2\,e}{1-e^2} \;.\; \frac{\sin\delta}{\cos 2\,\tau}$$

from which e can be readily calculated. Of the four variables in this equation, e is the unknown; the two angles τ and Θ can be determined from the positions of the major axes of the elliptical vibrations (E_1) and (E_2) and the position of the fast axis of the mica, which can be determined; and the retardation, δ, can be determined for each wavelength from the above equation if plane polarized light is used (e = 0) by the expression:

$$\cos\delta = \frac{\tan 2\,\Theta}{\tan 2\,\tau}$$

Data obtained for trisethylenediamminecobalt(III)ion agree quite closely with values obtained by MASON (1), and it is felt the values presently obtainable are reproducible to three significant figures at the wavelength of maximum circular dichroism for $[Co(en)_3]^{3+}$, 490 mμ.

The advantages and disadvantages of this technique will be discussed and the experimental methods necessary to obtain the maximum accuracy will be described. A comparison with existing data will be shown and new circular dichroism curves or inorganic complexes will be presented.

Reference

1. McCAFFREY, A.J., and MASON, S.F., Mol.Phys., 6, 359 (1963).

11 A 2

ROTATORY STRENGTH AND ABSOLUTE CONFIGURATION OF TRIS(DIAMINE) TRANSITION METAL COMPLEXES.

Flemming Woldbye
Chemistry Department A, The Technical University of Denmark,
Copenhagen K, Denmark.

In order to deepen our understanding of the optical activity of octahedral

metal complexes it appears necessary to correlate experimentally the structure
- or preferably the absolute configuration - and the rotatory strength of a
number of typical representatives for this class of compounds.

A determination of the absolute configuration by the ultimate means,
X-rays, has been carried out only for the complexes [Co en$_3$]$^{3+}$, [Co l-pn$_3$]$^{3+}$
and trans-[Co l-pn$_2$Cl$_2$]$^+$ (en = ethylenediamine, pn = 1,2-diaminopropane)
(1-3). Besides this rather cumbersome method a number of more or less
plausible experimental criteria for correlating absolute configurations of
structurally related complexes has been advocated.

Most prominent among those are arguments based on the solubility of
diastereomers, e.g. (-)$_D$-[Co(+)$_D$-chxn$_3$]$^{3+}$ and (+)$_D$-[Co(+)$_D$-cptn$_3$]$^{3+}$
(chxn = 1,2-diaminocyclohexane, cptn = 1,2-diaminocyclopentane) are
believed to have analogous absolute configurations around the central atom
because their diastereomers with D-tartrate - actually their halogenide
tartrates - are less soluble than those of their antipodes (4,5) and criteria
derived from the formation of "active racemates", e.g. the solubility of the
bromide of rac.-[Co en$_3$]$^{3+}$ is much lower than that of the single enantiomers.
The analogous statement about the corresponding Rh(III) complexes is also
true. (-)$_D$-[Co en$_3$]$^{3+}$ and (-)$_D$-[Rh en$_3$]$^{3+}$ form a sparingly soluble bromide
double salt, (-)$_D$-[Co en$_3$](-)$_D$-[Rh en$_3$]Br$_6$.5 H$_2$O (termed "active racemate"
by DELEPINE (6,7)), while no double bromide with a combination of a (+)$_D$-
and a (-)$_D$-complex appears to be less soluble than the single enantiomers.
(+)$_D$-[Co en$_3$]$^{3+}$ and (-)$_D$-[Rh en$_3$]$^{3+}$ are, therefore, believed to have
analogous absolute configurations.

A very promising short-cut to reliable absolute configurations of certain
complexes has been indicated by COREY and BAICAR (8). The basis for this
is the fact originally found by THEILACKER (9) that the five-membered rings
formed in ethylenediamine-metal complexes are not planar but twisted somewhat like cyclopentane. As pointed out (8) this means that in a tris(en) metal
complex the C-C bond in each of the en molecules may occupy one of two
positions, namely nearly parallel ("lel") to or obliquely slanted ("ob") with
respect to the trigonal axis (there is, of course, strictly speaking a trigonal
axis only if all of the three en molecules have the same conformation).
COREY and BAILAR (8) showed that the tris("lel") complex would be more
stable than any of the other possible conformations.

If a C-substituted ethylenediamine, e.g. 2,3-diaminobutane (2,3-bn),
is used as a ligand it is easily seen that the substituents may be classified as
equatorial* or axial*, equatorial* substitution leading to the more stable
compound (an asterisk was suggested by COREY and BAILAR as a reminder

that due to the ca. $90°$ N-M-N angle in the ring the situation is not rigorously the same as in a substituted cyclopentane). We can thus visualize four possible situations for a $(+)_D$-2,3-bn molecule (the absolute configuration of which is known through synthetic correlation (10)) in, say, $[Co(+)_D$-2,3-bn$_3]^{3+}$ as illustrated in fig. 1. Only the molecule labeled "a" having the λ absolute configuration around the central atom satisfies both the requirement of being "lel" and that of having exclusively equatorial* substituents, and this con-figuration is, therefore, believed to be the stable form of the species. This explains in a straightforward way the very pronounced tendency for optically active 2,3-bn (and several analogous optically active ligands) to form tris complexes by way of dissymmetric synthesis (11). Note that in the meso-form of the 2,3-bn as a ligand one methyl group will be equatorial* while the other must be axial*; this probably accounts for the lesser stability of the meso-2,3-bn complexes (12).

Clearly some of the configurations shown in fig. 1 must be quite unstable, but the difference in stability between the most stable configuration and the less stable ones cannot easily be ascertained. However, the calculations by

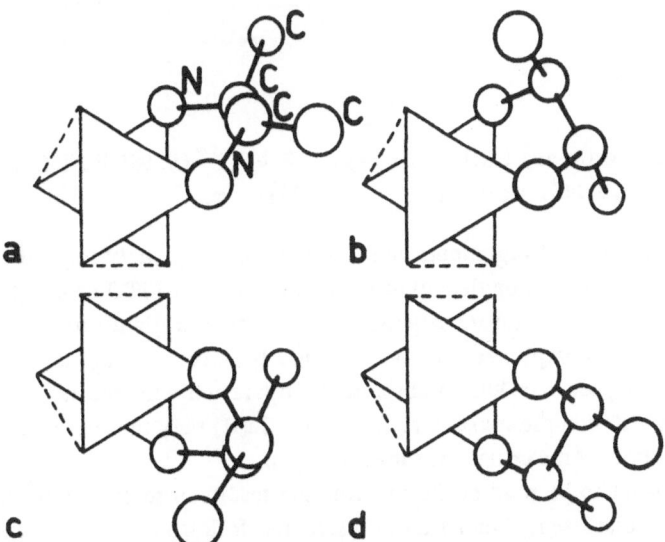

Fig. 1.

Four $(+)_D$-2,3-bn conformations in chelate rings of an octahedral complex. a:Λ, "lel", CH$_3$-groups equatorial*. b:Λ, "ob", axial*. c:Δ, "lel", axial*. d:Δ, "ob", equatorial*.

COREY and BAILAR as well as consideration of molecular models indicate that this difference suffices for making the most stable configuration the dominating species in an equilibrium between the possible forms. This, of course, requires at room temperature only 2 - 3 kcal/mole.

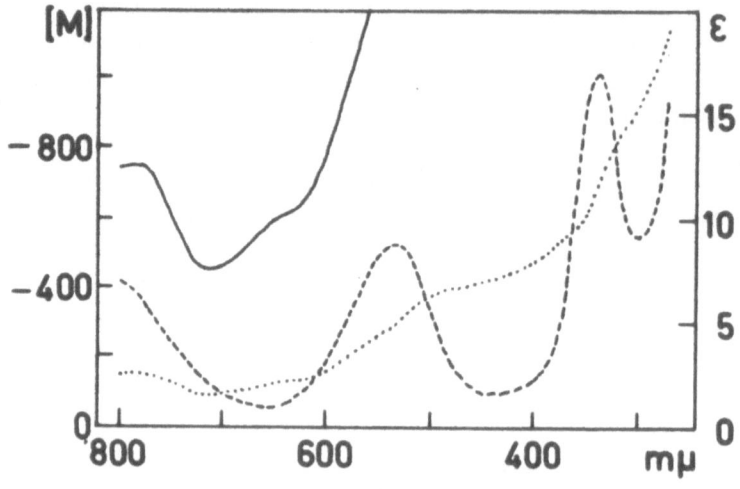

Fig. 2.

Absorption spectrum and ORD of $[Ni((+)_D-2,3-bn)_3]^{2+}$ (0, 085 F $NiSO_4$, 0, 424 F $(+)_D-2,3-bn$). ϵ :---. [M]:$\cdots$. 5x[M]:——.

If one accepts this argumentation which has been strongly reinforced by the investigations of 1-propylenediamine complexes by SAITO et al. (2,3). A study of the optical rotatory dispersion and of circular dichroism of tris($(+)_D-2,3-bn$) complexes become particularly interesting in that rotatory strengths of complexes of known absolute configuration with central ions not forming "robust" complexes such as Ni(II) and Cu(II) may thus be obtained for the first time. An example is shown in fig. 2.

In the paper to be read at the meeting the results of such a study will be presented and discussed. Among these results the following may be mentioned:

a) Co(III) and Cr(III) complexes of the same absolute configuration show a predominant Cotton effect corresponding to identical sign of the net rotatory strength of the "first" absorption band.

b) All of the $(+)_D-2,3-bn$ complexes studied so far have the same sign of the rotatory strength of the nearest UV charge transfer band.

c) tris($(+)_D-2,3-bn$) complexes seem to have a considerably higher rotatory strength than mono- or bis-complexes (also if calculated per metal-ligand ring).

References

1. SAITO, Y., NAKATSU, K., SHIRO, M., and KUROYA, H., Acta Cryst., 7, 638 (1954); 8, 729 (1955); Bull.Chem.Soc.Japan, 30, 158, 795 (1957); 35, 832 (1962).
2. SAITO, Y., and IWASAKI, H., Bull.Chem.Soc., Japan, 35, 1131 (1962).
3. SAITO, Y., IWASAKI, H., and OTA, H., Bull.Chem.Soc. Japan, 36, 1543 (1963).
4. JAEGER, F.M., and BIJKERK, L., Proc.Acad.Sci.Amsterdam, 40, 246 (1937).
5. JAEGER, F.M., Proc.Acad.Sci.Amsterdam, 40, 2, 108 (1937).
6. DELEPINE, M., Bull.Soc.Minéral.Français, 53, 73 (1933).
7. DELEPINE, M., Bull.Soc.Chim.France, [4] 29, 656 (1921).
8. COREY, E.J., and BAILAR, Jr., J.C., J.Am.Chem.Soc., 81, 2620 (1959).
9. THEILACKER, W., Z.anorg.allg.Chem., 234, 161 (1937).
10. DICKEY, F.H., FICKETT, W., and LUCAS, J.H., J.Am.Chem.Soc., 74, 944 (1952).
11. WOLDBYE, F., "Optical Rotatory Dispersion of Transition Metal Complexes", European Research Office, U.S.Army, Frankfurt a.M., 1959.
12. BASOLO, F., CHEN, Y.T., and MURMANN, R.K., J.Am.Chem.Soc., 76, 956 (1954).

11 A 3

CIRCULAR DICHROISM OF COBALT(III) COMPLEXES OF OPTICALLY ACTIVE AMINO ACIDS.

Bodie E. Douglas and Chong Tan Liu
The University of Pittsburgh, Pittsburgh, Pennsylvania, U.S.A.

O'BRIEN and TOOLE (1) studied the optical rotatory dispersion (ORD) of complexes formed by several colorless metal ions with optically active amines. They concluded that the effects were independent of the nature of the central metal ion and of the nature of the active amines, but dependent on the number of ligand molecules coordinated. SHIMURA (2) showed that for $[Co(NH_3)_4 L\text{-leuc}]\,(ClO_4)_2$ [leuc = $NH_2CH(CO_2^-)C_2H_3(CH_3)_2$] the presence of the optically active ligand caused an anomalous dispersion curve in the region of the first absorption band of Co(III). The present paper reports the study of the contributions to the optical activity from the dissymmetric spiral configuration of chelate rings about a metal ion and the vicinal effect of optically active amino acids. Circular dichroism (CD) was used since it is difficult to separate the effects using ORD curves because of the serious overlap of the individual components (3). The CD curves are much more easily resolved and reveal more splitting of bands than can be seen from absorption curves or ORD curves (4).

The complexes studied were of the type $[Co(en)_2aa]^{2+}$ (aa = amino acid anion) using glycine, D- and L-alanine, D- and L-leucine, and D- and L-phenylalanine. The compounds $[Co(en)_2pic](ClO_4)_2$ (pic = picolinate ion) and $[Co(NH_3)_4L-palan]I_2$ (palan = phenylalanine anion) were studied for comparison.

The absorption curves in the visible and near ultraviolet regions for complexes of the type $[Co(en)_2aa]^{2+}$ are nearly identical for all of the amino acids studied. The two absorption bands appear to be symmetrical so that one would conclude that the effective symmetry is cubic. The ORD curves for the resolved complexes show only one prominent Cotton effect in the low frequency band. Since there is no clear indication of splitting into two or more components, this would also indicate effectively cubic symmetry. The CD curves of the complexes of the type $(+)_{546}-[Co(en)_2L-aa]^{2+}$ reveal three components within the high frequency band, but only two components are seen in the low frequency region, presumably because of the dominating intensity of one peak. The presence of three components within a band indicates that the symmetry is only C_1.

The CD curves of the resolved complexes of the type $[Co(en)_2aa]^{2+}$ are similar for all of the optically active amines, but the curves for the corresponding glycine and picolinic acid complexes are significantly different. It was found that the differences between the CD curves of the complexes $(+)_{546}-[Co(en)_2L-aa]^{2+}$ and $(-)_{546}-[Co(en)_2L-aa]^{2+}$ could be accounted for quite well from the CD of the racemic $-[Co(en)_2L-aa]^{2+}$. The optical activity of the racemic complex containing an optically active amino acid is the result of the vicinal effect of the optically active amino acid which causes the electronic transitions of the metal ion to become optically active. The vicinal contribution and the larger contribution from the spiral configuration of the three chelate rings about the metal ion in the resolved complex are approximately additive. The CD curve of the racemic complex reveals three components even in the low frequency band since there is no dominating component. By making allowance for the vicinal effect of the optically active ligand, the CD curves of the optically active amino acid complexes could be related to that of the glycine complex. The CD curve of a racemic $[Co(en)_2aa]^{2+}$ complex of an optically active amino acid was found to be very similar to that of the corresponding complex $[Co(NH_3)_4aa]^{2+}$ where the optical activity arises only because of the active amino acid. Three CD peaks are present within each absorption band although their positions are shifted slightly in accordance with the shifts in the absorption bands relative to those of the ethylenediamine complex.

References

1. O'BRIEN, T.D., and TOOLE, R.C., J.Am.Chem.Soc., 77, 1368 (1955).
2. SHIMURA, Y., Bull.Chem.Soc. Japan, 31, 315 (1958).
3. BRUSHMILLER, J.G., AMMA, E.L., and DOUGLAS, B.E., J.Am.Chem. Soc., 84, 3227 (1962).
4. DOUGLAS, B.E., HAINES, R.A., and BRUSHMILLER, J.G., Inorg.Chem., 2, 1194 (1963).

11 A 4

THE ELECTRONIC SPECTRA, OPTICAL ROTATORY POWER, AND ABSOLUTE CONFIGURATION OF O-PHENANTHROLINE AND 2,2'-DIPYRIDYL COMPLEXES.

A.J. McCaffery, S.F. Mason and B.J. Norman
Chemistry Department, University of Exeter, U.K.

The present work reports the circular dichroism spectra and the unpolarized light absorption of the tris-2,2'-dipyridyl and the tris-1,10-phenanthroline complexes of Ni(II), Fe(II), Ru(II), Os(II), Ru(III) and Os(III). The particular enantiomers studied were those forming the less soluble (-)-Co(ox)$_3^{3-}$ salt and the less soluble antimonyl-(+)-tartrate. On solubility grounds these enantiomers are expected to have the same absolute configuration as (+)-Co(en)$_3^{3+}$, since the latter complex ion preferentially precipitates the (-)-isomer from a solution of racemic K$_3$Co(ox)$_3$.

In both the 2,2'-dipyridyl and the 1,10-phenanthroline series the major optical rotatory power is associated with the high-intensity long wavelength $\pi \rightarrow \pi^*$ absorption band of the ligand, the p-band, which lies at 3000 Å in dipyridyl and at 2700 Å in o-phenanthroline, and in each complex it consists of two main circular dichroism bands with nearly equal areas and opposed signs ($|\epsilon_l - \epsilon_r| \sim 500$). These two circular dichroism bands are due to transitions with the symmetries A$_2$ and E in D$_3$, the transition deriving in the dihedral complex from the coupling between the excitation moments which give the p-band in the isolated ligand. The strong circular dichroism of the complex indicates that the p-band excitation moment is orientated along the long-axis of the ligand.

Of the two major circular dichroism bands in the p-band absorption region the one at longer wavelengths always has a positive rotation strength in the o-phenanthroline series of complexes, whereas in the 2,2'-dipyridyl series it has invariably a negative rotational strength, the shorter wavelength band having a negative and a positive sign in the two series, respectively. This observation suggests that either the solubility criterion for absolute configuration is not well-founded, or that the A_2 and the E transitions derived from the ligand p-band excitations interchange their relative energies on passing from the o-phenanthroline to the 2,2'-dipyridyl series.

The second alternative is considered to be the more probable, since corresponding π-orbitals of o-phenanthroline and of 2,2'-dipyridyl have different symmetry properties with respect to reflection in the mirror plane which bisects the ligand perpendicular to the molecular plane. In particular, the lowest unoccupied π-orbital is symmetric in 2,2'-dipyridyl and anti-symmetric in o-phenanthroline, so that metal-ligand π-bonding in the complexes gives the A_2 transition the higher energy in the o-phenanthroline series and the E transition the higher energy in the 2,2'-dipyridyl series. The relative energies of the A_2 and the E transition are governed additionally by dipole-dipole interaction between the excitation moments of the three ligands in the complex, and by inter-ligand π-bonding.

2 A 1

OPTICAL ACTIVITY IN DIGONAL DIHEDRAL TRANSITION METAL COMPLEXES.

Andrew D. Liehr

Mellon Institute, 4400 Fifth Avenue, Pittsburgh, Pennsylvania, USA.

Although ROSENFELD, BORN and JORDAN correctly described the origin of optical rotatory power in molecular systems quantum mechanically as early als 1929 [its classical explanation had been well-understood since the work of FRESNEL (1822), CAUCHY (1842), GIBBS (1882), DRUDE (1892), NATANSON (1908), BORN (1915), OSEEN (1915), GRAY (1915), de MALLEMANN (1924), KUHN (1929), BOYS (1934), and their comtemporaries and followers], the explicit use of their equations [CONDON, ALTAR and EYRING (1937), KIRK-WOOD (1937), GORIN, WALTER, KAUZMANN and EYRING (1938-40), MOFFITT (1956-58), FITTS and KIRKWOOD (1956-58), MOSCOWITZ (1956-), WASSERMAN (1956-), de HEER and MOSCOWITZ (1956-), de HEER and WASSERMAN (1956-), MOSCOWITZ and SNYDER (1960-), JULG (1961) and their comtemporaries and followers] to elucidate the observed activity in specific compounds has proved disappointingly difficult because of the rigorous limitations placed upon the type of approximations that may be feasibly made. For example, SUGANO (1960), PIPER (1961), HAMER (1962), POULET (1962) and LIEHR (1956-), using first order perturbation theory, have recently shown that electrostatic fields cannot give rise to rotatory power in inorganic complexes(1). We have now been able to exhibit that this difficulty can be overcome by the inclusion of covalency and stereochemical forces, and a comparison between theory and experiment where circumstances permit has been made for trigonal dihedral compounds(1). It was displayed how the simultaneous use of spectra, electron spin resonance, and rotatory measurements unambiguously fixes all parameters occurring in the theory, and how numerous connective relationships tie the signs and magnitudes of the rotational and spectral strengths of different transition metal compounds one to the other. In the present paper we wish to extend these results to digonal dihedral compounds, and to demonstrate the universality of the spin free d^n, d^{n+5}, ($n = 0, 1, 2, 3, 4, 5$) relationship, the spin free $d^{1,4,6,9}$ relationship, and the numerous other indirect relationships between these and other spin free and spin paired d^n, ($n = 0, 1, 2, 3, 4, 5, 6, 7, 8, 9, 10$), configurations.

Reference

1. LIEHR, A.D., J.Phys.Chem., 68 (March 1964).

12 A 2

REACTION KINETICS AND STEREOCHEMISTRY OF SOME OPTICALLY ACTIVE TRANSITION METAL COMPLEXES.

Ronald D. Archer
Tulane University, New Orleans, Louisiana, U.S.A.

Reaction of cobalt(III) complexes in anhydrous liquid ammonia under a variety of reaction conditions has led to the conclusion that the bimolecular inversion reaction reported earlier in ammonia for $D^* \text{-}[Coen_2Cl_2]^+$ (in which D^* indicates the absolute configuration of the optically active cis ion and en = ethylenediamine) (1) must compete with other reaction paths. The inversion path appears to be insignificant in cobalt(III)ions possessing large ligand field stabilizations. For example retention of configuration is almost complete for substitution of ammonia in $D^* \text{-}[Coen_2NH_3Cl]^{2+}$ and the corresponding bromo complex. The rates of all of the reactions are catalyzed by hydroxide or amide ion eventually yielding racemic products unless the reactions are quenched with ammonium ion almost immediately.

The overall rate of the reaction may be considered as made up of a first order solvation reaction plus a second order base solvolysis reaction. Precise rate constants are difficult to obtain for reactions of cobalt(III) complexes in ammonia, as has been noted by other investigators (2, 3). Addition of amide ion yields measurable quantities of the conjugate base complex of the product, which is then converted to the amine (or ammine depending on the complex) by the addition of ammonium ion. This observation along with the retardation of the reaction by ammonium ion suggests the importance of conjugate base reactions in liquid ammonia.

Inhibition of the reaction is also abserved for large concentrations of the complex, apparently due to ion pairing. The importance of considering ion-pairing in liquid ammonia reactions is evident since the conductivity of the solutions pass through a maximum and actually decrease during the two-step reaction

$$[Coen_2Cl_2]^+ + Cl^- \xrightarrow{\ NH_3\ } [Coen_2NH_3Cl]^{2+} + 2\ Cl^-$$

$$\xrightarrow{\ NH_3\ } [Coen_2(NH_3)_2]^{3+} + 3\ Cl^-$$

This reversal is apparently caused by ion multiplets which diffuse more slowly through ammonia than the simple ions.

Improvement in the preparation of the trans-diamminebis-(ethylene-diamine)cobalt(III) ion has allowed a quantitative spectral evaluation of the products of the ammonation (acid ammonolysis). For example, one product of an inversion reaction has been found to consist of about 65% trans-diammine, 22% L*-cis-diammine, and 13% D*-cis-diammine. Extension of these studies to chromium complexes in ammonia has given no indication of comparable inversions.

The base hydrolysis of this D*-[Coen_2Cl_2]^+ ion in water at 0° gives interesting stereochemistry. CHAN and TOBE (4) as well as DWYER and co-workers (5) have found a slight excess of retention relative to inversion, whereas an inversion occurs under essentially the same conditions if silber ion is added (5). Inversion also is possible under these conditions without the addition of silver ion if a stoichiometric quantity of hydroxide ion is used (6). All studies indicate a predominance of the trans-isomer in the products. In all cases it has been found that the inversion, when it occurs, is operative during the replacement of the first chloro ligand, and that the second step primarily yields a product with retention of configuration.

No other cobalt(III) complexes show such inversions or predominance of the trans-isomer except possibly the $[Coen_2BrCl]^+$ and $[Coen_2Br_2]^+$ ions (4, 7). All the other bis(ethylenediamine) complexes studied have at least one ligand higher in the spectrochemical series than the chloro ligand. Therefore it seems possible that a mechanism is operative in the dihalo complexes that is slower and swamped by another mechanism for complexes of higher ligand field stability.

One such mechanism involves a spin-free intermediate. Calculations for $[Coen_2Cl_2]^+$ suggest that the spin-paired ground state is only a few kilocalories per mole below the spin-free state, and seven-coordinate intermediates for d^6 ions in spin-free states would not have the high activation energy associated with the spin-paired d^6 ions (8). This leads to inversion rather than a racemic product because of the large dipoles of these ions, the repulsion of basis groups near the halo groups, and the short time that the complex stays in the spin-free state. This intermediate explains several other apparent anomalies which have been observed in the first row transition metal complex ion substitution reactions. Alternatively, a converted reaction in which a base abstracts a proton from a chelated ethylenediamine allowing the advantages of a conjugate base reaction, but with a basic group entering the backside simultaneously. In the presence of silver ion a push-pull mechanism is undoubtedly operative. Calculations using a point-charge model suggest reasonable activation energies for either transition state.

Complexes containing ligand higher in the spectrochemical series all show retention to inversion ratios of about 2 : 1 even though the amount of trans product and the rates appear to vary greatly without much regard to ligand field strengths. The orientations of the hydroxide ions relative to the complex are important (1, 4), but an explanation of the partial racemization observed when no or very little trans product is formed suggests trigonal bipyramidal intermediates as has been suggested before (8). To obtain inversion it is necessary for the basic group to enter while the chloro group is leaving before the trigonal bipyramid can be formed, i.e., the reactions must be S_N2 or S_N2IP in nature.

References

1. ARCHER, R.D., and BAILAR, Jr., J.C., J.Am.Chem.Soc., <u>83</u>, 812 (1961).
2. WIESENDANGER, H.U.D., JONES, W.H., and GARNER, C.S., J.Chem. Phys., <u>27</u>, 558 (1957).
3. HUNT, J.P., paper presented at Am.Chem.Soc.Meeting, Atlantic City, Sept. 1962.
4. CHAN, S.C., and TOBE, M.L., J.Chem.Soc. (London), <u>1962</u>, 4531.
5. DWYER, F.P., SARGESON, A.M., and REID, I.K., J.Am.Chem.Soc., <u>85</u>, 1215 (1963).

6. BAILAR, J.C., private communication.

7. PEPPARD, D.F., and BAILAR, Jr., J.C., J.Am.Chem.Soc., 62, 821 (1940).

8. BASOLO, F., and PEARSON, R.G., "Mechanism of Inorganic Reactions", N.Y.Wiley, 1958; Advances in Inorganic Chemistry and Radiochemistry, 3, 1 (1961).

2 A 3

THE CRYSTAL STRUCTURES OF SOME MERCURIC COMPLEXES.

Carl-Ivar Brändén

Department of Agricultural Chemistry I, Royal Agricultural College,
Uppsala 7, Sweden.

As part of an investigation of complex mercuric compounds the crystal structures of four complexes were studied.

The structure of $HgCl_2 \cdot 2(C_6H_5)_3AsO$ comprises discrete adduct molecules $HgCl_2 \cdot 2(C_6H_5)_3AsO$. Mercury exhibits a highly distorted tetrahedral coordination, the ligands being two chlorine atoms and two oxygen atoms, one from each triphenylarsineoxide molecule. The bond angles subtended at the mercury atom show large deviations from the regular tetrahedral bond angle, the values being $Cl-Hg-Cl = 146,5°$, $Cl-Hg-O = 101,5°$ and $O-Hg-O = 92,5°$. The mean value of the two Hg-Cl bond lengths is 2,33 Å and that of the Hg-O bond lengths is 2,35 Å. The Hg-O-As bond angle is 135,5° and the As-O bond distance is 1,69 Å.

The structure of $[HgCl_2 \cdot (C_6H_5)_3AsO]_2$ comprises discrete dimeric centro-symmetrical molecules of $[HgCl_2 \cdot (C_6H_5)_3AsO]_2$. The oxygen atom of each donor molecule is bonded to two acceptor atoms forming oxygen bridges between the two mercury atoms of the dimer. Mercury thus attains a highly distorted tetrahedral coordination, the ligands being two chlorine atoms and two oxygen atoms. Oxygen is bonded to arsenic and to the two mercury atoms in a flat triangular pyramidal configuration with oxygen at the apex of the pyramid at a distance of 0,31 Å from the plane defined by the other three atoms. The mean bond lengths are: Hg-Cl = 2,32 Å, Hg-O = 2,47 Å and As-O = 1,66 Å. The bond angles subtended at the mercury atom are: $Cl-Hg-Cl = 144,8°$, $Cl-Hg-O = 103,4°$ and $O-Hg-O = 78,7°$.

The structure of $2\,HgCl_2 \cdot (C_2H_5)_2S$ is conveniently described as being built up from $[Cl-Hg-S(C_2H_5)_2]^+$ ions, Cl^- ions and unsubstituted $HgCl_2$ molecules linked together in a two-dimensional network by chlorine bridges. The mercury atom of the $[Cl-Hg-S(C_2H_5)_2]^+$ ion show a coordination inter-mediate in character between octahedral and tetrahedral coordination. One

chlorine atom and the sulphur atom of the thioether group are covalently
bonded to this mercury atom, the bond lengths being 2,35 and 2,41 Å,
respectively. The bond angle Cl-Hg-S is 158°. The coordination is completed
by two Cl⁻ ions at distances of 2,70 and 2,85 Å and by two very long bonds
of length 3,55 and 3,64 Å to two chlorine atoms of adjacent $HgCl_2$ molecules.
The bond angle Cl⁻-Hg-Cl⁻ is 83,4°.

The mercury atom of the unsubstituted $HgCl_2$ molecule exhibits the
distorted octahedral coordination characteristic of mercury forming two short
colinear Hg-Cl bonds of length 2,30 and 2,33 Å and four longer Hg-Cl bonds,
the lengths of which vary from 2,9 - 3,1 Å. The chloride ion is not covalently
bonded to anyone mercury atom but is instead associated with three different
mercury atoms at distances 2,70; 2,85 and 2,88 Å.

The strucutre of $HgCl_2 \cdot C_4H_8S$ (tetra-hydrothiophene) can be described
as composed of $[Cl-Hg-SC_4H_8]^+$ ions and Cl⁻ ions coupled together to form
infinite double chains. Within these chains each mercury atom is coordinated
by one covalently bonded chlorine atom at a distance of 2,30 Å and one
covalently bonded sulphur atom at a distance of 2,40 Å in a highly distorted
linear arrangement, the angle Cl-Hg-S being 143°. The mercury atom is
further bonded to two different Cl⁻ ions at distances of 2,64 and 2,83 Å, the
angle Cl⁻-Hg-Cl⁻ being 86°. These four atoms are thus arranged at the
corners of a highly distorted tetrahedron about the mercury atom. A third
Cl⁻ ion is, however, also bonded to mercury at a distance of 3,06 Å. The
configuration of ligands around mercury may alternatively be described as
highly distorted octahedral with the sixth corner of the octahedron occupied
by a sulphur atom of an adjacent chain, the Hg-S distance in this instance
being 3,90 Å. The chloride ion is bonded by three long bonds to three
different mercury atoms in a manner resembling the bonding of the Cl⁻ ion
in the structure of $2\ HgCl_2 \cdot (C_2H_5)_2S$.

These structure determinations show that thioethers and saturated cyclic
sulphur compounds do not form addition compounds with $HgCl_2$ as has previously
been believed. The complexes formed are the result of a substitution reaction,
in which one of the chlorine atoms in $HgCl_2$ has been replaced by the sulphur
atom of the donor molecule giving rise to positively charged mercuric complexes and negative chloride ions. The configuration of ligands around mercury
in these complexes is intermediate in character between that of tetrahedrally
and octahedrally coordinated mercury.

The covalent mercury-sulphur bond distances are correlated to the coordination of mercury, the distances being 2,34 Å for octahedrally and
2,55 Å for tetrahedrally coordinated mercury. In the structures previously
described the Hg-S bond distances are intermediate in length between these

two values. The mercury-chlorine bond length, on the other hand, is always approximately 2,32 Å regardless of the coordination of mercury.

Strong oxygen donor molecules like triphenylarsineoxide form addition compounds with mercuric chloride. In the dimeric adduct molecule $[HgCl_2.(C_6H_5)_3AsO]_2$ each oxygen atom forms two acceptor-donor bonds, in contrast to all other similar adducts so far investigated in which each donor atom is bonded to only one acceptor atom.

3 A 1

THE NATURE OF BONDING IN SOME PENTACOORDINATED MOLECULES.

Robert R. Holmes

Bell Telephone Laboratories, Incorporated, Murray Hill, New Jersey, U.S.A.

Elements present in groups III through VII as well as some of the transition elements form pentacoordinated molecules; although at present the group V elements make up the major fraction of such substances. The most prevalent geometry assigned to these molecules is the trigonal bipyramidal structure. In general, differences are expected in bonding properties in any reasonable pentacoordinated structure purely on the basis of geometrical considerations. Specifically, in the trigonal bipyramidal model, bonding in the trigonal plane should differ from that in the axial positions. Variations in bonding resulting from changes in the central atom as well as ligand changes are also expected.

Although the nature of bonding in pentacoordinated molecules at present is little understood, certain gross structural features recently have revealed themselves in the phosphorus(V)chlorofluorides. Results of studies involving nuclear quadrupole resonance spectroscopy (1), dipole moment measurements (1), IR- and Raman spectroscopy (2), NMR (3), and calorimetric measurements of complex formation (4) coupled with related information on the series $POCl_xF_{3-x}$ and PCl_xF_{3-x} have provided the necessary information to allow an initial correlative interpretation concerning stereochemical aspects and the nature of the nonequivalence of bonds in these molecules.

The phosphorus(V)chlorofluorides are especially interesting because they exist in molecular forms which slowly rearrange at room temperature to solid modifications. The molecular forms are trigonal bipyramids and have the following symmetries (1-3): PCl_4F, C_{3v}; PCl_3F_2, D_{3h}; PCl_2F_3, C_{2v}; PF_5, D_{3h}. The fluorine atoms occupy axial positions in PCl_4F and PCl_3F_2. In PCl_2F_3, the axial positions are again occupied by fluorine atoms while the remaining fluorine atom is located in an equatorial site. These molecules behave as typical Lewis acids in molecular addition compound formation with pyridine (4).

F^{19} NMR, chlorine quadrupole resonance and dipole moment studies as well as IR and Raman spectroscopy provide information on the differences in axial versus equatorial bonding. For example, axial fluorine atoms in the series PCl_xF_{5-x} are progressively less shielded as x increases while the equatorial fluorine atom in PCl_2F_3 appears at considerably higher fields. In PCl_4F there are three equatorial chlorine atoms and one axial chlorine atom. The | eQq | values for the Cl^{35} nucleus reflect this difference giving 65.08 Mc/sec and 57.96 Mc/sec respectively.

Assuming that bonding in the trigonal plane involves central atom orbitals which are largely sp^2 hybridized and axial orbitals that are largely d-p hybridized overlapping with p orbitals or the halogen atoms (having some s character), symmetry considerations suggest (5) that if π_{d-p} bonding is present in the phosphorus(V)chlorofluorides, axial bonds are expected to be stabilized to a greater extent than equatorial bonds. Fluorine atoms are assumed to enter into π bonding more effectively than chlorine atoms. Such a process would compensate for the large s to d promotional energy required in forming the d-p hybrids and makes the C_{2v} symmetry observed for PCl_2F_3 compared to the more symmetrical D_{3h} model seem more reasonable. Further, the trend

TABLE

BOND DISTANCES (Å) FOR TRIGONAL BIPYRAMID MOLECULES (6)

	Equatorial M-X	Axial M-X	Calc[a] M-X	Axial-Calc Δ	Method[b]
$H_3Al[N(CH_3)_3]$		2.18	1.84	+0.34	X
$(CH_3)_3SnClNC_5H_5$		2.26 Sn-N	2.00	+0.26	X
		2.42 Sn-Cl	2.37	+0.05	X
PF_5	$(1.57)^c$	(1.57)	1.59	-0.02	E
PCl_5	2.04	2.19	2.04	+0.15	E
$SbCl_5$	2.31	2.43	2.38	+0.05	E
$SbCl_5$	2.29	2.34	2.38	-0.04	X
$(CH_3)_3SbCl_2$		$[2.49]^d$	2.38	+0.11	X
$(ClCH=CH)_3SbCl_2$		2.45	2.38	+0.07	X
$(CH_3)_3SbBr_2$		[2.63]	2.53	+0.10	X

TABLE (contd.)

SF$_4$	1,545	1,646	1,56	+0,09	M
SeF$_4$	(1,77)	(1,77)	1,68	+0,09	E
φ$_2$SeCl$_2$		2,30	2,14	+0,16	X
φ$_2$SeBr$_2$		2,52	2,29	+0,23	X
TeCl$_4$	(2,33)	(2,33) or	2,34	-0,01	E
	or 2,27	2,40	2,34	+0,06	
ClF$_3$	1,598	1,698	1,54	+0,16	M
φICl$_2$		2,45 I-Cl	2,32	+0,13	X
PCl$_4^+$ICl$_2^-$		2,36 I-Cl	2,32	+0,04	X
C$_9$H$_7$NICl		2,51 I-Cl	2,32	+0,19	X
NH$_4^+$BrICl$^-$		2,38 I-Cl	2,32	+0,06	X
		2,50 I-Br	2,47	+0,03	
Fe(CO)$_5$ [e]	1,79-1,84	1,79-1,84	1,89	-0,05 to	X
				-0,10	
NbCl$_5$	(2,29)	(2,29)	2,30	-0,01	E
NbBr$_5$	(2,46)	(2,46)	2,45	+0,01	E
TaCl$_5$	(2,30)	(2,30)	2,30	0,00	E
TaBr$_5$	(2,45)	(2,45)	2,45	0,00	E

[a] Calculated by the SHOMAKER-STEVENSON equation corrected for electro-negativity differences (PAULING).

[b] X, X-ray; E, electron diffraction; M, microwave.

[c] Parantheses mean equal distances were assumed for axial and equatorial bonds.

[d] Brackets mean error uncertain. Other distances in the table are good to about ± 0,02 Å except the M values which are more accurate, ± 0,003 Å.

[e] In Fe(CO)$_5$ the orbital hybridization would be dsp^3 rather than sp^{3}d and since the energies of the individual orbitals are close together, the difference between equatorial and axial bonds should be at a minimum. Assuming the isoelectronic HCo(CO)$_4$ is a distorted trigonal bipyramid, likewise, little difference should be observable among the CO groups.

in magnetic shielding for axially oriented fluorine atoms, opposite from that expected on the basis of group electronegativity considerations, is explainable on this basis. Using the quadrupole coupling data, calculations indicate that little if any π bonding exists for equatorially oriented chlorine atoms; the situation is less certain for an axial chlorine atom.

Examination of available bond distance values (6) for known trigonal bipyramid molecules (Table) shows that, in general, axial bonds are considerably longer than equatorial bonds reflecting the use of higher energy axial orbitals. Axial bonds up to 0.3 Å longer than that calculated from the sum of the covalent radii corrected for electronegativity differences are observed while equatorial distances more closely approximate single bond values. However, the bond distance parameter may not be used to estimate stabilization due to π bonding effects that might be present since no reference distances are available.

The occupancy of equatorial positions by lone electron pairs in substances such as SF_4, ClF_3 and $[ICl_2]^-$ is reasonable on the basis of the lower promotional energy involved. For example in SF_4 with the electron pair in an equatorial position, a sulfur atom p electron must be promoted to a d orbital to give the necessary sigma bonding atom orbitals. If the electron pair is axial then a larger energy s to d electron transition is necessary. Partial energy compensation is achieved, however, by the formation of an additional sulfur-fluorine equatorial bond having higher stability than the sulfur-fluorine axial bond. An explanation based on lone electron pair repulsions leads to the same conclusion (7). Molecules having one or two lone electron pairs in equatorial positions are expected to have bent axial bonds because of the repulsion between the electron pairs, and the electronegative ligands in the axial positions (7). This is observed in ClF_3 (F_a-Cl-F_e angle is 87^0 29') and SF_4 (F_a-S-F_e angle is 86^0 32') (6) where precise microwave data are available. Any axial π bonding effects should be reduced in these molecules because of the loss of axial symmetry.

Alternate explanations of the suspected sigma bonding features in trigonal bipyramidal molecules are available but present data do not allow a definitive choice to be made among many of the approaches (8, 9). Information from further microwave studies, X-ray studies, and studies of the Zeeman splitting of nuclear quadrupole resonances of oriented crystals appears to be the most promising in offering conclusive evidence for the extent of π contributions, particularly in central atom - axial fluorine linkages, and in allowing the further unravelling of the versatile bonding in pentacoordinated molecules.

References

1. HOLMES, R.R., CARTER, Jr., R.P., and PETERSON, G.E., unpublished work.

2. GRIFFITHS, J.E., CARTER, Jr., R.P., and HOLMES, R.R., presented at the 147th ACS-meeting, Philadelphia, Pa., April 1964.

3. HOLMES, R.R., and GALLAGHER, W.P., Inorg.Chem., 2, 433 (1963) and unpublished work. See also MUETTERTIES, E.L., MAHLER, W., and SCHMUTZLER, R., Inorg.Chem., 2, 613 (1963).

4. HOLMES, R.R., GALLAGHER, W.P., and CARTER, Jr., R.P., Inorg. Chem., 2, 437 (1963).

5. KIMBALL, G.E., J.Chem.Phys., 8, 188 (1940).

6. Except as noted below bond distances are taken from L.E.SUTTON, Edit., "Tables of Interatomic Distances and Configuration in Molecules and Ions", The Chemical Society, London, 1958. $H_3Al \cdot 2N(CH_3)_3$: HEITSCH, C.W., NORDMAN, C.E., and PARRY, R.W., Inorg.Chem., 2, 508 (1963). $(CH_3)_3SnCl \cdot NC_5H_5$: HULME, R., J.Chem.Soc., 1963, 1524. SF_4: TOLLES, W.M., and GWINN, W.D., J.Chem.Phys., 36, 1119 (1962). C_5H_5N-ICl: HASSEL, O., and ROMMING, C., Acta Chem.Scand., 10, 696 (1956). $Fe(CO)_5$: HANSON, A.W., Acta Cryst., 15, 930 (1962).

7. GILLESPIE, R.J., Can.J.Chem., 39, 318 (1961).

8. CRAIG, D.P., NYHOLM, R.S., MACCOLL, A., ORGEL, L.E., and SUTTON, L.E., J.Chem.Soc. (London), 1954, 332.

9. MATWIYOFF, N.A., and DRAGO, R.S., Inorg.Chem., 3, 337 (1964) and references cited therein.

3 A 2

FIVE CO-ORDINATION IN TIN COMPOUNDS.

H.C. Clark and R.G. Goel
University of British Columbia, Vancouver, Canada.

The occurrence of five coordination in a wide range of organo-tin compounds has been firmly established in recent years. Of such compounds, the most surprising are those of the type R_3SnX, where X is an inorganic anion (e.g. F^-, ClO_4^-, BF_4^-). The crystallographic and spectroscopic evidence for five coordination in three such compounds, $(CH_3)_3SnF$, $(CH_3)_3SnBF_4$, and $(CH_3)_3SnSbF_6$ will be briefly reviewed, since the latter cases provide the first evidence for such bridging interaction by the BF_4^- and SbF_6^- ions. Detailed discussions of these compounds have already been published (1, 2).

These previous results clearly indicate the non-existence of the free $(CH_3)_3Sn^+$ cation in the solid state, and we have now extended our studies to related $(C_6H_5)_3SnX$ derivatives in particular the perchlorate and nitrate, since greater stability might be predicted for the $(C_6H_5)_3Sn^+$ ion. Infra-red studies show the perchlorate to be similar in structure to trimethyltin perchlorate, the ClO_4 group having C_{2v} symmetry, and the tin atom presumably being five coordinate. The nitrate may have a similar structure with bridging NO_3 units. However, whereas $(CH_3)_3SnClO_4$ and $(CH_3)_3SnNO_3$ readily form di-adducts with ammonia and pyridine, e.g. $[(CH_3)_3Sn(NH_3)_2]^+ClO_4^-$ so as to maintain the five coordinate structure, the triphenyltin compounds do not give stable adducts with nitrogen bases. Instead, the following typical reaction occurs:

$$(C_6H_5)_3SnClO_4 + NH_3 \xrightarrow[\text{(e.g. MeOH)}]{\text{solvent}} [(C_6H_5)_3Sn]_2O + NH_4ClO_4$$

This agrees with other evidence, which confirms the instability of nitrogen base adducts of triphenyltin derivatives, and distinguishes such derivatives from trialkyl tin compounds. This difference is clearly important in interpreting the interactions which can occur between R_3Sn groups and anions such as ClO_4^-, BF_4^- etc. on the one hand, and between R_3Sn group and nitrogen bases on the other. The possible nature of these interactions is then examined with reference to the use of filled 4d and unfilled 5d orbitals on the tin atom.

Because of the ability of R_3Sn groups to interact with, and distort groups such as ClO_4^- and BF_4^-, which are normally considered incapable of such interaction, these studies have also been extended to the examination of compounds such as $[(CH_3)_3Sn]_2CrO_4$. Such compounds may well be the first instances of complexing action by transition metal oxy-anions and the results of infra-red and ultra-violet spectroscopic studies will be presented.

Bis(trimethyltin)chromate is readily obtained from trimethyltin bromide and silver chromate in dry methanol. It is an orange powder, whose infra-red spectrum shows the presence of planar $(CH_3)_3Sn$ and tetrahedral CrO_4 groups. It thus resembles anhydrous $[(CH_3)_3Sn]_2SO_4$. The ultra-violet and visible spectrum shows appreciable shifts to higher wavelengths, compared with solid potassium chromate, and possible interpretations are considered. Finally, attempts to obtain trialkyltin derivatives of other transition-metal oxy-ions will be reported.

References

1. CLARK, H.C., O'BRIEN, R.J., and TROTTER, J., Proc.Chem.Soc. 1963, 85; J.Chem.Soc. (London) 1964, in press.
2. CLARK, H.C., and O'BRIEN, R.J., Inorg.Chem. 2, 1020 (1963).

3 A 3

FIVE COORDINATE COMPOUNDS OF COBALT.

A. Earnshaw, P.C. Hewlett and L.F. Larkworthy
Battersea College of Technology, Battersea Park Road, London, U.K.

Five coordinate compounds of cobalt are now comparatively common, though only on nitrosyl (dimethyldithiocarbonato) cobalt(I) has a detailed X ray examination been performed (1). This has been shown to have a square pyramidal structure with the N-O bond inclined at 139° to the pyramidal axis, forming an unsymmetrical π bond with the cobalt atom.

We have prepared the parent and several substituted Schiff's base complexes of cobalt and reacted them with nitric oxide. The mononitrosyls so formed give non-conducting solutions in nitrobenzene and molecular weight determinations, where they are possible, show them to be monomeric in nitrobenzene. Thus, unless the solvent is coordinated, they may be considered to be five coordinate with a structure as shown:

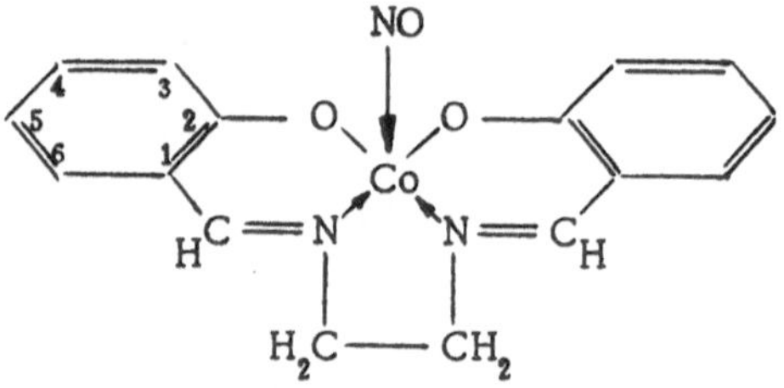

All have at room temperature magnetic moments of about 0.5 B.M. That this is temperature independent paramagnetism is indicated by the fact that, on four of the compounds which were measured over the range 90° to $300^O K$, the susceptibility did not vary significantly. The compounds therefore have no unpaired electrons and may be regarding as containing NO^- coordinated to Co(III) or as NO^+ coordinated to Co(I).

Measurement of the IR spectra suggest the latter formulation since absorptions occur in the range $1624 - 1724 \, cm^{-1}$ which is within the range ascribed to coordinated NO^+. Isoelectronic nitrosyl (dimethyldithiocarbonato) cobalt(I) has a band at $1626 \, cm^{-1}$ and a similar five coordinate tetragonal pyramidal structure seems likely for the present series of compounds.

Variations in the N-O stretching frequencies (Table) in these nitrosyls have been related to changes in double bond character in the Co-NO bond, consequent on electron withdrawal from, or release to, the metal by substituents in the benzene ring.

TABLE

VARIATION OF NO STRETCHING FREQUENCY
WITH SUBSTITUENT

Substituent	NO stretching frequency (cm^{-1})
unsubstituted	1624
5-methyl	1614
5-methoxy	1631
5-chloro	1638
5-nitro	1696
4-chloro	1686
3-methoxy	1635
3-nitro	1667
3,5-dinitro	1724
acacen	1660
phenen	1618

where acacen is acetylacetoneethylenediimine and phenen is o-hydroxy-acetophenonethylendiimine.

A preliminary investigation of the charge transfer spectra of these compounds has shown that the bands obtained are in general agreement with the predictions of GRAY et al (2) for a five coordinate compound in which an NO$^+$ group dominates the ligand field.

References

1. ALDERMAN, P.R.H., OWSTON, P.G., and ROWE, J.M., J.Chem. Soc. (London), 1962, 668.
2. GRAY, H.B., BERNAL, I., and BILLIG, E., J.Am.Chem.Soc., 84, 3404 (1962).

14 A 1

THE INFLUENCE OF STERIC AND ELECTRONIC EFFECTS UPON
THE STEREOCHEMISTRY OF GROUP VIII METAL COMPLEXES.

A.B.P. Lever
Department of Chemistry, Faculty of Technology,
University of Manchester, U.K.

Complexes of Iron, Cobalt and Nickel occur in a variety of stereochemistries. Nickel(II) is particularly interesting in that apparently small differences in the nature of the ligands around the metal ion can alter the preferred stereochemistry. Much of the published work in this connection has dealt with substituted salicylaldimines (1) and substituted phosphines (2). Some of the factors responsible for the variations in stereochemistry are discussed briefly below. It is proposed that electronic distortion which may lead to changes in stereochemistry occurs when dissimilar ligands are bonded to the metal ion. This distortion may be detected by magnetic and/or spectral studies, but not necessarily by crystallographic studies. This is well known with ions such as Co(III), but there has been little extension previously to nickel compounds. The Table illustrates the stereochemical variations occurring in amine nickel complexes when the halogen is changed and also indicates how such parameters as the magnetic moment, crystal field energy and Racah parameter vary.

A stereochemistry may be preferred for one or more of a number of reasons.

a) For a given stereochemistry, the lattice energy of the compound may be sufficiently high so as to favour the isolation of a solid in a form which does not exist appreciably in solution e.g. ethylenediamine will form low spin square planar (or tetragonal) complexes (3,4) of Fe(II) and Co(II) of the empirical formula $Men_2 . 2 AgI_2$, which only exist as such in the solid state.

b) Steric hindrance may preclude the formation of a particular stereo-chemistry. In general, this usually means that a coordination number greater than four cannot be achieved. Whether a four coordinate tetrahedral, or four coordinate square planar arrangement is then stabilised depends upon other factors. Many examples are known (5).

c) The electronic nature of the ligands may favour a particular stereo-chemistry. Highly electronegative ligands are likely to favour octahedral coordination whilst covalent ligands will probably prefer tetrahedral co-ordination. Interligand repulsion must be larger in a square complex than in a tetrahedral complex so that sterically hindered ligands should only form square planar complexes when there is obtained in this form some extra electronic stabilisation which compensates for the extra interligand repulsion energy. This extra stabilisation may take the form of π-back donation which is generally believed to be more important in square complexes than in tetra-hedral complexes. Strong π-acceptor ligands may therefore favour a square rather than a tetrahedral symmetry. The fact that dibromodiquinolinenickel(II) is tetrahedral whilst di-iododiquinolinenickel(II), which undoubtably has greater interligand repulsion energy, is square (6) may be due to the more polarisable iodide ion allowing the nickel ion to back donate electrons to the quinoline more effectively than it can in the bromide complex. It is interesting that the corresponding pyridine derivative is tetrahedral (7), perhaps because pyridine must be a poorer π-acceptor than quinoline. Increasing π-back donation should cause an increase in the observed crystal field strength, and this has been demonstrated in pyrazine nickel chemistry (8). Increasing π-back donation has also been used to explain the thermodynamic properties of some cobalt(II) complexes (9).

When dissimilar ligands are bonded to the metal ion, the degeneracy of the e_g and t_{2g} levels in the octahedral molecule may be lifted. We consider here the special case of tetragonal distortion. Weak tetragonal distortion is difficult to detect. Indeed it is reasonable to suppose that there may be some splitting of the e_g and t_{2g} levels before any alteration in bond lengths can be observed crystallographically. It is shown here that this is the case but that certain types of weak tetragonal distortion can be detected by both magnetic and spectral methods.

If X and Y, in a complex trans NiY_4X_2, are very similar, the splitting will be very small. If however there are differences between X and Y, particularly if one ligand is a π-acceptor, and the other is a π-donor, or if one is a strong base and the other is a weak base, or if one ligand hinders the other, then the splitting may be important (see figure), and may lead to change of stereochemistry.

Strong Tetragonal Distortion. In the case of nickel(II), the energy difference between the $d_{x^2-y^2}$ orbital and the next highest level may become sufficiently large for spin pairing to occur, and a diamagnetic product to result. The electronic spectra of such complexes are of course readily distinguishable from the spectra of regular octahedral nickel derivatives.

Medium Tetragonal Distortion. The spectra of the paramagnetic compounds dichloro-bis-N, N'-diphenylethylenediaminenickel(II), dichloro-bis--o-phenylenediaminenickel(II) and dichlorotetrapyridinenickel(II) have been interpreted in terms of a fairly strong tetragonal distortion (10, 11) but evidently not strong enough to cause spin pairing. The spectra of these complexes do not fit the Tanabe-Sugano equations for octahedral nickel (12). Recently dichloro-tetrakis-N, N'-diethylthioureanickel(II) has been shown to have a magnetic moment indicative of singlet and triplet magnetic states lying close together, i.e. this is a complex lying intermediate to the strong and medium cases cited above (13).

Weak Tetragonal Distortion. The order of the dxy and dxz, dyz levels depends upon the nature of the ligands. For complexes of the type NiY_4X_2

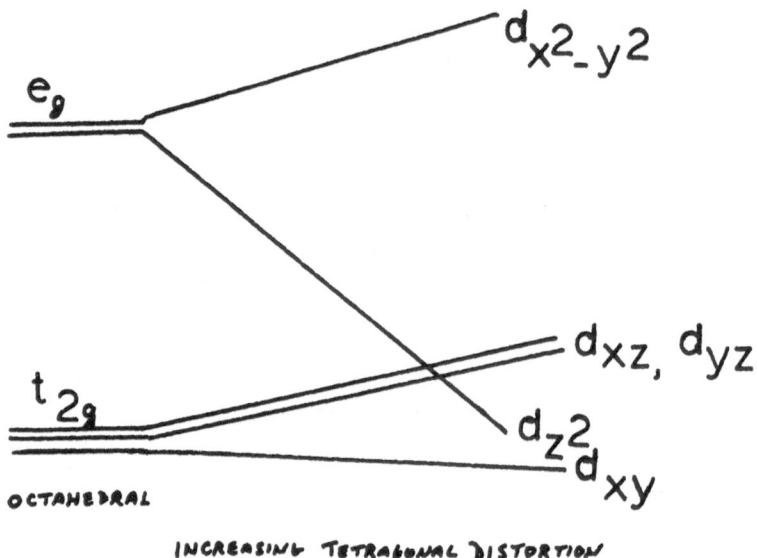

OCTAHEDRAL

INCREASING TETRAGONAL DISTORTION
⟶

where Y is a π-donor, and X is a π-acceptor amine, it is proposed that, assuming the amine lies in the xz plane, the order is as shown in the figure since only one of the original t_{2g} orbitals has the correct symmetry to form a π bond to the acceptor ligands, and hence to be stabilised. The other two orbitals will be destabilised by the filled π orbitals of Y. If we assume that the latter remain essentially degenerate then we may expect a magnetic moment which would be temperature independent and somewhat above the normal octahedral range for nickel(II), i.e. greater than 3.3 B.M. The tetragonal chloro and bromo pyridine, quinoline, quinoxaline and pyrazine complexes listed in the Table are of this type. (4 π-donor halides, and 2 π-acceptor unsaturated amines).

It is seen that the magnetic moments in all cases are significantly higher than the maximum theoretically possible for a regular octahedral derivative and lie at the lower end of the region generally considered to be derived from tetrahedral nickel complexes (14). No explanation has been offered previously to explain the high magnetic moments of this type of complex.

The spectra of these complexes may still be fitted to the Tanabe-Sugano matrices but give rise to values of 10Dq which are slightly lower, and of the Racah parameter B which are usually higher, than would be expected by the 'average environment rule' (15) for analogous undistorted octahedral derivatives. Although the values so obtained will not have an exact meaning they

TABLE

Complex	ref	Colour	μ_e (B.M.)	10Dq*)	B*)
φ NiPy$_2$Cl$_2$	(16)	Yellow green	3,37	8500	840
φ NiPy$_2$Br$_2$	(18)	Yellow	3,35	8200	829
* NiPy$_2$I$_2$	(7)	Deep green	3,44		
φ Ni(Quinoline)$_2$Cl$_2$	(6)	Yellow	3,41	7940	810
* Ni(Quinoline)$_2$Cl$_2$	(6)	Blue	3,54		
* Ni(Quinoline)$_2$Br$_2$	(6)	Blue	3,48		
Ni(Quinoline)$_2$I$_2$	(6)	Olive Green	0		
φ Ni(2,5-Dmp)Cl$_2$	(19)	Yellow	3,4	7500	805
** Ni(2,5-Dmp)Br$_2$	(19)	Purple	0		
** Ni(2,5-Dmp)I$_2$	(19)	Black	0		
φ Ni(2,6Dmp)$_2$Cl$_2$	(19)	Pale Green	3,38	8200	845
φ Ni(2,6Dmp)$_2$Br$_2$	(19)	Pale Blue	3,32	8100	825
** Ni(2,6Dmp)I$_2$	(19)	Black	0		
φ Ni(Quinoxaline)Cl$_2$	(20)	Yellow	3,44	7850	880
φ Ni(Quinoxaline)Br$_2$	(20)	Yellow	3,43	8050	650
φ Ni(Quinoxaline)Br$_2$	(20)	Brown	3,51		
Ni(Quinoxaline)$_2$I$_2$	(20)	Green	1,6 (3,0**)		
Ni(α-dimen)$_2$Cl$_2$	(21)	Blue	3,21	9710	
Ni(α-dimen)$_2$Br$_2$	(21)	Green	3,23	9000	
Ni(α-dimen)$_2$I$_2$	(21)	Orange	0		

Py = pyridine; 2,5-Dmp = 2,5-dimethylpyrazine; α-dimen = N,N-dimethyl-
ethylenediamine; 2,6Dmp = 2,6-dimethylpyrazine; **) Magnetic moment in

methanol solution. ø Halogen bridged polymers. *) Calculated on the
assumption that the complexes are regular octahedral. * Tetrahedral
** Square planar (23). All other complexes are believed to be tetragonal.

do act as a guide to the existence of weak tetragonal distortion (see Table).
The increase in magnetic moment should be emphasized if the d_{z^2} level drops
steeply, as will happen if there is any steric hindrance or if the base strengths
of the ligands differ markedly. Using this argument the tetragonal distortion
will increase with increasing halide size (and with decreasing halide base
strength), thus leading to diamagnetism in the heavier halides (see Table).
In the unhindered complexes the distortion may be purely electronic since
certainly in the case of $NiPy_2Cl_2$, no definite tetragonal distortion has been
observed crystallographically (16,17).

Complexes with 4 π-acceptor and 2 π-donor ligands, though still tetra-
gonally distorted, will be expected to have the energy order dxz, dyz < dxz
(arguing on the basis of π bonding as before) and should have moments within
the octahedral range $NiPy_4X_2$ X = Cl, Br, and I. μ_e = 3,11; 3,22; and 3,21 BM
respectively (18). In the case of the substituted ethylenediamine derivatives,
steric hindrance will lengthen the nickel to halide bond giving rise eventually
to either tetrahedral paramagnetic or tetragonal diamagnetic complexes,
strong σ-bonding favouring the latter. The magnetic moments of the paramagnetic tetragonal derivatives should not be noticeably enhanced.

References

1. FRASSON, PANATTONI and SACCONI, J.Phys.Chem., _63_, 1908 (1959);
   HOLM and SWAMINATHAN, Inorg.Chem., _1_, 599 (1962); _2_, 181 (1963);
   HOLM, J.Am.Chem.Soc., _82_, 5632 (1960), _83_, 4683 (1961);
   SACCONI, PAOLETTI and DEL RE, J.Am.Chem.Soc., _79_, 4062 (1957).
2. COUSSMAKER, HUTCHINSON, MELLOR, SUTTON and VENANZI,
   J.Chem.Soc. (London), _1961_, 2705;
   SCATTURIN, J.Inorg.Nucl.Chem., _8_, 447 (1958);
   GIACOMETTI and TURCO, J.Inorg.Nucl.Chem., _15_, 242 (1960).
3. LEVER, LEWIS and NYHOLM, J.Chem.Soc. (London), _1963_, 2552.
4. LEVER, Unpublished.
5. e.g. COTTON and FACKLER, J.Am.Chem.Soc., _83_, 2818 (1961);
   FACKLER and COTTON, J.Am.Chem.Soc., _83_, 3775 (1961).
6. GOODGAME and GOODGAME, J.Chem.Soc. (London), _1963_, 207.
7. GLONEK, CURRAN and QUAGLIANO, J.Am.Chem.Soc., _84_, 2014 (1962).
8. LEVER, LEWIS and NYHOLM, J.Chem.Soc. (London), _1964_, in press.
9. KING, KÖRÖS and NELSON, J.Chem.Soc. (London), _1963_, 5449.
10. MAKI, J.Phys.Chem., _29_, 162 (1952).

11. BOSTRUP and JØRGENSEN, Acta Chem.Scand., **11**, 1223 (1957).
12. TANABE and SUGANO, J.Phys.Soc.Japan, **9**, 753, 766 (1954).
13. HOLT Jr., BOUCHARD and CARLIN, J.Am.Chem.Soc., **86**, 519 (1964).
14. LEWIS, Science Progress, **1963**, 452.
15. JØRGENSEN, "Absorption Spectra and Chemical Bonding". Pergamon Press, London (1962).
16. GILL, NYHOLM, BARCLAY, CHRISTIE and PAULING, J.Inorg.Nucl. Chem., **18**, 88 (1961).
17. DUNITZ, Acta Cryst., **10**, 307 (1957).
18. NELSON, Private Communication.
19. LEVER, LEWIS and NYHOLM, J.Chem.Soc. (London), **1963**, 5042.
20. LEVER, In Press.
21. GOODGAME and VENANZI, J.Chem.Soc. (London), **1963**, 5909.

14 A 2

DISTORTIONS OF COORDINATION POLYHEDRA.

E.C. Lingafelter and H. Montgomery
University of Washington, Seattle, Washington, U.S.A., and
Canadian Services College, Royal Roads, Victoria, B.C., Canada.

The equilibrium configuration of a set of 4 or 6 point charges about an isotropic attracting center is a regular tetrahedron or octahedron, respectively. In many determinations of crystal structures of compounds containing coordination complexes exhibiting coordination numbers 4 and 6, the coordination polyhedra are found to be more or less distorted from these regular configurations. There are many factors which may influence these distortions: differing sizes of the ligand atoms; differing "bonding strength" of the ligand atoms; steric effects of other atoms bonded to the ligand atoms; crystal packing effects; hydrogen bonding; Jahn-Teller or other crystal field effects.

In a number of cases, observed distortions have been discussed and interpreted in terms of Jahn-Teller effects without consideration of the possibility of explanation in terms of one or more of the other factors listed above. It should be recognized that interpretation of distortions as illustrating a consequence of electronic structural influence of the metal ion is only valid if it can be assured that no other effect affords a possible explanation of the distortion. These are two ways in which the possibility of explanation by means of the other effects can be eliminated.

One way is through the consistent observation of a particular type of distortion of the coordination polyhedron of a particular metal ion in a large number of examples in which widely varying types of ligands and crystal structures are involved. This is probably best illustrated by the Cu(II) ion. A large number of structures of compounds of Cu(II) have been determined, and in most cases the Cu(II) ion is surrounded by a strongly distorted octahedron, with 4 ligands (in a plane) at distances which are comparable with those found for neighboring metal ions, and 2 ligands in the directions normal to this plane but at considerably greater (by 0,3 - 0,7 Å) distances.

Another way is through a comparison of the structures of isomorphous compounds which differ from one another only in the metal ion. It is the purpose of this paper to present and discuss two examples of isomorphous sets of compounds, one involving tetrahedra and one involving octahedra.

A number of compounds of the general type A_2MX_4, where A is Cs or $[N(CH_3)_4]$, have been investigated and found to be isomorphous (1-8). The structure consists of discrete A ions and $[MX_4]^{2-}$ tetrahedra. The space group is Pnma, the unit cell contains four molecules and the dimensions of those which have been investigated are given in Table I. Although there are fairly

TABLE I

UNIT CELL DIMENSIONS OF A_2MX_4

Compound	a_0	b_0	c_0	Ref.
Cs_2CoCl_4	9,74	7,39	12,97	1
Cs_2CuCl_4	9,72	7,66	12,36	2,7
Cs_2ZnCl_4	9,74	7,39	12,97	3
Cs_2CuBr_4	10,20	7,97	12,94	4
Cs_2ZnBr_4	10,20	7,77	13,52	5
$[N(CH_3)_4]_2CoCl_4$	12,24	8,92	15,39	6
$[N(CH_3)_4]_2CuCl_4$	12,13	9,04	15,16	7
$[N(CH_3)_4]_2ZnCl_4$	12,27	8,96	15,52	8

large differences in cell dimensions, they are due primarily to differences in the sizes of the A and X ions, and secondarily to differences in M-X bond distances and (for Cu) in regularity of the MX_4 tetrahedra.

The bond distances and angles in the MX_4 tetrahedra are given in Table II.

TABLE II

DISTANCES AND ANGLES IN MX_4 TETRAHEDRA

Compound	$M-X_1$	$M-X_2$	$M-X_3$ $M-X_4$	X_1-M-X_2	X_3-M-X_4	X_1-M-X_3 X_1-M-X_4	X_2-M-X_3 X_2-M-X_4
Cs_2CoCl_4	2,26	2,26	2,26	116,3	108,8	109,3	107,3
Cs_2CuCl_4	2,25	2,18	2,18	124,9	123,3	102,9	102,5
Cs_2CuBr_4	2,39	2,38	2,35	130,4	126,4	101,9	99,9
Cs_2ZnBr_4	2,39	2,41	2,38	114,9	109,6	109,7	106,4
$[N(CH_3)_4]_2CoCl_4$	2,20	2,25	2,26	113,2	110,0	109,2	107,6
$[N(CH_3)_4]_2CuCl_4$	2,25	2,22	2,23	130,6	127,2	99,5	101,8
$[N(CH_3)_4]_2ZnCl_4$	2,25	2,24	2,24	111,7	110,4	109,1	108,3

It is apparent that in all cases the tetrahedra are distorted. One type of distortion is exhibited by the Zn and Co compounds, in all cases showing the angle X_1-M-X_2 larger than $109,5^\circ$ and the angle X_2-M-X_3 smaller than $109,5^\circ$. Examination of intermolecular approach distances and individual atom temperature factors in the case of the Cs_2ZnBr_4 has shown that this distortion is

TABLE III

UNIT CELL DIMENSIONS OF $(NH_4)_2M(II)(SO_4)_2 \cdot (H_2O)_6$

M(II)	a	b	c	β	Ref.
Mg	9,38	12,67	6,22	107,1	9,10,6
Mn	9,35	12,74	6,26	107,0	6
Fe	9,32	12,65	6,24	106,8	6
Co	9,20	12,56	6,24	107,2	6
Ni	9,26	12,54	6,24	107,0	11,6
Cu	9,27	12,44	6,30	106,0	6
Zn	9,28	12,57	6,25	106,8	6
Cd	9,37	12,73	6,28	106,7	6

due to a crystal packing·effect (5). The Cu compounds, on the other hand, exhibit a quite different, and considerably greater, distortion which is not explicable on the basis of steric effects. The distortion of the CuX_4 tetrahedron has been satisfactorily explained by FELSENFELD in terms of the balancing of the mutual repulsion of the ce⁻ and the crystal field stabilization of the Cu^{2+}

The Tutton salts, $M(I)_2M(II)(SO_4)_2 \cdot (H_2O)_6$, form an isomorphous series containing the octahedral complex $M(II)(H_2O)_6$. The space group is $P2_1/a$ and the unit cell contains two formula units (6, 9-11). The cell dimensions of a number of members of this series with $M(I) = NH_4$ are given in Table III.

The bond distances and angles in the MO_6 octahedra are given in Table IV. Here, again, all of the cases show some distortion. In all cases

TABLE IV

DISTANCES AND ANGLES IN $M(H_2O)_6$ OCTAHEDRA

Metal	$M-O_7$	$M-O_8$	$M-O_9$	O_7-M-O_8	O_7-M-O_9	O_8-M-O_9
Mg	2,09	2,10	2,05	89,8	91,4	90,3
Mn	2,18	2,20	2,15	89,3	91,4	90,5
Fe	2,17	2,14	2,10	89,4	91,2	90,9
Co	2,11	2,09	2,08	87,4	91,9	88,9
Ni	2,09	2,08	2,04	88,5	90,4	89,3
Cu	2,20	2,10	1,97	88,8	90,3	88,6
Zn	2,14	2,12	2,07	88,7	90,8	89,4
Cd	2,26	2,24	2,21	89,0	91,6	90,8

except Cu, the $M-O_7$ and $M-O_8$ distances are the same within the experimental accuracy. However, in all cases the $M-O_9$ distance is shortest. Since this shortening of $M-O_9$ compared with $M-O_7$ and $M-O_8$ is found for all metals, it is presumably due to some effect other than a "crystal field" effect.

Similarly, in all cases except Cu, the bond angles are distorted in the same manner, with $O_7-M-O_8 < O_8-M-O_9 < O_7-M-O_9$, and even for Cu the deviation from this order is not significant.

Thus,in the octahedral coordination polyhedra of all of the metal ions studied, except for Cu, there are slight but significant deviations from regularity. Since these deviations are of exactly the same nature for all of the metal ions, they must be ascribed to effects of the crystal structure. The coordination polyhedron of the Cu(II) ion, on the other hand, shows a much larger and unique distortion, which is therefore to be ascribed to "crystal field" effects. It is interesting to note that the distortion of the Cu(II) coordination octahedron in the present case is not merely tetragonal but definitely orthorhombic in character.

References

1. PORAI-KOSHITS, M.A., Kristallografya, 1, 291 (1956).
2. HELMHOLTZ, L., and KRUH, R.F., J.Amer.Chem.Soc., 74, 1176 (1952).
3. BREHLER, B., Z.Krist., 109, 68 (1957).
4. MOROSIN, B., and LINGAFELTER, E.C., Acta Cryst., 13, 807 (1960).
5. MOROSIN, B., and LINGAFELTER, E.C., Acta Cryst., 12, 744 (1959).
6. Unpublished results, University of Washington.
7. MOROSIN, B., and LINGAFELTER, E.C., J.Phys.Chem., 65, 50 (1961).
8. MOROSIN, B., and LINGAFELTER, E.C., Acta Cryst., 12, 611 (1959).
9. HOFMANN, W., Z.Krist., 78, 279 (1931).
10. MARQULIS, T.N., and TEMPLETON, D.H., Z.Krist., 117, 344 (1962).
11. GRIMES, N.W., KAY, H.F., and WEBB, M.W., Acta Cryst., 16, 823 (1963).

14 A 3

ELECTRONIC STRUCTURES OF SQUARE PLANAR METAL COMPLEXES.

Harry B. Gray

Department of Chemistry, Columbia University, New York, New York, U.S.A.

The relative energies of the molecular orbitals in square planar metal complexes is a problem which has received considerable attention in the last few years (1-10). Of the antibonding molecular orbitals derived from the d valence orbitals, it is generally accepted that the $\sigma^*(x^2-y^2)$ orbital is much less stable than the other four levels $[\pi^*(xy), \sigma^*(z^2), \pi^*(xz, yz)]$. This level scheme is consistent with the fact that almost all known square planar metal complexes contain a d^8 central metal ion and are diamagnetic, since eight valence electrons can be exactly accomodated in the four reasonably stable orbitals.

In D_{4h} symmetry, the exact ordering of $\pi^*(xy)$, $\sigma^*(z^2)$, and $\pi^*(xz, yz)$ is a tough problem; for complexes such as $[PtCl_4]^{2-}$ and $[Ni(CN)_4]^{2-}$, the current state of friendly disagreement follows:

a) CHATT, GAMLEN and ORGEL have suggested (1) $xy > xz$, $yz > z^2$; $[PtCl_4]^{2-}$;

b) MARTIN and coworkers have proposed two possibilities,
$xy > z^2 > xz, yz$ (6) and $z^2 > xy > xz, yz$ (10); $[PtCl_4]^{2-}$;

c) PERUMAREDDI, LIEHR and ADAMSON have suggested (7)
$z^2 > xy > xz, yz$; $[Ni(CN)_4]^{2-}$;

d) We have suggested (9) $xy > z^2 > xz, yz$; $[PtCl_4]^{2-}$, $[Ni(CN)_4]^{2-}$.

It is evident from this sampling that, in all probability, the solution to the problem has been found. All we have to do now is find who found it.

The first part of the lecture will consist of a summary of the facts relating to the principal d orbital ordering in square planar complexes. A short discussion of the main features of intramolecular charge transfer in these complexes will also be presented.

As mentioned above, the square planar geometry has been restricted mainly to the diamagnetic, d^8 complexes of Ni(II), Pd(II), Pt(II), Rh(I), Ir(I), and Au(III). Recently, however, it has been possible to stabilize the planar geometry using sulfur donor ligands which possess extensive π-orbital systems; the result has been a considerable number of new electronic structures for the planar situation (11-16). We have investigated the complexes containing the ligands maleonitriledithiolate {MNT} and toluene-3,4-di-thiolate {TDT}, which are shown below:

{MNT}   {TDT}

Examples of the new square planar complexes are as follows:

1. a square planar complex with S = 1, $[Co(TDT)_2]^-$;

2. a square planar rhodium complex with S = 1/2, $[Rh(MNT)_2]^{2-}$;

3. square planar copper complexes with S = 0, $[Cu(TDT)_2]^-$, $[Cu(MNT)_2]^-$;

4. square planar nickel, palladium, and platinum complexes with S = 1/2, $[Ni(TDT)_2]^-$, $[Pd(TDT)_2]^-$, $[Pt(TDT)_2]^-$, $[Ni(MNT)_2]^-$, $[Pd(MNT)_2]^-$, and $[Pt(MNT)_2]^-$;

5. square planar iron complexes with S = 1/2 and 3/2, $[Fe(TDT)_2]^-$, $[Fe(MNT)_2]^-$.

The last half of the lecture will deal with the electronic structures of these interesting new complexes.

It is a pleasure to acknowledge the National Science Foundation for support of this research.

References

1. CHATT, J., GAMLEN, G.A., and ORGEL, L.E., J.Chem.Soc. (London), 1958, 486.
2. MAKI, G., J.Chem.Phys., 28, 651 (1958); 29, 162, 1129 (1958).
3. BALLHAUSEN, C.J., and LIEHR, A.D., J.Am.Chem.Soc., 71, 538 (1959).
4. BAN, M.I., Acta Chim.Acad.Sci.Hung., 19, 459 (1959).
5. KIDA, S., FUKITA, J., NAKAMOTO, K., and TSUCHIDA, R., Bull. Chem.Soc. (Japan), 31, 79 (1958).
6. FENSKE, R.F., MARTIN, D.S., and RUEDENBERG, K., Inorg.Chem., 1, 441 (1962).
7. PERUMAREDDI, J.R., LIEHR, A.D., and ADAMSON, A.W., J.Am. Chem.Soc., 85, 249 (1963).
8. FERGUSON, J., J.Chem.Phys., 34, 611 (1961).
9. GRAY, H.B., and BALLHAUSEN, C.J., J.Am.Chem.Soc., 85, 260 (1963).
10. MARTIN, D.S., and LENHARDT, C.A., to be published.
11. SCHRAUZER, G.N., and MAYWEG, V., J.Am.Chem.Soc., 84, 3221 (1962).
12. GRAY, H.B., WILLIAMS, R., BERNAL, I., and BILLIG, E., J.Am. Chem.Soc., 84, 3596 (1962).
13. GRAY, H.B., and BILLIG, E., J.Am.Chem.Soc., 85, 2019 (1963).
14. DAVISON, A., EDELSTEIN, N., HOLM, R.H., and MAKI, A.H., Inorg.Chem., 2, 1227 (1963).
15. EISENBERG, R., IBERS, J.A., CLARK, R.J.H., and GRAY, H.B., J.Am.Chem.Soc., 86, 113 (1964).
16. BILLIG, E., SHUPACK, S.I., WATERS, J.H., WILLIAMS, R., and GRAY, H.B., J.Am.Chem.Soc., 86, 926 (1964).

15 A 1

NOVEL STEREOCHEMICAL EFFECTS IN CERTAIN COORDINATION COMPOUNDS.

J.L. Hoard
Cornell University, Ithaca, N.Y., U.S.A.

This paper is concerned with the stereochemical properties of certain generically related mononuclear complexes in which the coordination number of the central atom or cation is 7, 8, 9, or 10. New results are reported for amine-polycarboxylate chelates of Zr (IV), high spin Fe^{3+}, and the M^{3+} rare earth ions; the chelating agents are ethylendiaminetetraacetic acid (EDTA;

H$_4$Y), 1,2-diaminocyclohexane-N,N'-tetraacetic acid (DCTA, H$_4$Z), and nitrilotriacetic acid (NTA; H$_3$X). X-ray diffraction analyses of three-dimensional data from single crystals provide quantitative descriptions of the chelates for theoretical discussion and correlation with other experimental observations.

We first record certain structural principles which are obeyed by all of the amine-polycarboxylate chelates which have been investigated in our laboratory during the past six years. Ten different structural types of crystalline arrangements containing such chelates have been determined. No instance of an uncomplexed carboxylate group - a free ·CH$_2$COO$^-$ arm - is found in any crystal. Thus the Y^{4-} and Z^{4-} anions from EDTA and DCTA are actively sexadentate; when the coordination number of the central atom exceeds the number of "teeth", the coordination group is filled out with water molecules. Familiar examples of sexadentate aquo complexes are the seven-coordinate [Mn(OH$_2$)Y]$^{2-}$ (1,2) and [Fe(OH$_2$)Y]$^-$ (1,3) ions (which have different configurations); new examples (vide infra) are the seven-coordinate [Fe(OH$_2$)Z]$^-$ (4) and the nine-coordinate [La(OH$_2$)$_3$Y]$^-$ (5) ions. Weak chelation through carbonyl oxygen of a carboxylic acid arm, ·CH$_2$COOH, to the central atom, with the acid hydrogen used to form an external stabilizing hydrogen bond, is observed for three of five crystalline acid complexes of proved structure; thus the quinquedentate octahedral Ni(OH$_2$)YH$_2$ molecule (6) has just one unchelated ·CH$_2$COOH arm in the crystal and the sexadentate tencoordinate La(OH$_2$)$_4$YH molecule (7), described herein, has none.

The severe geometrical limitations attendant upon formation of a sexadentate octahedral EDTA or DCTA complex were set forth in the original structural investigations of the octahedral CoY$^-$ (8) and Ni(OH$_2$)YH$_2$ (6). Briefly, the ring span of a five-membered glycinate ring, i.e., the separation of the nitrogen and oxygen atoms defining an edge of the coordination polyhedron, is limited at best, and is substantially decreased whenever the ring departs markedly from planarity. Inasmuch as the ethylenediamine ring always must be strongly puckered, and bond angles at nitrogen - the common junction for three rings - are preserved within ±5° of the regular tetrahedral value, two of the four glycinate rings in a sexadentate octahedral complex (including CoY$^-$) are markedly folded. It is, moreover, just these rings that need maximum ring span because they involve complexing M-O bonds which should lie approximately in the same plane as, and opposite to, the M-N bonds. The difficulties of ring closure to achieve quasi-octahedral coordination are accentuated if, as is true of all complexes given detailed structural study excepting those of Co(III) and Ni(II), M-N bonds are much longer (0,15 - 0,30 Å) than are M-O bonds. Sexadentate octahedral coordination is limited to small central atoms, and is dependent upon the M-O and M-N bond lengths.

Given primarily electrostatic bonding, as with high spin Mn^{2+} or Fe^{3+}, or a rare earth M^{3+}, as the central ion, the chelate more readily accepts a co-ordination polyhedron with stronger ligand repulsions than those of the octahedron in order to minimize ring strain. The dominant role of ring constraints is more openly displayed as the central atom increases in size. Complexes with coordination numbers above six generally retain two structural classes of glycinate rings which, in contrast to the octahedral case, differ only moderately in accumulated ring strain.

A nearly completed structure determination (4) of crystalline $Ca[Fe(OH_2)Z]_2 \cdot 9 H_2O$ reveals a sexadentate seven-coordinate $[Fe(OH_2)Z]^-$ which resembles that described (3) for $[Fe(OH_2)Y]^-$ as closely as one could expect for analogous DCTA and EDTA chelates. Excepting the four outer carbon atoms in the cyclohexane ring skeleton of the DCTA complex, this latter is described quite adequately in respect to effective symmetry, shape, and bond parameters by the published data (3) for the EDTA complex to which the Reader is referred. The greater stability of the DCTA chelate is not reflected in shorter complexing bonds, but is attributable rather to a more favorable entropy of formation. $[Fe(OH_2)Z]^-$, as with $[Mn(OH_2)Y]^{2-}$ (2) and $[Fe(OH_2)Y]^-$ (3), is expected to be the dominant species throughout an appropriate range of pH in aqueous solution.

The eight-coordinate bisnitrilotriacetatozirconate (IV) ion, $[ZrX_2]^{2-}$, as it exists in both the potassium and the rubidium salts (9), utilizes a co-ordination polyhedron of the dodecahedral $[Mo(CN)_8]^{4-}$ type (10). Four oxygen atoms occupy positions B (cf. Fig. 1, Ref. 10), two oxygen and two nitrogen atoms occupy positions A. $Zr-O_B$ bond lengths are 2.13 Å and, as anticipated (10), $Zr-O_A$ bond lengths are longer at 2.25 Å; $Zr-N$ links, 2.44 Å, are very much longer still, a result seemingly not required by strain inasmuch as the glycinate rings are of "standard" geometry.

Fig. 1 illustrates the sexadentate ten-coordinate $[La(OH_2)_4Y]H$ molecule as given by a very satisfactory determination of the crystalline arrangement (7). The four chelated oxygen atoms, O_C, form a nearly planar trapezoidal array; the lanthanum ion is 0.77 Å from this plane on the side opposite to the nitrogen atoms with $La-N = 2.86$ Å. Three $La-O_C$ links to carboxylate oxygen average to 2.54 Å, and the $La-O_C$ link to carbonyl oxygen is 2.61 Å. The four water molecules attached to La^{3+} at an average distance of 2.60 Å lie in the hemisphere opposite to that which contains the chelating agent.

The sexadentate nine-coordinate $[La(OH_2)_3Y]^-$, as this exists in the potassium salt (5), differs in degree rather than in kind from the $[La(OH_2)_4Y]H$ molecule. Deprotonation of the latter produces a shrinkage of the four $La-O_C$

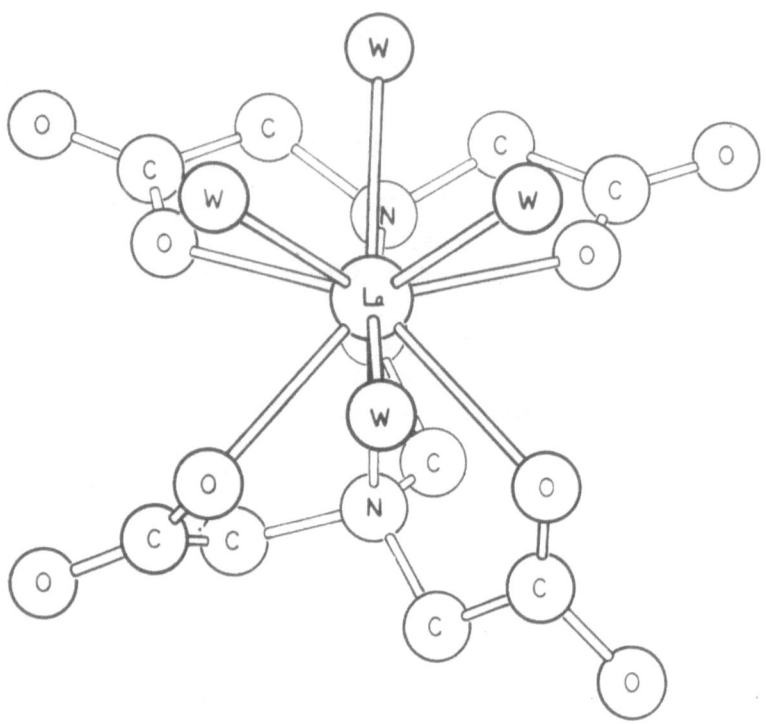

Fig. 1

Model of the $[La(OH_2)_4Y]H$ molecule. The water molecule
at the bottom of the diagram is missing in $[La(OH_2)_3Y]^-$.

bonds to 2, 44 Å, approach of La^{3+} to within 0, 60 Å of the mean plane of the
4 O_C atoms, and the ejection of one water molecule from the coordination
group; the average length for three La-OH$_2$ bonds which remain is 2, 59 Å.
We then hypothesize that replacement of La^{3+} by successively smaller rare
earth M^{3+} ions with accompanying shrinkage of the complex should result,
in one such step, in the ejection of a second water molecule to leave an
eight-coordinate $[M(OH_2)Y]^-$ with the superior coordination geometry (of
$[Mo(CN)_8]^{4-}$ type) needed for a relatively small M^{3+} ion. Further, this transition
should occur upon substitution of Tb^{3+} for Gd^{3+}; we would have then a natural
explanation for the jump in the entropy of formation reported (11) to occur at
this point in the EDTA-rare earth series. Our hypothesis, which demands a
rather minor alteration in the bulky chelate, seems likely to be compatible
with the observation that whole classes of rare earth salts AMY.XH$_2$O,
$A^+ = Na^+$, K^+, NH_4^+; $M^{3+} = La^{3+}$ ..... Er^{3+} (at least); $X \sim 8$, are closely

related in structure to the point of apparent isomorphism (5). Further experimental studies are in hand.

We thank the National Science Foundation, the National Institutes of Health, the Army Research Office (Durham), and the Advanced Research Projects Agency for support of our work in this field.

References

1. HOARD, J.L., SMITH, G.S., and LIND, M.D., in "Advances in the Chemistry of the Coordination Compounds", the Macmillan Company, New York, 1961, pp. 296-302.
2. RICHARDS, S., PEDERSEN, B., SILVERTON, J.V., and HOARD, J.L., Inorg.Chem., 3, 27 (1964).
3. LIND, M.D., HAMOR, M.J., HAMOR, T.A., and HOARD, J.L., Inorg.Chem., 3, 34 (1964).
4. COHEN, G.H., and HOARD, J.L., to be published.
5. LEE, B., LIND, M.D., and HOARD, J.L., in progress.
6. SMITH, G.S., and HOARD, J.L., J.Am.Chem.Soc., 81, 556 (1959).
7. LIND, M.D., LEE, B., and HOARD, J.L., to be published.
8. WEAKLIEM, H.A., and HOARD, J.L., J.Am.Chem.Soc., 81, 549 (1959).
9. WILLSTADTER, E., and HOARD, J.L., to be published.
10. HOARD, J.L., and SILVERTON, J.V., Inorg.Chem., 2, 235 (1963).
11. BETTS, R.H., and DAHLINGER, O.F., Can.J.Chem., 37, 91 (1959).

15 A 2

STEREO-CHEMICAL FACTORS IN COORDINATION COMPOUNDS OF NICKEL: STRUCTURE OF COMPLEX WITH TRIMETHYLENE DINITRAMINE.

J.H. Robertson and D.M. Liebig
School of Chemistry, University of Leeds, Leeds, 2, England, U.K.

Ni(II) readily forms complexes with the following three, simply related, polymethylene dinitramine ligands, as follows:

$$O_2N.NH.(CH_2)_2.NH.NO_2 \longrightarrow Ni[(CH_2)_2(N.NO_2)_2].4 H_2O$$

bluish, microcrystalline, paramagnetic;

$$O_2N.NH.(CH_2)_3.NH.NO_2 \longrightarrow K_2Ni[(CH_2)_3.(N.NO_2)_2]_2.4 H_2O$$

red, very well crystalline, diamagnetic;

$O_2N.NH.(CH_2)_4.NH.NO_2 \longrightarrow Ni[(CH_2)_4(N.NO_2)_2].H_2O$

green, well crystalline, paramagnetic.

Clearly, the nickel atom is quite differently coordinated in these complexes, despite the family resemblance of the ligands. The magnetic properties and the colours in the first place, and also the spectra, indicate that
in the first and third the metal is octahedrally coordinated, while in the
second it is in square-planar coordination. We have begun an X-ray structural
investigation on these three compounds.

The second complex listed, viz., the potassium salt of trimethylene
dinitramine nickel(II) of the composition, $K_2Ni[(CH_2)_3(N.NO_2)_2]_2.4H_2O$ is
especially interesting because, although it is a deep red colour in the solid
state, its aqueous solution is dark green. Our X-ray work on this compound
has been completed so that we are now able to describe the configuration of
the complex ion in this particular case. The details are as follows.

The complex ion $Ni[(CH_2)_3(N.NO_2)_2]_2$ has 2/m symmetry in the crystal;
it is of an overall flattish ellipsoidal shape and is formed of two ligand
molecules, like plates, with the nickel atom held at the centre. Fig.1 shows

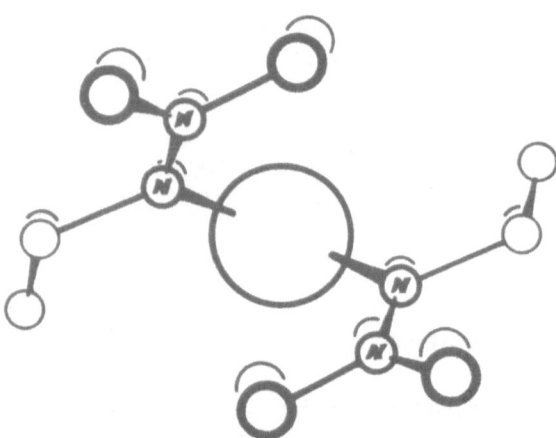

Fig. 1

the ion as it is seen looking down the 2-fold axis between the two ligands.
The nickel is square-planar bonded to the amine nitrogen atoms of the ligands.
It is not bonded to the nitro group at all, nor to the water molecules in the
crystal. (Instead, the water molecules are found associated with the potassium
ions in hydrogen-bonded layers.) The octahedral position on each side of the
nickel atom is unoccupied and it is effectively barred sterically by the
nearness of the nitro group oxygen atoms (4,42 Å apart and 2,95 Å distant

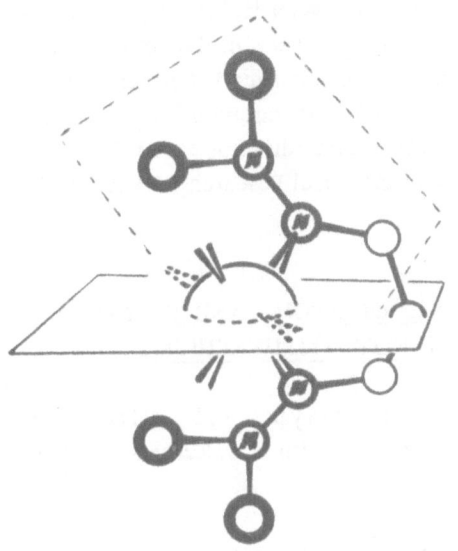

Fig. 2

from the nickel). The situation is depicted in the perspective sketch of Fig.2, where, for clarity, only one ligand is drawn. The nitro group might be assumed to enjoy free rotation about the N-N bond, but in the crystal structure as we have found it this is not the case, for the N-N bond, with the length, 1,275 Å, is almost as short as a full double bond (ef. 1,25 ± 0,02 Å in N_2F_2), and the six atoms,

$$\underset{Ni}{\overset{C}{\diagdown}} N = N \overset{\diagup O}{\underset{\diagdown O}{}}$$

are found to lie in a plane. This plane is indicated by the broken line in Fig.2.

It would therefore appear that sp^2 hybridization of the amine nitrogen forces the nitro group to the particular orientation which obstructs the octahedral position of the nickel, making the nickel square-planar coordinated despite the presence of water of crystallization in the structure.

We suppose that, on dissolution, the molecular motion and impact of solvent molecules suffices to rotate the nitro groups out of the way and to permit coordination of the octahedral positions by solvent. Work is continuing on the tetramethylene complex.

We should like to acknowledge the kindness of Dr.S.W.HAWKINS (of I.C.I., Billingham) who gave us our specimens and who provided our information on the magnetic properties; Professor H.IRVING, for suggesting this problem to us; Dr.M.R.TRUTER, for guidance in the computations; the Director and staff of the Leeds University Computing Laboratory, Dr.B.J. HATHAWAY (of the University of Hull) for measurement of and comments up the reflectance spectra; Mr.J.LYDON and other colleagues for assistance; and the Department of Scientific and Industrial Research for a maintenance grant.

5 A 3

ON MUTUAL INFLUENCE OF ATOMS AND GROUPS IN COMPLEXES OF NITROSYLRUTHENIUM.

O.E. Zvjagintsev, N.M. Sinitsyn, V.N. Pitchkov
Institut Obshchey i Neorganitcheskoy Khimii im N.S.Kurnakova AN SSSR,
Moscow, USSR.

The question of trans-effect in the complex compounds Pt, Co, Pd, Rh is widely discussed in literature. The effects of mutual influence of atoms and groups in complexes of nitrosylruthenium were not as yet observed.

In investigating the complexes of ruthenium in solutions as well as in polycrystals an experimental evidence was obtained to prove the existence of trans-influence in complexes of nitrosylruthenium. It was found out that the nitrosyl group has a strong trans-influence manifesting itself in labilization of ligand bond being situated in trans-position to it on the coordinate A-Ru-NO, with A being a neutral or acid ligand. Thus in hydrolysis or titrating the nitrosylruthenium pentahalogens aquo solutions with silver nitrate, the halogen being in trans position to nitrosyl group was found to be the most reactionable. This gives rise to forming the trans-configuration nitrosylruthenium tetra-halogens.

The rise in mobility is observed in $[RuNO(NH_3)_5]Cl_3 \cdot H_2O$ for ammonia being in trans position to nitrosyl group on the coordinate NH_3-Ru-NO. In cultivating this compound with alkali or with ammonia, trans-$[RuNO(NH_3)_4OH]Cl_2$ (1) results; and in thermal decomposing, trans-$[RuNO(NH_3)_4Cl]Cl_2$ results. X-ray structural investigations by KCHODASHEVA (2) have shown the distance Ru-N (from NH_3) on the coordinate NH_3-Ru-NO is longer than that on the coordinate NH_3-Ru-NH.

A comparison of the X-ray structural data obtained by BOKIY and his collaborators to those being IR-spectroscopic (Table I) shows that both distances Ru-N (from NO) r_{RuN} and N-O (from NO) r_{NO} change symbatically in passing from one compound to another.

TABLE I

Compound	ν_{NO}, cm^{-1}	Force constant of the bond NO, f_{NO}, milli-dyne/Å	Calcula-ted accor-ding to ν_{NO}, r_{NO}, Å	Found with the X-ray structural method r_{RuN}, Å
$[RuNO(NH_3)_5]Cl_3 \cdot H_2O$	1935 1922	16,3	$1,14_5$	$\leq 1,70\pm0,07\geq$
$K_2[RuNOCl_5]$	1919 (3)	16,2	$1,14_7$	$1,70\pm0,07$
trans-$[RuNO(NO_2)_4OH]K_2$	1900 (4)	15,9	$1,15_0$	$1,85\pm0,05$
trans-$[RuNOCl_4OH](NH_4)_2$	-	-	-	$2,04\pm0,05$
trans-$[RuNO(NH_3)_4OH]Cl_2$	1845 (4)	15,0	$1,15_9$	$2,07\pm0,04$

This gave us a chance to demonstrate, according to the change of valence vibration frequency ν_{NO}, that diverse ligands in ruthenium compounds which are in trans-position to the nitrosyl group on the coordinate A-Ru-NO could be arranged in the series as follows: NH_3, $NO_3^- < Cl^- < Br^- < J^-$ (Table II).

TABLE II

Coordinate in tetra-ammine	ν_{NO}, cm^{-1}	f_{NO}, millidyne/Å	r_{NO}, Å
NH_3-Ru-NO	1922, 1935	17,2	1,13
Cl-Ru-NO	1893	17,0	
Br-Ru-NO	1885 (4)	16,9	
OH-Ru-NO	1845 (4)	16,2	1,14

In passing along the series from NH_3 to J^-, r_{RuN}, r_{NO} increase, but ν_{NO}, f_{NO} and the bond order decrease.

The methods of obtaining, chemical and physical-chemical properties of the ruthenium complex compounds with general formula $[RuNO(NO_2)_2(Am)_2OH]$ (where Am = NH_3, py etc.) are described. The structure of these compounds is interpreted as follows:

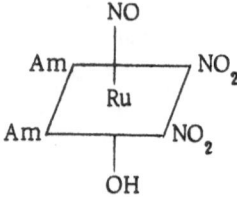

In these compounds the cis-effect was observed which manifested itself in the weakening of the between nitrosyl group and ruthenium, depending on the amine electronodonor properties in cis-position to the coordinate HO-Ru-NO. The increase of amine electronodonor properties leads to displacement of electron atmosphere to ruthenium, diminishes its effective positive charge. This, in turn, stimulates the decrease of ν_{NO} in IR absorption spectrum.

A certain influence on the redistribution of electronic density in the molecule of nitrosylruthenium complexes is caused as well by outside sphere substituents which lead to the increase of displacement of electronic density from ligands to the central atom in the series:

$$Na^+ < K^+, NH_4^+ < Rb^+ < Cs^+ \quad \text{(Table III)} ;$$

it is due to this fact that frequency of valence vibration ν_{NO} decreases.

TABLE III

Outside sphere substituent	ν_{NO}, cm^{-1}	
	$[RuNOCl_5]^{-2}$	$[RuNO(NO_2)_4OH]^{-2}$
Na^+		1907 (3)
K^+	1919 (3)	1900 (4)
NH_4^+	1916, 1907	
Rb^+	1899	1890
Cs^+	1880 (4)	

This additive contribution of outside sphere substituents to the redistribution of electronic density is to be taken into account in regarding the questions on mutual influence of atoms and groups in complex compounds.

References
1. PARPIEV, N.A., BOKIY, G.B., J.neorgan.kchimii, 4, 2452 (1959).
2. KCHODASHEVA, T.S., Autoreferat dissertatsii IONCh AN SSSR, 1963.
3. BROWN, G.M., J.Inorg.Nucl.Chem., 13, 71 (1960).
4. LEWIS, J., IRVING, R., WILKINSON, E., J.Inorg.Nucl.Chem., 7, 32 (1958).

16 A 1

DECOMPOSITION OF SUBSTITUTED π-ALLYL PALLADIUM HALIDE COMPLEXES.

E.J. Smutny
Koninklijke/Shell Laboratorium, Amsterdam-N, The Netherlands.
Present address: Shell Development Company, Emeryville, California.

The interaction between diolefins and palladium salts in hydroxylic and non-hydroxylic solvents leads to π-allylic complexes in yields of 80% or greater. These complexes have the empirical formula [(diolefin)(alcohol)PdCl] and their structure has been established (1) as the π-allyl type {I}.

These complexes have been decomposed in alcoholic media by sodium acetate and by hydrogen gas. 1-Methoxymethyl-2-methyl-π-allyl palladium chloride {I}, derived from isoprene, methanol and sodium chloropalladite, was dissolved in methanol and decomposed with sodium acetate at reflux. The products were separated by repeated GLC fractionation and identified by chemical and spectrochemical means. The scheme summarizes the products identified and accounts for approximately a 95% material balance. The decomposition of complex {I} with hydrogen gas was run at room temperature in methanol. Similar means were used for product separation and identification; and 95% of the reduction products were identified as outlined in the scheme. The π-allyl complex from butadiene, methanol and sodium chloropalladite shows similar behavior with sodium acetate and with hydrogen reduction. The product distribution is solvent dependent. These reactions provide a chemical proof for structure {I}.

The mechanism of decomposition will be discussed and properties of the π-allyl metal complexes described.

Complexes have been prepared from tetramethylallene and palladium chloride. The composition and structure will be discussed.

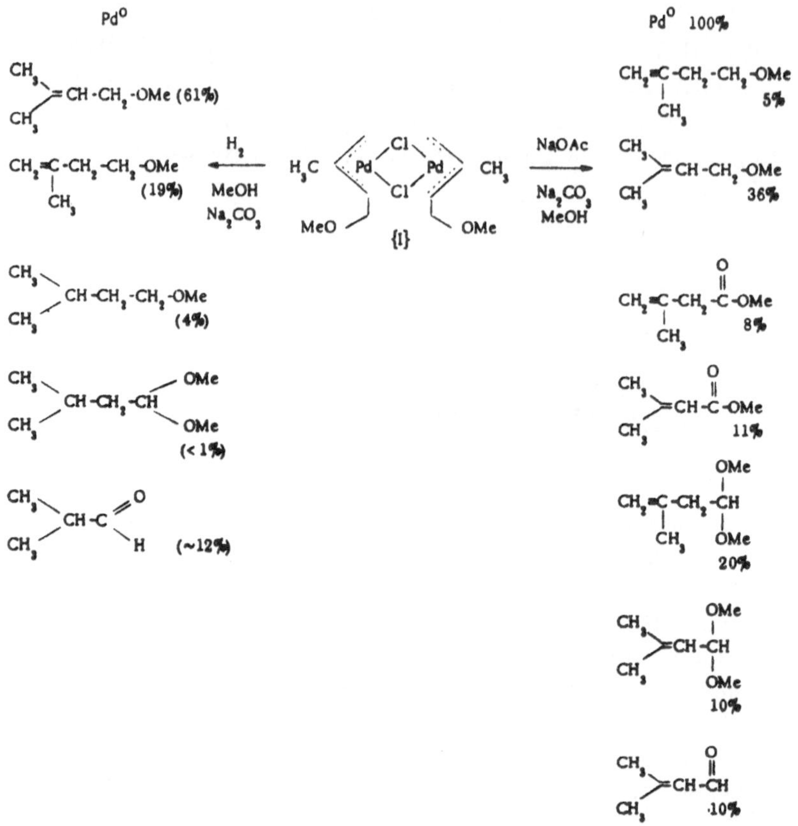

Pd⁰ → Pd^0 100%

(CH₃)₂=CH-CH₂-OMe (61%)

CH₂=C(CH₃)-CH₂-CH₂-OMe (19%) ← H₂, MeOH, Na₂CO₃

(CH₃)₂CH-CH₂-CH₂-OMe (4%)

(CH₃)₂CH-CH₂-CH(OMe)₂ (<1%)

(CH₃)₂CH-C(=O)H (~12%)

H₃C–[Pd(Cl)₂Pd]–CH₃ with MeO and OMe {I}

NaOAc, Na₂CO₃, MeOH →

CH₂=C(CH₃)-CH₂-CH₂-OMe 5%

(CH₃)₂=CH-CH₂-OMe 36%

CH₂=C(CH₃)-CH₂-C(=O)-OMe 8%

(CH₃)₂=CH-C(=O)-OMe 11%

CH₂=C(CH₃)-CH₂-CH(OMe)₂ 20%

(CH₃)₂=CH-CH(OMe)₂ 10%

(CH₃)₂=CH-CH=O 10%

Reference

1. ROBINSON, S.D., and SHAW, B.L., J.Chem.Soc. (London), 1963, 4806.

6 A 2

COMPLEX FORMATION BY MOLYBDENUM(II)CHLORIDE.

J.E. Fergusson, B.H. Robinson and C.J. Wilkins
University of Canterbury, Christchurch, New Zealand, U.K.

Comparison of the ultra-violet reflectance spectrum of Mo(II)chloride with that of the acid $H_2[Mo_6Cl_8.Cl_6].8 H_2O$ (1) which has an Mo_6Cl_8 cage confirms the close structural similarity of the chloride. It seems clear that in the "dichloride" each molybdenum of the cage is coordinated externally with a chlorine, the six-fold coordination being completed through inter-cage bridging. The structural evidence, including the diamagnetism, indicates that all compounds containing the Mo_6Cl_8 unit are electronically saturated.

Molybdenum(II)chloride reacts with varied monodentate ligands (pyridine, pyridine-N-oxide, triphenylphosphine, and alcohol) to give complexes of the type $Mo_6Cl_8.Cl_4.L_2$ whose electrolytic conductivities are low and comparable with those of the chloride itself in the same solvents. It is inferred that the two additional ligands complete the coordination octahedron without the need for inter-cage bridging. The relevance of six-coordination is underlined by the existence of complexes $[(Mo_6Cl_8)Cl_3(Ph_3P)_3]Cl$ and $[(Mo_6Cl_8)Cl_2(Ph_3P)_2(EtOH)_2]Cl_2$ which are respectively 1:1 and 2:1 electrolytes in nitrobenzene. Molecular weight values accord with these formulations.

Molybdenum(II)chloride is more selective in its behaviour towards bidentate ligands. It does not react with N,N,N',N'-tetramethylethylenediamine and does not yield the normal type of chelate complex with ethylenediamine or acetylacetone. However, with 2,2'-bipyridyl and o-phenylenebisdimethylarsine it gives complexes of the type $[Mo_6Cl_8.Cl_2.L_2]Cl_2$. These give conductivities characteristic of 2:1 electrolytes. Bisdiphenylphosphinethane (with pyridine) gives a complex $[(Mo_6Cl_8)Cl_3(Diphos)(Py)]Cl$, and 2,2',6',2"-terpyridyl forms a compound $[Mo_6Cl_8.Cl_3(Terpy)]Cl$ in which six-coordination is again maintained. In the UV spectra of these complexes, as compared with those containing only monodentate ligands, there is a small shift of two of the three bands to lower frequency, and the appearance of a fourth band. The essential cage structure undoubtedly persists but may undergo some bonding modification. Models show the molybdenum-molybdenum bond distance, 2.62 Å, could be spanned by the bidentate ligands if 'bent' bonds are involved, but this description is not adequate for terpyridyl. The rigid adherence to six-coordination irrespective of the geometry of the ligand suggests the cage may be able to provide a delocalised molecular orbital capable of accepting electron pairs which are not disposed symmetrically.

Reference

1. BROSSET, Arkiv Kemi, 1, 353 (1949).

16 A 3

IODINETRIAMMINES AND TETRAMMINES(CIS) OF Pt(IV).

I.I. Chernjaev and V.S. Orlova
Institute of Inorganic Chemistry, Academy of Science,
Leninskii Prospect 31, Moscow, UdSSR.

According to the trans-effect principle the diammines of Pt(IV) with iodine in the inner sphere, are capable to replace in a reversible way some ligands occupying the trans-position in relation to iodine. The cis-diammines of Pt(IV) replace one atom of iodine, the trans-diammines two, forming accordingly face triiodinetriammines and cis-diiodinetetrammines. The latter differ quite distinctly from the trans-tetrammines by their capacity to transfer into iodinetriammines and then to tetraiodinediammines interacting with potassium iodide.

Trans-diiodine-tetrammines do not interact with potassium iodide, but react with ammonia. Cis-diiodine-tetrammines do not enter into any reaction with ammonia.

The face structure of the triiodinetriammines has been proved by their reduction to cis-diammines and that of the diiodinetetrammines to tri-ammines of Pt(II). Besides, the cis-tetrammines react with alkali to yield amido hydroxo products and interact with concentrated alkali-diamidodi-ammine of platinum by elimination of the second proton. The elimination of the two protons and the formation of hydroxoamidoderivates are in accord with the low pH of the iodinederivates of Pt(IV). The easy formation of the iodinehydroxo-coordinate with a slight increase in pH of the solutions for all iodinederivates is quite remarkable.

1 B 1

METAL-METAL BONDS IN COMPLEX COMPOUNDS.

R.S. Nyholm

University College London, Department of Chemistry, London W.C.1, U.K.

This paper summarises some recent investigations concerned with the synthesis and structure of compounds containing metal-to-metal (M-M) bonds. It extends the earlier work described at the VII. I.C.C.C. in Stockholm (1) and is complimentary with the paper by Dr. K. VRIEZE given in the following abstract. Developments in our understanding of the nature of the M-M bond and some new types of compounds are reported.

The obvious way to classify substances containing M-M links is in terms of the nature of the bond present, i.e. multicentre, ionic or covalent. However, the distinction between these is not always clear and the following division into four types of substances is more convenient.

Substances Containing Metal-Metal Bonds

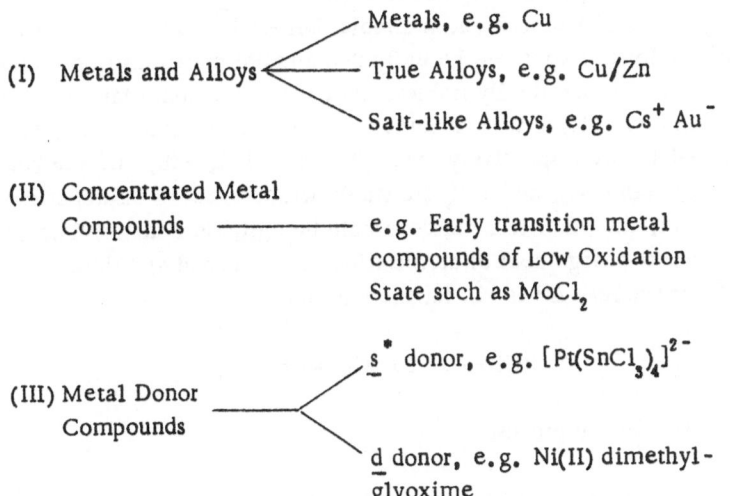

(I) Metals and Alloys
- Metals, e.g. Cu
- True Alloys, e.g. Cu/Zn
- Salt-like Alloys, e.g. $Cs^+ Au^-$

(II) Concentrated Metal Compounds —— e.g. Early transition metal compounds of Low Oxidation State such as $MoCl_2$

(III) Metal Donor Compounds
- $\underline{s}^*$ donor, e.g. $[Pt(SnCl_3)_4]^{2-}$
- $\underline{d}$ donor, e.g. Ni(II) dimethyl-glyoxime

$$\text{(IV) Covalent Bond Type} \begin{cases} \underline{s}^*/\underline{s}^*, \text{ e.g. } Hg_2Cl_2 \\ \underline{s}^*/\underline{d}, \text{ e.g. } Ph_3P \longrightarrow AuMn(CO)_5 \\ \underline{d}/\underline{d}, \text{ e.g. } Mn_2(CO)_{10} \end{cases}$$

* Strictly $\underline{s}/\underline{p}$

Type I. The structures of metals and alloys are mentioned briefly as a background for the discussion of concentrated metal compounds. We take the view that in metals and alloys the valency displayed (used here to imply the number of electrons per atom used for binding) is commonly that shown in usual chemical compounds rather than the high values assigned by PAULING (2). For example the latter attributes a valency of 5,56 to Cu, Ag and Au whereas we consider it should be unity as for the alkali metals. Similarly the alkaline earth metals form multicentre bonded lattices in which the metal is bivalent. Proceeding across the transition series Sc, Y, La are ter- and Ti, Zr, and Hf quadri-valent. The later transition elements provide major problems of interpretation but will not concern us here. The big increase in heats of atomisation as we pass from the alkali metals to Cu, Ag and Au can be understood solely in terms of an increase in electronegativity rather than of bond order (3).

Type II. Multicentre binding between metal atoms also occurs in concentrated metal compounds such as $MoCl_2$. The unusual properties (e.g. diamagnetism) of compounds such as $MoCl_2$ and Ta_6Cl_{14} etc, as compared with the first transition series, can be understood in terms of (i) the small number of $\underline{d}$ electrons present; (ii) the large number of vacant $\underline{d}$ orbitals available; (iii) the large heats of atomisation. A simple Born-Haber calculation shows that most of the lower halides of the early 2nd and 3rd non-transition metals would not be thermodynamically stable if they existed as ionic lattices. The heats of formation ($-\Delta H$) of the (hypothetical) ionic dichlorides of Cr, Mo, W, V, Nb, and Ta are respectively -96, -7, +98, -121, -48, and +33 kcal/mole. Whereas both $CrCl_2$ and VCl_2 are stable with respect to disproportionation into MCl_3 and M, the other four halides would be unstable if ionic. The controlling factor is the large heat of atomisation of the second and third row elements. Thus whereas ΔH is +27 for the reaction

$$3\ CrCl_2 \longrightarrow 2\ CrCl_3 + Cr \quad ,$$

the value is -147 for the process

$$5\ "Mo^{2+}\ 2\ Cl^-" \longrightarrow 2\ MoCl_5 + 3\ Mo$$

The above compounds are stable only because intra-nuclear multicentre bonds are formed. These often involve six octahedral metal atoms as in $MoCl_2$ containing the $[Mo_6Cl_8]^{4+}$ unit. Other arrangements are known; thus in the $[Re_3Cl_{12}]^{3-}$ ion there are 3 metal atoms in a triangle (4).

   Type III. Many compounds are now known in which one metal atom acts as a donor to another metal atom. Perhaps the simplest are those in which the $[SnCl_3]^-$ ion, which is isoelectronic with SbR_3, forms bonds with Rh, Pt etc. Undoubtedly the donor capacity of the $[SnCl_3]^-$ group is enhanced by the fact that it has a negative charge.

   Type IV. The factors governing the formation of the various kinds of covalent bonds (s-s, s-d, d-d) are being studied in these laboratories. General methods of preparation used are: (i) Reduction of a suitable halide, e.g. the action of reducing agents on $HgCl_2 \longrightarrow Hg_2Cl_2$. (ii) The reaction involving metathesis between a coordinated metal halide (e.g. $Ph_3P \longrightarrow AuCl$) and the sodium salt of a carbonyl metal anion (e.g. $Na^+[Mn(CO)_5]^-$). This gives the compound $Ph_3P \longrightarrow ^-Au-Mn(CO)_5$. Work on this reaction has been extended to the preparation of bonds between copper and transition metals and silver and transition metals. Thus KASENALLY (5) has shown that the reaction

yields the yellow complex Tridentate $\rightleftharpoons$ $Cu-Mn(CO)_5$. This is monomeric in organic solvents, diamagnetic and a non-electrolyte in nitrobenzene. The corresponding silver complex decomposes in a few minutes in daylight and is difficult to characterise. This work has been extended to other triarsines and to different transition metals and will be discussed. (iii) Treatment of the hydride of a transition metal with a post-transition metal halide yields a M-M bond and hydrogen halide (see VRIEZE below) (6). (iv) In certain cases direct addition to a square complex gives a metal-metal bond, e.g. the reaction

$$(Ph_3P)_2COIrCl + HgCl_2 \longrightarrow (Ph_3P)_2COCl_2Ir-HgCl$$

gives an octahedral complex of tervalent iridium (see VRIEZE below).

Factors governing the stability of metal-metal bonds have been investigated. So far as the post-transition metal is concerned it is clear that high electronegativity is an important factor. Thus the stability of the Group I diatomic molecules parallels electronegativity, viz. Au_2 (88,3 kcal) > Cu_2 (81,1 kcal) > Ag_2(68,4). Similarly Hg-Hg bonds are more stable than those of Zn and Cd. For the transition metal expansion of the $\underline{d}$ orbital used binding is important and this is favoured by low or negative oxidation states (e.g. by using carbonyl anions). A study of the infra-red spectra of compounds of the type $L \rightarrow Au-Mn(CO)_5$, reveals that for $L = Ph_3P$, Ph_3As, $(PhO)_3P$, $(CH_3O)_3P$ the mean C-O stretching frequency is constant at 2015 ± 3 cm^{-1} but this changes to 1979 cm^{-1} for $L = Ph_3Sb$. This decrease in C-O stretching frequency is taken to indicate that the Au^+-Mn^- bond polarity is greater in the Ph_3Sb compounds. Attempts to prepare the Ph_3Bi compound have been unsuccessful. We conclude that the Au-Mn bond becomes less stable as the Au atom becomes more electropositive, and it seems clear that instability results if the effective electronegativity of the metal atom attached to the Mn atom changes more than slightly. Complexes of the type $Ph_3P \rightarrow Au-Mn(CO)_4L$ have been prepared and studied; when $L = Py$, Ph_3As or PPh_3 the mean C-O stretching frequency is practically constant at about 1935 cm^{-1} but this increases to 1969 cm^{-1} for $L = P(OPh)_3$. This is attributed to the greater π bonding capacity of the $P(OPh)_3$ group competing with the CO ligands for π bonding from the metal.

The preparation and properties of various new compounds containing metal-metal bonds will be discussed.

References

1. COFFEY, LEWIS and NYHOLM, Proc. of VIIth I.C.C.C. (Stockholm), p.66.
2. PAULING, "Nature of the Chemical Bond", Ithaca Cornell Univ. Press, 3rd Ed. 1960.
3. NYHOLM, Unpublished.
4. ROBINSON-WARD, FERGUSSON and PENFOLD-BRUCE, Nature, 201, 181 (1964), and Proc. Chem. Soc., 1963, 116. See also BERTRAND, COTTON and DOLLASE, J. Amer. Chem. Soc., 85, 1349 (1963), and COTTON and HAAS, Inorganic Chem., 3, 10 (1964).
5. KASENALLY, NYHOLM and STIDDARD, J. Am. Chem. Soc. 86, in the press (1964).
6. NYHOLM and VRIEZE, Proc. Chem. Soc., 1963, 138.
7. NYHOLM and VRIEZE, Chem. and Ind., 1964, 318.

COMPOUNDS CONTAINING RHODIUM-MERCURY, OSMIUM-MERCURY OR IRIDIUM-MERCURY BONDS.

K. Vrieze

University College London, Gower Street, W.C.1, U.K.

In the preceding paper presented by R.S. NYHOLM various methods which can be used to prepare covalent metal-to-metal bonded compounds were summarised. The work described in this investigation employs two new essentially different reactions (see I and II below) to produce complexes containing transition metal-to-mercury bonds; these are the first complexes containing a metal-metal bond between a post-transition metal and a group VIII metal in the second and third rows.

I. Hydride Reactions (see figure). These involve a reaction of the type:

$$(Ph_2MeAs)_3Cl_2RhH + HgCl_2 \longrightarrow (Ph_2MeAs)_3Cl_2RhHgCl + HCl$$

The same product is obtained if one uses Hg_2Cl_2, Hg as well as HCl being formed, or by using PhHgCl.

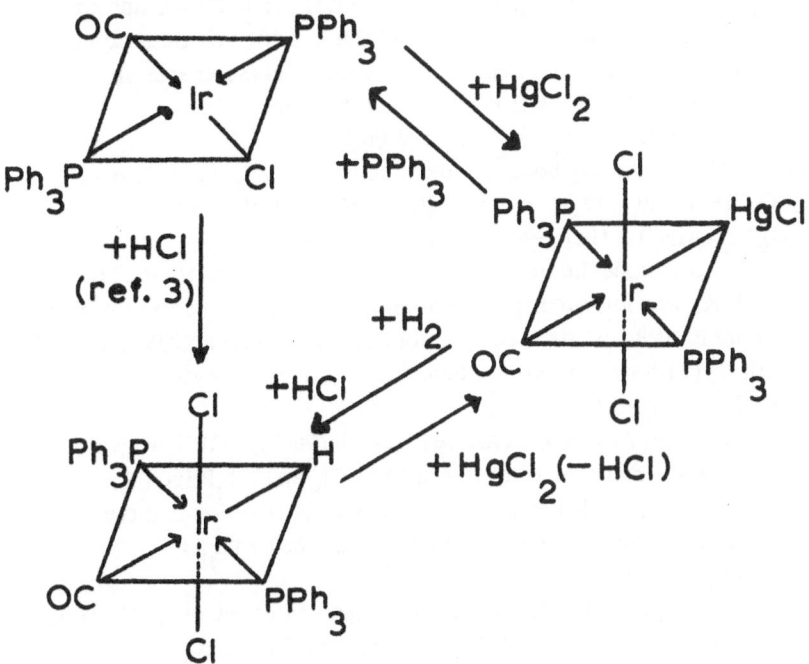

In this way we have prepared the rhodium complexes L_3X_2RhHgY (1) ($L = Ph_2MeAs$; $X = Cl$, Br; $Y = F$, Cl, Br, I, OAc, CN, SCN) which are monomeric for $Y =$ Halogen and OAc and dimeric for $Y = CN$ and SCN. The Os-compound $(Ph_3As)_3(CO)ClOsHgSCN$ was shown to be dimeric too; however the Ir-complexes $(Ph_3P)_2Cl_2IrHgY$ ($Y = Cl$, Br, SCN) made from HgY_2 and $(Ph_3P)_2(CO)Cl_2IrH$ (3) are all monomeric in solution (5).

II. <u>Oxidation-Reduction Reactions</u> (see figure). It was discovered that the easily oxidisable $IrCl(CO)(Ph_3P)_2$ (3) reacts at room temperature in solution with mercury compounds (2) as follows:

$$IrCl(CO)(Ph_3P)_2 + HgCl_2 \longrightarrow (Ph_3P)_2(CO)Cl_2IrHgCl$$

This compound is also formed by using Hg_2Cl_2 instead of $HgCl_2$, Hg being precipitated.

By adding HgY_2 to $IrCl(CO)(Ph_3P)_2$ in solution we isolated, in quantitative yield, the <u>monomeric</u> complexes $(Ph_3P)_2(CO)ClYIrHgY$ ($Y = Cl$, Br, I, OAc, CN, SCN). $RhCl(CO)(Ph_3P)_2$, isomorphous with $IrCl(CO)(Ph_3P)_2$, is <u>not</u> oxidised by $HgCl_2$ <u>nor</u> by $AgNO_3$ (4) whereas $IrCl(CO)(Ph_3P)_2$ reduces $AgNO_3$ in acetone solution to Ag.

<u>Properties</u>.

The complexes are diamagnetic, non-electrolytes in $PhNO_2$ and soluble in suitable organic solvents. The products of the hydride reactions do not show metal-hydrogen stretching and bending frequencies characteristic of the original hydrides. All compounds are stable to air and moisture.

Reactions with Halogens, HCl, H_2 and Ph_3P.

The metal-to-mercury bond is broken by Cl_2, Br_2, I_2 and HCl e.g.
$(Ph_3P)_2(CO)ClBrIrHgBr + Br_2 \longrightarrow (Ph_3P)_2(CO)ClBr_2Ir + HgBr_2$ and
$(Ph_2MeAs)_3Cl_2RhHgCl + HCl \longrightarrow (Ph_2MeAs)_3Cl_2RhH + HgCl_2$
In both type of reactions the products were isolated and identified. The reaction with Br_2 was followed spectrophotometrically; the reaction is quantitative, one mole of Br_2 being used per mole of the complex $(CO)(Ph_3P)_2Cl_2IrHgCl$ treated with H_2 (without catalyst) in benzene for about 20 h yields the hydride $(CO)(Ph_3P)_2Cl_2IrH$, Hg and Hg_2Cl_2.

Ph_3P, but <u>not</u> Ph_3As or Ph_2MeAs, reacts with $(Ph_3P)_2(CO)Cl_2IrHgCl$ to give $IrCl(CO)(Ph_3P)_2$ and a mixture of $(Ph_3P)_2HgCl_2$ and $(Ph_3PHgCl_2)_2$.

The Rh-Hg-compounds behave similarly but the reaction is more complicated because Ph_2MeAs is replaced by Ph_3P but not by Ph_3As.

<u>Spectral Observations</u>.

It was observed that the CO-stretching frequency of the only CO-group present in the Ir-complexes (in $CHCl_3$) decreases in the series $(Ph_3P)_2(CO)Cl_2IrX$; $X = Cl$ $(2080 \text{ cm}^{-1}) > H$ $(2045 \text{ cm}^{-1}) > HgCl$ (2030 cm^{-1})

in keeping with the decreasing electronegativity in the order Cl > H > Hg.
The CO frequency of $IrCl(CO)(Ph_3P)_2$ is 1944 cm$^-$ (Nujol).

In the case of the Rh-compounds L_3X_2RhZ we find that the group Z causes
a change in the visible spectrum. It was found that the longer wavelength
absorption $^1A_{1g} \rightarrow {}^1T_{1g}$ ($t_{2g}^6 \rightarrow t_{2g}^5e_g$ estimated epsilon 200-400) shifts
to higher frequency in the order: Br < Cl < H < HgI < HgBr < HgCl < HgF <
HgOAc (5) (the higher frequency absorption $^1A_{1g} \rightarrow {}^1T_{2g}$ ($t_{2g}^6 \rightarrow t_{2g}^5e_g$)
is not visible because of strong electron transfer). The relative shifts are
bigger for the $L_3Br_2RhZ -$ than for the L_3Cl_2RhZ compounds.

If we assume that the ligand field parameter $10D_q$ is proportional to the
frequency of the longer wavelength transition (supposing that the electron
repulsion is essentially constant) we can rationalise the high ligand field of
the HgY group by assuming a strong covalent bond between the Rh- and the
Hg atom which raises the energy of the antibonding e_g-orbitals.

A rationalisation for the increasing ligand field from HgI to HgOAc could
be that the ligand Y contracts the 6s6p-hybrid of the Hg- atom causing
energetically a more favourable interaction between the 6s6p (Hg) and the
4d (Rh) orbitals, which in turn raises the antibonding e_g-orbitals of the Rh
along the HgY series.

Acknowledgement: The author is indebted to the Koninklijke Shell Laboratorium
in Amsterdam for leave ob absence.

References

1. NYHOLM, R.S., and VRIEZE, K., Proc.Chem.Soc., 1963, 138.

2. NYHOLM, R.S., and VRIEZE, K., Chem.and Ind., 1964, 318.

3. VASKA, L., and DILUZIO, J.W., J.Amer.Chem.Soc., 83, 2784 (1961);
   84, 679 (1962).

4. VALLARINO, L., J.Chem.Soc. (London), 1957, 2287.

5. NYHOLM, R.S., and VRIEZE, K., Unpublished work.

6. LEWIS, J., NYHOLM, R.S., and REDDY, G.K.N., Chem.and Ind.,
   1960, 1386.

1 B 3

CRYSTAL STRUCTURES OF SOME COMPOUNDS CONTAINING BONDS BETWEEN DISSIMILAR METAL ATOMS.

H.M. Powell, K. Mannan, B.T. Kilbourn and P. Porta
Chemical Crystallography Laboratory, Oxford University, England, U.K.

In the last few years many compounds have been synthesised in which two, or more, metal atoms are linked. It has been known for some time that atoms of the same metallic element are joined together in some compounds, but the more recent discoveries include examples of bonding between dissimilar atoms. Thus many compounds containing an Au-Mn bond have been made (1). Molecular weights, magnetic and other properties establish the nature of these compounds, but it is desirable with any new type of compound to determine molecular structure by the direct method of crystal structure analysis. This has been done for several compounds. In each case a metal-metal bond is disclosed by the interatomic distance.

In $(C_6H_5)_3P$-Au-Mn(CO)$_5$ there is an approximately linear bond system P-Au-Mn. This is the normal stereochemical form that would be expected for an Au(I) atom forming two bonds. Considerable difficultiy arose in the structure determination. The crystal is non-centrosymmetric and, as a further complication, contains two crystallographically different molecules with a pseudosymmetry of the heavy atom positions higher than that of the space-group. In this situation the common methods of crystal structure refinement are severely hampered. Nevertheless all atoms have been located. The distance Au-P may be used to get, by difference, a covalent radius applicable to the gold atom and hence, from the Au-Mn distance, a radius for Mn is found. A more suitable compound is the variant

$$(C_6H_5)_3P\text{-Au-Mn(CO)}_4\text{-P(OC}_6H_5)_3$$

which forms a centrosymmetric structure with one molecule per asymmetric unit. From the Au-P distance observed in the first of these two compounds and a tetrahedral phosphorus radius of 1,1 Å the Au contribution to the single bond length is found to be 1,2 Å. From the Au-Mn separation observed as 2,47 Å a manganese radius of 1,27 Å is deduced. This is considerably less than 1,46 Å, half the Mn-Mn distance in $Mn_2(CO)_{10}$ (2). According to DAHL and RUNDLE (2) this Mn-Mn distance is 0,5 Å longer than the "normal". If the radius of gold in the present compound is 1,2 Å, the expected Au-Mn distance will be 2,66 Å if the radius is taken from $Mn_2(CO)_{10}$ and 2,41 Å if DAHL and RUNDLE's "normal" radius is used.

In $(C_6H_5)_2Sn[Mn(CO)_5]_2$, refined to an R value of 11,35%, two crystallographically independent Sn-Mn distances provide an internal check on accuracy by their agreement to within 0,01 Å. These distances 2,70 Å lead to a Mn radius of 1,31 Å, which is more accurate than the value derived above from the gold compound because of greater ease of refinement. In this case the appropriate Sn contribution to the metal-metal bond length is taken as 1,39 Å, half the Sn-Sn distance in the ring polymer $[(C_6H_5)_2Sn]_6$, which contains tin atoms in a chemical state closely resembling that in the carbonyl.

In both these compounds the Mn radius deduced is neither the high value found in the decacarbonyl nor the "normal" radius cited by DAHL and RUNDLE. More detailed discussion is perhaps best deferred until further results are available from structural analyses of related compounds.

The complex bis(cyclo-octa-1,5-diene)iridiumtrichlorotin, of previously uncertain constitution, is found to contain the $SnCl_3$ group linked through tin to the iridium átom. The distance Ir-Sn, 2,73 Å (at a stage of refinement R = 12,9%), is fairly close to a value derived from normal radii applicable to cómpounds not containing metal-metal bonds.

References

1. COFFEY, C.E., LEWIS, J., and NYHOLM, R.S., J.Chem.Soc. (London), 1964, in press.
2. DAHL, L.F., and RUNDLE, R.E., Acta Cryst., <u>16</u>, 419 (1963).

1 B 4

PLATINUM COMPLEXES WITH Pt-Pt INTERACTION IN CRYSTAL AND SOLUTION.

Klaus Krogmann, Peter Dodel and Hans Dieter Hausen
Laboratorium für Anorganische Chemie der T.H. Stuttgart,
Schellingstr.26, 7 Stuttgart-N, Germany.

d^8-metal complexes under strong field conditions stabilize the d_{z^2}-electrons by lowering their coordination symmetry from octahedral to tetragonal or even square planar. That means, two negative ligands are removed until they disappear from the coordination sphere. But in the crystal lattice, there may be a further increase of d_{z^2}-stabilization by approaching positive charges. It is, in fact, a continuous process from lifting two opposite negative charges from the octahedral complex to infinity, till bringing back two positive charges on the same way. In this picture, the "free" square planar complex is only an intermediate form in the series

octahedral - negative tetragonal - square planar - positive tetragonal

coordination. Thus, under strong field conditions, the most stable stacking of square planar complexes may be the columnar type as in the Magnus' green salt and in the vic-dioximes.

Since the work of GODYCKI and RUNDLE (1) about the structure of Ni-dimethylglyoxime, there was much discussion about a metal-to-metal bond in these compounds (2), which often display unusual optical properties (3,4). We prefer to regard this as a second question, not as a necessary condition for columnar stacking.

As MILLER (2) has pointed out, the Magnus salt structure is even favoured electrostatically, with alternating cations and anions

$$[(NH_3)_2 \, Pt \, (NH_3)_2]^{2+}$$

$$[Cl_2 \, Pt \, Cl_2]^{2-}$$

at a distance of 3,21 Å (5,6).

In the vic-dioximes of Ni, Pd and Pt and in the α-N-methylsalicyl-aldimin complexes, the columnar type of stacking is the one in which the Van der Waals and ligand field stabilization energies combine to the lowest sum. In the "square planar" structures of K_2PdCl_4 and K_2PtCl_4 there is still some sort of columnar stacking (7,8), but the Me-Me distance is very long, 4,1 Å, since the Coulomb repulsion of the chlorine atoms is large, and the Δ value is relatively small (9). On the other hand, the cyanide complexes of Ni, Pd, and Pt show distances of about 3,3 Å (10,11). Here, the less localized negative charge on the smaller ligands diminishes the Coulomb repulsion, whereas the ligand field splitting effect of cyanide is very strong (9).

If the ligands remain small, and their charges are more dispersed, as in oxalate, for example, the columnar stacking distance can drop to 3,04 Å, as in $CaPt(C_2O_4)_2 \cdot 4\,H_2O$ (12). Thus, the controling factor for columnar stacking seems to be the ligand-ligand interaction more than ligand field splitting, providing the latter exceeds the "square-planar limit".

The shorter the metal-to-metal distance becomes, the more the d_{z^2}-electrons of different central atoms will disturb each other and raise the question for another distribution of orbitals and charges. Little or no bond will be produced by simply combining all d_{z^2}-orbitals in the crystal to a band (2). But if, by careful oxidation, some electrons could be drawn off from this band, a bonding effect should result, leading to shorter distances and other physical effects. This can be observed in the violet crystals with analytical formula $K_{1,6}Pt(C_2O_4)_2 \cdot 2,5\,H_2O$ and a Pt-Pt distance of only 2,76 Å (12). The oxidation state of this compound is now well established and can be reached from the (2+) as well as the (4+) state. Similarly, by oxidation of $K_2Pt(CN)_4$, compounds like $K_2Pt(CN)_4 \cdot Cl_{0,2}$ with 2,83 Å as stacking distance, and Cl^- (or Br^-, etc.) bound in cavities, are found (12). No distinction can be made about different sorts of Pt atoms in the unit cells. Therefore, the additional charge must be smeared out in the solid.

It is clear, however, that too much oxidation will favour disproportionation in the normal Pt(2+) and Pt(4+) compounds, the latter tending to negative octahedral coordination. The same occurs in the reversible solutions. But at least in the case of the dioxalato complex, there are deeply colored polymeric species at high concentrations, as can be shown by osmotic, conductivity, pH, and spectral measurements. The equilibrium is shifted towards polymerization by addition of H^+-ions, while protons are generated by the depolymerization process, according to the overall equation

$$[Pt(C_2O_4)_2]_x^{1,6\ x-} + 0,4 \times H_2O \rightleftharpoons$$
$$0,8 \times [Pt(C_2O_4)_2]^{2-} + 0,2 \times [Pt(C_2O_4)_2(OH)_2]^{2-} + 0,4 \times H^+$$

The oxalato complexes form true equilibria of this sort with variable x in water. With the cyanides, we found kinetic intermediates only.

References

1. GODYCKI, L.E., and RUNDLE, R.E., Acta Cryst. 6, 487 (1953).
2. MILLER, J.R., in "Advances of Inorganic and Radiochemistry", Vol. 4, p. 133-195 (1962).
3. YAMADA, S., and TSUCHIDA, R., J. Am. Chem. Soc., 75, 6351 (1953).
4. YAMADA, S., J. Am. Chem. Soc., 73, 1579 (1951).
5. COX, E.G., PINKARD, F.W., WARDLAW, W., and PRESTON, G.H., J. Chem. Soc. (London), 1932, 2527.
6. ATOJI, M., RICHARDSON, J.W., and RUNDLE, R.E., J. Am. Chem. Soc., 79, 3017 (1957).
7. DICKINSON, G.E., J. Am. Chem. Soc., 44, 2404 (1922).
8. THEILACKER, W., Z. anorg. allg. Chemie, 234, 161 (1937).
9. GRAY, H.B., and BALLHAUSEN, C.J., J. Am. Chem. Soc., 85, 260 (1963).
10. BRASSEUR, H., and DE RASSENFOSSE, A., Bull. Soc. franc. Mineral., 61, 129 (1938).
11. MONFORT, F., Bull. Soc. roy. Sci. Liege, 11, 567 (1942).
12. KROGMANN, K., to be published.

2 B 1

MACROCYCLIC COMPLEXES OF GERMANIUM.

J.N. Esposito, R. Rafaeloff, A.J. Starshak,
L.E. Sutton and M.E. Kenney
Case Institute of Technology, Cleveland 6, Ohio, U.S.A.

Recently ELVIDGE and LINSTEAD (1) reported the macrocyclic, tetra-dentate ligand, hpH_2, and several of its divalent metal complexes. A structure determination has been carried out on the non-planar nickel complex by SPEAKMAN (2).

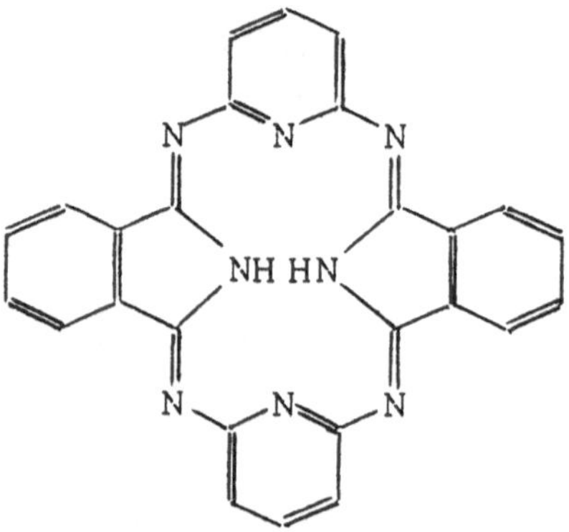

Hemiporphyrazine, hpH_2

Now a series of tetravalent germanium complexes of this ligand has been synthesized and at the same time the series of germanium phthalocyanines (3, 4) has been extended. Accordingly comparisons between the two series have been made possible and these enable the chemistry of octahedral germanium to be more clearly delineated.

The metal free ligand, hpH_2, was prepared by the reaction of phthalo-nitrile with 2, 6-diaminopyridine in 1-chloronaphthalene (5). Treatment of the free ligand with $GeCl_4$ in quinoline gave $hpGeCl_2$ in good yield. The $hpGeCl_2$ was converted into $hpGe(OH)_2$ by hydrolysis with pyridine and aqueous ammonia and this in turn was converted to $hpGeF_2$ by aqueous HF in pyridine. In each case satisfactory elemental analyses were obtained on pure samples.

The extension of the studies on the germanium phthalocyanines has included work on the dihalides. Reaction of phthalocyanine (PcH_2) with $GeCl_4$ in quinoline by the method already reported was used to prepare $PcGeCl_2$. The difluoride was synthesized by treating $PcGe(OH)_2$ (made by hydrolyzing $PcGeCl_2$)

with aqueous HF. The dibromide and diiodide, $PcGeBr_2$ and $PcGeI_2$, were made by reacting $PcGe(OH)_2$ with Et_3SiBr and Et_3SiI respectively. It is of interest that the diiodide can be formed because it shows that germanium can accomodate at least two iodines in stable octahedral coordination in spite of the fact that the iodines are large and of low electronegativity.

Dipseudohalides, $PcGe(NCO)_2$, $PcGe(NCS)_2$ and $PcGe(NCSe)_2$ were also prepared. They were synthesized by the action of the corresponding silver pseudohalide on $PcGeCl_2$ or $PcGeI_2$. It seems probable that in each case the iso or N bonded isomers were obtained. The pseudohalides thus constitute examples of germanium octahedrally bonded to six nitrogen atoms.

A series of germanium alkoxides was obtained by reacting $PcGe(OH)_2$ with the corresponding alcohols. These include $PcGe(OC_2H_5)_2$, $PcGe(OC_4H_9)_2$ and $PcGe(OC_8H_{17})_2$. Similarly the diacetate, $PcGe(OAc)_2$ was prepared from $PcGe(OH)_2$ and acetic acid. These and the previously reported siloxides and phenoxides illustrate the variety obtainable in germanium phthalocyanines having trans oxygen atoms.

Attempts to produce germanium phthalocyanines containing trans sulfur atoms by standard methods were unsuccessful. In view of the existence of $PcGeI_2$ this behaviour cannot be steric in origin and must be attributed to changes in bonding in going from tetrahedral to octahedral germanium, perhaps to a reduction in the importance of π-bonding.

Comparison of the IR spectra of the germanium hemiporphyrazine and phthalocyanine complexes made possible clear identification of the germanium halogen and oxygen absorptions. In the phthalocyanine series strong bands which were not attributable to the ligand were found at 647, 607, 313, 227 cm^{-1} in $PcGe(OH)_2$, $PcGeF_2$, $PcGeCl_2$ and $PcGeBr_2$ respectively. No similar absorption was found for $PcGeI_2$ above 220 cm^{-1}. Analogous strong non-ligand bands were found in the spectra of $hpGe(OH)_2$, $hpGeF_2$ and $hpGeCl_2$ at similar but somewhat higher frequencies. These strong non-ligand bands were assigned to the asymmetric Ge-O or Ge-X stretching modes in each case. The fact that the absorptions appeared at higher frequencies in the hemiporphyrazines is of interest and may indicate that stronger bonds are formed in this case.

References

1. ELVIDGE, J.A., and LINSTEAD, R.P., J.Chem.Soc. (London), 1952, 5008.

2. SPEAKMAN, J.C., Acta.Cryst., 6, 784 (1953).

3. JOYNER, R.D., and KENNEY, M.E., J.Am.Chem.Soc., 82, 5790 (1960).

4. JOYNER, R.D., LINCK, R.G., ESPOSITO, J.N., and KENNEY, M.E., J.Inorg.Nucl.Chem., 24, 299 (1962).

5. CAMPBELL, J.B., U.S. 2,765,308.

2 B 2

EXCHANGE OF MOLECULAR PARTS BETWEEN COORDINATION COMPOUNDS AT EQUILIBRIUM.

John R. Van Wazer

Monsanto Chemical Company, St. Louis/Miss., U.S.A.

Because of the availability of modern analytical and computative methods, particularly nuclear magnetic resonance and high-speed computers, it is now possible to investigate chemical systems consisting of various sizes and shapes of molecules at equilibrium with each other. It is common in inorganic chemistry for molecules to be sufficiently labile so that during the standard preparative laboratory operations, an equilibrium mixture is obtained via scrambling reactions. In this kind of reaction, the total number of bonds of any type remains constant but the atoms are rearranged. Thus, to the extent that bond energies are additive, scrambling reactions should have zero enthalpy and hence be entropy controlled.

In our laboratory, an extensive study is underway on the scrambling reactions of those molecules the backbones of which contain elements representative of different parts of the Periodic Table. In this work, two types of systems have been investigated. One type consists of the scrambling of two kinds of monofunctional substituents on a given central atom. These systems will be exemplified in the talk by equilibrium data on the groups of compounds $OVCl_i(OCH_3)_{3-i}$ and $OVCl_i[OSi(CH_3)_3]_{3-i}$, where $i = 0, 1, 2,$ or 3. In the other more interesting type of reaction, scrambling on a given central atom occurs between a bridging bifunctional substituent (such as -O- or -N(CH$_3$)-) and a monofunctional substituent. This type of work will be exemplified by studies on several new families of arsenic compounds (1). Thus the polyarsenous oxychlorides making up the system $AsCl_3$-As_2O_3 and their fluorine analogs making up the system AsF_3-As_2O_3 will be discussed briefly, along with the polyarsenous N-methylimidochlorides, making up the $AsCl_3$-$As_2(NCH_3)_3$ system of compounds.

It has been found that the deviations from random scrambling of substituents are generally attributable to nonadditivity of bond energies; i.e. nonzero enthalpies of the scrambling reactions. In the case of complicated molecules resulting from scrambling of bifunctional bridging groups with monofunctional substituents, consideration of the enthalpy involved in exchanging a monofunctional substituent for any bridging atom or vice versa is not sufficient to explain the observed equilibrium distribution of molecular species so that it is necessary to consider higher-order effects. This has been done by including contributions to the free energy of scrambling due to the arrangement of atoms beyond the particular bridging atom in the molecule. As expected, such effects are found to attenuate rapidly with increasing distance from a given position in a molecule.

To a first approximation, the deviations from randomness in the scrambling of a given pair of substituents on a central atom is independent of the central atom and of whether or not either or both of the substituents are polyfunctional and hence act as bridging atoms or groups. This finding, which has a quantum-mechanical explanation, represents a major theorem of inorganic structural chemistry - the theorem which determines the kind(s) of molecules obtained as a single-phase fluid from a given set of reaction ingredients and the size distribution of these molecules.

References

1. Van WAZER, J.R., MOEDRITZER, K., and MATULA, D.W., J.Am. Chem.Soc., 86, 807 (1964).
2. Van WAZER, J.R., and MAIER, L., J.Am.Chem.Soc., 86, 811 (1964).

2 B 3

ZUR KRISTALLCHEMIE VON DIMETATETRAFLUOROARSENATEN; EINIGE NEUERE ERGEBNISSE DER RÖNTGENSTRUKTURANALYSE.

H. Dunken und W. Haase

Institut für Physikalische Chemie der Universität Jena, Germany.

Bei der thermischen Behandlung von Alkalipentafluoromonohydroxo-arsenaten $M[AsF_5OH]$ im Vakuum erhielten KOLDITZ und Mitarbeiter (1,2,3) unter HF-Abspaltung Reaktionsprodukte, aus denen sich sowohl kettenförmig als auch ringförmig vernetzte Polytetrafluoroarsenate der Formel $(MAsF_4O)_n$ iso-lieren ließen. Während von den kettenförmigen Produkten keine Einkristalle zu erhalten waren, kristallisierten die ringförmigen Produkte in schönen, bemerkenswert luftbeständigen Blättchen. Von diesen Ringverbindungen mit den Kationen $M = NH_4^+$, K^+ und Rb^+ wurden Röntgenstrukturuntersuchungen durchgeführt (4), die zu dem Ergebnis gelangten, daß es sich bei dieser Substanzklasse um Dimetatetrafluoroarsenate der Formel $M_2[As_2F_8O_2]$ handelt. Die Strukturuntersuchung wurde durch die Anwendbarkeit der Methode des isotypen Ersatzes und der Schweratomtechnik wesentlich erleichtert. Alle drei Verbindungen kristallisieren in der monoklinen Raumgruppe $P2_1/n - C_{2h}^5$. Die Abmessungen der Elementarzellen sind:

	a	b	c	∢ β
$(NH_4)_2[As_2F_8O_2]$	5,09	6,26	13,89 Å	103,9°
$K_2[As_2F_8O_2]$	5,19	5,83	13,53 Å	92,5$_5$°
$Rb_2[As_2F_8O_2]$	5,33	5,98	13,94 Å	92,4$_5$°

Die Anzahl n der Moleküle $M_2[As_2F_8O_2]$ in der Elementarzelle beträgt 2.

Zur Erweiterung der Erkenntnisse über die Kristall-und Molekülstruktur dieser Verbindungsklasse wurden die Röntgenstrukturuntersuchungen auf Verbindungen mit großvolumigen Kationen - M = Pyridinium bzw. Nitron - ausgedehnt (5, 6). Die Pyridiniumverbindung $(C_5H_6N)_2[As_2F_8O_2]$ kristallisiert in der triklinen Raumgruppe P$\bar{1}$ mit

$$a = 8,32 \qquad b = 7,33 \qquad c = 8,38 \text{ Å}$$

$$\alpha = 93° \qquad \beta = 130° \qquad \gamma = 91°$$

Die Solvat-Acetonitrilmoleküle enthaltende Nitronverbindung $(H\text{-Nitron})_2[As_2F_8O_2] \cdot 2 \, CH_3CN$ kristallisiert in der monoklinen Raumgruppe $P2_{1/n}\text{-}C_{2h}^5$ mit

$$a = 17,29 \qquad b = 7,44 \qquad c = 18,49 \text{ Å} \qquad \beta = 93,1°$$

Die Untersuchungen an den beiden letztgenannten Verbindungen ergaben übereinstimmend mit den Ergebnissen bei den vorstehend beschriebenen Alkaliverbindungen das Vorliegen eines komplexen Dimetatetrafluoroarsenat-Anions. Dieses konnte aus Elektronendichteprojektionen, bei der Nitronverbindung nur aus Pattersonprojektionen, geschlossen werden.

Die bei den Alkaliverbindungen sehr stark ausgeprägte Schichtenstruktur in der Ebene (001) geht aus dem Strukturmodell zwanglos hervor. Auch alle übrigen, bei den Verbindungen auftretenden Spaltbarkeiten finden eine plausible Erklärung in der strukturellen Anordnung. Bei allen Verbindungen besetzen die Mittelpunkte der Molekülanionen die durch die Raumgruppen-symmetrie geforderten Symmetriezentren. Danach enthält der aus zwei $(AsF_4O)^-$-Einheiten - mit der Verbindungslinie zweier Sauerstoffatome als gemeinsamer Kante - gebildete Dioctaeder einen vollkommen planaren Ring, der von zwei Arsen-und zwei Sauerstoffatomen gebildet wird. In Abb. 1 ist ein Modell des Anions dargestellt. Jedes Arsenatom ist oktaedrisch umgeben von 4 Fluor-und 2 Sauerstoffatomen entsprechend einer oktaedrischen sp³d²-Hybridisierung. Interessant ist besonders der geringe Bindungswinkel am Sauerstoff von 94,5°. Danach wird die Bindung vom Sauerstoff fast ausschließlich

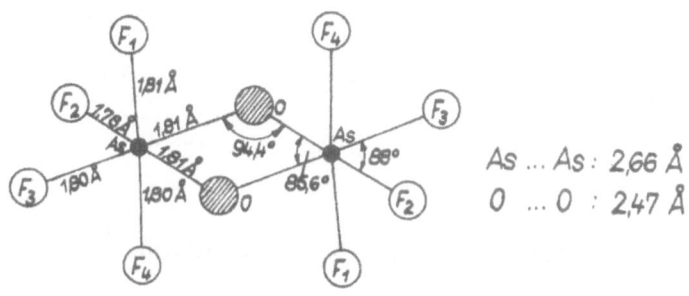

Abb. 1 Strukturmodell des $[As_2F_8O_2]^{2-}$-Dioktaeders

As ... As : 2,66 Å
O ... O : 2,47 Å

unter Beteiligung der p_x-, p_y-Orbitale gebildet, wobei nur ein sehr geringer s-Beitrag beigemischt sein dürfte. Entsprechend dem Winkel am Sauerstoff muß der Bindungswinkel ⊰ OAsO 85,6° betragen. Die Verzerrung der Oktaedersymmetrie am Arsen ist gering. Alle rechten Bindungswinkel liegen zwischen 85-95°. Die durchschnittlichen Abstände As-F betragen übereinstimmend bei allen Verbindungen 1,79-1,80 Å. Die Abstände As-O sind innerhalb der Meßgenauigkeit gleich und betragen im Mittel 1,81 Å.

Der planare Ring ist etwas verzerrt, bedingt durch die enge Nachbarschaft zweier Oktaedergruppierungen. Hier beträgt der Abstand As...As = 2,66 Å und der Abstand O...O = 2,46 Å.

Die Länge der Oktaederkanten beträgt im Durchschnitt 2,55 Å. Sie stimmt damit gut überein mit den aus anderen oktaedrischen Fluorkomplexen bekannten Daten. Der nächste Abstand zwischen zwei Fluoratomen am Dioktaeder beträgt 2,85 Å, er ist damit nur etwas größer als die Summe der Ionenradien (2,72 Å).

Den Dioktaeder könnte man grob angenähert als zwei sich intensiv durchdringende Kugeln auffassen. Dadurch werden die empfindlichen As-O-Bindungen gut abgeschirmt, und die bemerkenswerte Stabilität des Komplexes, insbesondere gegenüber hydrolytischen Einflüssen, wird verständlich.

Die Kationen bilden um sich Koordinationspolyeder mit 12 nächsten Fluor- bzw. Sauerstoffnachbarn. Die durchschnittlichen Abstände stimmen auch hier sehr gut mit den aus der Summe der Ionenradien für 12er-Koordination berechenbaren Werten überein.

Literatur

1. KOLDITZ, L., und NUSSBÜCKER, B., Z. anorg. allgem. Chem., 312, 299 (1961).
2. KOLDITZ, L., und RENNO, D., Z. anorg. allgem. Chem., 315, 46 (1962).

3. KOLDITZ, L., Z.Chem., 2, 186 (1962).
4. DUNKEN, H., und HAASE, W., Z.Chem., 3, 433 (1963).
5. DUNKEN, H., und HAASE, W., Z.Chem., im Druck.
6. DUNKEN, H., und HAASE, W., Z.Chem., im Druck.

3 B 1

SOME NEW ADDUCTS OF IODOSILANE.

B.J. Aylett and R.A. Sinclair
Chemistry Department, Westfield College, Hampstead,
London, N.W.3., U.K.

Monohalosilanes are known to combine very readily with tertiary amines to yield 1:1 adducts (1). In the case of iodosilane, a 1:2 adduct with tri-methylamine has also been reported, made under forcing conditions (2).

In the present study, pyridine, 2,4,6-trimethylpyridine (collidine), and 2-hexylpyridine all reacted readily at low temperatures with iodosilane, either alone or in an inert solvent. With excess base, a 1:2 adduct was formed in each case, with almost no side reactions; with excess iodosilane, some hydrogen and silane were produced, but no evidence for a 1:1 adduct was found.

The white solid pyridine adduct was stable in vacuo at room temperature, and dissolved in excess pyridine or in acetonitrile to give moderately stable solutions. The specific conductivity of these solutions has been compared with that of tetrapropylammonium iodide in the same solvents and it is clear that the adduct is considerably dissociated as $[SiH_3py_2]^+I^-$. In air, the adduct fumed and was partly oxidised to iodine, while it reacted slowly with water to yield:

$$SiH_3I, \ 2 \ py + (2+n) \ H_2O \longrightarrow SiO_2, \ n \ H_2O + 3 \ H_2 + pyH^+I^- + py$$

$$\{1\}$$

Pyrolysis at 100° or more led to two reactions: (a) disproportionation to give silane and adducts of diiodosilane, and (b) abstraction of HI from iodo-silane to yield pyH^+I^- and adducts of polymeric silicon hydrides.

A comparison of the pyridine (py) adduct with those of collidine (coll) and 2-hexylpyridine (hexpy) shows the following orders:

Volatility	coll	>	hexpy	>	py
Thermal stability	hexpy	>	py	>	coll
Reactivity	hexpy	>	py	>	coll

These orders are discussed in terms of the steric requirements of the bases and the adducts. It appears that the 2-hexylpyridine adduct is the most promising compound for use as a silylating agent.

The collidine adduct reacted with excess trimethylamine with complete displacement of collidine to give a trimethylamine adduct of iodosilane. The nature of this adduct is discussed in conjunction with that of similar previously-reported compounds.

Diglyme reacted readily with the hexylpyridine adduct at 40° to yield the theoretical amount of silane according to the equation:

$$2 \, [SiH_3I.(hexpy)_2] + (MeOCH_2CH_2)_2O \longrightarrow$$

$$SiH_4 + MeI + 4 \, hexpy + ISiH_2OCH_2CH_2OCH_2CH_2OMe \qquad \{2\}$$

The amount of iodomethane recovered suggested that some cleavage also occurred at the central oxygen of the polyether to yield $MeOCH_2CH_2I$ and $ISiH_2OCH_2CH_2OMe$. Dimethyl ether reacted even more readily with the hexyl-pyridine adduct to give, after one hour at 0°, silane, iodomethane, and tri-methoxysilane. This is consistent with an initial cleavage to yield methoxy-silane (3), which then disproportionates in a base-catalysed reaction (4).

As an example of the preparative use of these complexes, the reaction of phosphine with the hexylpyridine complex was studied. With no solvent, and with petroleum ether or excess hexylpyridine, no Si-P compounds were formed, but when dibutyl ether or, better, tributylamine was used as solvent, trisilylphosphine, $P(SiH_3)_3$, was produced in moderate yield.

$$3 \, [SiH_3I.(hexpy)_2] + PH_3 \longrightarrow P(SiH_3)_3 + 3 \, hexpyH^+I^- + 3 \, hexpy$$

$$\{3\}$$

The chelating base, 2,2'-dipyridyl, was found to react with iodosilane to form a yellow-brown 1:1 adduct, similar in properties to the pyridine adducts. Dipyridyl was displaced from the adduct by excess trimethylamine, especially when the reaction was carried out in acetonitrile solution.

The IR spectra of the dipyridyl adduct from $4000 - 200 \, cm^{-1}$, and of the pyridine adducts in the NaCl region have been recorded. These spectra are discussed, and suggestions are made concerning the nature of the adducts and the mechanism of some of their reactions.

References

1. EMELÉUS and MILLER, J.Chem.Soc. (London) 1939, 819.
2. AYLETT, EMELÉUS and MADDOCK, J.Inorg.Nucl.Chem., 1, 187 (1955).
3. STERNBACH and MacDIARMID, J.Am.Chem.Soc., 83, 3384 (1961).
4. MacDIARMID, J.Inorg.Nucl.Chem., 25, 1534 (1963).

3 B 2

COORDINATION CHEMISTRY OF ORGANOHETEROSILOXANES AND RELATED COMPOUNDS.

Hubert S c h m i d b a u r and Ingeborg R u i d i s c h
Institut für Anorganische Chemie der Universität Marburg, Germany.

(to be presented in German)

It is well known, that the structural unit $\geqslant Si-O-Si\leqslant$ in siloxane compounds has very weak donor properties and that only few and unstable coordination compounds have been reported. The now well established principle of $d_\pi p_\pi$ -multiple bonding is a reasonable explanation for this abnormal behaviour of the "silico-ethers". It was the purpose of our current investigations to examine the change of the donor properties of the bridging oxygen atoms on substitution of one silicon atom of the siloxane unit by certain hetero-atoms of group I-IV b, leading to the $\geqslant Si-O-X\leqslant$ hetero-siloxane structure (where X = hetero-atom).

Chemical and spectroscopic studies on germano-, stanno- and plumbo-siloxane model compounds $R_3Si-O-XR_3$, (X = Ge, Sn, Pb; R = organyl) have shown, that the bridging oxygen atoms therein are somewhat better, but still weak donors because of increasing multiple bonding in the $Si-O$ bond and increasing polarity in the $X-O$ bonds, following the sequence X = Si, Ge, Sn, Pb. On complex formation with strong acceptors all compounds undergo rapid heterolytic cleavage of the heterosiloxane bridges, e.g.:

$$2\ R_3Si-O-XR_3\ +\ 2\ AlCl_3\ \longrightarrow\ [R_3SiOAlCl_2]_2\ +\ 2\ R_3XCl$$

$$\{1\}$$

In all cases only the $X-O$ bonds are cleaved, whereas the $Si-O$ bonds remain intact in the course of the reaction. Analoguous organoheterogermoxanes $\geqslant Ge-O-X\leqslant$ in most cases appeared to be better donor molecules, the adducts of which, however, were even less stable to subsequent heterolytic cleavage and rearrangement than the related heterosiloxane derivatives. In contrast to their stannoxane analogues no coordination oligomers could be observed with the stannosiloxane and stannogermoxane compounds.

Heterosiloxanes of the group III elements Al and Ga, however, exist as coordination dimers throughout, having the following general formula:

$$\begin{array}{c} Z \diagdown \diagup Z \\ Al \\ R_3Si-O \diagup \diagdown O-SiR_3 \\ Al \\ Z \diagdown \diagup Z \end{array}$$

instead of $R_3SiOAlZ_2$.

(R = organyl, Z = halogen, hydrogen, organyl, siloxy etc.)

The dimeric structures are characterized by a planar $d_\pi p_\pi$-stabilysed four-membered ring structure

$$\diagup Si-O \begin{array}{c} \diagup Al \diagdown \\ \diagdown Al \diagup \end{array} O-Si \diagdown$$

of relatively high thermal and chemical stability. Some methods of preparation for these compounds are represented by the following equations:

$$2\ R_3Si-O-SiR_3 + 2\ AlZ_3 \longrightarrow 2\ R_3SiZ + [R_3SiOAlZ_2]_2 \qquad \{2\}$$

$$[R_3SiOAlZ_2]_2 + 4\ LiR \longrightarrow 4\ LiZ + [R_3SiOAlR_2]_2\ [I] \qquad \{3\}$$

$$[R_3SiOAlZ_2]_2 + 2\ LiAlH_4 \longrightarrow 2\ LiZ + 2\ AlH_3 + [R_3SiOAlZH]_2 \qquad \{4\}$$

$$[R_3SiOAlZ_2]_2 + 4\ LiAlH_4 \longrightarrow 4\ LiZ + 4\ AlH_3 + [R_3SiOAlH_2]_2 \qquad \{5\}$$

$$[R_3SiOGaZ_2]_2 + 4\ NaOSiR_3 \longrightarrow 4\ NaZ + [(R_3SiO)_3Ga]_2\ [II] \qquad \{6\}$$

Starting from the dimers [I] and [II] on addition of stoichiometric amounts of alkali triorganosilanolate salt-like heterosiliconate compounds are obtained according to the equations (R = CH$_3$):

$$\begin{array}{ccc} & SiR_3 & \\ & | & \\ R \diagdown \quad O \quad \diagup R & & SiR_3 \\ Al \quad Al & + 2\ R_3SiONa \longrightarrow & 2\ Na \cdots \begin{array}{c} O \diagdown \diagup R \\ Al \\ O \diagup \diagdown R \end{array} \qquad [III] \\ R \diagup \quad O \quad \diagdown R & & | \\ & | & & SiR_3 \\ & SiR_3 & & \{7\} \end{array}$$

$$\begin{array}{ccc} & SiR_3 & \\ & | & \\ R_3Si-O \diagdown \quad O \quad \diagup O-SiR_3 & & \left[R_3SiO \diagdown \diagup OSiR_3 \right] \\ Ga \quad Ga & + 2\ KOSiR_3 \longrightarrow 2\ K & Ga \\ R_3Si-O \diagup \quad O \quad \diagdown O-SiR_3 & & \left[R_3SiO \diagup \diagdown OSiR_3 \right] \\ & | & & [IV] \\ & SiR_3 & & \{8\} \end{array}$$

[III] , monomeric in benzene, m.p. 215°, and K(R$_3$SiOAlR$_2$OSiR$_3$), dimeric in benzene, m.p. 125°, contain the octamethyldisiloxalanat anion, an iso-ester of octamethyltrisiloxane. The anion of [IV] is isosteric to the branched dodecamethylgermanotetrasiloxane.

With few exceptions no heterosiloxanes of definite composition could be obtained from the group II elements Mg, Ca, Sr, and Ba. Trimethylsiloxy derivatives of all alkali metals, however, are now known and an examination of their physical and spectroscopic properties has shown, that they are coordination oligomers with multiple-coordinating oxygen and alkali atoms, the degree of association depending on the nature of the alkali metal and the solvent.

Pair-wise reaction of different alkali-trimethylsilanolates yields new coordination compounds of a similar type:

$$Na[Li(OSiR_3)_2] \quad m.p. \ 233\text{-}5^{o} \qquad K[Li(OSiR_3)_2] \quad m.p. \ 258\text{-}62^{o}$$

The chemical, IR- and NMR-spectroscopic properties of these novel compounds, which readily crystallise from inert organic solvents, were investigated.

NMR-spectroscopic studies of trimethylsilanol Me$_3$SiOH proved a similar principle of oligomerisation via strong hydrogen bonds instead of the alkali bridges. Trimethylgermanol could not be prepared.

When compared with the analoguous properties of trimethyltin fluoride Me$_3$SnF, (a high melting solid), the physical constants of trimethylfluorosilane Me$_3$SiF, (a highly volatile gas) prove, that only very weak intermolecular interactions are present in this compound. Trimethylfluorogermane Me$_3$GeF, however, appeared by its specific properties to undergo intermolecular interactions, too, and was shown by NMR-spectroscopy to undergo rapid fluorine exchange in the liquid state. This reaction, which probably involves the coordination intermediate

$$R_3Ge \overset{\displaystyle \diagup F \diagdown}{\underset{\displaystyle \diagdown F \diagup}{}} GeR_3$$

with pentacovalent germanium atoms, disappears on dilution with various solvents.

3 B 3

DARSTELLUNG UND KOMPLEXCHEMISCHES VERHALTEN VON SILYLAMIDEN VERSCHIEDENER ÜBERGANGSMETALLE.

Hans Bürger

Institut für Anorganische Chemie, Technische Hochschule Graz, Austria.

Ausgehend von Natrium-bis-(trimethylsilyl)-amid $NaN(SiR_3)_2$; $R = CH_3$ (1) wurden durch Reaktion mit verschiedenen Übergangsmetallhalogeniden in Äther, THF, Pyridin oder Petroläther die folgenden, destillierbaren oder sublimierbaren Übergangsmetall-disilylamide dargestellt:

TABELLE I

SILYLAMIDO-VERBINDUNGEN VON ÜBERGANGSMETALLEN

Verbindung	Farbe	Sdp.	Torr	Schmp.
$Cl_3TiN(SiR_3)_2$ (2)	orange	$104\text{-}105^0$	1	$75\text{-}77^0$
$(C_3H_7O)_3TiN(SiR_3)_2$ (2)	farblos	93^0	1	-78^0
$(C_3H_7O)_2VON(SiR_3)_2$ (3)	gelb	81^0	0,5	-15^0
$[(R_3Si)_2N]_2V(NSiR_3)(OSiR_3)$ (3)	olivgrün	130^0	0,5	68^0
$Cr[N(SiR_3)_2]$ (4)	grün	135^0	0,8	120^0
$Mn[N(SiR_3)_2]_2$ (4)	rosa	100^0	0,2	
$Fe[N(SiR_3)_2]$ (5)	grün	130^0	1*	135^0
$Co[N(SiR_3)_2]_2$ (5)	grün	101^0	0,6	73^0
$Ni[N(SiR_3)_2]_2$ (4)	rot	80^0	0,2	
$[CuN(SiR_3)_2]_2$ (4)	farblos	180^0	0,2*	
$Zn[N(SiR_3)_2]_2$	farblos	82^0	0,5	$12,5^0$
$Cd[N(SiR_3)_2]_2$	farblos	93^0	0,5	8^0
$Hg[N(SiR_3)_2]_2$	farblos	78^0	0,15	11^0

* sublimiert

Alle Verbindungen sind äußerst hydrolyseempfindlich, lösen sich leicht in organischen Lösungsmitteln und liegen darin monomer vor. Lediglich die Cu-Verbindung scheint wie die Alkali-Verbindungen (1) dimer zu sein.

Lewis-Basen lassen sich nur an $Cu[N(SiR_3)_2]_2$, $Ni[N(SiR_3)_2]_2$ und $ClCoN(SiR_3)_2$ (5) anlagern. Chlorhaltige Titan- und Vanadium-silylamide spalten hingegen bei der Einwirkung von Pyridin (Py) oder Chinolin zuerst R_3SiCl ab und addieren dann die Base:

$$Cl_3TiN(SiR_3)_2 + 2\ Py \longrightarrow ClSiR_3 + Cl_2TiNSiR_3 \cdot 2\ Py\ (2)$$

sowie

$$Cl_2VON(SiR_3)_2 + VOCl_3 + 4\ Py \longrightarrow 2\ ClSiR_3 + Cl_3V_2NO_2 \cdot 4\ Py\ (3)$$

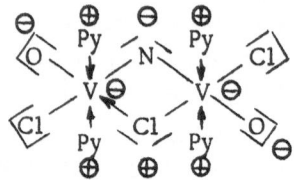

Alle anderen Verbindungen können keine weiteren Liganden an das Zentralatom anlagern. Hierfür ist neben sterischen Gründen die ausgeprägte Fähigkeit der Disilylamino-Gruppe, über die kovalente Einfachbindung hinaus noch partielle Doppelbindungen auszubilden, verantwortlich. Dabei teilt das Übergangsmetall Me seinen Anspruch auf das freie Elektronenpaar des Stickstoffs mit den beiden Si-Atomen

IR-, Raman- und Protonenresonanz-Untersuchungen zeigen, daß z.B. für Me = Zn, Cd und Hg die Bindungsverhältnisse eher durch (a), für Co, Fe, Ti und besonders für V besser durch (b) wiedergegeben werden.

So steigen ν_{as} und ν_s SiNSi bei $Zn[N(SiR_3)_2]_2$ und den entsprechenden Cd- und Hg-Verbindungen auf 990 (ν_{as}) bzw. 630 cm^{-1} (ν_s), während die Me-N-Valenzschwingung auf ca. 400 cm^{-1} absinkt; die Protonensignale liegen bei 0,03 bis 0,1 ppm Niedrigfeld gegen TMS. Hingegen sinkt ν_{as} SiNSi in V-, Fe- und Ti-Verbindungen auf 900 - 915 cm^{-1}, während ν_{as} Me-N (bei gleichzeitiger Kopplung mit ν SiN) auf 800 cm^{-1} steigt; die Protonensignale (V- und Ti-Verbindungen) zeigen chemische Verschiebungen von 0,22 - 0,35 ppm. Zum Vergleich liegen diese Werte für $(R_3Si)_3N$ (gleiche Anteile von (a) und (b)) bei 916 cm^{-1} (ν_{as} SiNSi) (6) bzw. 0,18 ppm (7).

Bei ausgesprochen elektropositiven Elementen (Me = Na, K) geht Form (a) unter Streckung des SiNSi-Bindungswinkels in eine salzartige Struktur (c)

$$R_3Si \diagdown \atop R_3Si \diagup N-Me \quad \longleftrightarrow \quad R_3Si \overset{\ominus}{\underset{\ominus}{\overset{\|}{\underset{\|}{N^{\oplus}}}}} R_3Si \quad + Me^+$$

(a)   (c)

über,aus der heraus die Verbindungen als ausgesprochene Lewis-Säuren reagieren. Eine solche Struktur ließ sich jedoch bei Übergangsmetall-Verbindungen, in denen der Stickstoff stets sp^2-hybridisiert vorliegt, nicht beobachten.

Im Gegensatz zu den Amiden selbst zeigen die Silylamide der Übergangs-metalle eine unerwartet geringe Neigung, Lewis-Basen anzulagern. Vielmehr kann das Zentralatom innere Elektronenbahnen durch zusätzliche Wechselwirkungen mit dem Elektronenpaar des kovalent gebundenen Amid-Stickstoffs auf-füllen.

Literatur

1. WANNAGAT, U., und NIEDERPRÜM, H., Chem.Ber., 94, 1540 (1961).

2. BÜRGER, H., und WANNAGAT, U., Mh.Chem., 94, 761 (1963).

3. BÜRGER, H., SMREKAR, O., und WANNAGAT, U., Mh.Chem., 95, 282 (1964).

4. BÜRGER, H., und WANNAGAT, U., Mh.Chem. (im Druck).

5. BÜRGER, H., und WANNAGAT, U., Mh.Chem., 94, 1007 (1963).

6. GOUBEAU, J., und JIMENEZ-BARBERA, J., Z.anorg.allg.Chem., 303, 217 (1960).

7. SCHMIDBAUR, H., J.Am.Chem.Soc., 85, 2336 (1963).

3 B 4

ÜBER ORGANOFLUOROSILIKATE.

R. Müller

Institut für Silikon- und Fluorkarbon-Chemie, Radebeul-Dresden, Germany.

Leitet man in Wasser ziemlich schwer verseifbares Methyltrifluorsilan (Sdp. -30,2°) in eine möglichst konzentrierte Lösung von Ammonium- oder Kaliumfluorid (1), so werden Komplexsalze der Organopentafluorokieselsäure von der Art $(NH_4)_2[CH_3SiF_5]$ gebildet. In ähnlicher Weise konnten $K_2[CH_3SiF_5]$, $K_2[(CH_2=CH)SiF_5]$, $K_2[C_2H_5SiF_5]$ oder $(NH_4)_2[C_6H_5SiF_5]$ dargestellt werden. Auch die Natriumsalze können - wenn auch wegen der schweren Löslichkeit des NaF nicht unmittelbar - gewonnen werden. Mit ebenfalls schwer hydro-lysierbaren $(CH_3)_2SiF_2$ (Sdp. $\pm2,7°$) bzw. $(CH_3)_3SiF$ (Sdp. +16,4°) waren dagegen keine entsprechenden Komplexe darzustellen. Beide gingen unverändert gasförmig z.B. durch Ammoniumfluoridlösung hindurch. Man kann infolgedessen diese beiden Fluorsilane von dem ersten über dessen Komplex trennen. Die Salze des CH_3SiF_3 sind in Wasser ziemlich gut löslich. Die Ammoniumsalze fallen erst bei genügend hoher Konzentration aus der wäßrigen Lösung des Fluorides aus und können abfiltriert werden. Sie stellen Gegenstücke zum Hexafluorosilikat K_2SiF_6 usw. dar; im Anion ist offenbar wie bei diesem sechsbindiges Silizium enthalten. Da man Chlorsilane nicht nur mit wäßriger Flußsäure, sondern auch mit wäßrigen Lösungen von KF oder NH_4F fluorieren und in Fluorsilane überführen kann (1), bilden sich obige Komplexe auch aus den entsprechenden Chlorsilanen.

$$CH_3SiCl_3 + 5\ NH_4F \xrightarrow{\ H_2O\ } (NH_4)_2[CH_3SiF_5] + 3\ NH_4Cl \qquad \{1\}$$

Das Kaliumsalz kann man auch bei Zimmertemperatur durch Überleiten von Methyltrifluorsilan über festes KF erhalten

$$2\ KF + CH_3SiF_3 \xrightarrow{\ 20^0\ } K_2[CH_3SiF_5] \qquad \{2\}$$

Aus den Lösungen konnte CH_3SiF_3 entweder durch Erhitzen zum Sieden unter teilweiser oder durch Ansäuern ohne nennenswerte Hydrolyse rückgewonnen werden. Auf Grund der in diesen Fällen nur dem trifunktionellen Fluorsilan eigenen Fähigkeit zur Komplexbildung konnte ein Verfahren zur Aufarbeitung der bei der direkten Synthese anfallenden Methylchlorsilane über deren Fluorsilane auf "nassem" Wege ausgearbeitet werden (1). Man leitet das Gemisch der Methylchlorsilane in wäßrige Flußsäure. Die danach daraus entweichenden Methylfluorsilane geben in Ammonfluoridlösung $(NH_4)_2[CH_3SiF_5]$. $(CH_3)_2SiF_2$ und $(CH_3)_3SiF$ bleiben unverändert gasförmig. Das Komplexsalz wird durch HF in NH_4F und CH_3SiF_3 gespalten. Die Komplexe können sowohl aus polaren als auch unpolaren Lösungsmitteln (Aceton, Acetonitril, Benzol oder Petroläther) gewonnen werden.

Es besteht Aussicht, viele andere Verbindungen, die $-SiCl_3$ -Gruppen enthalten, nach Fluorierung in entsprechende Komplexe überzuführen. Z.B. konnte die Verbindung $(NH_4)_2SiF_5CHClSiF_5(NH_4)_2$ dargestellt werden.

Beim Versuch, entsprechende quarternäre Ammoniumsalze durch Einleiten von CH_3SiF_3 in wäßrige oder alkoholische Lösungen von Tetraäthylammoniumfluorid herzustellen, entstand das $(C_2H_5)_4N[CH_3SiF_4]$, also ein Salz der bisher ebenfalls unbekannten Organotetrafluorokieselsäure.

Einfache Komplexsalze z.B. $(NH_4)_2[CH_2=CHSiF_5]$ bilden sich nur bei Einhaltung ganz bestimmter Versuchsbedingungen. Oftmals fallen Doppelsalze z.B. $(NH_4)_2[F_5SiCH_2-CH=CH-CH_2SiF_5](NH_4)_2 \cdot 2\ NH_4F$ an.

In manchen Fällen sind die Komplexe allerdings so unbeständig, daß sie sich praktisch nicht darstellen lassen. So zerfallen in H_2O die des $HSiF_3$ und $CF_2=CFSiF_3$ fast augenblicklich unter Entwicklung von H_2 und $CF_2=CFH$. Das $CH_2=CH-CH_2SiF_3$ spaltet sich schon mit wäßriger Flußsäure teilweise nach der Gleichung:

$$CH_2=CH-CH_2SiF_3 + 3\ HF \longrightarrow CH_2=CH-CH_3 + H_2SiF_6 \quad \{3\}$$

Die erhaltenen Komplexe vermögen als Carburierungsmittel verwendet zu werden. Z.B. wird in wäßriger Lösung Silbernitrat in Silberphenyl verwandelt, das durch Anlagerung von Silbernitrat so stabilisiert ist, daß es sich eine Zeit lang als zitronengelbe bzw. orangefarbene Verbindung hält, schließlich aber unter Abscheidung von metallischem Silber zerfällt. Die Verbindung $(C_6H_5Ag)_2 \cdot AgNO_3$ wurde erstmalig von KRAUSE und SCHMITZ (2) aus Triphenyläthylblei bzw. Triphenyläthylzinn mit Silbernitrat in wasserfreiem Alkohol als prachtvoll leuchtend gelbes, auf dunkler Unterlage fluoreszierendes Pulver dargestellt. Mit dem Salz des $CH_2=CHSiF_3$ entsteht vorübergehend eine tiefblaue Färbung; schließlich scheidet sich daraus Silber in Form eines Spiegels ab.

Man kann die Vorgänge für die Vinylverbindung in folgender Weise deuten:

$$AgNO_3 + NH_4F \longrightarrow Ag^+ + [:F]^- + NH_4NO_3 \quad \{4\}$$

$$(NH_4)_2CH_2=CHSiF_5 \longrightarrow [(NH_4)_2CH_2=CHSiF_5] \longrightarrow$$
$$\uparrow$$
$$[:F]^-$$

$$(NH_4)_2SiF_6 + [CH_2=CH:]^- \quad \{5\}$$

Es ist möglich, daß dabei siebenbindiges Silizium entsteht (3) und der Komplex dadurch so unbeständig wird, daß die Vinylgruppe abgespalten wird:

$$[CH_2=CH:]^- + Ag^+ \longrightarrow Ag:CH=CH_2(\cdot AgNO_3) \quad \{6\}$$
$$\text{(blau)}$$

$$Ag:CH=CH_2 \longrightarrow Ag\cdot + \cdot CH=CH_2 \quad \{7\}$$
$$\text{(schwarz)}$$

$$2 \cdot CH=CH_2 \longrightarrow CH_2=CH-CH=CH_2 \quad \{8\}$$

Butadien entsteht aber nicht allein, in den meisten Fällen überwiegt in Abhängigkeit von der Silberionenkonzentration in den untersuchten Abgasen Äthylen, so daß offenbar auch folgende Umsetzung stattfindet:

$$Ag:CH=CH_2 + H_2O \longrightarrow CH_2=CH_2 + OH^- + Ag^+ \qquad \{9\}$$

$$[CH_2=CHSiF_5]^{2-} + Ag^+ \longrightarrow CH_2=CHAg + [SiF_5OH]^{2-} \qquad \{10\}$$
$$\uparrow OH^-$$

Aus $(NH_4)_2[CH_3SiF_5]$ bildet sich in entsprechender Weise C_2H_6 und CH_4. Bei Radikalresten mit π-Elektronen z.B. $CH_2=CH-$ oder C_6H_5- wäre u.U. auch an einen gleichzeitigen elektrophilen Angriff der Silberionen auf die Alkenylgruppe usw. zu denken, wie dies von MILLER (4) für den Angriff von Hg^{2+} auf Perfluorpropen angenommen wird.

Mit Hilfe dieser als Carburierungsmittel wirkenden Komplexe konnten auch quecksilberorganische Verbindungen in wäßriger Lösung dargestellt werden:

$$HgCl_2 + (NH_4)_2[CH_3SiF_5] \xrightarrow{H_2O} (NH_4)_2SiF_5Cl + CH_3HgCl \qquad \{11\}$$

Das Methylquecksilberchlorid fällt aus der Lösung aus. Auf ähnliche Weise wurden gewonnen:

$$CH_2=CHHgCl, \quad C_6H_5HgNO_3, \quad C_6H_5HgCl.$$

Mit Brom finden folgende Umsetzungen statt:

$$K_2[CH_3SiF_5] + Br_2 \xrightarrow{H_2O} 2\,KF + CH_3Br + BrSiF_3 \qquad \{12\}$$

$$BrSiF_3 + 3\,KF \longrightarrow K_2SiF_6 + KBr \qquad \{13\}$$

Literatur

1. MÜLLER, R., und DATHE, Chr., J.Prakt.Chem. [4], 22, 232 (1963).
2. KRAUSE, E., und SCHMITZ, M., Ber., 52, 2159 (1919);
   SEMERANO, G., und RICCOBONI, L., Ber., 74, 1089 (1941);
   GILMAN, H., und WOODS, L.A., J.Am.Chem.Soc., 65, 435 (1943).
3. HOARD, J.L., und WILLIAMS, M.B., J.Am.Chem.Soc., 64, 633 (1942);
   PEARSON, R.G., EDGINGTON, D.N., und BASOLO, F., J.Am.Chem. Soc., 84, 3233 (1962).
4. MILLER, Jr., W.T., FREEDMAN, B.M., FRIED, J.H., und KOCH, H.F., J.Am.Chem.Soc., 83, 4105 (1961).

BORON HYDRIDE FRAGMENTS AS COORDINATING LIGANDS.

R.W. Parry, L.J. Malone and K.W. Morse
Department of Chemistry, University of Michigan,
Ann Arbor, Michigan, U.S.A.

The diammoniate of borane carbonyl was first prepared by BURG and SCHLESINGER in 1937 (1). Using chemical and spectroscopic methods CARTER and PARRY (2) showed that the closely related bis-methylamine adduct of borane carbonyl is a methyl ammonium salt of a new N-methyl borano-carbamate anion:

$$[CH_3NH_3]^+ \quad \left[\begin{array}{c} H_3C \\ \diagdown \\ N - C \diagup ^O \\ H \diagup \qquad \diagdown \\ HBH \\ H \end{array} \right]^-$$

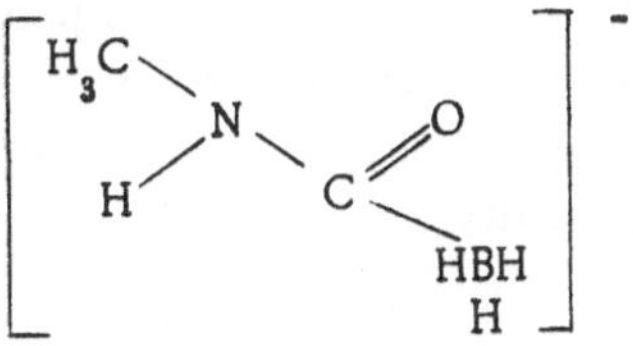

NORDMAN (3) confirmed the above structure by a single crystal X-ray diffraction study. All atoms in the anion are essentially planar. The key distances are C-N 1,45 Å; C=O 1,28 Å; C-B 1,60 Å; the angle C-N-C in the anion is 125,5°. If one considers that the oxyacids are coordination compounds in which oxygen atoms serve as ligands coordinated around a central atom, then the boranocarbamate anion represents a carbamate ion in which one coordinated oxygen atom is replaced by a BH_3 group. Using this model, a series of new oxyacid analogues can be visualized. Several members of the series have been prepared. The carbonate analogue is of interest.

Methyl ammonium N-methyl borano carbamate and excess KOH react in boiling alcohol in accordance with the equation:

$$[NH_3CH_3]^+[H_3BCONHCH_3]^- + 2\ KOH \longrightarrow K_2H_3BCO_2 + 2\ NH_2CH_3 + H_2O$$

The potassium boranocarbonate precipitates from the ethyl alcohol solution as a fine crystalline powder. The compound is stable in dry air indefinitely and in water for several hours. When heated in the absence of air and moisture, sodium and potassium salts show little decomposition below a temperature of 400°. The potassium salt can be titrated with dilute hydrochloric acid. From the titration curve the ionization constant for the process:

$$H[H_3BCO_2]^- \rightleftarrows H^+ + [H_3BCO_2]^{2-} \quad \text{is} \quad K \approx 5,5 \cdot 10^{-11}$$

The corresponding value for bicarbonate is $4,7 \cdot 10^{-11}$. With 85% H_3PO_4 potassium boranocarbonate liberates borane carbonyl in excellent yield; in dilute acid solution hydrolysis of the liberated borane carbonyl occurs.

The proton NMR spectrum is surprisingly sharp showing clearly splittings due to both B^{10} and B^{11}.

The boranocarbonate anion, like its carbonate analogue, can act as a ligand in forming coordinating compounds with transition metal ions. Some complexes formed will be discussed.

References

1. BURG, A.B., and SCHLESINGER, H.F., J.Am.Chem.Soc., 59, 780 (1937).
2. CARTER, J.C., and PARRY, R.W., to be published; CARTER, J.C., Doctoral Dissertation, University of Michigan (1960).
3. NORDMAN, C.E., to be published. University of Michigan (1962); Boron-Nitrogen Chemistry, A.C.S. Advances in Chemistry Series (1964).

4 B 2

NOVEL COORDINATION COMPOUNDS OF BORON.

James C. Carter
Department of Chemistry, University of Pittsburgh,
Pittsburgh, Pennsylvania, U.S.A.

The characteristic reactions of many boron compounds as Lewis acids have been extensively investigated. This has resulted in the characterization of salts of usual Lewis bases such as $(CH_3)_3NBH_3$, H_3NBH_3, $OCBH_3$ and $(C_2H_5)_2OBF_3$. The higher boranes and borane fragments have demonstrated similar properties in the formation of compounds such as $H_3NB_3H_7$, OCB_4H_8 and $(CH_3)_3NB_4H_6$. Studies of decaborane (10) have clearly established its properties as a Lewis acid, or perhaps more appropriately, the acid nature of the hypothetical $B_{10}H_{12}$ group. Compounds such as $(CH_3CN)_2B_{10}H_{12}$ and $(C_5H_5N)_2B_{10}H_{12}$ are examples of this behavior.

The same bases have produced the extensive variety of metal complexes with the transition metals, the amines and carbonyls serving as representative examples. Recently several transition metal complexes have also been shown to exhibit Lewis base properties to BF_3 or BH_3 with the formation of metal-boron bonds. The first example of this was SHRIVER's preparation of $(C_5H_5)_2WH_2BF_3$ which contains a tungsten-boron bond (1). The $(C_5H_5)_2WH_2$ contains an unshared pair of electrons which are donated to the electron deficient BF_3. The protonation of the lone pair of electrons to form $[(C_5H_5)_2WH_3]^+$ is a related reaction and also demonstrates the basic nature of $(C_5H_5)_2WH_2$.

The addition of BH_3 to the anions $[Re(CO)_5]^-$, $[Mn(CO)_5]^-$ and $[(C_6H_5)_3PMn(CO)_4]^-$ has been reported by PARSHALL (2). As in the case of $(C_5H_5)_2WH_2$, an unshared pair of electrons is available on the metal and the complexes such as $[H_3BRe(CO)_5]^-$ have a metal-boron bond. DOBROTT and LIPSCOMB have reported the existence of strong covalent metal-boron bonds in $Cu_2B_{10}H_{10}$. They suggest three center bonds between copper(I) and edge boron-boron bonds of the $[B_{10}H_{10}]^{2-}$ ion on the basis of their single crystal X-ray study (3).

The present work describes the formation of titanium-boron bonds. Comparing titanocene dichloride with $(C_5H_5)_2WH_2$, one major difference is the presence of an unshared electron pair on the tungsten but not on the titanium. It is unlikely, therefore, that $(C_5H_5)_2TiCl_2$ can act as a Lewis base to borane fragments. The possibility exists that it could act as an acid. Several reactions are known in which the titanium-chlorine bond is replaced by a titanium-carbon linkage. For instance, titanocene dichloride reacts with phenyl lithium to form $(\pi-C_5H_5)_2Ti(\sigma-C_6H_5)_2$ (4). This has now been extended to prepare $(C_5H_5)_2Ti(B_{10}H_{13})_2$ from the reaction of $(C_5H_5)_2TiCl_2$ with $NaB_{10}H_{13}$ in tetrahydrofuran. The other anticipated product, sodium chloride, was recovered and identified by its X-ray powder diffraction pattern. The compound is found to have limited stability and decomposes with the evolution of hydrogen. It appears that a $B_{10}H_{12}$ derivative is formed during the decomposition. The decaboranate reacts rapidly with air giving products which have not been completely identified. Reactions have also been observed with triethylamine and ethylenediamine. Titanocene dichloride was not formed when the material was allowed to react with HCl in absolute ethanol.

No reaction was observed between $(C_5H_5)_2TiCl_2$ and $B_{10}H_{14}$ in ethyl ether at room temperature. Addition of trimethylamine resulted in an immediate reaction; the titanocene dichloride was consumed. A similar reaction took place between $(C_5H_5)_2TiCl_2$, $B_{10}H_{14}$ and $(C_2H_5)_3N$ in tetrahydrofuran and the anticipated $[(C_2H_5)_3NH]^+Cl^-$ was precipitated and identified by its X-ray powder diffraction pattern. However, $(C_5H_5)_2Ti(B_{10}H_{13})_2$ was not obtained because of further reaction with $(C_2H_5)_3N$.

The nature of the above products and extensions of the $NaB_{10}H_{13}$ reactions to other systems will be discussed.

References

1. SHRIVER, D., J.Am.Chem.Soc., __85__, 3509 (1963).

2. PARSHALL, G., J.Am.Chem.Soc., <u>86</u>, 361 (1964).

3. DOBROTT, R.D., and LIPSCOMB, W.N., J.Chem.Phys., <u>37</u>, 1779 (1962).

4. SUMMERS, L., ULOTH, R.H., and HOLMES, A., J.Am.Chem.Soc., <u>77</u>, 3604 (1955).

4 B 3

<u>KOORDINATIONSVERBINDUNGEN VON PHOSPHINO- UND AMINO-BORANEN SOWIE VON BORAZOLDERIVATEN.</u>

<u>H. Nöth</u>, G. Schmid und Y. Chung

Institut für Anorganische Chemie der Universität München, Germany.

(to be presented in English)

Polymere Phosphinoborane $(X_2B-PR_2)_n$ (X = H, R, Cl, Br, n = 2, 3, 4) enthalten Bor und Phosphor der Koordinationszahl 4 und sind deshalb als Liganden zum Aufbau von Komplexverbindungen ungeeignet, es sei denn, es gelingt, sie thermisch zu monomerisieren und das Monomere in Form von Koordinationsverbindungen z.B. mit PdJ_2, $FeCl_2$ etc. abzufangen. Versuche in dieser Richtung blieben ohne Erfolg.

Die von uns in jüngster Zeit dargestellten monomeren Dialkylamino-diäthylphosphino-borane $(R_2N)_{3-n}B(PÄt_2)_n$ (n = 1, 2) verfügen über dreibindige P-Atome (1). Da elektrophile Agenzien die B-P- bevorzugt vor der B-N-Bindung angreifen, erwarteten wir, daß diese Verbindungen als Liganden zur Synthese neuer Koordinationsverbindungen geeignet sind.

Die Umsetzung von $[(CH_3)_2N]_2B-P(C_2H_5)_2$ [I] mit HgJ_2 und CdJ_2 lieferte relativ instabile 1:1-Verbindungen. Insbesondere bei der Umsetzung von HgJ_2 mit [I] konkurriert die Reaktion

$$HgJ_2 + [(CH_3)_2N]_2B-P(C_2H_5)_2 \longrightarrow (C_2H_5)_2PHgJ + [(CH_3)_2N]_2BJ$$

$$\{1\}$$

mit der zur Komplexbildung führenden Reaktion

$$HgJ_2 + [(CH_3)_2N]_2B-P(C_2H_5)_2 \longrightarrow HgJ_2 \cdot [(CH_3)_2N]_2B-P(C_2H_5)_2$$

$$\{2\}$$

Hiebei wird das P-Atom als Donor, nicht jedoch das N-Atom, benutzt.

Die Spaltung der B-P-Bindung in [I] erfolgt über die Komplexbildung. Sie erfolgt bei Umsetzungen von Metallchloriden mit [I] leichter als bei den

Metalljodiden, da dem Angriff des Cl^- auf das B-Atom in [I] ein geringerer Widerstand entgegengesetzt wird als dem sperrigeren Jod-Ion. Reaktionen dieser Art werden durch die Systeme von [I] mit $CuCl_2$, $CuCl$, $NiCl_2$ und $FeCl_3$ repräsentiert.

Eindeutiger und weniger durch Nebenreaktionen gestört verlaufen Umsetzungen von [I] mit Nickelcarbonyl. Nach

$$Ni(CO)_4 + (C_2H_5)_2P\text{-}B[N(CH_3)_2]_2 \longrightarrow (CO)_3Ni\{(C_2H_5)_2P\text{-}B[N(CH_3)_2]_2\} + CO$$
$$\{3\}$$

$$Ni(CO)_4 + 2\,(C_2H_5)_2P\text{-}B[N(CH_3)_2]_2 \longrightarrow (CO)_2Ni\{(C_2H_5)_2P\text{-}B[N(CH_3)_2]_2\}_2 + 2\,CO$$
$$\{4\}$$

werden stufenweise 2 CO-Moleküle je $Ni(CO)_4$ durch [I] substituiert. Erwartungsgemäß fungiert das P-Atom von [I] als Koordinationspartner, wie IR- und H^1 NMR-spektroskopisch gezeigt wurde. Substitution über das 1:2-Verhältnis hinaus ist nicht möglich, da bei Temperaturen über 80^0, die für eine weitere Substitution mindestens aufzuwenden wären, Zerfall der genannten Phosphinoboran-nickel-carbonyle stattfindet.

Umsetzungen von $Ni(CO)_4$ mit $(CH_3)_2NB(Cl)P(C_2H_5)_2$ [II] oder $(C_2H_5)_2NB[P(C_2H_5)_2]_2$ [III] entsprechen im Prinzip den Gl. {1} und {2}. Mit [III] wird ein polymeres $[(C_2H_5)_2N\text{-}B[P(C_2H_5)_2]_2Ni(CO)_2]_n$ als gelb gefärbtes, zähöliges Produkt erhalten.

Gleiches gilt auch für Substitutionsreaktionen mit der Dibor-Verbindung $[(C_2H_5)_2N](C_2H_5)_2PB\text{-}BP(C_2H_5)_2[N(C_2H_5)_2]$, jedoch läßt sich aus dem Rohprodukt die monomere Verbindung [IV] in etwa 10% Ausbeute isolieren. Das schwach gelb gefärbte Öl löst sich in Benzol monomer. Seine geringe thermische Stabilität erlaubt allerdings keine Reinigung durch Destillation, so daß bisher nur das nicht ganz reine [IV] untersucht werden konnte.

[IV]

Die Umsetzungen B-halogenierter Phosphinoboran-nickel-carbonyle mit z.B. $NaMn(CO)_5$ stellt den Prototyp eines neuen Typs von anorganischen Polymeren dar. In THF-Lösung erfolgt die Umsetzung

$$(CO)_3Ni[(C_2H_5)_2P\text{-}B(Cl)N(C_2H_5)_2] + NaMn(CO)_5 \longrightarrow$$

$$NaCl + (CO)_3Ni[(C_2H_5)_2P\text{-}B\text{-}N(C_2H_5)_2]\text{-}Mn(CO)_5 \qquad \{5\}$$

quantitativ unter Aufbau einer Ni-P-B-Mn-Kette. Polymere Produkte werden gemäß Reaktion {6} erhalten:

$$R_2NB(Cl)PR_2' + NaMn(CO)_5 \xrightarrow{\quad -NaCl \quad} R_2NB(PR_2)\text{-}Mn(CO)_5 \xrightarrow{\quad -CO \quad}$$

$$[R_2NB(PR_2)Mn(CO)_4] \qquad \{6\}$$

[V]

Das Diäthylamino-diäthylphosphinoboryl-mangantetracarbonyl [V] ist polymer.

Der Ersatz von CO durch [I] oder [II] in $Fe(CO)_5$, $Cr(CO)_6$, $Mo(CO)_6$ und $Mn_2(CO)_{10}$ ließ sich bei Temperaturen bis zu 100^0 nicht erzwingen.

Im Laufe der Untersuchung der Einwirkung von [I] auf Metallhalogenide wurde insbesondere bei Verwendung von $TiCl_4$ festgestellt, daß dieses mit dem bei der B-P-Spaltung von [I] resultierenden $[(CH_3)_2N]_2BCl$ eine 1:1-Additionsverbindung bildet. Auch im System $B[N(CH_3)_2]_3/TiCl_4$ existiert das sehr hydrolyseempfindliche 1:1-Addukt $TiCl_4 \cdot B[N(CH_3)_2]_3$, das bei Raumtemperatur allerdings nicht allzu stabil ist und sich in noch unbekannte, schwarzbraun gefärbte Verbindungen zersetzt, während das reine Addukt eine hellgelbe, kristalline Verbindung ist.

Nach bisherigen Untersuchungen liegen in den 1:1-Addukten Nichtelektrolytkomplexe vor, denen auf Grund ihrer IR-Spektren die Struktur [VI] zuzuordnen ist.

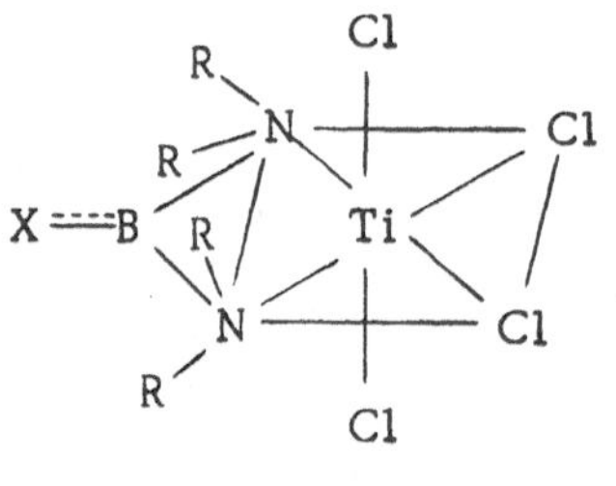

[VI]

Die Möglichkeit, daß Borazol und seine Derivate zur Bildung von Koordinationsverbindungen geeignet sein müsse, sei es als π-Komplexligand oder als einfaches Donorsystem über die N-Atome, wurde bereits oft diskutiert. Die Isolierung derartiger Komplexe gelang unseres Wissens aber noch nicht. Ausgehend von der Tatsache, daß B-alkylierte Borazole zur charge-transfer-Komplexbildung geeignet sind und daß bei Verwendung von $TiCl_4$, $VOCl_3$ oder $AlBr_3$ Aromaten-Metallhalogenid-Komplexe zu isolieren sind, untersuchten wir das Verhalten von $TiCl_4$ und $AlBr_3$ gegenüber einigen Borazolderivaten.

Das Hexamethylborazol lieferte, unabhängig vom Molverhältnis mit $TiCl_4$ einen sublimierbaren, gelben, kristallisierten 1:1-Komplex $TiCl_4 \cdot (CH_3B\text{-}NCH_3)_3$, der sich in Benzol monomer löst und in diesem Lösungsmittel keine elektrische Leitfähigkeit zeigt. Thermische Zersetzung des Komplexes liefert unverändertes Hexamethylborazol. Da der Borazolring intakt bleibt, wenn die Verbindung aufgebaut wird, schlagen wir die Struktur [VII] für diese Verbindung vor. Wir hoffen, bis zum Berichtszeitpunkt über weitere Daten zu verfügen, damit dieser Vorschlag zur Struktur von $TiCl_4 \cdot (CH_3B\text{-}NCH_3)_3$ ausführlich diskutiert werden kann.

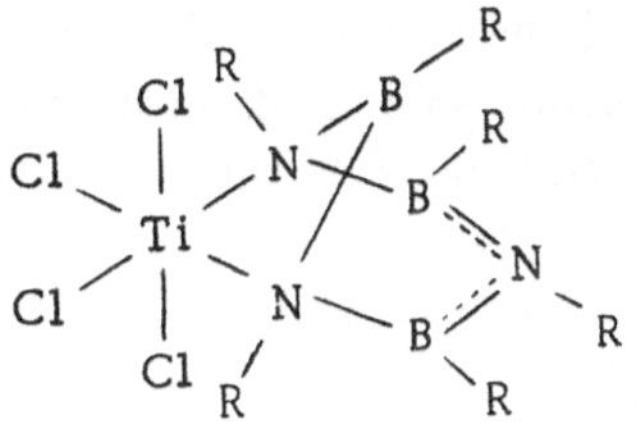

[VII]

Literatur

1. NÖTH, H., und SCHRÄGLE, W., Angew.Chem., _74_, 718 (1962).

5 B 1

REACTIONS BETWEEN NITRILES AND SOME ORGANIC COMPOUNDS OF BORON AND ALUMINIUM.

J.E. Lloyd and K. Wade
University of Durham, U.K.

Diborane is known to react reversibly with nitriles to form adducts $RCN.BH_3$ {1}, which may also rearrange irreversibly by a two-stage hydrogen transfer process (1):

$$\frac{1}{2} B_2H_6 + RC\ N \ \rightleftharpoons \ RC\vdots N.BH_3 \ \longrightarrow \ (RCH\vdots N \cdot BH_2)_n \ \longrightarrow \ (RCH_2NBH)_m$$

$$\{1\} \qquad\qquad \{2\} \qquad\qquad \{3\}$$

The products generally obtained {3} are N-trisubstituted borazines, $(RCH_2NBH)_3$, and related polymers in which borazine rings are linked by groups $-N(CH_2R)-$ (2), although in some systems the intermediates {2} are sufficiently resistant to further rearrangement to be isolated. For example, an explosive crystalline solid $(Cl_3CCH:N \cdot BH_2)_n$ has been prepared from diborane and trichloroacetonitrile (3), and various butyl derivatives $(RCH:NBHBu^t)_2$ have been obtained by the hydroboration of nitriles with the complex $Me_3N.BH_2Bu^t$ (4).

In the present work, dialkylboron derivatives of type {2}, the dimeric ethylideneaminodialkylboranes, $(MeCH:N \cdot BR_2)_2$ (R = Me or Et) have been prepared from methyl cyanide and the appropriate tetra-alkyldiborane, R_2BHBHR_2, or unsymmetrical dialkyldiborane R_2BHBH_3, in the latter case N-triethylborazine being formed also. Isomeric forms of the methyl compound, $(MeCH:N \cdot BMe_2)_2$, have been separated, one a crystalline solid, m.p. 76°, v.p. 0,6 mm at 20°, the other a liquid, f.p. ca. -5°, v.p. 1,8 mm at 20°. Their infrared and H^1 nuclear magnetic resonance spectra, and a preliminary X-ray crystallographic study of the solid, indicate the solid isomer to have the trans structure {4}, the liquid the cis structure {5} (5).

{4} {5}

Such structures are supported by the detection of dimethylborinic acid, acetaldehyde and ammonia, but not ethylamine, among their hydrolysis products, and the formation of dimethylchloroborane by reaction with hydrogen chloride.

Aluminium compounds with related structures result from reactions between nitriles and organo-aluminium compounds, when adducts $RCN.AlR_3'$ rearrange by transfer of organic groups R' from aluminium to the carbon of the cyanide group:

$$RCN + AlR_3' \longrightarrow RCN.AlR_3' \xrightarrow{150-250^o} (RR'C:N \cdot AlR_2')_2$$

Thus, adducts of phenyl cyanide with trimethyl-, triethyl- and triphenyl-aluminium rearrange at temperatures between 150 and 250° to form dimeric substituted benzylideneamino-aluminium compounds, $(R'PhC:N \cdot AlR_2')$, crystalline solids which are readily hydrolysed to the ketone PhCOR', ammonia, aluminium hydroxyde and the hydrocarbon R'H.

Adducts of methyl or ethyl cyanide with organo-aluminium compounds can rearrange when heated to form the related dimers $(MeR'C:N \cdot AlR_2')_2$ and $(EtR'C:N \cdot AlR_2')_2$, although these are obtained in low (< 20%) yield because at the same or lower temperatures the adducts also evolve hydrocarbon R'H through loss of one of the acidic hydrogen atoms on the alpha carbon atom of the nitrile, leaving polymeric solids containing such groups as ($\longrightarrow$ $AlMe_2 \cdot CH_2 C:N \longrightarrow$) and ($-AlMe \cdot CH_2 CMe:N-$), e.g.:

$$MeCN \cdot AlMe_3 \quad \xrightarrow{150^0} \quad \begin{cases} \xrightarrow{-MeH} (Me_2AlCH_2C:N)_n(-MeAlCH_2CMe:N-)_m \\ \\ \longrightarrow 15\% (Me_2C:N \cdot AlMe_2)_2 \text{ m.p. } 96^0 \end{cases}$$

m.p. 73^0

Further variations on these reactions are illustrated by the reaction of nitriles with dimethylaluminium chloride or hydride:

$$PhCN + Me_2AlCl \longrightarrow PhCN \cdot AlMe_2Cl \longrightarrow (PhMeC:N \cdot AlMeCl)_2$$

$$PhCN + Me_2AlH \longrightarrow PhCN \cdot AlMe_2H \longrightarrow (PhHC:N \cdot AlMe_2)_2$$

These reactions establish the nature of the intermediates which were involved in reactions reported by other workers (6), who have shown that nitriles can be converted into ketones, aldehydes or amines by reaction with suitable organo-aluminium compounds. Some aspects of the chemistry of these compounds will be discussed.

References

1. BURG, A., Record of Chemical Progress (Kresge-Hooker Science Library), 15, 159 (1954), EMELÉUS, H.J., and WADE, K., J.Chem.Soc., 1960, 2614.

2. JENNINGS, J.R., and WADE, K., unpublished observations.

3. LEFFLER, A.J., Inorg.Chem., 3, 145 (1964).

4. HAWTHORNE, M.F., Tetrahedron, 17, 117 (1962).

5. LLOYD, J.E., and WADE, K., J.Chem.Soc., 1964, in the press.

6. ZAKHARKIN, L.I., and KHORLINA, I.M., Doklady Akad.Nauk, 116, 422 (1957), ZIEGLER, K., SCHNEIDER, K., and SCHNEIDER, J., Ann.Chem., 623, 9 (1959), PASYNKIEWICZ, S., DAHLIG, W., and TOMASZEWSKI, B., Roczniki Chem., 36, 1383 (1962).

5 B 2

THE COORDINATION CHEMISTRY OF BORON-NITROGEN COMPOUNDS.

J.J. Lagowski
The University of Texas, Austin, Texas, U.S.A.

The borazines and aminoboranes are examples of compound types that are isoelectronic with benzene and ethylene, respectively. Although the preparation, reaction (predominantly substitution and addition reactions), and physical properties of these boron-nitrogen systems have been extensively studied (1,2) little has been done toward investigating their electronic nature with respect to the reactions or subtle interactions that are characteristic of the corresponding carbonoid systems where electron delocalization is known to occur.

Theoretically, borazines could form addition compounds with electron-pair acceptors whether the nitrogen lone-pair electrons are localized or de-localized as in benzene. In the former case the borazine could be considered as a cyclic triamine and hence might act as a Lewis base. On the other hand, a delocalized electron system in the borazine ring could lead to the formation of "sandwich-type" compounds similar to dibenzenechromium. It is apparent that the isolation of a "sandwich-type" compound containing the borazine ring can not be taken as evidence that a delocalized π system of electrons is involved. Cyclic triamines can act as tridentate ligands as exemplified by the reaction of hexahydro-s-triazine with chromium hexacarbonyl to form $Cr(CO)_3(C_3N_3H_6)$ (4). In view of the possible electron-pair donating (at the nitrogen sites) and accepting (at the boron sites) ability of the borazine ring an interesting possibility arises for the formation of addition compounds with transition metals. If the electron pair is localized on the nitrogen atoms of the borazine ring, the borazine can behave as a tridentate ligand, donating its electrons to available vacant d orbitals of a suitable transition metal atom, and it can also accept electron pairs at the boron sites from filled d orbitals of the transition metal atom. A compound formed in this manner might be difficult to distinguish from a "sandwich-type" compound resulting from the interaction of delocalized π electrons on the borazine ring.

Similar considerations are valid for the aminoboranes. Localization of the nitrogen electron pair leads to an amine structure while delocalization yields a formal B-N double bond. In the first case aminoboranes might be expected to yield complex compounds (with, for example, transition metal ions) by coordination through the nitrogen atom, while in the latter instance coordination might be expected to occur through the "double bond" as in $(PtCl_2 \cdot C_2H_4)_2$ (3).

Although no coordination compounds containing the borazine nucleus have been characterized yet, charge-transfer complexes involving borazines have been reported (5-7). Charge-transfer complexes are well known for benzenoid systems, but the charge-transfer spectrum observed with the borazine ring does not necessarily mean that a delocalized π system of electrons is present in this ring because amines have also been reported to give charge transfer complexes. In addition, the observed dipole moments of several borazines, measured in benzene solution, could be interpreted as arising from solute-solvent association in these systems as well as from deviations from planarity of the borazine ring, as was originally suggested (8). A careful study (9) of the polarization of pure hexamethylborazine as well as of benzene, carbon tetrachloride, and n-heptane solutions of this compound indicates that the apparent dipole moments of the borazines arise from large atom polarization in these systems.

Aminoboranes interact with iodine in inert solvents to yield red-brown solutions (λ_{max} 360 to 390 mμ); the spectra of solutions containing varying mole ratios of reactants exhibited one isosbestic point and the method of continuous variations indicates the presence of a 1:1 complex in these systems. The equilibrium constants determined from the spectral data for the formation of these complexes are numerically similar to those observed for the corresponding amine complexes; the equilibrium constants for olefin-iodine complexes are orders of magnitude smaller (10). The products obtained from the reaction of aminoboranes with transition metal ions also suggest that the former compounds act as amines.

The preparation (11) of compounds containing a sigma-bonded borazine moeity will be discussed.

References

1. MELLON, Jr., E.K., and LAGOWSKI, J.J., Advan. Inorg. Chem. and Radiochem., 5, 259 (1963).
2. a) NIEDENZU, K., and DAWSON, J.W., J. Am. Chem. Soc., 81, 5553 (1959).
   b) COATES, G.E., and LIVINGSTON, J.G., J. Chem. Soc. (London), 1961, 1000.
   c) BARFIELD, P.A., LAPPERT, M.F., and LEE, J., Proc. Chem. Soc., 1961, 421.
   d) NIEDENZU, K., and DAWSON, J.W., J. Am. Chem. Soc., 82, 4223 (1960).

3. ANDERSON, J.S., J.Chem.Soc. (London), 1934, 971.
4. LÜTTRINGHAUS, A., and KULLICK, W., Tetrahedron Letters, No.10, 13 (1959).
5. MELLON, Jr., E.K., and LAGOWSKI, J.J., Nature, 199, 997 (1963).
6. CHAMPION, N.E.S., FOSTER, R., and MACKIE, R.K., J.Chem.Soc. (London), 1961, 5060.
7. MUSZKAT, K.A., and KUSON, B., Israel J.Chem., 1, 150 (1963).
8. WATANABE, H., and KUBO, M., J.Am.Chem.Soc., 82, 2428 (1960).
9. MELLON, Jr., E.K., Ph.D.Dissertation, The University of Texas, 1963.
10. EUBANKS, I.D., Ph.D.Dissertation, The University of Texas, 1963.
11. KÖHL, H., and COMPTON, R.C., unpublished results.

5 B 3

SOME NEW COMPLEXES OF RHENIUM.

F. Nyman and E.S. Stern
I.C.I. Ltd., Petrochemical and Polymer Laboratory,
The Heath, Runcorn, England, U.K.

Octahedral complexes of rhenium(V), rhenium(III), and rhenium(II) derived from ditertiary phosphines have been prepared by reaction of the phosphine with potassium perrhenate or rhenium heptoxide in the presence of a halogen acid and an organic solvent. These include complexes of the types $[(R_2PCH_2CH_2PR_2)_2ReO_2]^+X^-$, $[(R_2PCH_2CH_2PR_2)_2ReX_2]^+X^-$ and $[(R_2PCH_2CH_2PR_2)_2ReX_2]$ where $R = C_6H_5$ or CH_3 and $X = Cl$, Br, or I.

Zero-valent complexes of rhenium have been prepared by reacting dirhenium decacarbonyl with mono- and ditertiary phosphines and amines in the presence of an organic solvent; heat or UV light can be used to produce reaction.

The structures of many of these complexes have been established by the application of various physical techniques and will be discussed.

6 B 1

DIALKYLAMINOFLUOROPHOSPHINES AND THEIR COORDINATION CHEMISTRY WITH TRANSITION METALS.

Reinhard Schmutzler
E.I.duPont de Nemours and Co., Inc., Explosives Department,
Experimental Station Laboratory, Wilmington 98, Del., U.S.A.

First reference to a dialkylaminofluorophosphine, $(C_2H_5)_2NPF_2$, was made

25 years ago, but it was not until recently when this class of compounds of the composition $(R_2N)_nPF_{3-n}$ (n = 1, 2) attracted further interest. The compound $(CH_3)_2NPF_2$ has been prepared by the dimethylaminolysis of PF_3, and some of its reactions have been studied (1-3).

We have obtained several dialkylaminofluorophosphines from the corresponding chloro compounds, using sodium fluoride in a tetramethylene sulfone medium or antimony trifluoride (in the case of dialkylaminodifluoro-phosphines) as a fluorinating agent. A number of dialkylaminofluorophosphines of the composition R_2NPF_2, $(R_2N)_2PF$, and,most recently, also $RPFN(CH_3)_2$, were readily obtained as stable, distillable liquids.

As there are two potential bonding sites in dialkylaminofluorophosphines, their coordination chemistry with transition metals was deemed particularly interesting.

A series of coordination compounds of zerovalent nickel and molybdenum, involving the novel ligands, was obtained in accordance with:

$$Ni(CO)_4 + 2L \longrightarrow Ni(CO)_2L_2 + 2CO$$

(stoichiometric quantities of reactants);

$$Ni(CO)_4 + 4L \longrightarrow NiL_4 + 4CO$$

(excess L) ;

$$Mo(CO)_3(C_7H_8) + 3L \longrightarrow Mo(CO)_3L_3 + C_7H_8$$

(C_7H_8 = cycloheptatriene).

The exchange reactions proceeded readily at room temperature and were completed by gentle heating, if necessary, e.g. in the case of the tetra-substituted derivatives of $Ni(CO)_4$.

Further reaction of a dicarbonyl-nickel(O) derivative, $Ni(CO)_2(C_5H_{10}NPF_2)_2$, with $(C_2H_5)_2NPF_2$ gave the mixed compound, $Ni[(C_2H_5)_2NPF_2]_2 [C_5H_{10}NPF_2]_2$.

Especially noteworthy is the substitution of four CO groups in $Mo(CO)_6$ upon heating the hexacarbonyl with $C_5H_{10}NPF_2$ in a molar ratio of ca. 1:5 under autogenous pressure. Tetrasubstitution of CO in a group VI hexacarbonyl has previously been described only with the bidentate ligand, o-phenylene--bis-dimethylarsine, where the chelating properties of the ligand may be a driving force in the replacement of CO. No monodentate ligand has been reported to replace directly more than three CO groups in a group VI hexa-carbonyl.

The novel coordination compounds are mostly colorless, crystalline solids, and, in some instances, liquids which could not be induced to crystallize. The properties of the compounds are generally comparable to those containing fluorophosphite, $(RO)_n PF_{3-n}$, ligands (4).

Observations have been made which support coordination of the dialkyl-aminofluorophosphine ligands to the transition metal through phosphorus, rather than through nitrogen, whereas in the reaction of $(CH_3)_2 NPF_2$ with some boron acceptor molecules coordination both through phosphorus or nitrogen was observed (1).

In general, p_π-d_π interaction between nitrogen and phosphorus (promoted by the presence of the electronegative fluorine atoms on the phosphorus) is expected to reduce the basicity of nitrogen and to increase the tendency of coordination through phosphorus. Strong evidence for co-ordination via phosphorus, rather than nitrogen in the case of $Ni(C_5 H_{10} NPF_2)_4$ is provided by the investigation of its crystal structure, which is currently in progress (5).

Especially informative as to the bonding site in $(CH_3)_2 NPF_2$ is a comparison of the C-H stretching region in the IR spectra both of uncoordinated $(CH_3)_2 NPF_2$ and of some of its coordination compounds. An absorption around 2800 cm^{-1} in N-CH$_3$ compounds, where the nitrogen atom retains its lone pair of electrons, is characteristically affected upon quaternization of the nitrogen atom, e.g. by complex formation. In case of coordination through nitrogen, this absorption would be expected to disappear.

The IR spectra of the novel coordination compounds were recorded. The CO stretching absorptions both for the dicarbonyl-nickel(O) and tricarbonyl-molybdenum(O) derivatives were found in the region typical of terminal CO groups. A pronounced increase in the CO stretching frequencies of phosphine-substituted metal carbonyls, as compared to those of derivatives containing nitrogen donor molecules, has been attributed to the π-character of the metal-ligand bond in the phosphine complexes (6). Reasonably, the multiple bond character of the metal-phosphorus bond is expected to increase with increasing availability of donor atom d-orbitals, due to the influence of the electro-negative fluorine substituents. This effect is reflected in the trends observed in the CO stretching region of both series, $Ni(CO)_2 L_2$, and $Mo(CO)_3 L_3$.

References

1. TerHAR, G., FLEMING, M.A., and PARRY, R.W., J. Am. Chem. Soc., **84**, 1767 (1962).
2. NÖTH, H., and VETTER, H.J., Chem. Ber., **96**, 1298 (1963).
3. CAVELL, R.G., University Chemical Laboratory, Cambridge, England, personal communication.

4. SCHMUTZLER, R., Chem.Ber., 96, 2435 (1963).

5. GREENBERG, B., Hunter College, New York, N.Y., U.S.A.; personal communication; c.f. also GREENBERG, B., AMENDOLA, A., and SCHMUTZLER, R., Naturwiss., 50, 593 (1963).

6. ABEL, E.W., BENNETT, M.A., and WILKINSON, G., J.Chem.Soc. (London), 1959, 2323.

6 B 2

TRANSITION METAL COMPLEXES OF CAGED PHOSPHITE ESTERS.

J.G. Verkade, T.J. Huttemann, B.M. Foxman, R.E. McCarley and D.G. Hendricker

Chemistry Hall, Iowa State University, Ames, Iowa, U.S.A.

The unusual donor abilities of caged phosphites {Ia} and {III} have been the subject of recent investigations in these laboratories. The low steric requirements and high dipole moments of {Ia} and {II} are undoubtedly significant

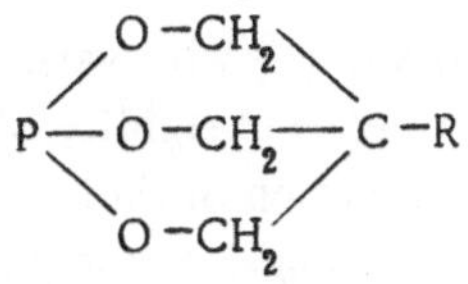

{Ia} (R = CH_3)
{Ib} (R = CH_2CH_3)

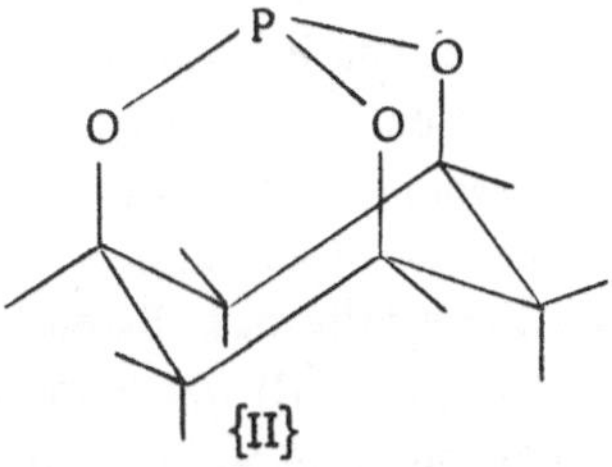

{II}

factors in the formation of several transition metal ion complexes exhibiting maximum coordination numbers with {Ia} (1,2) and in the formation of stable adducts of {Ia} and {II} with Lewis acids such as the BH_3 and B_3H_7 groups (3,4). A measure of the ligand field strength of {Ia} was obtained from the UV spectrum of the colorless $[Co\{Ia\}_6]^{3+}$ ion which indicated that D_q is 3320 cm^{-1}; a value very close to that for D_q in the $[Co(CN)_6]^{3-}$ ion which is 3350 cm^{-1}(2).

A series of new complexes of {Ia}, {Ib} and {II} with transition metal ions and transition metal carbonyls have been characterized and are here reported. Previously published procedures were used to prepare {Ia}, (5) {Ib}, (5) and {II} (6).

When {II} is allowed to react with $Co(ClO_4)_2 \cdot 6 H_2O$ in acetone, diamagnetic crystalline complexes of the indicated stoichiometries are formed in near quantitative yield according to the following equation:

$$2 \, Co(ClO_4)_2 \cdot 6 \, H_2O + 11\{II\} \longrightarrow [Co\{II\}_5]ClO_4 + [Co\{II\}_6] \, [ClO_4]_3$$

yellow   colorless

In addition to the diamagnetic properties of $[Co\{II\}_6] \, [ClO_4]_3$, the high ligand field exerted by $\{II\}$ is substantiated by a D_q value of 3246 cm^{-1} for this complex. Reaction of $\{II\}$ with $Ni(ClO_4)_2 \cdot 6 \, H_2O$ in acetone quantitatively yields yellow diamagnetic crystals having a stoichiometry $[Ni\{II\}_5 H_2O] \, [ClO_4]_2$ which upon heating in water produces a colorless diamagnetic solid possessing the Formula $Ni\{II\}_4$. Similar procedures with $[Ni\{Ia\}_5 H_2O] \, [ClO_4]_2$ and $[Ni\{Ib\}_5 H_2O] \, [ClO_4]_2$ form $Ni\{Ia\}_4$ and $Ni\{Ib\}_4$, respectively. The following equation is postulated to account for this unusual reduction. The stoichiometric liberation of the protons was shown by pH measurements. The product complex of nickel was shown to be non-conducting in polar solvents and dielectric constant measurements indicated no appreciable dipole moment.

$$[Ni\{II\}_5 H_2O] \, [ClO_4]_2 \longrightarrow Ni\{II\}_4 + \{II\}oxide + 2 \, H^+ + 2 \, [ClO_4]^-$$

It is interesting that these compounds showed no discernable visible or UV absorption. Colorless transition metal ion complexes of manganese and iron with these ligands have also been isolated.

A study of the H^1 and P^{31} NMR spectra of $\{Ia\}$ and $\{II\}$ with various boron Lewis acids has been carried out (7). It was hoped that the rigidity of these cage-like ligands would tend to minimize rehybridization of their atomic orbitals upon coordination and that conformational changes accompanying coordination would be minimal. Extension of this study to the transition metal ion complexes of $\{Ia\}$ and $\{II\}$ was hampered by their low solubility in suitable NMR solvents. This problem was circumvented by the use of $\{Ib\}$. Analogous complexes are thus obtained whose drastically increased solubility in organic solvents is apparently due to the substitution of an ethyl for a methyl group in the ligand. P^{31} and H^1 NMR chemical shifts and coupling constants for transition metal ion complexes of $\{Ib\}$ will be presented and discussed in terms of the possible nature of the P-metal link. Ligand $\{Ib\}$ proved particularly useful in the assignment of the configuration of $[Pd\{Ia\}_2 Cl_2]$. That the configuration is trans is supported by the lack of any appreciable dipole moment or conductivity and the appearance of a triplet methylene absorption in the NMR (8).

A series of mono- and di-substituted carbonyl complexes of $Ni(CO)_4$, $Fe(CO)_5$, $Cr(CO)_6$, $Mo(CO)_6$ and $W(CO)_6$ with $\{Ia\}$ have been prepared by refluxing stoichiometric quantities of ligand and metal carbonyl under nitrogen flush in ethyl benzene. The colorless diamagnetic complexes are soluble in acetone and chloroform. It is significant that the IR carbonyl stretching frequencies occur at relatively high values (ca. 2000 cm^{-1}). The indicated increase in C-O bond order suggests the possibility of substantial doublebond character of the P-M link. That the di-substituted compounds have the tráns configuration is supported by the IR spectra (9). Further support stems from the proton NMR spectra of the methylene groups wherein the triplet has been ascribed to a "virtual coupling" involving the protons and the trans P^{31} nuclei in similar complexes (9). The mono-substituted complexes exhibit only a methylene doublet in the NMR spectra arising from spin-spin inter-action with the single P^{31} nucleus. The variation in relative intensity of the middle peak of the triplet, magnitude of the coupling constants and values of the P^{31} and H^1 chemical shifts for these complexes will be discussed in terms of the ligand properties of {Ia}.

References

1. VERKADE and PIPER, Inorg.Chem., 1, 453 (1962).
2. VERKADE and PIPER, Inorg.Chem., 2, 944 (1963).
3. HEITSCH and VERKADE, Inorg.Chem., 1, 392 (1962).
4. HEITSCH and VERKADE, Inorg.Chem., 2, 512 (1963).
5. WADSWORTH and EMMONS, J.Am.Chem.Soc., 84, 610 (1962).
6. BERLIN, HILDEBRAND, VERKADE and DERMER, Chem.and Ind., 291 (1963).
7. VERKADE, KING and HEITSCH, Inorg.Chem., in press.
8. JENKINS and SHAW, Proc.Chem.Soc., 1963, 279.
9. COTTON and KRAIHANZEL, J.Am.Chem.Soc., 84, 432 (1962).

6 B 3

NEUERES ÜBER P-SUBSTITUIERTE SCHWERMETALLPHOSPHIDE.

K. Issleib

Anorganisch-Chemisches Institut der Universität Halle/Saale, Germany.

P-substituierte Schwermetallphosphide bilden sich in der Regel aus was-serfreien Metallsalzen und Alkaliphosphiden des Typs $MePR_2$, MePHR bzw. $R(Me)P-[CH_2]_n-P(Me)R$. Je nach der Natur des Metallsalzes und des Alkali-phosphids wird unter Verwendung beispielsweise von $KP(C_6H_5)_2$ bzw. $LiP(c-C_6H_{11})_2$ neben einer Substitution, wobei u.a. Redoxreaktionen gemäß

$$Me'X_n \xrightarrow{MePR_2} \begin{cases} Me'(PR_2)_n & \{1\} \\ Me' = V(III),\ Cr(III),\ Fe(II),\ Co(II),\ Ni(II),\ Cu(I) \\ Me'(PR_2)_n + R_2P-PR_2 & \{2\} \\ Me' = Ti(II),\ Cu(I) \end{cases}$$

auftreten, auch das Entstehen komplexer Phosphide nach

$$Me'(PR_2)_n + MePR_2 \longrightarrow MeMe'(PR_2)_{n+1} \qquad \{3\}$$

$$Me = Li, K$$

$$Me' = Mn(II), Fe(II), Ni(II), Cu(I)$$

beobachtet (1-4). Während letztere mit Ausnahme von $KMn[P(C_6H_5)_2]_3$, $K_2Mn[P(C_6H_5)_2]_4$ bzw. $LiMn[P(c-C_6H_{11})_2]_3$ nach Solvolyse Phosphid-phosphin-komplexe $(c-C_6H_{11})_2PHFe[P(c-C_6H_{11})_2]_2$, $\{(c-C_6H_{11})_2PHNi[P(c-C_6H_{11})_2]_2\}_2$ und $[R_2PHCuPR_2]-R=C_6H_5$, $c-C_6H_{11}$ – liefern, verhalten sich die einfachen Phosphide $Me'(PR_2)_n$ zu komplexaktiven N- bzw. P-haltigen Liganden meist indifferent. Die nach $\{2\}$ gebildete Menge Diphosphin R_2P-PR_2 entspricht dem Wertigkeitswechsel von Titan(IV) bzw. (III) zu Titan(II) oder Kupfer(II) zu Kupfer(I). Im Falle der Umsetzung von $FeBr_2$ mit $KP(C_6H_5)_2$ verläuft die Redoxreaktion nach

$$3\,FeBr_2 + 6\,KP(C_6H_5)_2 \longrightarrow P[FeP(C_6H_5)_2]_3 + 2\,P(C_6H_5)_3 + 6\,KBr$$

$$\{4\}$$

unter Bildung von Triphenylphosphin und Tris-[diphenylphosphido-eisen(II)]-phosphid. Tetraphenyldiphosphin und Eisen, die analog der Bildung von $\{Ni[P(C_6H_5)_2]_2\}_2$ (2) hätten entstehen sollen, ließen sich nicht identifizieren. Als Beweis für die Struktur des in THF monomolekular löslichen $P[FeP(C_6H_5)_2]_3$ ist nicht nur die Bildung von Na_3PO_4, sondern auch die von PH_3 gemäß

$$P[FeP(C_6H_5)_2]_3 \begin{array}{l} \xrightarrow{\;NaOH/H_2O_2\;} Na_3PO_4 + 3\,(C_6H_5)_2POONa + 3\,Fe(OH)_3 \qquad \{5\} \\[2em] \xrightarrow{\;HCl\;} PH_3 + 3\,(C_6H_5)_2PH + 3\,FeCl_2 \qquad \{6\} \end{array}$$

anzusehen. Das Entstehen von $P[FeP(C_6H_5)_2]_3$ läßt auf eine Wechselwirkung des Phosphors mit $FeP(C_6H_5)_2$ - beide durch Zersetzung von intermediär gebildetem $Fe[P(C_6H_5)_2]_2$ und $(C_6H_5)_2P-P(C_6H_5)_2$ entstanden - schließen. Diese Annahme wird durch die Umsetzung von P_4 mit $Ti[P(c-C_6H_{11})_2]_2$ in THF zu $P\{Ti[P(c-C_6H_{11})_2]_2\}_3$, das mittels Salzsäure gleichfalls PH_3 und Titan(III) liefert, gestützt. Auch andere Phosphide $Me'(PR_2)_n$, in denen ein Übergang zur höheren Oxidationsstufe von Me' leicht möglich ist, zeigen analoges Reaktionsverhalten.

Eine Oxydation des Schwermetalls mit Jod läßt sich auch in einigen

Phosphiden Me'$(PR_2)_n$ erreichen. Je nach den Mengenverhältnissen resultieren aus Ti$[P(c-C_6H_{11})_2]_2$ bzw. V$[P(c-C_6H_{11})_2]_3$ die Oxydationsprodukte TiJ$_2[P(c-C_6H_{11})_2]_2$, TiJ$_3P(c-C_6H_{11})_2$ bzw. VJ$[P(c-C_6H_{11})_2]_3$. Überschüssiges Jod spaltet die Metall-Phorphor-Bindung.

Die Schwermetallphosphide können zur Metall-Phorphor-σ-Bindung eine koordinative π-Bindung gemäß Me'$\rightleftharpoons PR_2$ ausbilden, die die Stabilität der Metall-Phorphor-Bindung erhöhen und die Spaltung mit CH$_3$J erschweren. Aus entsprechenden Versuchen mit P.P-cyclohexylsubstituierten Phosphiden resultiert für die Stabilität der Metall-Phosphor-Bindung gegenüber CH$_3$J folgende Abstufung:

$$Ti > V > Cr > Mn < Fe < Co < Ni < Cu \quad.$$

Die abnehmende Stabilität in Richtung Mangan(II) und die zunehmende zum Kupfer(I) ist einmal durch einen geringer werdenden π-Bindungsanteil, zum andern durch die Ausbildung von d_π-d_π-Bindungsanteilen zwischen Metall und Phosphor zu erklären. Dies äußert sich auch in den Löslichkeitseigenschaften der Phosphide.

Im Vergleich zu Alkaliphosphiden MePR$_2$ zeigen auch die des Typs MePHR bzw. R(Me)P$-[CH_2]_n-$P(Me)R gegenüber Schwermetallsalzen ein z.T. analoges Reaktionsverhalten. So bildet sich beispielsweise aus KPHC$_6$H$_5$ und CoBr$_2$ das Kobalt(II)-bis-monophenylphosphid und aus C$_6$H$_5$(Li)P$-$CH$_2$·CH$_2-$P(Li)C$_6$H$_5$ und CoBr$_2$ sowie NiBr$_2$ resultieren [NiC$_6$H$_5$P$-$CH$_2$·CH$_2-$PC$_6$H$_5$] und [CoC$_6$H$_5$P$-$CH$_2$·CH$_2-$PC$_6$H$_5$], für die eine Ring- bzw. Kettenstruktur zu diskutieren ist.

Eine weitere Möglichkeit zur Synthese der Schwermetallphosphide besteht in der Umsetzung von Metallorganylen mit arylsubstituierten prim., sek. bzw. disek. Phosphinen, wie es die Beispiele der Bildung von Zn$[P(C_6H_5)_2]_2$ bzw. Cr(II)$[P(C_6H_5)_2]_2$ aus Zn(C$_2$H$_5$)$_2$ bzw. Cr(C$_6$H$_5$)$_3$·3 THF und (C$_6$H$_5$)$_2$PH zeigen. CuPR$_2$ wie auch C$_6$H$_5$(Cu)P$-$CH$_2$·CH$_2-$P(Cu)C$_6$H$_5$ sind auf Grund der Schwerlöslichkeit und der relativ großen Beständigkeit gegenüber solvolysierenden Lösungsmitteln als Koordinationspolymere

$$\left\{ Cu \leftarrow \overset{R_2}{P} \!\!-\!\!-\!\!-\! Cu \leftarrow \overset{R_2}{P} \right\}_n$$

aufzufassen. Andere Schwermetallphosphide hingegen lösen sich mono- bzw. dimolekular in indifferenten Lösungsmitteln, was einen teilweisen Komplexcharakter vom Typ des

$$R_2P-Co \underset{PR_2}{\overset{PR_2}{<\,>}} Co-PR_2$$

erkennen läßt. Die Zink- und Manganphosphide gleichen in ihren Reaktionen und in den Solvolyseeigenschaften den Alkaliphosphiden.

Die magnetischen Eigenschaften der Schwermetallphosphide sind recht unterschiedlich. Häufig werden stark reduzierte Momente beobachtet, während andere die gleichen magnetischen Eigenschaften entsprechender Metall-salze aufweisen.

Literatur

1. ISSLEIB, K., und FRÖHLICH, H.-O., Chem.Ber., 95, 375 (1962).
2. ISSLEIB, K., FRÖHLICH, H.-O., und WENSCHUH, E., Chem.Ber., 95, 2742 (1962).
3. ISSLEIB, K., und WENSCHUH, E., Z.Naturforschg. 17b, 778 (1962).
4. ISSLEIB, K., und WENSCHUH, E., Chem.Ber. 97, 715 (1964).

7 B 1

COMPLEXES OF XENON HEXAFLUORIDE *

Henry Selig

Argonne National Laboratory, Argonne, Illinois, U.S.A.

It has recently been reported (1) that XeF_2 reacts with SbF_5 or TaF_5 to form complexes with the composition $XeF_2 \cdot (SbF_5)_2$ or $XeF_2 \cdot (TaF_5)_2$. However, BF_3 does not appear to form complexes with XeF_2 or XeF_4 up to $200°$ (1, 2). Complexes with XeF_6 have not yet been described.

It has now been found that XeF_6 will form addition compounds with BF_3, AsF_5, and SbF_5.

BF_3 and XeF_6 combine at room temperature to form an adduct of composition $XeF_6 \cdot BF_3$. At room temperature the adduct is a white solid which melts at $90°$ to a pale-yellow viscous liquid. The liquid has a marked tendency to supercool and eventually crystallizes in the form of colorless long needles. The adduct has a vapor pressure of less than 1 mm at $20°$. However, it can be sublimed under vacuum at this temperature. The infrared spectrum of the solid at $-195°$ is complex, but shows a broad absorption centered at about 1025 cm^{-1} which is characteristic of BF_4^- (3). The infrared spectrum of the vapor was obtained in a reflection cell of 60 cm path length (4). This shows absorptions characteristic of BF_3 and XeF_6, indicating that the adduct is at least partially dissociated in the vapor phase.

---

* Based on work performed under the auspices of the U.S. Atomic Energy Commission.

$XeF_6 \cdot BF_3$ reacts with NaF at 100° to form $NaBF_4$ and an illdefined addition product between XeF_6 and NaF. The XeF_6 can be recovered from the latter by pumping at 100° (5).

XeF_6 also reacts with AsF_5 at room temperature to form a 1 : 1 addition compound. This is a white solid which cannot be sublimed at room temperature. Raman spectra of solid $XeF_6 \cdot BF_3$ and $XeF_6 \cdot AsF_5$ prepared in situ in a sapphire cell show scattering peaks at 612 cm^{-1} and 620 cm^{-1}, respectively, which cannot be ascribed to any of the components (3, 6, 9). It is interesting that solutions of XeF_6 in anhydrous HF show a scattering peak at 620 cm^{-1} (7), and possibly the species in HF may be similar to that in the adducts.

Similarly to XeF_2, XeF_6 forms a complex with SbF_5 of composition $XeF_6 \cdot (SbF_5)_2$. This complex, which is a white solid at room temperature, possesses an extremely low vapor pressure. When introduced into a Bendix time-of-flight mass spectrometer, no fragmentation pattern could be observed at room temperature. $XeF_6 \cdot (SbF_5)_2$ appears to be insoluble in SbF_5 at room temperature. However, its solubility increases with temperature. A 0.19 M solution of $XeF_6 \cdot (SbF_5)_2$ in SbF_5 at 35° showed no increase in conductivity over that of pure SbF_5.

The adducts of XeF_6 compare in chemical reactivity with that of XeF_6 itself (8). They are extremely hygroscopic. Although they can be stored in nickel containers, they appear to react with pyrex and more slowly with Kel-F. The reactivity of these compounds and their low volatility present formidable obstacles to the determination of their structures.

PEACOCK (1) has suggested that the complex $XeF_2 \cdot (SbF_5)_2$ may be covalent with the possibility of fluorine bridges linking the heavy atoms. The XeF_6 adducts may have similar structures, particularly in the case of the SbF_5 complex. Ionic complexes of the type $[XeF_5]^+[AsF_6]^-$ cannot be ruled out, however. The $[AsF_6]^-$ configuration is known to be particularly stable (9), while the stability of a possible $(XeF_5)^+$ may be enhanced by the favorable octahedral geometry predicted for one lone and five bonded electron pairs (10).

Although $[SbF_6]^-$ is even more stable (11) than $[AsF_6]^-$, the XeF_6 complex with SbF_5 may represent a special case. The high viscosity of pure SbF_5 is associated with the formation of polymeric species. These may play a role in the different stoichiometry of the SbF_5 complex as compared with that of the AsF_5 complex.

References

1. EDWARDS, A.J., HOLLOWAY, J.H., and PEACOCK, R.D., Proc.Chem. Soc. 235 (1963).

2. BARTLETT, N., Chem.Eng.News, <u>41</u>, 36 (1963).

3. SHARP, D.W.A., "Advances in Fluorine Chemistry", Vol.I, p.79.

4. WEINSTOCK, B., CLAASSEN, H.H., and CHERNICK, C.L., J.Chem. Phys. <u>38</u>, 1470 (1963).

5. SHEFT, I., private communication.

6. SMITH, D.F., "Noble Gas Compounds", HYMAN, H.H., Editor, University of Chicago Press, p.301 (1963).

7. HYMAN, H.H., and QUARTERMAN, L.A., ibid., p.275.

8. MALM, J.G., SHEFT, I., and CHERNICK, C.L., J.Am.Chem.Soc. <u>85</u>, 110 (1963).

9. HYMAN, H.H., LANE, T.J., and O'DONNELL, T.A., Presented at 145th Meeting of American Chemical Society, New York, Sept.1963.

10. GILLESPIE, R.J., ref. (6), p.333.

11. HYMAN, H.H., QUARTERMAN, L.A., KILPATRICK, M., and KATZ, J.J., J.Am.Chem.Soc. <u>65</u>, 123 (1961).

7 B 2

COORDINATION OF SILICON AND GERMANIUM TETRAFLUORIDE WITH SOME OXYGEN DONORS.

R.C. Aggarwal, J.P. Guertin and <u>M. Onyszchuk</u>
Department of Chemistry, McGill University, Montreal 2, Canada.

Few coordination compounds of silicon and germanium tetrafluoride have been prepared and characterized. MUETTERTIES (1) has described $SiF_4[(CH_3)_2SO]_2$ and $SiF_4[(CH_3)_2NCHO]_2$ and the corresponding complexes of germanium tetra-fluoride. ISSLEIB and REINHOLD (2) have prepared $SiF_4[(C_6H_5)_3PO]_2$, $SiF_4[(C_6H_{11})_3PO]_2$, $SiF_4[(CH_3)_3NO]_2$ and $SiF_4[C_5H_5NO]_2$. The interaction of silicon tetrafluoride with methanol has been shown recently to involve strong hydrogen bonds between fluorine atoms and hydroxyl protons rather than octa-hedral coordination of silicon (3). We have now examined the relative tendency of silicon and germanium tetrafluoride to complex with simple ethers, acetone, and methanol.

Ethylene oxide reacted with silicon tetrafluoride at -78° in a vacuum to produce a white solid complex, $SiF_4[(CH_2)_2O]_2$, the composition of which was confirmed by quantitative synthesis and tensimetric titration. Attempts to measure dissociation pressures of the adduct were unsuccessful because it dis-sociated and decomposed slowly above -78° and rapidly at -45° into 1,4-di-oxane, a non-volatile polymer, and silicon tetrafluoride. This type of dimerization and polymerization of ethylene oxide has previously been reported in reactions of boron trifluoride and trichloride with ethylene oxide. Trimethylene oxide also reacted with silicon tetrafluoride at -78° in a 1:2 mole ratio to form a white solid complex which decomposed above -78° into a clear viscous polymer and silicon tetrafluoride. An infrared spectrum of the complex, measured at -185°, contained a strong band at 720 cm^{-1} which is characteristic of the Si-F octahedral bond stretching vibration.

Tetrahydrofuran and tetrahydropyran reacted at -78° with silicon tetrafluoride to give white solid 1:2 adducts. These melted at about -10° without polymerization and were completely dissociated in the gas phase at 25°. Heats of dissociation were estimated from dissociation pressures measured in the range -70 to -30°. Dimethyl ether formed a 1:2 complex which melted with dissociation at about -45°. Surprisingly, 1,4-dioxane did not react with silicon tetrafluoride in the range 25 to -115°, as evident from the fact that total pressures of mixtures of the two gases obeyed Dalton's Law of Partial Pressures. There was also no reaction between acetone and silicon tetrafluoride, even at -115°. The analytical data and stabilities of the new silicon tetrafluoride-ether complexes are summarized in the following table.

TABLE 1

Complex	Ratio of ether to SiF$_4$		Decomp. and/or Dissoc.		$\Delta H_{dissoc.}$ kcal. mole^{-1}
	Tensimetric titration	Quantitative synthesis	$^\circ$C	Products	
SiF$_4$[(CH$_2$)$_2$O]$_2$	2,02	2,01	-78	SiF$_4$, (CH$_2$)$_2$O, C$_4$H$_8$O$_2$, polymer.	-
SiF$_4$[(CH$_2$)$_3$O]$_2$	2,27	2,09	-64	SiF$_4$, polymer.	-
SiF$_4$[(CH$_2$)$_4$O]$_2$	2,00	1,91	-78	SiF$_4$, (CH$_2$)$_4$O	11,5
SiF$_4$[(CH$_2$)$_5$O]$_2$	2,08	2,04	-95	SiF$_4$, (CH$_2$)$_5$O	11,0
SiF$_4$[(CH$_3$)$_2$O]$_2$	2,01	2,01	-110	SiF$_4$, (CH$_3$)$_2$O	9,0

Combining ratios of germanium tetrafluoride with ethers, acetone, and methanol were obtained by tensimetric titration and quantitative synthesis. These values are shown in the following table together with analytical data and melting points.

The ethylene oxide complex decomposed, like SiF$_4$[(CH$_2$)$_2$O]$_2$, into 1,4-dioxane and a purple polymer. The other ether complexes were stable, white, highly hygroscopic solids.

TABLE II

Complex	Combining ratio Donor/GeF$_4$	Analysis, %				m.p. °C
		Germanium		Fluorine		
		found	calc.	found	calc.	
GeF$_4$[(CH$_2$)$_2$O]$_2$	1,98	-	-	-	-	dec. -78
GeF$_4$[(CH$_2$)$_4$O]$_2$	1,97	24,7	24,79	26,3	25,96	152
GeF$_4$[(CH$_2$)$_5$O]$_2$	1,98	24,0	24,52	24,0	23,69	115
GeF$_4$[(CH$_3$)$_2$O]$_2$	1,96	30,2	30,16	31,5	31,57	subl. 90
GeF$_4$[(CH$_3$)$_2$CO]$_2$	2,08	27,4	27,42	28,6	28,70	dec. 65
GeF$_4$[CH$_3$OH]$_2$	2,04	34,8	34,13	35,4	35,68	dec. 145
GeF$_4$[O(CH$_2$)$_4$O]	1,01	30,4	30,67	31,9	32,11	subl. 125

The following striking differences between silicon and germanium tetra-fluoride are apparent: (i) germanium tetrafluoride forms more stable complexes with cyclic ethers and dimethyl ether than does silicon tetrafluoride; (ii) silicon tetrafluoride does not coordinate with 1,4-dioxane whereas germanium tetrafluoride forms a stable 1:1 adduct; (iii) germanium tetrafluoride but not silicon tetrafluoride gives a 1:2 complex with acetone and methanol, both of which are very weak donors. These differences are less evident in the complexes of silicon and germanium tetrafluoride with nitrogen donors, such as ammonia, trimethylamine, and pyridine, all of which are stronger donors than ethers, acetone, or methanol.

The greater electron pair acceptor power of germanium tetrafluoride compared with silicon tetrafluoride suggests that the reorganization energy required to convert sp^3-tetrahedral bonding in silicon and germanium tetra-fluoride to sp^{3}d^2-octahedral bonding in their 1:2 complexes is less for germanium than for silicon. As in the case of the boron trihalides, the reorganization energy is propably mainly that necessary to break supplementary π-bonding in silicon and germanium tetrafluoride, although π-bonding probably also occurs in the octahedral complexes but to a lesser degree because the bond distances are greater. It appears, therefore, that the d$_\pi$-p$_\pi$ bond energy is greater in silicon tetrafluoride than in germanium tetrafluoride. This lesser tendency of germanium to use its vacant 4d-orbitals in d$_\pi$-p$_\pi$ supplementary bonding is also indicated by nuclear quadrupole resonance measurements, basicity of oxygen in Ge-O-Ge and Ge-O-C linkages, and dipole moments and bond shortening in germyl halides.

References

1. MUETTERTIES, E.L., J.Am.Chem.Soc., **82**, 1082 (1960).
2. ISSLEIB, K., and REINHOLD, H., Z. anorg.Chem., **314**, 113 (1962).
3. GUERTIN, J.P., and ONYSZCHUK, M., Can.J.Chem., **41**, 1477 (1963).

7 B 3

NEUE VERBINDUNGEN MIT DER F_5Te-GRUPPE.

A. Engelbrecht und F. Sladky
Institut für Anorganische und Analytische Chemie der
Universität Innsbruck, Austria.

Vor kurzem berichteten wir über die Darstellung von Pentafluoro-ortho--tellursäure, $HOTeF_5$, als Hauptprodukt der Reaktion von Bariumtellurat mit Fluorsulfonsäure (1). Die Analyse und Eigenschaften dieser Verbindung zeigen, daß es sich um eine starke, leichtflüchtige Säure handelt, welche mit Fluoriden, Chloriden und Karbonaten unter Entwicklung von HF, HCl bzw. CO_2 reagiert.

Durch geringe Änderungen der Bedingungen bei der Umsetzung von Bariumtellurat mit Fluorsulfonsäure lassen sich andere, ebenfalls leichtflüchtige und gleichfalls die F_5Te-Gruppe enthaltende Verbindungen darstellen.

So konnte in einem der Versuche die Verbindung F_5TeOSO_2F mit einer Ausbeute von über 60% (berechnet auf $BaTeO_4$) gewonnen werden. Dieses Pentafluor-tellur-fluorsulfonat ist eine leicht bewegliche Flüssigkeit (Sdp. 58°), welche gegen Wasser und verdünnte Laugen bemerkenswert stabil ist.

Eine weitere, mit etwa 15% Ausbeute erhaltene, bei 118° siedende Flüssigkeit ergab eine, mit der Formel eines Bis-pentafluorotellur-sulfates, $(F_5Te)_2SO_4$ übereinstimmende Analyse. Auch diese Verbindung ist überraschend inert und braucht mehrere Stunden, um mit 3 N Lauge vollständig zu hydrolysieren.

Während uns vom Tellur nur eine, auf gänzlich andere Art erhaltene, ähnliche Verbindung $(F_5TeOTeF_4OTeF_5)$ bekannt ist (2), wurden die unseren Verbindungen analogen Schwefelverbindungen kürzlich dargestellt (3), allerdings mit Ausnahme des $HOTeF_5$, deren Schwefel-Analogon wohl nicht stabil sein dürfte.

Weitere sehr reaktionsfähige Produkte der Reaktion von $BaTeO_4$ mit HSO_3F sind in Untersuchung.

Literatur

1. ENGELBRECHT, A., und SLADKY, F., Angew.Chem., im Druck.
2. CAMPBELL, R., und ROBINSON, P.L., J.chem.Soc. (London), 1956, 3454.
3. MERRILL, C.I., und CADY, G.H., J.Am.Chem.Soc., 85, 909 (1963);
   PASS, G., und ROBERTS, H.L., Inorg.Chem., 2, 1016 (1963);
   PASS, G., J.chem.Soc. (London), 1963, 6047.

7 B 4

THE PREPARATION AND PROPERTIES OF RARE EARTH HEXAFLUOROACETYLACETONATES.

M. Steinberg, J. Mashall and A. Glasner
Department of Inorganic and Analytical Chemistry,
The Hebrew University, Jerusalem, Israel.

Reports on the preparation of metal hexafluoroacetylacetonates (hfaa) have recently been published in which these chelates were proposed as a possible means for the separation of metal cations (1,2). The work presented here deals with the preparation of rare earth chelates of hfaa.

Preliminary experiments proved that the unsubstituted acetylacetonates of the rare earths were thermally unstable, as already noted (3), so that the chelates would not be suitable for their separation by gas chromatography. On the other hand, as suggested by SIEVERS et al. (1), the hfaa chelates seemed to be more promising. In order to prevent thermal hydrolysis, the hfaa chelates had to be obtained in an anhydrous state. The hfaa chelates prepared from aqueous solution of the cation in the usual manner are hydrated compounds, and the tenaciously held water could not be removed even after drying for months under high vacuum in the presence of phosphorus pentoxide. Attempts to prepare the chelates by the direct reaction of pure and dry hfaa with the metals were unsuccessful. The reaction began only after the addition of small amounts of water, which set off a very vigorous and uncontrolled reaction. Dry hfaa also reacted very slowly with calcined oxide, especially when the calcination temperature was high ($\sim 1000^\circ$).

A suitable method for the preparation of the anhydrous chelate was found to be the reaction of a suspension of anhydrous chloride in dry acetone with an excess of redistilled anhydrous hfaa. The elementary analysis showed that an addition product with acetone was obtained, in the ratio of 1:1 (e.g. in the case of $Gd(hfaa)_3(CH_3)_2CO$: C found 25,65; calc. 25,84%; H found 1,35; calc. 1,08%; Gd found 18,8; calc. 18,74%).

IR spectra of the chelates in NaCl discs and in CCl_4 solutions were recorded. In the case of the NaCl discs the free carbonyl stretching band in the 1700 cm^{-1} region was not observed. However, the IR spectra of CCl_4 solutions showed a band at 1692 cm^{-1}, in addition to the carbonyl-metal band in the 1640 cm^{-1} region.

The hfaa chelates of Sc, Y, Pr, Nd, Eu and Gd were prepared by the above method. The absorption bands observed in the NaCl region in the IR spectra indicated that the chelates were cyclic (4). Apparently their stability constants are nearly the same, the carbonyl stretching bond peaking nearly at the same frequencies:

hfaa (in CCl_4)	hfaa (gas)	Sc	Y	La	Pr	Nd	Eu	Gd
1682	1682	1643	1650	1640	1637	1640	1640	1644 cm^{-1}

For comparison the IR spectrum of the chromium hfaa chelate was taken, when the carbonyl shifted to 1610 cm^{-1}, indicating the formation of a more stable chelate (the value reported in the literature is 1612 cm^{-1} (5)).

The volatility and ease of oxidation of the solid chelates seems to be independent of the stability constants. Differential thermal analysis and thermo-gravimetric runs showed that the Sc-chelate sublimed at about 100° both in pure argon and in air. The Gd-chelate sublimed at about 150° in argon, but decomposed in air. The weight of the residue conformed fairly well with the oxide.

Gas chromatography experiments (injection port 150°, column 60°) on the chelates in hfaa solutions gave two distinct peaks, the first presumably due to the solvent, and the second that of the chelate. However, the retention times of the individual chelates (with the columns so far used) did not differ sufficiently to make quantitative determinations possible. No third peak which might be attributed to acetone from the addition product was observed.

References

1. SIEVERS, R.E., PONDER, B.W., MORRIS, M.L., and MOSHIER, R.W., Inorg.Chem., 2, 693 (1963).
2. ROSS, W.D., and WHEELER, G., Anal.Chem., 36, 266 (1964).
3. BAILAR, Jr., J.C., "The Chemistry of the Coordination Compounds", Reinhold Publ.Corp., New York, 1956, p.42.
4. NONHEBEL, D.C., J.Chem.Soc. (London), 1963, 738.
5. MORRIS, M.L., MOSHIER, R.W., and SIEVERS, R.E., Inorg.Chem., 2, 411 (1963).

S B 1

MULTIPLY BONDED OXO, IMIDO AND NITRIDO COMPLEXES OF RHENIUM(V).

J. Chatt, J.D. Garforth, N.P. Johnson and G.A. Rowe
Imperial Chemical Industries Limited, Heavy Organic Chemicals Division,
Akers Research Laboratories, The Frythe, Welwyn, Herts., U.K.

Transition metals (M), especially those around the vanadium and chromium Groups in the Peridic Table have a strong tendency to form oxy-cations of the type MO^{n+} in which one oxygen atom is strongly bonded to the metal and remarkably resistant to protonation. This behaviour is to be associated with strong π-bonding between the metal and the oxygen, so that the M—O bond has considerable triple bond character (M≡O) (1, 2). The oxygen atom thus resembles that of carbon monoxide rather than that of the ketones, and should carry considerably less negative charge than would a truly doubly bonded oxygen (M=O).

It is difficult to obtain exact data relating to the polarity of the M—O grouping in complex ions, but the discovery of complexes of the type trans-$[ReOCl_3(PR_3)_2]$ and related arylimido and nitrido complexes which are soluble in benzene and resistant to protonation gives some data about the possible polarity of the Re-O and Re-N bonds by the measurement of dipole moments.

TABLE I

DIPOLE MOMENTS (DEBYE UNITS)

trans-$[ReOCl_3(PR_3)_2]$		trans-$[Re(N-4-C_6H_4X)Cl_3(PEt_2Ph)_2]$	
Phosphine		X	
PPh_3	2,25	OMe	7,2
PEt_2Ph	1,7	Me	6,5
PEt_3	1,7	H	5,9
$P(Pr^n)_3$	1,35	Br	5,2
		Cl	5,0
		F	4,6
		COMe	4,5

Owing, however, to the discovery of considerable distortion from the octahedral angles in trans-[ReOCl$_3$(PEt$_2$Ph)$_2$] (3) the information is not so useful as
we originally expected when the moments were measured. Relevant dipole
moments are listed below.

TABLE II

DIPOLE MOMENTS (DEBYE-UNITS)

[ReCl$_3$(PEt$_2$Ph)$_3$]	6,3	[ReNCl$_2$(PEt$_2$Ph)$_3$]	6,4
[ReBr$_3$(PEt$_2$Ph)$_3$]	6,5	[ReNBr$_2$(PEt$_2$Ph)$_3$]	5,8 *
carbon monoxide	0,13 **		
benzoisonitrile	3,56 **		
p-toluisonitrile	4,0 **		

*   unstable in solution
** taken from Tables of Experimental Dipole Moments.
    A.L.McCLELLAN, Freeman and Company, San Francisco
    and London 1963.

The p-substituted phenylimido series of complexes is most instructive.
The more electron releasing substituents give the higher dipole moments,
indicating that they lie at the positive end of the dipole. The formally double
negative arylimido ligands thus carry considerable positive charge. Mesomeric
effects influence the dipole moments more than inductive effects, as would be
expected if the moment arises mainly from strong N-Re π-bonding. The seat
of positive charge responsible for the dipole moments of the arylimido series
of complexes is thus the nitrogen atom and its mesomerically induced positive
charges in the aryl system. The seat of negative charge will be the chlorine
atom trans to the nitrogen atom. The rather high moment of e.g. 5,9 D for
the phenylimido complex indicates the presence of somewhat strong π-bonding
of the type Ph—N≋Re , rather as in benzoisonitrile Ph—N≋C which has a
moment of 3,5 D.

The dipole moments of the oxo complexes also provide evidence of strong
Re≋O bonding in keeping with the short Re—O distance (1,6 Å) found by
X-ray structure analysis (3). In the solid trans-[ReOCl$_3$(PEt$_2$Ph)$_2$] has a distorted
octahedral structure in which the P—Re—P angle is 162°, the O—Re—Cl(2)
angle 180° and Cl(1)—Re—Cl(3) angle 174,6°. If this shape is retained in
solution and the reasonable but approximate values of 6 D and 2 D assumed for
the P—Re and Re—Cl bond moments respectively, the Re=O bond moment can

be calculated as approximately 1,3 D with oxygen positive or 2,1 D with oxygen negative.

All evidence points to the oxygen being positive rather than negative; (a) it is not easily protonated; (b) the more electronegative triphenylphosphine gives a complex, with the phosphine molecule in partial opposition to the oxygen, and of higher moment than that of the more aliphatic phosphines; (c) the short Re—O distance. The estimated bond moment of 1,3 D is very approximate but is probably within 1 D of the true value putting the oxygen atom definitely on the positive side of neutrality.

The complex $[ReX_3(PEt_2Ph)_3]$ and $[ReNX_2(PEt_2Ph)_3]$ must have the same cis-meridial arrangement of phosphine molecules. The alternative cis-facial arrangement would demand much higher dipole moments. Comparing these two series of complexes it is evident that replacement of chlorine by nitrogen, with appropriate change in oxidation number of the rhenium has little effect on the dipole moment. It appears, therefore, that the Re(V)≡N and Re(III)—Cl bond moments are about equal.

This comparison may, however, be spurious. Since the oxo-complex is considerably distorted so might the nitrido complex. However, distortion in the same sense as in the oxo complex would enhance the apparent negative charge carried by the nitrogen atom, and we may safely conclude that the actual negative charge carried by the coordinated N^{3-} ion is not greater than that carried by the coordinated Cl^- in spite of its greater formal negative charge. This again indicates strong triple bonding of the nitrogen to the metal, Re≡N.

References

1. e.g., see BALLHAUSEN, C.J., and GRAY, H.B., J.Inorg.Chem., 1, 111 (1962).
2. CHATT, J., and ROWE, G.A., J.Chem.Soc. (London), 1962, 4019.
3. EHRLICH, H.W.W., and OWSTON, P.G., J.Chem.Soc., (London), 1963, 4368.

B 2

SOME NOVEL COORDINATION COMPOUNDS OF GROUP IV METALS.

A.F. Reid and P.C. Wailes
Division of Mineral Chemistry and Division of Organic Chemistry,
C.S.I.R.O., Australia.

In investigations to determine whether ligand groups can be added directly to the lower halides of zirconium and titanium it has been found that cyclopentadiene vapour reacts with these compounds at 200 - 300° to put a C_5H_5 ring in each of the vacant valency positions according to the reaction:

$$n\ C_5H_6 + (Zr,Ti)X_{4-n} \longrightarrow (C_5H_5)_n(Zr,Ti)X_{4-n} + \frac{n}{2}H_2$$

$$\{1\}$$

with the hydrogen apparently reacting with excess cyclopentadiene. From the appropriate lower halides the previously undescribed compounds $C_5H_5ZrCl_3$, $C_5H_5ZrBr_3$ and $C_5H_5ZrI_3$ and the known compounds $(C_5H_5)_2ZrCl_2$, $(C_5H_5)_2ZrBr_2$ and $C_5H_5TiCl_3$ were made. The known compounds had physical properties and X-ray powder patterns identical with samples made by other routes. The zirconium trihalogeno-compounds were also synthesized by the action of magnesium cyclopentadienide in xylene solution on zirconium tetrahalides, and these compounds were identical with those from the lower halides reactions. In all cases a single sharp proton magnetic resonance peak was obtained from the compounds in solution, showing them to contain π-bonded C_5H_5 rings. This was confirmed by the near infra-red reflectance spectra of the solids (1), which correspond closely with that of ferrocene.

In attempts to prepare a direct carbon zirconium bond CF_3I was reacted with ZrI_3 at 210°. A copious evolution of iodine occurred and reaction products included an extremely unreactive black, involatile residue containing Zr, C and F, and a very small yield of clear yellow hexagonal platelets subliming at 50-80°. Mass spectral examination showed these to contain CF_3 groups and iodine, but their characterization as a CF_3-zirconium compound is incomplete as yet.

In a separate study we have been investigating the product of the reaction of aniline and other bases with $(C_5H_5)_2ZrCl_2$, reported by SAMUEL and SETTON (2) as "$C_{10}H_9ZrCl$". Proton magnetic resonance and near infrared reflectance spectra show the C_5H_5 rings to be unchanged and all the protons to be in identical situations. The compound is diamagnetic and M.W. measurements and mass spectra show the M.W. to be 490 (510 is required). The parent mass peak is not obtained, but the isotope ratio distributions show two Zr atoms to be present in a number of fragments above mass 290. Thus the compound is a dimer and should be written $[(C_5H_5)_2ZrCl]_2$. It was found to be orthorhombic, with a = 13,69 Å, b = 8,09 Å, c = 36,16 Å, each ± 0,2 Å, and to have a density of 1,700 ± 0,005, and thus with good accuracy to have 8 molecules (16 Zr atoms) per unit cell.

The compound is rapidly and quantitatively reconverted to the starting material by the action of HCl, and thus the reversible system exists:

$$2\,(C_5H_5)_2ZrCl_2 + 2\,C_6H_5NH_2 \rightarrow [(C_5H_5)_2ZrCl]_2 + 2\,C_5H_6NH_2HCl \quad \{2\}$$

$$[(C_5H_5)_2ZrCl]_2 + HCl \longrightarrow 2\,(C_5H_5)_2ZrCl_2 \quad\quad\quad \{3\}$$

No hydrogen is evolved in reaction $\{3\}$, and for $\{2\}$ mass balances for aniline, aniline hydrochloride and dimer have been obtained, and the reaction can occur in carbon tetrachloride solution. Thus the hydrogen does not appear to come from the base of the solvent. Use of DCl in $\{3\}$ shows that no C-D bonds are formed on the cyclopentadienyl rings, and the possibility appears to exist that $(C_5H_5)_2ZrCl_2$ contains a "hidden" proton.

The final novel zirconium compound we wish to report is $(C_5H_5)_4Zr$, made by the action of excess sodium cyclopentadienide solution on $ZrCl_4$, followed by vacuum separation and sublimation. The yellow compound is very reactive in air, contains π-C_5H_5 rings but no chlorine, and is diamagnetic.

A novel and simple reaction process will also be mentioned which is particularly suited to the preparation of the tetracyclopentadienyl compounds of thorium and uranium, and the tricyclopentadienides of scandium and the rare earths.

References

1. REID, A.F., SCAIFE, D.E., and WAILES, P.C., Spectrochimica Acta (1964), in press.
2. SAMUEL, E., and SETTON, R., Compt.Rend., 256, 443 (1963).

8 B 3

SOME COMPLEX SULPHUR COMPOUNDS OF TITANIUM AND VANADIUM.

G.W.A. Fowles
Department of Chemistry, University of Southampton, U.K.

The reactions between $TiCl_4$, $TiBr_4$, $TiCl_3$, VCl_4, VCl_3, and VBr_3 with a range of monodentate sulphur ligands, have been investigated. The ligands used include the aliphatic thioethers, R_2S, for R = Me, Et, Pr^n, and Bu^n, and tetrahydrothiophen and pentamethylenesulphide. Notable differences have been observed in the reactions of the halides of the two elements. Thus, whereas the tetrahalides of titanium form simple adducts with all ligands tried, vanadium(IV) chloride is reduced rapidly and quantitatively to the tervalent state. Furthermore, tervalent vanadium forms a series of well-defined complexes with sulphur ligands, but the complexes formed by tervalent titanium are very much less stable and readily break down with the formation of complexes of quadrivalent titanium.

With $TiCl_4$ and $TiBr_4$, all the monodentate sulphur ligands (L) mentioned above, form complexes analysing to $TiX_4.L_2$, except for $TiBr_4$ with SPr_2^n and SBu_2^n where the products appear to be $TiBr_4.L$. The 1:2 adducts are soluble in non-polar solvents, and molecular weight studies indicate that while the chloro compounds are monomeric in solution, with the bromo compounds there is extensive dissociation probably of the kind $TiBr_4.L_2 \rightleftharpoons TiBr_4.L + L$. The available spectroscopic data (i.r. and u.v.) will be discussed. $TiCl_3$ and $TiBr_3$ both appear to give compounds of the type $TiX_3.L_2$ with dialkyl sulphides, but these compounds break down readily on gentle warming with the formation of $TiX_4.L_2$ and a so-far unidentified residue. The analogous complexes formed with tetrahydrothiophen are rather more stable, and behave in every way as complexes of tervalent titanium (e.g. visible spectra shows typical d-d transition), but they have low magnetic moments (e.g. $TiCl_3(THT)_2$, $\mu = 1.05$ B.M.).

The products formed in the reactions of the vanadium chlorides with the dialkyl sulphides (R = Me and Et) have the composition $VCl_3(SR_2)_2$. These compounds are monomeric in benzene solution and have dipole moments of about 2.5 D. On the basis of the spectroscopic evidence, and by analogy with the trimethylamine adducts ($VX_3(NMe_3)_2$) these compounds have been assigned trans-trigonal-bipyramidal structures. Their visible and ultra-violet spectra have been measured and the significance of these results will be discussed. With the higher members of the thioether series, VCl_3 forms rather indeterminate compounds, the thermal decomposition of which indicates (for the VCl_3-SPr_2^n systems) some stability for the composition $VCl_3(SPr_2^n)_2$.

Several complex compounds of titanium with thioxan and dithian have been prepared. Thus, with $TiBr_4$ and $TiCl_4$, thioxan(L) gives compounds $TiX_4.L_2$; on the basis of infra-red spectra the ligands have been shown to coordinate through sulphur and not oxygen. For tervalent titanium the following three thioxan compounds have been prepared: $TiCl_3.L_2$, $TiCl_2.L$ and $TiBr_3.L_2$. Spectra and magnetic susceptibility data have been obtained and will be discussed.

A limited number of compounds of the type $TiX_4.B$ have been prepared, where B represents a number of bidentate sulphur ligands. Structural information is not yet available on these compounds.

B 1

HYDRIDO AND CARBONYLHYDRIDOIRIDIUM COMPOUNDS.

L. Malatesta, M. Angoletta and G. Caglio
Istituto di Chimica Generale e Inorganica dell'Università,
Via Saldini, 50, Milano, Italy.

The iridium complexe hydrides so far reported, belong to the type IrH_3L_3, where L may be triphenylphosphine (1) or the diethylphenylphosphine (2). Using as ligand triphenylphosphine we have now prepared the unsaturated hydride IrH_3L_2, as well as the hydridocarbonylphosphineiridium compounds $IrHCOL_2$, IrH_3COL_2, $IrH(CO)_2L_2$ and $IrHCOL_3$.

The trihydridobistriphenylphosphineiridium IrH_3L_2 [I] was prepared by reduction with both sodium tetrahydridoborate and lithium tetrahydrido-aluminate, of IrI_3L_2 or IrI_2HL_2. It is a white powder, stable to air and moisture, almost insoluble in all solvents, diamagnetic, m.p. 145^0 (dec.). The number of hydridic hydrogen atoms, which can not be determined by analysis only, was proved by the reaction with triphenylphosphine, which gave place very rapidly in the cold to the two isomeric forms of the well known trihydride IrH_3L_3 (1), and by the following quantitative reactions:

$$IrH_3L_2 + HClO_4 = [IrH_2L_2]ClO_4 + H_2$$
$$[IrH_2L_2]^+ + I^- = IrH_2IL_2$$

The very low solubility of compound [I], could have been ascribed to polymerisation, through hydrogen bonds, but this was disproved by the fact that it has only one very strong absorption band in the IR at 1945 cm^{-1}, that is in the region characteristic for normal metal-hydrogen bonds. This band could be unequivocally assigned to Ir-H stretching, because it moves to 1360 cm^{-1} in the corresponding deuteride, prepared by using LiAlD$_4$ as reducing agent. The presence of only one hydrogen stretching band is consistent with a trigonal structure with the two phosphine at the apices and the hydrogen atoms on the equatorial plane of the bipyramid.

Compound [I] reacted in very mild condition with pyridine giving the saturated hydride IrH_3L_2Py, m.p. 134^0, which has two strong bands in the IR at 1700 and 2120 cm^{-1}. These two bands are very similar to the bands of the α-form of IrH_3L_3 (1), for which we proposed a structure with 2 hydrogen atoms trans each other. We suggest that in this case too the trans isomer is formed.

Compound [I] absorbed carbon monoxide yielding a soluble carbonyl-hydride, m.p. 132°, whose analytical composition agreed fairly well to IrH_xCOL_2 [II]. When we tried to prepare the same compound by reduction of $IrClCOL_2$ and of $IrHI_2COL_2$ with $NaBH_4$, we obtained a different carbonyl-hydride [III], in two isomeric forms, melting at 138° and 146° resp., which however showed, within experimental error, the same analytical composition as [II]. Compounds [II] and [III] (isomeric) reacted with perchloric acid giving the same perchlorate, the former without any evolution of gas, the latter with the evolution of one mole hydrogen. We therefore formulated the two hydride as monohydride and trihydride resp.:

$$IrHCOL_2 \text{ [II]} + HClO_4 = [IrH_2COL_2]ClO_4$$

$$IrH_3COL_2 \text{ [III]} + HClO_4 = [IrH_2COL_2]ClO_4 + H_2$$

This was confirmed by hydrogenating in solution compound [II] to compound [III] with hydrogen at atmospheric pressure, and by accomplishing the inverse reaction on [III], with nitrogen. This was the first example of a complex iridium hydride adding reversibly hydrogen analogously as VASKA's $IrClCOL_2$ (3).

The perchlorate $[IrH_2COL_2]ClO_4$ behaved in acetone as a uni-univalent electrolyte and showed in the IR spectrum three bands at 2165, 2085 and 2050 cm^{-1}. With ethanolic potassium hydroxide it immediately gave the monohydride [II]:

$$[IrH_2COL_2]^+ + OH^- = IrHCOL_2 + H_2O$$

With halide ions X$^-$ it gave the non-electrolyte $IrXH_2COL_2$, and with triphenylphosphine the cation $[IrH_2(CO)L_3]^+$ (see below).

The monohydride [II] has two strong IR bands at 1960 and 2000 cm^{-1}, as expected. The high melting form of the trihydride has three strong bands at 1785, 1965 and 2080 cm^{-1}. The first of these bands, which is also the strongest, falls in the region which seems to be characteristic of compounds having two hydrogens in trans to each other (1). This trihydride therefore, deriving from a d^6, low spin, hexacoordinated iridium(III) can be considered to have an octahedral structure, analogous to that of the trans-trihydridotris-triphenylphosphineiridium(III), with a CO in the place of a triphenylphosphine. The low melting form has three bands at 2118, 2080, 1960 cm^{-1} and should be considered the cis isomer.

The monohydride [II] derives from a d^8, tetracoordinated, low spin, iridium(I) atom, which is considered to give square planar coordination. Its electric moment of about 6,3 D and the position of the hydrogen stretching band seem to indicate that it is the cis isomer.

Compound [III] (either isomer) reacted with carbon monoxide at 200 atm. and 50-60° to give a compound [IV] melting above 300° (dec.) corresponding in composition to $[Ir(CO)_3L]_2$ or to $IrH(CO)_3L$ and showing a very strong unique band, in the CO and Ir-H stretching region, at 1950 cm^{-1}. It does not react with triphenylphosphine in boiling benzene, thus favouring the first structure, which, being diamagnetic, has been formulated dimeric.

The perchlorate deriving from compound [III] reacted with carbon monoxide to give the perchlorate of a saturated cation:

$$[IrH_2COL_2]^+ClO_4 + CO \rightleftharpoons [IrH_2(CO)_2L_2]ClO_4$$

showing in the IR a strong band at 2002 cm^{-1}, with a shoulder at 1978 cm^{-1} This compound by action of lithium chloride or bromide LiX gave the following reaction:

$$[IrH_2(CO)_2L_2]^+ + X \rightleftharpoons IrXCOL_2 + H_2 + CO$$

while with potassium hydroxide it gave a new carbonylhydride m.p. 123° which we temptatively formulated as $IrH(CO)_2L_2$ [V]. We have still doubts on the formula of [V], which showed a molecular weight in accordance to the monomeric structure, especially because of its IR spectrum. In fact the two strong bands at 2000 and 1945 cm^{-1} could very well be assigned to CO stretching, while the strong band at 1634 cm^{-1} has a wave number much lower than usually observed for metalhydrogen stretching.

Another compound belonging to these series, is the hydridocarbonyl-tristriphenylphosphineiridium $IrHCOL_3$ [VI], m.p. 145°. This could be obtained from many inorganic iridium carbonyl derivatives as $K_2[IrCOBr_5]$, $K[IrCOI_5]$, $Ir(CO)_2I_3$ (4), by reduction with hydridoborate, in presence of an excess of triphenylphosphine, as well as by action of a mole of triphenylphosphine on the hydrides [II] and [III]. It was also obtained in very little amount when $[IrH_2L_3]ClO_4$ (1) was reacted with ethanolic potassium hydroxide. The elemental analysis of compound [VI] was indicative of the $IrCOL_3$ group, the presence of a CO ligand was confirmed by the IR spectrum and by the fact that [VI] could be obtained only from iridium carbonyl compounds. The number of hydridic hydrogen was deduced from the reaction with acids (see below) and confirmed by the IR spectrum. [VI] is a yellow crystalline substance m.p. 145⁰, stable to air and moisture, soluble in benzene and chloroform and almost insoluble in ethanol. It shows two bands in the IR region characteristic for Ir-H and CO stretching at 1920 and 2120 cm^{-1}.

Compounds [VI] and [II] react reversibly with strong acids adding a proton. With perchloric acid the reaction is:

$$IrHCOL_3 + HClO_4 \quad = \quad [IrH_2COL_3]ClO_4$$

$$[IrH_2COL_3]ClO_4 + KOH \quad = \quad [IrHCOL_3] + KClO_4 + H_2O$$

This perchlorate which could be also abtained from $[IrH_2COL_2]ClO_4$ and triphenylphosphine, is a white crystalline substance, soluble in ethanol and benzene, behaving in solution as a uni-univalent electrolyte. It shows in the IR three bands at 2155, 2188 and 2080 cm^{-1} as expected. The cation $[IrH_2COL_3]^+$ is coordinatively saturated and does not react in solution either with neutral or anionic ligands, behaving in this respect differently from the cation $[IrH_2COL_2]^+$. Its constitution was also confirmed by the preparation from $[IrH_2L_3]^+[ClO_4]^-$ and CO at ordinary temperature and pressure. Fig. 1 indicates the relationship among these new hydrido- and hydridocarbonylcompounds.

This research has been sponsored by the Office of Scientific Research through the European Office of Aerospace Research United States Air Force.

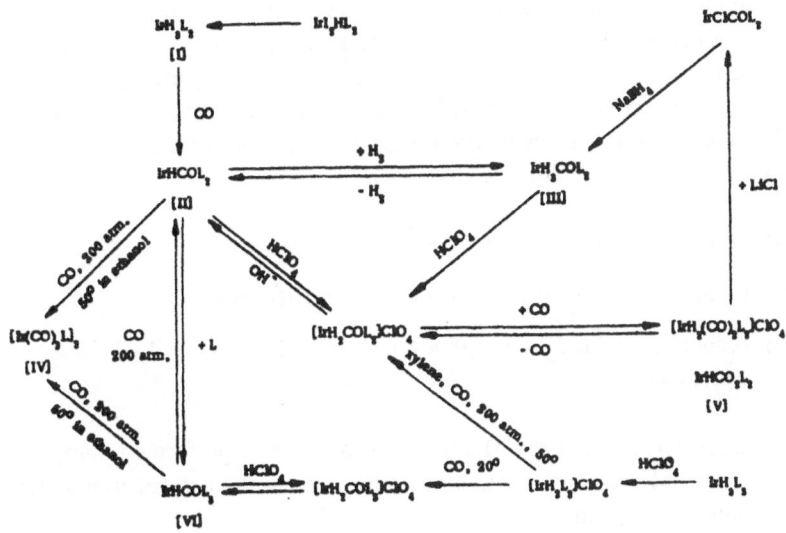

Fig. 1

References

1. ANGOLETTA, M., Gazz.Chim.It., 92, 811 (1962).
2. CHATT, J., Unpublished work.
3. VASKA, L., DILUZIO, J.W., J.Am.Chem.Soc., 83, 2784 (1961).
4. MALATESTA, L., NALDINI, L., CARIATI, E., J.Chem.Soc., (London), 1964, 961.

B 2

THE PREPARATION OF TRIS-2,2'-BIPYRIDYL-IRON(O) [Fe(bipy)$_3$].

S. Herzog and H. Präkel
Ernst Moritz Arndt Universität, Greifwald, Germany.
(to be presented in German)

Iron relatively often approves the oxidation state zero, in most cases, however, it has the tendency to avoid the coordination number six.

Especially no case of zero-valent iron with sixfold nitrogen coordination has been known so far. The action of 1,10-phenanthroline (phen) or 2,2'-bi-pyridyl (bipy) on Fe(CO)$_5$ had brought about a transient blue coloring, which was interpreted as hinting at an adduct Fe(CO)$_5 \cdot$phen, but it was only possible to isolate preparations of the type of [Fe(phen)$_3$][Fe$_2$(CO)$_8$], i.e. the iron atom had been encased by three molecules of phenanthroline, but at the same time there had happened a valence disproportioning (HIEBER and coworkers).

After we had succeeded, with some metals, in preparing mostly hexa-coordinated complexes of bipyridyl with lower oxidation states, respectively with oxidation state zero, by using the three main methods of addition a), of reduction b), resp. of reducing addition c):

a) addition $Mg + 2\ bipy_0 \longrightarrow Mg(bipy)_2$

b) reduction $[Cr(bipy)_3]^{3+} + 3\ominus \longrightarrow Cr(bipy)_3$

c) reducing addition $TiCl_4 + 3\ bipy + 3\ominus \longrightarrow Ti(bipy)_3 + 4\ Cl^-$

we asked ourselves if perhaps the electronic charging of a preformed complex ion up to the neutral complex would be possible also in the case of iron under carefully planned conditions.

For the experimental test we brought together a suspension of the red [Fe(bipy)$_3$]J$_2$, which is practically insoluble in absolute tetrahydrofurane, with a green solution of the bis-lithium-bipyridyl, Li_2bipy, in the same solvent and strictly excluding air and moisture (1). The shares of the reducing solution added successively were immediately discolored and the iron(II)-complex dissolved at the rate of dripping and its surface transiently taking a brown color. After adding two reducing equivalents the original crystals had dissolved completely, the color of the solution was deep violet red, and deep green, almost black crystals fell out, which on analysis showed themselves to be the expected $[Fe(bipy)_3]$.

Obviously the reaction can be expressed by the equation

$$[Fe(bipy)_3]J_2 + Li_2bipy \xrightarrow{\text{THF}} [Fe(bipy)_3] + 2\ LiJ + bipy$$

The substance got in this way still contained, however, changing quantities of lithium and iodine from the preparation. Experiments for recrystallization failed, since always a partial or complete decomposition took place and metallic iron was separated.

Eventually we succeeded in preparing the $[Fe(bipy)_3]$ in a pure state by the direct reaction of anhydrous iron(III)-chloride with bipyridyl and Li_2bipy in tetrahydrofurane by the "reducing addition":

$$FeCl_3 + 1,5\ bipy + 1,5\ Li_2bipy \longrightarrow Fe(bipy)_3 + 3\ LiCl.$$

The new compound is soluble in tetrahydrofurane, dioxane, benzene etc. with a violet color. The solutions as well as the solid substance oxidized immediately in the presence of air. When the $[Fe(bipy)_3]$ is titrated with iodine solution under exclusion of air, the red iron(II) complex is quantitatively formed again under consuming the amount of iodine calculated according to

$$[Fe(bipy)_3] + J_2 \longrightarrow [Fe(bipy)_3]J_2$$

The quantity of the iron(II) complex as analyzed colorimetrically was equivalent to the original quantity of $[Fe(bipy)_3]$. A corresponding oxidation takes place as well, when $[Fe(bipy)_3]$ is brought together with water, alcohol, or dilute acids.

The thermal stability of $[Fe(bipy)_3]$ is not sufficient for sublimation in high vacuum, just as it happens with its neighbors $[Mn(bipy)_3]$ and $[Co(bipy)_3]$. On the other hand $[Ti(bipy)_3]$, $[V(bipy)_3]$, and $[Cr(bipy)_3]$ can easily be sublimed in a high vacuum.

According to these qualities, the $[Fe(bipy)_3]$ can easily be grouped with the series of complexes of bipyridyl rich in electrons, which were found and investigated more closely by our team (2-10). In the 3d-series, which is the best investigated one so far, the results for type [Me(bipy)$_3$] are summarized in Table I.

TABLE I - COMPLEXES OF THE TYPE [Me(bipy)$_3$]

Total number
of electrons

42 ————————————————	Zn(0)
41	
40 ————————————————	Zn(+2)
39 ——————————— Co(0) ——— Cu(+2)	
38 ——— Mn(-1)—Fe(0)——Co(+1)—Ni(+2)	
37 ——— Mn(0)——— Co(+2)	
36 —— V(-1)—Cr(0)——— Fe(+2)—Co(+3)	
35 —— Ti(-1)—V(0)—Cr(+1)—Mn(+2)—Fe(+3)	
34 —— Ti(0)—V(+1)—Cr(+2)	
33 —Sc(0)——— V(+2)—Cr(+3)	
32	
31	
30 — Sc(+3)	

For the time being, a general synopsis of the neutral type [Me(bipy)$_n$] (n = 1, 2, 3, 4, depending on the central atom), which is relatively preferred, results in Table II.

TABLE II - NEUTRAL TYPE [Me(bipy)$_n$]

																H	He
Li	Be											B	C	N	O	F	Ne
Na	Mg											Al	Si	P	S	Cl	Ar
K	Ca	Sc	Ti	V	Cr	Mn	Fe	Co	Ni	Cu	Zn	Ga	Ge	As	Se	Br	Kr
Rb	Sr	Y	Zr	Nb	Mo	Tc	Ru	Rh	Pd	Ag	Cd	In	Sn	Sb	Te	J	Xe
Cs	Ba	La a	Hf	Ta	W	Re	Os	Ir	Pt	Au	Hg	Tl	Pb	Bi	Po	At	Rn
Fr	Ra	Ac b															

a	=	Ce	Pr	Nd	Pm	Sm	Eu	Gd	Tb	Dy	Ho	Er	Tm	Yb	Lu
b	=	Th	Pa	U	Np	Pu	Am	Cm	Bk	Cf	Es	Fm	Md	No	Lr

References

1. HERZOG, S., and DEHNERT, J., Z.Chem., 4, 1 (1964).

2. HERZOG, S., and TAUBE, R., Z.Chem., 2, 208 (1962).
3. TAUBE, R., and HERZOG, S., Z.Chem., 2, 225 (1962).
4. HERZOG, S., and GRIMM, U., Z.Chem., 3, 31 (1963).
5. HERZOG, S., and KUBETSCHEK, E., Z.Naturforsch., 18b, 162 1963).
6. HERZOG, S., GEISLER, K., and PRÄKEL, H., Angew.Chem., 75, 94 (1963).
7. HERZOG, S., and KREBS, F., Naturwissenschaften, 50, 1 (1963).
8. HERZOG, S., and SCHMIDT, M., Z.Chem., 3, 392 (1963).
9. HERZOG, S., and OBERENDER, H., Z.Chem., 3, 429 (1963).
10. HERZOG, S., KLAUSCH, R., and LANTOS, J., Z.Chem., in press.

9 B 3

COLOR AND VARIABLE VALENCE IN NIOBIUM AND TANTALUM HALIDES.

M.B. Robin and N.A. Kuebler

Bell Telephone Laboratories, Incorporated, Murray Hill, New Jersey, U.S.A.

The very stable polynuclear halide ions $[M_6X_{12}]^{2+}$ (M = Nb or Ta, X = Cl or Br) are unique in that though the X-ray structure analyses of the complex (Fig. 1) demonstrate that the metal ions are all in equivalent environments, the net charge of +2 on the $[M_6X_{12}]^{2+}$ ion implies that two of the M ions have formal charge +3 (nd^2) and four have the formal charge +2 (nd^3). Actually, quantum mechanics demands that these formal valences oscillate equally among all six metal ions with the result that states arise for the polynuclear ion which have only a distant relationship to those of either the M^{2+} or M^{3+} ions.

The $[M_6X_{12}]^{2+}$ ion is treated by first considering each M ion as subject to a halide ion induced tetragonal field. As a result of this field, the nonbonding metal d orbitals d_{xz} and d_{yz} lie lowest and carry the valence shell electrons. Molecular orbitals are then constructed on the assumption that there is strong M-M bonding involving d_{xz} and d_{yz} a.o.'s. From this molecular orbital picture of the valence shell structure, the magnetic and optical (absorption and luminescence) properties of these ions can then be explained.

We have found that the paramagnetism of these complexes is temperature independent as expected for the singlet ground state predicted by the MO theory with M-M bonding. The assignment of the bands in the optical spectrum is complicated because there are three types of transitions possible and these must be sorted out first. The ligand $\longrightarrow$ metal charge transfer bands are identified by comparing the spectra of $[M_6Cl_{12}]^{2+}$ and $[M_6Br_{12}]^{2+}$ and assigning those bands which shift to lower frequency as $L \rightarrow M$. The unshifted bands do not involve ligand orbitals and the strong ones of these are assigned as metal $\rightarrow$ metal, oscillating valence transitions. At the moment, there is an ambiguity concerning the assignment of the remaining weak bands, which can be either $M \rightarrow M$, oscillating valence, but symmetry forbidden or transitions to m.o.'s composed of metal-ligand antibonding d orbitals. Transitions of the latter type are forbidden regardless of the M.O. symmetries.

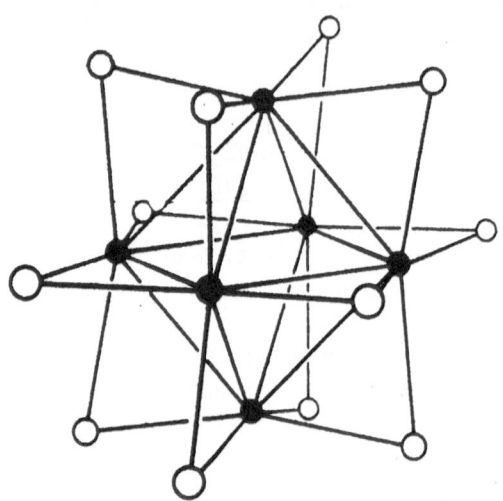

Fig. 1: The structure of the $[M_6X_{12}]^{2+}$ cation.

0 B 1

TETRAPHENYLCYCLOBUTADIENE-TRANSITION METAL COMPLEXES.

P.M. Maitlis, (Mrs.) M.L. Games and A. Efraty
McMaster University, Hamilton, Ontario, Canada.

The prediction, in 1956 by LONGUET-HIGGINS and ORGEL (1), that while cyclobutadienes are likely to be unstable, their complexes with transition metals would be stable, has been amply confirmed. In 1959 CRIEGEE and SCHRÖDER prepared tetramethylcyclobutadiene nickel chloride (2) and this was quickly followed by the preparation of tetraphenylcyclobutadiene $[Ph_4C_4]$ complexes of iron, [II] (3), nickel, [III] (4), and cobalt, [VI] (5). We have, following the work of MALATESTA et al. (6), recently prepared the tetraphenylcyclobutadiene palladium halides [I] by simple routes from diphenylacetylene (7,8) and characterized these compounds. All of these cyclobutadiene-metal complexes however, were prepared by unique routes, not readily applicable to the synthesis of cyclobutadiene derivatives of other transition metals. We report here the first general syntheses of tetraphenylcyclobutadiene-transition metal complexes from [I] by ligand-exchange and ligand-transfer reactions.

Reaction of [I] with R_3P at moderate temperatures causes replacement of the $[Ph_4C_4]$ group by phosphines giving a transient paramagnetic species (presumably related to the unknown tetraphenylcyclobutadiene) which underwent further reaction to give octaphenylcubane (9):

$$2\,[Ph_4C_4]PdBr_2 + 4\,R_3P \longrightarrow 2\,[R_3P]_2PdBr_2 + Ph_8C_8$$
$$[I]$$

[I] also reacted with phosphine-substituted nickel bromide to give nearly quantitative exchange of the ligands (10):

$$[I] + (R_3P)_2NiBr_2 \longrightarrow [R_3P]_2PdBr_2 + [Ph_4C_4]NiBr_2$$

Since it had been shown that the halogens in [I] were easily removed, the reactions of [I] with various halogen-acceptors such as metal carbonyls were investigated and it was found that transfer of the $[Ph_4C_4]$ ligand occurred in nearly all cases (11, 12):

$$[I] \;\; + Fe(CO)_5 \longrightarrow [Ph_4C_4]Fe(CO)_3 \qquad [II]$$

$$[I] \;\; + Ni(CO)_4 \longrightarrow [Ph_4C_4]NiBr_2 \qquad [III]$$

$$[III] + Fe(CO)_5 \longrightarrow [II]$$

$$[I] \;\; + Mo(CO)_6 \longrightarrow \{[Ph_4C_4]Mo(CO)_3Br\}_2 \;\; [IV]$$

$$[I] \;\; + Co_2(CO)_8 \longrightarrow [Ph_4C_4]Co(CO)_2Br \;\; [V]$$

$$[I] \;\; + (C_5H_5)_2Co \longrightarrow [Ph_4C_4]Co[C_5H_5] \qquad [VI]$$

Reaction of [I] with $\{[C_5H_5]Mo(CO)_3\}_2$ occurs analogously with retention of the cyclopentadienyl group on the molybdenum:

$$[Ph_4C_4]PdCl_2 + \{[C_5H_5]Mo(CO)_3\}_2 \longrightarrow \{[Ph_4C_4]Mo[C_5H_5][CO]Cl\}$$

However, reaction of [I] with $[C_5H_5]Fe(CO)_2Br$ (or $\{[C_5H_5]Fe(CO)_2\}_2$, but giving lower yields) caused transfer of the cyclopentadienyl group from iron to palladium:

$$[I] + 2\,\{[C_5H_5]Fe(CO)_2Br\} \longrightarrow \{[Ph_4C_4]Pd[C_5H_5]\}^+\,FeBr_4^-$$
$$[VII]$$

[VII] was converted into the bromide [VIII] and this, on treatment with bases, gave substituted (π-tetraphenylcyclobutenyl) (π-cyclopentadienyl) palladium complexes [IX]

$$[VII] + K_4Fe(CN)_6 \longrightarrow \{[Ph_4C_4]Pd[C_5H_5]\}^+\,Br^- + 3\,KBr + KFeFe(CN)_6$$
$$[VIII]$$

$$[VIII] + OR^- \longrightarrow$$

[IX, R = H, Me, Et]

[IX], with HBr gave [I], probably via [VIII] since this also gave [I] under the same conditions.

[III] and 1,5-cyclooctadiene palladium bromide react with $[C_5H_5]Fe(CO)_2Br$ to give compounds analogous to [VII]. The cyclopentadiene ring has also been transferred onto cobalt:

$$[V] + [C_5H_5]Fe(CO)_2Br \longrightarrow [VI]$$

and the preparation of [VI] by simultaneous transfer of the cyclopentadienyl and the tetraphenylcyclobutadiene from [VIII] onto cobalt has also been accomplished.

The further reactions of these compounds and the mechanisms of the ligand-transfer reactions will also be discussed.

References

1. LONGUET-HIGGINS, H.C., and ORGEL, L.E., J.Chem.Soc. (London), 1956, 1969.
2. CRIEGEE, R., and SCHRÖDER, G., Annalen, 633, 1 (1959);

CRIEGEE, R., Angew.Chem. (Int.Ed.), 1, 519 (1962).

3. HÜBEL, W., and BRAYE, E.H., J.Inorg.Nucl.Chem., 10, 250 (1959).
4. FREEDMAN, H.H., J.Am.Chem.Soc., 83, 2194, 2195 (1961).
5. NAKAMURA, A., and HAGIHARA, N., Bull.Chem.Soc.Japan, 34, 452 (1961).
6. MALATESTA, L., et al., Angew.Chem., 72, 34 (1960).
7. BLOMQUIST, A.T., and MAITLIS, P.M., J.Am.Chem.Soc., 84, 2329 (1962).
8. MAITLIS, P.M., and GAMES, M.L., Canad.J.Chem., 42, 183 (1964).
9. MAITLIS, P.M., and STONE, F.G.A., Proc.Chem.Soc., 1962, 330.
10. MAITLIS, P.M., and POLLOCK, D., unpublished observations.
11. MAITLIS, P.M., and GAMES, M.L., J.Am.Chem.Soc., 85, 1887 (1963).
12. MAITLIS, P.M., and GAMES, M.L., Chem.and Ind., 1963, 1624.

10 B 2

AN INFRARED AND NMR STUDY OF THE BONDING BETWEEN OLEFINS AND SILVER IONS.

H.W. Quinn, J.S. McIntyre and D.J. Peterson
Dow Chemical of Canada Limited, Sarnia, Ontario, Canada.

An earlier study (1) of the stoichiometry and stability of crystalline complexes of some mono-olefins with anhydrous silver fluoroborate established that these were quite stable by comparison with those silver ion-olefin complexes which had been known and that, with the exception of ethylene, the most stable complex in every instance had a silver ion:olefin stoichiometric ratio of 1:2. The study established as well an essentially linear correlation between the ionization potential of the olefin and the infrared shift in the double bond stretch vibration frequency upon complexation with the silver salt.

The spectral study has now been extended to cover a greater variety of olefins and to include a broader investigation of the changes effected in the infrared and NMR spectra of the olefins when silver fluoroborate complexes are formed. The spectral determinations have employed chloroform or deuterochloroform solutions of the complexes and have established that in these solutions the silver ion-olefin ratio in the complexes is also 1:2.

Infrared investigations

In Table I are recorded infrared data for the carbon double bond stretch vibration and the vinyl proton out-of-plane deformation frequencies for a number of olefins and the corresponding $AgBF_4 . 2$ olefinates. There is in general shift of $\nu_{C=C}$ to lower frequencies and of γCH and γCH_2 to higher frequencies upon complexation with silver ion. The frequency shifts (particularly $\Delta\nu_{C=C}$) increase with increasing alkyl substitution at the double bond indicating stronger interaction with silver ion as a result of the increasing inductive effect although there is some evidence of steric effects. The infrared absorption bands attributed to the asymmetric carbon-hydrogen stretch vibration of the $=CH_2$ groups (at 3075 - 3095 cm^{-1}) and to the CH stretch vibration for the $=CH\diagdown_H$ group (at 3010 - 3040 cm^{-1}) are not distinctly observable in the complex. These bands have either been shifted to lower frequencies and are hidden by other CH stretch vibration absorption bands, (a shift of at least 80 cm^{-1} for $\nu_{=CH_2}$) or have diminished very markedly in intensity. The shifts of $\nu_{C=C}$ and possibly of $\nu_{=CH_2}$ to lower frequencies and of γCH_2 and γCH to higher frequencies is suggestive of a change from sp^2 to sp^3 character of the hybrid orbitals of the double bond carbon atoms as suggested by REEVES for the olefin in the ethylene platinous chloride dimer (2).

TABLE I - INFRARED FREQUENCY SHIFTS [a] FOR C=C STRETCH VIBRATIONS
AND CH AND CH$_2$ OUT-OF-PLANE DEFORMATION VIBRATIONS IN OLEFINS
UPON FORMATION OF AgBF$_4$.2-OLEFINATES IN SOLUTION [b] (cm^{-1})

Olefin	$\nu_{C=C}$		$\Delta\nu_{C=C}$	γCH		$\Delta\gamma CH$	γCH_2		$\Delta\gamma CH_2$
	Olefin	Complex		Olefin	Complex		Olefin	Complex	
Propylene	1642	1587	-55	991	1012	21	913	947	34
Isobutene	1647	1587	-60				885	918	33
Cis-2-butene	1653	1595	-58	670	715	45			
Trans-2-butene	1673(R)[c]	1614(R)	-59	967	996	29			
2-Methyl-2-butene	1669	1596	-73	800	839	39			
2,3-Dimethyl-2-butene	1672(R)	1599(R)	-73						

(a) $\Delta\nu = \nu$ complex - ν olefin.

(b) I.R. spectra of chloroform or deuterochloroform solution.

(c) Raman spectra of nitromethane solution.

NMR investigations

In Table II are recorded the chemical shifts for the protons in a number of methyl-substituted ethylenes and in the corresponding $AgBF_4$.2-olefinates. The proton chemical shifts in the complexed olefin are in every instance downfield from those of the uncomplexed olefin but the spectral pattern is essentially unchanged (some small changes of coupling constants observed) indicating retention of the olefinic character as observed by POWELL and SHEPPARD (3). The downfield change of chemical shift is attributed to proton deshielding as a result of the tendency of the silver ion to withdraw π electrons from the double bond. The dependence of the change in chemical shift for the methyl protons on the position of methyl substitution may be interpreted in terms of an asymmetric positioning of the silver ion with respect to the double bond and the effect which this has on the shielding of the protons.

TABLE II - CHANGE OF PROTON CHEMICAL SHIFTS [a] IN NMR SPECTRA OF OLEFINS

UPON FORMATION OF AgBF₄.2-OLEFINATES [b] (CYCLES PER SECOND)

Olefin	R_1	R_2	R_3	R_4	ν_1			ν_2			ν_3			ν_4		
					ν_1^o	ν_1^c	Δ [c]	ν_2^o	ν_2^c	Δ	ν_3^o	ν_3^c	Δ	ν_4^o	ν_4^c	Δ
Ethylene	H	H	H	H	323	344	21	$\approx \nu_1$			$\approx \nu_1$			$\approx \nu_1$		
Propylene	H	H	H	CH_3	352	400	48	295	324	29	300	319	19	104	123	19
Isobutene	CH_3	H	H	CH_3	$\approx \nu_4$			279	304	25	$\approx \nu_2$			103	128	25
Cis-2-butene	H	H	CH_3	CH_3	328	373	45	$\approx \nu_1$			$\approx \nu_4$			96	115	19
Trans-2-butene	H	CH_3	H	CH_3	326	371	45	$\approx \nu_4$			$\approx \nu_1$			100	117	17
2-Methyl-2-butene	CH_3	H	CH_3	CH_3	$\approx \nu_4$			311	350	39	90	107	17	95	120	25
2,3-Dimethyl-2-butene	CH_3	CH_3	CH_3	CH_3	$\approx \nu_4$			$\approx \nu_4$			$\approx \nu_4$			97	116	19

(a) Chemical shifts expressed in c.p.s. at 60 Mc/sec downfield with respect to internal Si(CH₃)₄.

(b) Stoichiometry of ethylene complex in solution not established.

(c) $\Delta = \nu^c - \nu^o$ where c and o designate complex and olefin respectively.

A further study of the effect of the nature of the alkyl substituent in monoalkyl ethylenes on the change in chemical shift of the vinyl protons upon complexation with silver ion will be discussed in terms of the change in the dipole moment of the molecule.

References

1. QUINN, H.W., and GLEW, D.N., Can.J.Chem., 40, 1103 (1962).
2. REEVES, R.W., Can.J.Chem., 38, 736 (1960).
3. POWELL, D.B., and SHEPPARD, N., J.Chem.Soc.(London), 1960, 2519.

10 B 3

OLEFIN AND π-ALLYLIC COMPLEXES OF PALLADIUM, PLATINUM AND IRIDIUM.

M.S. Lupin, S.D. Robinson and B.L. Shaw
Department of Inorganic and Structural Chemistry,
The University, Leeds 2, U.K.

We are making a study of compounds containing metal-carbon bonds with particular reference to complexes of the second and third transition series. Compounds of this type are known to be catalytic intermediates in several reactions of industrial interest; for example, palladium(II) which readily forms olefin and π-allylic complexes, is a particularly good catalyst for some reactions involving olefins. We have investigated reactions of palladium(II) compounds with diolefins and here describe some of this work.

Dichlorobis(benzonitrile)palladium(II) is known to react with butadiene to give a yellow complex, formulated by previous workers as a diolefin complex [PdCl$_2$(C$_4$H$_6$)] or [Pd$_2$Cl$_4$(C$_4$H$_6$)$_2$] (1). We have shown this compound to have the chloro bridged π-4-chlorobut-2-enyl structure [Pd$_2$Cl$_2$(C$_4$H$_6$Cl)$_2$] by means of NMR and chemical reactions (2, 3). The complex is rapidly solvolysed with methanol to give a more stable methoxy-derivative [Pd$_2$Cl$_2$(C$_4$H$_6$OCH$_3$)$_2$]. This and a large number of similar alkoxy-π-allylic-palladium complexes were readily prepared by treating sodium chloropalladite in an alcohol with conjugated dienes; this reaction probably constitutes the simplest synthesis of π-allylic complexes yet discovered. Dienes included, buta-1,3-diene, 2-methylbuta-1,3-diene, 4-methylpenta-1,3-diene, 2,3-dimethylbuta-1,3-diene, 2,4-dimethylpenta-1,3-diene, 2,5-dimethyl-hexa-2,4-diene, sorbic acid, methyl sorbate and a number of cyclic 1,3-dienes; alcohols were usually methanol or ethanol. Reactions involving 1,2-dienes (allenes) have also been investigated - Allene and dichlorobis-benzonitrilepalladium(II) in benzene gave the π-2-chloroallyl complex [Pd$_2$Cl$_2$(C$_3$H$_4$Cl)$_2$]; methylallene and 1,1-dimethylallene reacted similarly. Allene and sodium chloropalladite in methanol give a π-allylic complex [Pd$_2$Cl$_2$(C$_6$H$_8$Cl)$_2$] where the π-allylic ligand has the structure (I). The structures and stereochemistry of these π-allylic

(I)

complexes were confirmed by their NMR spectra.

These chloro-bridged π-allylic complexes gave the usual bridge splitting reactions with amines and the bridging chlorine could readily be replaced by bromine, iodine etc. Treatment with silver salts of carboxylic acids (RCOOH) gave a series of complexes of type [Pd$_2$(RCOO)$_2$(all)$_2$] (all = π-allylic ligand) containing carboxylate bridging groups. Mononuclear acetylacetonato complexes [Pd(acac)all], formed from the chloro-bridged complexes by treatment with thallous acetylacetonate, were very useful for NMR studies because of their solubility. Cyclopentadienyl complexes [Pd(all)(C$_5$H$_5$)] were also prepared, by treatment with cyclopentadienylsodium.

The alkoxy π-allylic derivatives underwent alkoxyl exchange very readily when treated with the appropriate alcohol containing a little mineral acid (10^{-3}M) (e.g. methoxy-ethoxy and vice versa). These reactions are believed to go via an intermediate carbonium ion stabilised by the delocalised electron system of the π-allylic group. Methoxy π-allylic complexes derived from 2,5-dimethylhexa-2,4-diene, cyclohepta-1,3-diene and cycloocta-1,3-diene on heating lost methanol irreversibly to give $\alpha\beta$-unsaturated π-allylic complexes e.g. cyclohepta-1,3-diene first gave a π-methoxycycloheptenyl and then a π-cycloheptadienyl complex.

In contrast with palladium(II), platinum(II) shows little tendency to form π-allylic complexes. Only one π-allylic platinum(II) complex, π-allyl-π-cyclopentadienylplatinum(II) has been prepared so far (4). We have now shown that several acyclic and cyclic 1,3-dienes react with sodium chloroplatinite in alcohols to give products of composition [PtCl$_2$(diene)]; these are very insoluble and do not appear to involve π-allylic bonding to the platinum i.e. they are true olefin complexes.

Halogen-bridged rhodium(I) complexes of the type [Rh$_2$Cl$_2$(diene)$_2$] have been prepared from chelating cyclic dienes e.g. cycloocta-1,5-diene (5) and bicycloheptadiene (6) although attempts to make analogous complexes of iridium have hitherto been unsuccessful. We have reinvestigated the reaction between cycloocta-1,5-diene and chloroiridic acid in ethanol and find that a hydridodiene iridium(III) complex of type [IrHCl$_2$(C$_8$H$_{12}$)]$_2$ is formed, v(Ir-H) = 2261 cm^{-1}, v(Ir-D) = 1613 cm^{-1}. This with a base in methanol loses hydrogen chloride to give a bridged methoxy iridium(I) complex [Ir(OCH$_3$)(C$_8$H$_{12}$)]$_2$, this reaction is reversed by hydrochloric acid. Mononuclear complexes of type [IrX(C$_8$H$_{12}$)] (X = acetylacetonato, cyclopentadienyl) have also been prepared. The hydrido diene [IrHCl$_2$(C$_8$H$_{12}$)]$_2$ is the first example of a stable compound containing both hydrogen and an olefin bonded to the same metal atom. Such complexes have been postulated as catalytic intermediates in the oxo and Fischer-Tropsch processes.

References

1. SLADE and JONASSEN, J.Am.Chem.Soc., _79_, 1277 (1957).
2. SHAW, Chem. and Ind., _1962_, 1190.
3. ROBINSON and SHAW, J.Chem.Soc. (London), _1963_, 4806.
4. SHAW and SHEPPARD, Chem.and Ind., _1961_, 517.
5. CHATT and VENANZI, J.Chem.Soc. (London), _1957_, 4735.
6. ABEL, BENNETT and WILKINSON, J.Chem.Soc. (London), _1959_, 3178.

π-CYCLOPROPENYL DERIVATIVES OF THE TRANSITION METALS.

E.W. Gowling and S.F.A. Kettle
Department of Chemistry, The University, Sheffield, England.

Complexes are known in which 4, 5, 6 and 7 membered carbon rings are symmetrically bonded to a transition metal atom. Two attempts to form similar complexes of three membered rings, using triphenylcyclopropenyl (1) salts, have been unsuccessful (2, 3). It is a fairly true generalisation that stable complexes of C_n conjugated systems with n small exist at the end of a transition series, those with n large at the beginning. Accordingly, we investigated the reaction between nickel carbonyl and triphenyl-cyclopropenyl bromide {I} in a mutual solvent, methanol. Fine red crystals of π-triphenylcyclopropenyl nickel carbonyl bromide {II} precipitated. The probable structure of this compound is shown in the Figure.

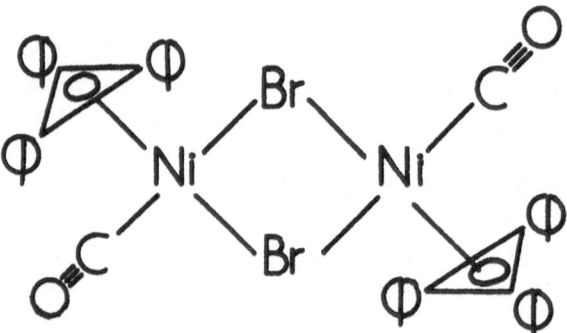

The reactions of this compound have been studied and will be discussed at the Conference. {II} is virtually insoluble in organic solvents unless it reacts with them; warming with potential ligands in the presence of donor solvents causes reaction both with ligand and solvent, the latter reaction frequently predominating. A structure determination of {II} is being presently commenced and it is hoped that preliminary results will be available for the Conference.

{I} reacts with dicobalt octacarbonyl to give the compound first prepared by COFFEY (2); with iron pentacarbonyl to give three products, one of which is the $FeBr_4^-$ salt of {I}, another is a hydrocarbon of empirical formula $(C_7H_5)_n$ but which is not hexaphenylbenzene. The third, purple, compound appears and then disappears at an early stage in the reaction. It has yet to be characterised.

References

1. COFFEY, C.E., J.Am.Chem.Soc., 84, 118 (1962).
2. CHATT, J., and GUY, R.G., Chem.and Ind. (London), 1963, 212.
3. BRESLOW, R., and CHANG, H.W., J.Am.Chem.Soc., 83, 2367 (1961).
   See also SUNDERALINGHAM, M., and JENSEN, C.H., J.Am.Chem.Soc.,
   85, 3302 (1963).

11 B 2

BISCYCLOPENTADIENYL ALLYL COMPLEXES OF
TITANIUM AND VANADIUM.

Hendrik A. Martin and F. Jellinek
Laboratorium voor anorganische chemie, Fijksuniversiteit,
Bloemsingel 10, Groningen, the Netherlands.

No allyl complexes of titanium or vanadium have yet been reported. We have prepared allyl-bis(cyclopentadienyl)-titanium [I] by various methods, of which the following is the most convenient one. Red bis(cyclopentadienyl)-titanium dichloride [II] in tetrahydrofuran solution was reduced to the green monochloride by means of zinc powder (1). When the reduction was completed, one equivalent of allylmagnesium chloride was added to the solution, the color of which changed to violet. The reaction mixture was evaporated in vacuo, and the residue extracted with pentane. When the violet pentane solution was concentrated to saturation and cooled to -80°, purple crystals of [I] precipitated. After two recrystallizations from pentane pure [I] was obtained in 35% yield. Analysis: found 21,8% Ti; 70,89% C; 7,01% H; calc. for $(C_5H_5)_2Ti(C_3H_5)$: 21,86% Ti; 71,24% C; 6,89% H. Melting point: 118°C (decomp.). Reaction of [I] with octanol gave propene in 80% of the theoretical yield.

The infrared spectrum of [I] in a KBr pellet has no absorption bands in the 1650 - 1600 cm^{-1} range, but a band at 1500 cm^{-1}; we conclude, therefore (2), that the allyl group is π-bonded to the metal. Another band at 1435 cm^{-1} is ascribed to the valence vibration of the π-bonded cyclopentadienyl ligands. Solutions of [I] in hexane show a rather sharp absorption maximum at 510 mμ; this is the only band between 400 and 1200 mμ. Beyond 400 mμ there is intense absorption. Compound [I] is paramagnetic.

Crystalline [I] and its solutions in pentane remained unchanged when kept under nitrogen for several months at room temperature. No reaction with triphenylphosphine was observed. By methylene dichloride or allyl chloride in excess, [I] is converted into bis(cyclopentadienyl)titanium di-chloride [II]. When ether/HCl was added to a solution of [I] in toluene, a green product was formed, which was slowly converted into red [II]; the latter product was formed instantaneously when air was passed through the solution. Solid [I] is rapidly oxidized in air, and solutions of [I] immediately turn yellow on exposure to air. This is in contrast to benzyl-bis(cyclopenta-dienyl)-titanium, which is stable towards air, both in the solid state and in ether solution. Benzyl-bis(cyclopentadienyl)-titanium was obtained by reaction of [II] with two to three equivalents of benzylmagnesium chloride in ether. Its absorption spectrum in hexane resembles that of [I]; the maximum is at 490 mμ.

[I] could also be prepared by reaction of [II] with two to three equivalents of allyl magnesium chloride in tetrahydrofuran at room temperature; pure [I] could be isolated from the violet reaction mixture in the same way as described above. If only one equivalent of allyl magnesium chloride was used, however, the solution turned brown; after evaporation a brown oil separated. This product may contain allyl-bis(cyclopentadienyl)-titanium monochloride, but so far all attempts to isolate this compound have failed, even when the solvent was varied and the temperature of the solution kept as low as -80°.

By reaction of bis(cyclopentadienyl)-vanadium monochloride [III] with allyl magnesium chloride, a brown solution was obtained. Its absorption spectrum shows bands at 475 mμ and 720 mμ and is closely similar to that of benzyl-bis(cyclopentadienyl)-vanadium (3). We assume, therefore, that our reaction product is allyl-bis(cyclopentadienyl)-vanadium. The compound is unstable and has not yet been isolated in pure form. The same compound is formed, together with [III], by reaction of bis(cyclopentadienyl)-vanadium with half an equivalent of allyl chloride. If an excess of allyl chloride is used, only [III] is obtained.

References

1. BIRMINGHAM, J.M., FISCHER, A.K., and WILKINSON, G., Natur-wissensch., 42, 96 (1955).
2. McCLELLAN, W.R., HOEHN, H.H., MUETTERTIES, E.L., and HOWK, B.W., J.Am.Chem.Soc., 83, 1601 (1961).
3. de LIEFDE MEIJER, H.J., JANSSEN, M.J., and van der KERK, G.J.M., Rec.trav.chim. Pays-Bas, 80, 831 (1961).

METALCOMPLEXES OF THE PENTALENE SYSTEM.

Michael Cais, A. Modiano, N. Tirosh and A. Eisenstadt
Technion - Israel Institute of Technology, Haifa, Israel.

In several recent communications (1-4) we have reported preliminary results of our attempts at the synthesis of organometallic π -complexes of the pentalene system. The key step involved solvolysis, in acid solution, of the alcohols {1}, {3} and {5} to obtain the metal π -complexes of the pentalenylium, methylpentalenylium and benzopentalenylium cations, {2}, {4} and {6} respectively.

$\{5\}$ HX $\{6\}$

In order to understand better what happens to the cations $\{2\}$, $\{4\}$ and $\{6\}$ produced in the first step of the solvolysis reaction it is necessary to consider a recently investigated aspect of metallocene chemistry, namely the role of the metal in the formation of α-metallocenyl carbonium ions. Several investigators (5, 6) have discussed the unusual stability possessed by carbonium ions adjacent to a metallocene nucleus.

Our work with ferrocenylphenylcarbinyl cations, $\{7\}$, has been directed towards differentiating between the several possible reaction paths, (i) (8), (ii) (7) and (iii) (6). Evidence will be presented that the reaction conditions are critical in determining which course will prevail.

$$C_5H_5FeC_5H_4\overset{\oplus}{C}\varnothing R$$

$$\{7\}$$

radical (i)   (ii) electrophilic
formation   attack

$$\overset{\oplus}{C_5H_5}FeC_5H_4\overset{\bullet}{C}\varnothing R \qquad\qquad C_5H_4FeC_5H_4\overset{\oplus}{C}\varnothing R$$

$$\overset{\oplus}{C_5H_5}FeC_5H_4\overset{|}{C}\varnothing R \qquad\qquad R\varnothing CC_5H_4FeC_5H_5$$

$$\overset{\oplus}{C_5H_5}FeC_5\overset{.}{H}_4\overset{|}{C}\varnothing R$$

                               (iii)   further electrophilic
                                        attack

trimer and higher polymers

The results obtained with the model compounds $\{7\}$ have been applied in interpreting the reactions of the cations $\{2\}$, $\{4\}$ and $\{6\}$. Some of the possibilities are shown in the reaction scheme below.

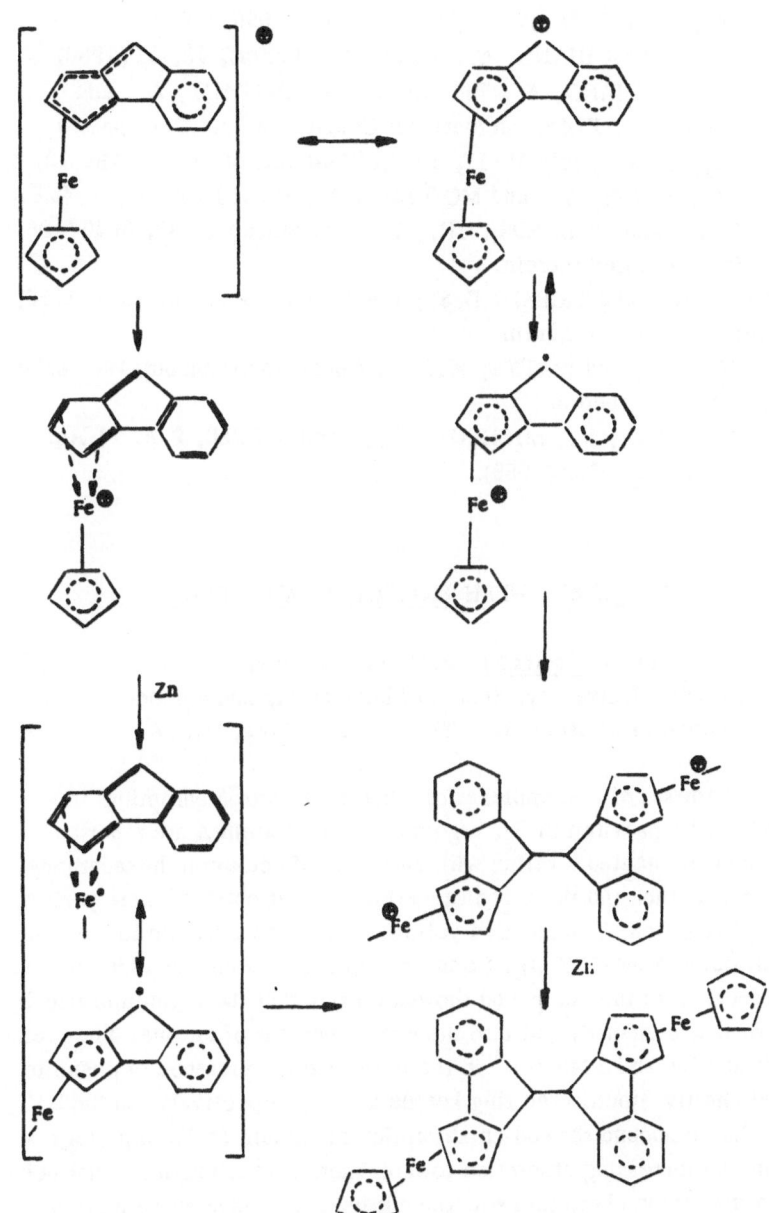

References

1. CAIS, M., and MODIANO, A., Chem. & Ind. 1960, 202.
2. CAIS, M., and MODIANO, A., Tetrahedron Letters, 18, 31 (1960).
3. CAIS, M., MODIANO, A., TIROSH, N., EISENSTADT, A., and
   RUBINSTEIN, A., XIXth International Congress of Pure and Applied
   Chemistry, London, July 10-17, 1963; cf Abstracts of Papers, AB4-12, p. 166.
4. CAIS, M., RAVEH, A., and MODIANO, A., Israel J. Chem., 1, 228 (1963).
5. HILL, E.A., and RICHARDS, J.H., J. Am. Chem. Soc., 83, 3840 (1961),
   and references cited therein.
6. NEUSE, E.W., and TRIFAN, D.S., J. Am. Chem. Soc., 85, 1952 (1963),
   and references cited therein.
7. PAUSON, P.L., and WATTS, W.E., J. Chem. Soc. (London), 1962, 3880, and
   references cited therein.
8. RINEHART, K.L., Jr., MICHEDA, C.J., and KITTLE, P.A., J. Am.
   Chem. Soc., 81, 3162 (1959).

1 B 4

π-COMPLEXES WITH BIOLOGICAL MATERIALS.

Minoru Tsutsui and Akira Nakamura
New York University, School of Engineering and Science,
Research Division, New York 53, New York, U.S.A.

The syntheses of π-complexes of a hormone, estron chromium tricarbonyl, and a provitamin D_2, ergosterol, and vitamin A are reported.

Estrone was heated a reflux with an excess of chromium hexacarbonyl in di-n-butyl ether and the greenish-yellow crystals obtained were purified by recrystallization. The purified yellow compound was shown to have the composition, estrone-$Cr(CO)_3$. Strong absorption at 1855 and 1950 cm^{-1} in the IR spectrum of the compound showed clearly that the chromium tricarbonyl group was incorporated into the phenol nucleus of estrone. The presence of bands at 3150 and 1748 cm^{-1} in the IR showed the presence of phenolic hydroxyl and five-membered ring ketone groups, respectively. In the UV region, the compound showed an absorption maximum at 315 mμ (log ϵ = 3.38) and an increasing absorption towards shorter wavelengths. It has been reported that arene chromium tricarbonyls generally have an absorption maximum at 319 mμ (log ϵ = 4).

The stereochemical problem of whether the chromium tricarbonyl group is attached on the center or the side of the molecule is not settled.

Acetylergosterol was found to yield a yellow iron carbonyl π-complex in reaction with an excess of tri-iron dodecacarbonyl.

The reaction was run in benzene under nitrogen at 90 - 100° for 18 hours, and the reaction mixture was separated by chromatography on alumina into two fractions. From one fraction 3,5-cyclo-Δ6,8,(14),22-ergostatriene was isolated in 3% yield and identified by its IR and UV spectra (2).

The other fraction, a yellow solid, was further purified by digestion with methanol. A methanol-insoluble colorless portion was found to be un-changed starting material by its IR spectrum. The yellow methanol-soluble portion crystallized on concentration and standing. The pure compound was isolated through recrystallization from methanol as yellow crystals in 16% yield and found to have the composition, acetylergosterol.Fe(CO)$_3$ by elemental analysis: mp 95 - 99°, sintering at 90°. The compound was stable in air at room temperature and was soluble in all organic solvents tried. The IR spectrum of the compound showed peaks at 1950 cm^{-1} due to the iron carbonyl group and at 1730 cm^{-1} due to the acetyl group. In the UV region, the compound had an absorption maximum at 235 mμ (logϵ = 3,86) and a shoulder at 295 mμ (log ϵ = 2,89). Comparison with acetylergosterol, which has absorption maxima at 271 and 281 mμ, shows clearly that a shift towards shorter wavelengths has occurred on π-complex formation. The same kind of shift has been observed with other diene iron tricarbonyl complexes (3).

Isomerization of the double bond has been observed in some cases on reaction with iron carbonyl compounds (4). However, no evidence for iso-merization was observed in this case. The starting material was recovered unchanged in 28% yield. This fact suggests that the position of the double bonds in erogosterol is a stable one. The compound was found to have optical activity, $[\alpha]_D^{22}$ = -77,2° (AcOEt); the value can be compared with that of ergosterol, $[\alpha]_D^{20}$ = -135°.

Although a number of iron carbonyl complexes have been prepared with cyclic conjugated polyene systems such as cyclooctatetraene, linear con-jugated polyene systems have been studied only up to 1,3,5-hexatriene. Linear polyenes containing more than four double bonds have not been in-vestigated for the formation of π-complexes. The reaction of vitamin A acetate, which contains five conjugated double bonds, with iron carbonyl compounds is thus of interest.

Vitamin A acetate (6,02 mmole) reacted with tri-iron dodecacarbonyl (6,6 mmole) in refluxing benzene (20 ml) under nitrogen for 6 hours. In the course of the reaction ca. 200 ml (8 mmole) of a gas was evolved. The initial green color of the reaction mixture faded as the reaction proceeded and finally turned to brown. The brown precipitate formed was removed by filtration and identified as ferrous acetate. The yellow filtrate was evaporated under reduced pressure below room temperature. Alumina chromatography of the yellow semi-solid from the filtrate gave two fractions. One fraction gave an orange semi-solid in ca. 20% yield on short pass sublimation in vacue at 120°.

The IR spectrum of the compound showed strong peaks at 1980 cm^{-1} and 2040 cm^{-1} due to the presence of a $Fe(CO)_3$ group π-bonded to the diene part of vitamin A acetate. The presence of strong peaks at 1745 cm^{-1} and 1232 cm^{-1} indicated an acetoxy group in the complex. The position of the $C=O$ and $C-O$ stretching frequencies in the complex is almost identical with that of vitamin A acetate, viz. 1743 cm^{-1} and 1230 cm^{-1}. The UV spectrum had absorption maxima at 332 mμ and 288 mμ. These values for the maxima may be compared with that of vitamin A acetate, 327 mμ. UV absorption maxima of conjugated dienes have been shown to shift toward shorter wavelengths on π-complex formation with the iron tricarbonyl group.

The splitting of the absorption maximum at 327 mμ of vitamin A acetate into two peaks on π-complex formation can be said to be due to a "π-complex effect". The NMR spectrum of the compound was found to be very complex. However, it was possible to assign all strong peaks to the proposed structure by a comparison with the NMR spectrum of vitamin A acetate. The peaks due to the protons around the six-membered ring of the vitamin A structure kept their position on π-complex formation. Therefore the six-membered ring may not participate in π-complex formation with the iron tricarbonyl group. However, the exact location of π-bonding in the vitamin A structure could not be determined unequivocally.

References

1. NICHOLLS, B., and WHITING, M.C., Proc.chem.Soc. (London), 1958, 152; J.Chem.Soc. (London), 1959, 551.
2. FIESER, M., ROSEN, W.E., and FIESER, L.F., J.Am.Chem.Soc., 74, 5397 (1952).
3. HALLAM, B.F., and PAUSON, P.L., J.Chem.Soc. (London), 1958, 642.
4. a) ARNETT, J.E., and PETTIT, R., J.Am.Chem.Soc., 83, 295 (1961).
   b) MANUEL, T.A., J.Org.Chem., 27, 3941 (1962).

ORGANIC RADICALS AS LIGANDS.

Raymond E. Dessy, Theodore Psarras and S.E.I. Green
Department of Chemistry, University of Cincinnati,
Cincinnati, Ohio, U.S.A.

In viewing the field of organometallic chemistry one is struck by the fact that, except for highly electronegative groups and unsaturated groups capable of accepting back-bonding, little use has been made of CHATT's original suggestion that much could be gained by considering organic radicals in organometallic compounds as ligands.

In a manner analogous to the corresponding areas involving inorganic ligands, the reactions of organometallic compounds recently studied in our laboratory will be discussed under three headings

A. Ligand Displacement
1. $Rhg + HA \longrightarrow hgA + RH$
2. $Rsn + HA \longrightarrow snA + RH$
3. $Rb + HA \longrightarrow bA + RH$

B. Ligand Transfer
1. $Rm + R'm' \rightleftarrows R'm + Rm'$
2. $Rm + m'X \longrightarrow$

C. Charge Transfer

$$Rm \xrightarrow{e}$$

In organic ligand displacement reactions the activity of replacing ligand is determined by solvent, and the central metal atom, as might be expected. Different sequences of activity, in the same solvent, are found for ligands such as Cl^-, Br^-, I^-, RO^-, $\geq S$, $\geq N$ acting on d^0, d^{10} and "no d" atoms.

In organic ligand exchange reactions a sequence of ligand preference for a given site have been arrived at, with particular emphasis on the equilibrium

$$Rmg + R'hg \underset{K}{\rightleftarrows} R'mg + Rhg$$

For the following groups the numbers below each symbol give the factors from which relative equilibrium constants can be calculated

i-Pr	Et	Me	Δ	Vi	$\emptyset$	Allyl	$\emptyset CH_2$
$< 10^{-4}$	$< 10^{-2}$	1	13	30	100	170	365

Finally, in electrode reactions of organometallic compounds the similarities between inorganic ligands and organic ligands will be emphasized.

2 B 2

RECENT STUDIES ON FULMINATE COMPLEXES.

Wolfgang B e c k , Erich S c h u i e r e r und Klaus F e l d l
Anorganisch-Chemisches Laboratorium der
Technischen Hochschule München, Germany.

The visible and ultraviolet spectra of various fulminate complexes have been determined in order to estimate the ligand field strength of the fulminate ion CNO$^-$. The absorption bands of the diamagnetic (1) octahedral fulminato complexes [Fe(CNO)$_6$]$^{4-}$, [Fe(CN)$_5$CNO]$^{4-}$, [Co(CNO)$_6$]$^{3-}$ have been assigned; from the wavenumbers of the two ligand field bands ($^1A_{1g} \rightarrow {}^1T_{1g}$ and $^1A_{1g} \rightarrow {}^1T_{2g}$ transitions) the d-d splitting parameter Δ and the RACAH parameter B have been calculated using a relation given by JØRGENSEN (2). It is seen from Table I

TABLE I

VALUES OF Δ AND B FOR HEXACOORDINATED FULMINATO
AND CYANO COMPLEXES

	Δ [cm^{-1}]	B [cm^{-1}]
[Fe(CNO)$_6$]$^{4-}$	27000	410
[Fe(CN)$_5$CNO]$^{4-}$	31000	400
[Fe(CN)$_6$]$^{4-}$	32200	400
[Co(CNO)$_6$]$^{3-}$	26100	450
[Co(CNO)$_6$]$^{3-}$	33500	460

that the ligand field strength of the fulminate ion is large, but less than that of the cyanide ion. From the high ligand field strength one can conclude that the CNO ligand is bonded through the carbon to the metal (3), since the position of a ligand in the spectrochemical series is determined mainly by the ligand atom nearest to the metal (2), the ligand field increasing in the order halogen < O < N < C . These results confirm the suggestion that all ligands with carbon as bonding atom have high ligand field strength (4). The almost identical B values of fulminate and cyanide (Table I) show that these ligands have the same nephelauxetic effect. The large ligand field strength of CNO⁻ may also be derived from the electronic spectra and the diamagnetism of the compounds Fe phen$_2$(CNO)$_2$ and Fe dipy$_2$(CNO)$_2$ (5) (phen = 1,10-phen-anthroline, dipy = α,α'-bipyridyl). The strong absorption bands of $[Ni(CNO)_4]^{2-}$, $[Pd(CNO)_4]^{2-}$, $[Pt(CNO)_4]^{2-}$, $[Cu(CNO)_2]^-$ and $[Cu(CNO)_3]^{2-}$ which are due to metal-ligand charge transfer transitions are observed at higher wavenumbers than in the corresponding cyano complexes (6).

In order to estimate the trans effect of the fulminate ion the platinum compound trans $PtH(CNO)[P(C_2H_5)_3]_2$ and the corresponding deuteride were prepared and the wavenumber of the Pt-H (Pt-D) stretch was compared with that of other complexes of the type $PtHX(PEt_3)_2$ (7) (X = CN, SCN, I, Br etc). According to CHATT (7) with decreasing values of ν (Pt-H) the trans effect of the group X increases; a similar relationship has been found in the case of the chemical shift of the hydridic hydrogen. For $PtH(CNO)(PEt_3)_2$ the observed values indicate that the fulminate ion has a large trans effect, though again less than that of cyanide ion.

Compounds of Ni(II), Pd(II) and Pt(II) - fulminates with trialkyl- or triaryl-phosphine are formed by the reaction of the tetrafulminato complexes with the corresponding phosphine:

$$[M(CNO)_4]^{2-} + 2 L \longrightarrow ML_2(CNO)_2 + 2 CNO^-$$

M = Ni, Pd, Pt; L = $P(C_2H_5)_3$, $P(C_6H_5)_3$, $P(C_6H_{11})_3$; 2 L=$[(C_6H_5)_2PCH_2CH_2P(C_6H_5)_2]$

The colorless complexes of Pd and Pt being stable to water are easily obtained by shaking the aqueous solution of the complex tetrafulminates with the corresponding phosphine; the phosphine complexes separate immediately. The yellow Ni compound $Ni(PEt_3)_2(CNO)_2$, however, which is decomposed by water has to be prepared in dried ethanol or chloroform. The thermal stability increases from the nickel to the platinum compound (Table II).

In the same manner the light yellow bistriethylphosphine Ni(II)cyanide is obtained

$$[Ni(CN)_4]^{2-} + 2 P(C_2H_5)_3 \longrightarrow [P(C_2H_5)_3]_2Ni(CN)_2 + 2 CN^-$$

All these tetracoordinated bisphosphine metal fulminates are, like the tetrafulminato complexes, diamagnetic and hence have planar structures. Also the cyanide compound $(PEt_3)_2Ni(CN)_2$ is planar whereas Ni(II) halides with triphenylphosphine were reported to be tetrahedral (8). The complexes with the monodentate ligands $P(C_2H_5)_3$, $P(C_6H_5)_3$, $P(C_6H_{11})_3$ have the trans configuration as is shown by their low dipole moments (Table II). Complexes with cis configuration should have moments of about 6 - 8 D. The non zero moments may be attributed to some dissociation giving the free triethyl-phosphine. As is required for the trans configuration these compounds show only one CN and NO stretching frequency in the infra red spectra (taken on Nujol). In the infra red spectra of the Pd(II) and Pt(II) fulminates with the bidentate ligand 1,2-bisdiphenylphosphinoethane, however, two NO stretching bands are observed (in CHCl$_3$ solution) and consequently here the two fulminate ligands are in cis positions. The absorption spectra of the planar complexes in the visible and ultraviolet region are discussed.

TABLE II

	Decomposed at	$\mu[D]$ in benzene	$\nu(CN)$ [cm^{-1}]	$\nu(NO)$ [cm^{-1}]
$(PEt_3)_2Ni(CNO)_2$	159°	1,75	2146	1150
$(PEt_3)_2Pd(CNO)_2$	193°	1,46	2160	1153
$(PEt_3)_2Pt(CNO)_2$	205°	1,31	2180	1152
$(Ph_2PCH_2CH_2PPh_2)Pd(CNO)_2$	224°	-	2167	1164, 1157
$(Ph_2PCH_2CH_2PPh_2)Pt(CNO)_2$	260°	-	2184	1173, 1154

Similarly to the Ni compounds bistriphenylphosphine mercury fulminate has been obtained:

$$[Hg(CNO)_4]^{2-} + 2\ P(C_6H_5)_3 \longrightarrow Hg[P(C_6H_5)_3]_2(CNO)_2 + 2\ CNO^-$$

Also alkyl- and arylmetal fulminates have been prepared, among them the triphenylmetal fulminates of the group IV b elements $(C_6H_5)_3MCNO$ (M = Si, Ge, Sn, Pb) by reaction of the corresponding halides with silver or alkali fulminate. The termal stability increases from the silicon to the lead compound (Table III).

TABLE III

$(C_6H_5)_3MCNO$	m.p. (with dec.)	$2\nu(NO)\,[cm^{-1}]$	$\nu(CN)\,[cm^{-1}]$	$\nu(NO)\,[cm^{-1}]$
M = C	154^0	2599 *	2283	1304
Si	105^0	2594	2200	1302
Ge	122^0	2540	2185	1276
Sn	146^0	2315	2156	1160
Pb	174^0	2298	2130	1149

* infra red spectra in nujol

Remarkable is the decrease of the CN frequency going from triphenyl-acetonitrile-N-oxide (9) to triphenyl lead fulminate. This lowering of the CN stretching frequency can be explained by increasing interaction of the π-electrons of the CN bond with the free d-orbitals of the central atom (Si, Ge, Sn, Pb). This may also be responsible for the increasing thermal stability. Similar to these compounds the diphenylthallium fulminate and the arylmercury fulminates RHgCNO (R = C_6H_5, $CH_3C_6H_4$ or α-naphthyl) are rather stable, whereas alkyl compounds such as $(CH_3)_2TlCNO$ or $(C_3H_7)_3PbCNO$, like the alkalifulminates, are very explosive.

References

1. BECK, W., Z.Naturforsch., 17b, 130 (1962); Z.anorg.allg.Chem., in the press.
2. JØRGENSEN, C.K., "Absorption Spectra and Chemical Bonding in Complexes", Pergamon Press, Oxford 1962; "Orbitals in Atoms and Molecules", Academic Press, London 1962.
3. BECK, W., and LUX, F., Chem.Ber., 95, 1683 (1962).
4. CHATT, J., and HAYTER, R.G., J.Chem.Soc. (London), 1961, 772.
5. BECK, W., and SCHUIERER, E., Chem.Ber., 95, 3048 (1962).
6. GRAY, H.B., and BALLHAUSEN, C.J., J.Am.Chem.Soc., 85, 260 (1963).
7. CHATT, J., and SHAW, B.L., J.Chem.Soc. (London), 1962, 5075.
8. VENANZI, L.M., J.Chem.Soc. (London), 1958, 719; 1961, 2705, 4816.
9. WIELAND, H., and ROSENFELD, Liebigs Ann.Chem., 484, 236 (1930).

2 B 3

ÜBER EINE σ-PHENYL-VANADIN(II)-VERBINDUNG.

E. Kurras

Forschungsstelle für Komplexchemie der Deutschen Akademie
der Wissenschaften zu Berlin, Jena, Germany.

In den letzten Jahren konnten HEIN et al.σ -Organochrom(III)verbindungen mit Phenyl- und substituierten Phenylgruppen durch Komplexbildung stabilisieren.

$RCrX_2L_3$	R_2CrXL_3	R_3CrL_3	
$C_6H_5CrCl_2(THF)_3$ * (1)	$(C_6H_5)_2CrCl(DMG)_2$ * (4)	$(C_6H_5)_3Cr(THF)_3$	(5)
$p\text{-}BrC_6H_4CrCl_2(THF)_3$ (2)		$(p\text{-}ClC_6H_4)_3Cr(THF)_3$	(3)
$p\text{-}ClC_6H_4CrCl_2(THF)_3$ (3)		$(p\text{-}BrC_6H_4)_3Cr(THF)_3$	(6)
$m\text{-}ClC_6H_4CrCl_2(THF)_3$ (3)		$(m\text{-}ClC_6H_4)_3Cr(THF)_3$	(6)
		$(m\text{-}ClC_6H_4)_3Cr(Py)_3$	(6)
		$(p\text{-}C_6H_5C_6H_4)_3Cr(THF)_3$	(3)
		$(p\text{-}(CH_3)_2NC_6H_4)_3Cr(THF)_3$	(7)
		$(o\text{-}CH_3OC_6H_4)_3Cr$	(8)
$R_4CrL_2Me(I)$	$R_5CrLMe(I)_2$	$R_6CrMe(I)_3$	
$(C_6H_5)_4CrLi(DMG)_4$ * (4)	$(C_6H_5)_5CrLi_2\ddot{A}_3$ (10)	$(C_6H_5)_6CrLi_3$	(11)
$(C_6H_5)_4CrNa(DMG)_4$ (4)	$(C_6H_5)_5CrNa_2\ddot{A}_3$ (10)	$(p\text{-}CH_3C_6H_4)_6CrLi_3$	(12)
$(C_6H_5)_3Cr(C_5H_5)Li$ (9)		$(p\text{-}C_6H_5C_6H_4)_6CrLi_3$	(13)
		$(p\text{-}CH_3OC_6H_4)_6CrLi_3$	(12)
		$(p\text{-}(CH_3)_2NC_6H_4)_6CrLi_3$	(8)
		$(p\text{-}BrC_6H_4)_6CrLi_3$	(14)
		$(m\text{-}ClC_6H_4)_6CrLi_3$	(14)

* THF = Tetrahydrofuran
  DMG = Dimethylglykoläther

Zur Darstellung wurden die Metallaustauschreaktionen verwendet:

$$6\ RLi \quad + CrX_3 \xrightarrow{\text{Äther}} Li_3CrR_6 \quad + 3\ LiX \quad (11)$$

$$3 \text{ RMgX} + \text{CrX}_3 \xrightarrow{\text{THF}} \text{R}_3\text{Cr(THF)}_3 + 3 \text{ MgX}_2 \qquad (5)$$

$$\text{R}_3\text{Al} + \text{CrX}_3 \xrightarrow{\text{THF}} \text{RCrX}_2(\text{THF})_3 + \text{R}_2\text{AlX} \qquad (1)$$

$$2 \text{ Li}_3\text{CrR}_6 + \text{CrX}_3 \xrightarrow{\text{DMG}} 3 \text{ LiCrR}_4(\text{DMG})_4 + 3 \text{ LiX} \qquad (4)$$

Nach den Ergebnissen der magnetischen Messungen ($\mu_{\text{eff}} = 3,65 - 3,95$ B.M.) sind es oktaedrisch koordinierte spin-free-Komplexe mit d^3-Konfiguration.

Derartige σ-Alkyl-chrom(III)verbindungen, wie das $\text{CH}_3\text{CrCl}_2(\text{THF})_3$ (15), $\text{Li}_3\text{Cr(CH}_3)_6(\text{Dioxan})_3$ (16) und das $(\text{C}_3\text{H}_5)_3\text{Cr}$ (17), das wegen der Allyl-Liganden eine Sonderstellung einnimmt, sind vor kurzem hergestellt worden.

Zu einer analogen σ-Metallorganoverbindung des V(II) mit d^3-Konfiguration, $\text{Li}_4\text{V(C}_6\text{H}_5)_6(\ddot{\text{A}})_{3,5}$ führte die Umsetzung ätherischer Phenyllithiumlösung mit $\text{VCl}_3(\text{THF})_3$ (18). Es sind äußerst oxydations- und feuchtigkeitsempfindliche, dunkelviolette Kristalle. Sie spalten bei etwa 50° Äther ab und zersetzen sich oberhalb 80° unter Diphenylbildung. Die Löslichkeit in Äther ist sehr gut, in Dioxan, Benzol, Cyclohexan mäßig. Die Verbindung ist paramagnetisch ($\mu_{\text{eff}} = 3,85$ B.M.), in Übereinstimmung mit der d^2sp^3-Hybridisierung für oktaedrische spin-free-Komplexe. Das $\text{Li}_4\text{V(C}_6\text{H}_5)_6$ gibt die typischen Organometallreaktionen, wie

$$(\text{C}_6\text{H}_5)_6\text{VLi}_4 + 6 \text{ H}_2\text{O} \longrightarrow 6 \text{ C}_6\text{H}_6 + \text{V(OH)}_2 + 4 \text{ LiOH}$$

$$(\text{C}_6\text{H}_5)_6\text{VLi}_4 + 6,5 \text{ J}_2 \longrightarrow 6 \text{ C}_6\text{H}_5\text{J} + \text{VJ}_3 + 4 \text{ LiJ}$$

$$(\text{C}_6\text{H}_5)_6\text{VLi}_4 + 7 \text{ HgCl}_2 \longrightarrow 6 \text{ C}_6\text{H}_5\text{HgCl} + 1/2 \text{ Hg}_2\text{Cl}_2 + \text{VCl}_3 + 4 \text{ LiCl}$$

Im Gegensatz zu der Umsetzung des $\text{VCl}_3(\text{THF})_3$ mit Phenyllithium bildet sich mit Phenylmagnesiumbromidlösung ein Gemisch von π-Aromaten-vanadin(O)-Komplexen (17% Ausbeute). Daraus konnte Dibenzolvanadin(O) und Bis-diphenylvanadin(O) abgetrennt werden. Sie sind in ihren chemischen und physikalischen Eigenschaften mit Vergleichssubstanzen, die durch reduzierende Friedel-Crafts-Reaktion dargestellt wurden, völlig identisch. Bei der Synthese des bisher unbekannten Bis-diphenyl-vanadin(O) ist ein modifiziertes Verfahren mit Chlorbenzol als Lösungsmittel angewendet worden.

Für die Bildung der π-Aromaten-vanadin(O)-Komplexe bei der Grignardreaktion müssen σ-Phenyl-vanadin-Verbindungen als Zwischenprodukte angenommen werden. Die Umsetzung von $\text{Li}_4\text{V(C}_6\text{H}_5)_6$ mit wasserfreiem Magnesiumchlorid liefert nach der Hydrolyse das gleiche π-Aromatenkomplexgemisch. Auf das zentrale V(II)-Atom bezogen, ergeben quantitative Untersuchungen eine Disproportionierung nach

$$3 \ V(II) \longrightarrow \ 2 \ V(III) \ + \ V(O)$$

Das Grignardsystem kann das im $Li_4V(C_6H_5)_6$ komplex maskierte "$V(C_6H_5)_2$" nicht stabilisieren; es bilden sich π-Aromatenkomplexe.

Literatur

1. KURRAS, E., Dissertation Universität Jena 1959.
2. HEIN, Fr., und STOLZE, G., unveröffentlicht.
3. HEIN, Fr., und HÄHLE, J., unveröffentlicht.
4. HEIN, Fr., und SCHMIEDEKNECHT, K., unveröffentlicht.
5. ZEISS, H.H., und HERWIG, W., J.Am.Chem.Soc., 79, 6561 (1957).
6. HEIN, Fr., HÄHLE, J., und STOLZE, G., Mber.Dt.Akad.Wiss., 5, 528 (1963).
7. SCHMIDT, K., Diplomarbeit Universität Jena 1962.
8. HEIN, Fr., und TILLE, D., Mber.Dt.Akad.Wiss., 4, 414 (1962).
9. HEIN, Fr., und HEYN, B., Mber.Dt.Akad.Wiss., 4, 220 (1962).
10. HEIN, Fr., HEYN, B., und SCHMIEDEKNECHT, K., Mber.Dt.Akad. Wiss., 2, 552 (1960).
11. HEIN, Fr., und WEISS, R., Z.anorg.allg.Chem., 295, 145 (1958).
12. HEIN, Fr., WEISS, R., HEYN, B., BARTH, K.H., und TILLE, D., Mber.Dt.Akad.Wiss., 1, 541 (1959).
13. HEIN, Fr., und TILLE, D., unveröffentlicht.
14. HEIN, Fr., HÄHLE, J., und STOLZE, G., Mber.Dt.Akad.Wiss., 5, 530 (1963).
15. KURRAS, E., Mber.Dt.Akad.Wiss., 5, 378 (1963).
16. KURRAS, E., und OTTO, J., Mber.Dt.Akad.Wiss., 5, 705 (1963).
17. KURRAS, E., und KLIMSCH, P., Mber.Dt.Akad.Wiss., im Druck.
18. KURRAS, E., Naturwiss., 46, 171 (1959).

3 B 1

STRUCTURES AND BONDING OF NEW TYPES OF TRANSITION ORGANOMETALLIC COMPLEXES.

Lawrence F. Dahl and Willi E. Oberhansli
Department of Chemistry, University of Wisconsin,
Madison 6, Wisconsin, U.S.A.

The results of a number of recent structural investigations have provided new insight into the nature of bonding of organometallic complexes. These studies include a systematic examination of several (π-allylic) transition metal compounds to determine the stereochemical nature of these complexes. The detailed molecular configuration of $[\pi\text{-}C_3H_5PdCl]_2$ was elucidated from a three-dimensional X-ray analysis (1). The dimeric molecule of idealized point symmetry C_{2h}-$2/m$ (i.e., the required crystallographic molecular symmetry is $C_i\text{-}\bar{1}$) consists of a Pd_2Cl_2 rhombus with each palladium symmetrically bonded to an allylic group. The overall crystallographic results of ROWE (2) based on two-dimensional data are substantiated with the notable exception that our more precisely determined atomic coordinates (which provide more refined molecular parameters) reveal that the plane of the three allylic carbons is not perpendicular to the plane of the Pd_2Cl_2 bridge system; the dihedral angle between these two planes is 118° (e.s.d. 4°) with the central allylic carbon tipped away from the palladium; the palladium, however, is closer to the central carbon (Pd-C = 2.01 ± 0.04 Å) than to the two terminal carbons (mean Pd-C = 2.15 ± 0.02 Å).

The structural determination (1) of a nickel complex of formula $C_{13}N_{17}NiC_5H_5$ prepared by the reaction of sodium cyclopentadienide with tetramethylcyclobutadiene nickel dichloride (3) has revealed an unexpected new type of molecular configuration consisting of a nickel coordinated to an allylic fragment which is part of a cyclobutenyl group. The resulting compound is π-cyclopentadienyl-1,2,3,4-tetramethyl-1-cyclopenta-1',3'-dienecyclobutenyl-nickel(II). The nickel is symmetrically bonded to the three allylic carbons of the non-planar cyclobutenyl group with the central carbon being closer to the nickel than are the two terminal carbon atoms ($1.89 \pm 0.01_4$ Å vs. 1.99 ± 0.01 Å (av.)).

The above work prompted our X-ray investigation (1) of two palladium isomers of empirical formula $C_{30}H_{25}OPdCl$ prepared by the reactions of palladium compounds with diphenylacetylene (4). Both structures were found to consist of dimeric molecules (each possessing a crystallographic center of symmetry) in which each palladium of a rhombic Pd_2Cl_2 fragment is bonded to the allylic part of a cyclobutenyl group. A phenyl group is attached to each of the four ring carbons with an ethoxy group being the other substituent attached to the tetrahedral carbon. These compounds are geometrical isomers (endo- and exo-forms) with the phenyl and ethoxy groups on the tetrahedral carbon interchanged.

The detailed molecular features of these four compounds will be compared with one another and with those of 3-chloro-1,2,3,4-tetraphenyl-cyclobutenium pentachlorostannate whose structure was recently determined (5).

The organometallic interactions of several other complexes will be discussed.

References

1. OBERHANSLI, W.E., and DAHL, L.F., submitted for publication.
2. ROWE, J.M., Proc.Chem.Soc. (London), 1962, 66.
3. CRIEGEE, R., and LUDWIG, P., Ber., 94, 2038 (1961); KING, R.B., Inorg.Chem., 2, 528 (1963).
4. BLOMQUIST, A.T., and MAITLIS, P.M., J.Am.Chem.Soc., 84, 2329 (1962).
5. BRYAN, R.F., J.Am.Chem.Soc., 86, 733 (1964).

3 B 2

THE REACTION OF HEXACARBONYLVANADIUM WITH AROMATIC HYDROCARBONS.

Fausto Calderazzo

Cyanamid European Research Institute, Cologny, Geneva, Switzerland.

It has been known since 1958 that the hexacarbonyls of the Group VI metals react with aromatic hydrocarbons to give tricarbonylarenemetal derivatives, $Me(CO)_3Arene$. It has now been found that hexacarbonylvanadium reacts at $20-50^{\circ}$, preferably at 35°, with aromatic hydrocarbons with evolution of carbon monoxide and precipitation of solid products. The reaction has been successfully carried out with benzene, toluene, p-xylene and mesitylene. Compounds of formula $[V(CO)_4Arene][V(CO)_6]$ have been isolated from the corresponding reaction mixtures as red crystalline air-sensitive solids.

The ionic character of the hexacarbonylvanadates $[V(CO)_4Arene][V(CO)_6]$ is evidenced by their non-volatility, by a negligible solubility in hydrocarbons, and by the presence of infrared bands at 1895 and 1859 cm^{-1} due (1,2) to the anion $[V(CO)_6]^-$. The corresponding hexafluorophosphates, prepared by the following metathetical reaction:

$$[V(CO)_4Arene][V(CO)_6] + NH_4PF_6 \rightarrow [V(CO)_4Arene]PF_6 + NH_4[V(CO)_6]$$

$$\{1\}$$

show only C-O stretching bands at higher frequencies (between 2070 and 1979 cm^{-1}) due to the cationic carbon monoxide.

The cation $[V(CO)_4Arene]^+$ is electronically analogous to the cyclopentadienyl complex $[V(CO)_4C_5H_5]$ (3). The infrared data suggest that they are also structurally similar. In addition to that, the infrared measurements

- 245 -

show that the stretching frequencies of the cationic carbon monoxide depend on the degree of substitution of the aromatic ring. The compounds $[V(CO)_4 Arene]X$ ($X = [V(CO)_6]^-$, PF_6^-, $[B(C_6H_5)_4]^-$) show in fact a regular shift to lower frequencies of 2-3 cm^{-1} per methyl group with increasing number of methyl substituents (tetrahydrofuran or acetone solutions). This is explained in terms of a corresponding increased d_π-p_π bonding from the metal to the CO ligands caused by the inductive effect of the methyl substituents. Similar effects have been found for the tricarbonylarenechromium derivatives (4, 5).

The precipitation of $[V(CO)_4 Arene][V(CO)_6]$ in the reaction medium is probably preceded by the formation of a soluble substitution product of $V(CO)_6$. In view of the known (1, 6) oxydising properties of hexacarbonylvanadium, the formation of the final ionic compounds clearly involves an oxidation-reduction step by a second molecule of $V(CO)_6$. The nature of the intermediate substitution product will be discussed.

References

1. CALDERAZZO, F., Inorg.Chem., in the press.
2. CALDERAZZO, F., and ERCOLI, R., Chim.Ing. (Milan), 44, 990 (1962).
3. FISCHER, E.O., and HAFNER, W., Z.Naturforsch., 9b, 503 (1954).
4. FISCHER, R.D., Chem.Ber., 93, 165 (1960).
5. BROWN, D.A., and SLOAN, H., J.Chem.Soc. (London), 1962, 3849.
6. CALDERAZZO, F., and BACCIARELLI, S., Inorg.Chem., 2, 721 (1963).

13 B 3

NOVEL ACETYLENE TUNGSTEN CARBONYL COMPLEXES.

D.P. Tate, J.M. Augl, W.M. Ritchey, B.L. Ross
and J.G. Grasselli

Research Department, The Standard Oil Company (Ohio), Cleveland, Ohio.

We recently reported the preparation of a new acetylenic complex of tungsten (1). This study has now been extended to other alkynes and the other subgroup VI-B metals.

Tris-hexyne tungsten (O) monocarbonyl $(C_2H_5C\equiv CC_2H_5)_3 W(CO)$, [I], was prepared by heating excess 3-hexyne with either monoacetonitrile tungsten pentacarbonyl or tris-acetonitrile tungsten tricarbonyl (1). Other disubstituted alkynes give analogous products. Thus, diphenylacetylene and methylphenyl-acetylene formed the stable complexes $(C_6H_5C\equiv CC_6H_5)_3 W(CO)$, [II], and $(C_6H_5C\equiv CCH_3)_3 W(CO)$, [III], respectively. The 3-hexyne may be displaced and recovered by treatment of [I] with CO at 90° and 3000 psi. Methylphenylacetylene may be recovered on a cold finger by vacuum pyrolysis of [III]. Alkynes that contain a terminal hydrogen tended to cause decomposition and polymerization and no stable organometallic compounds were isolated. Acetylene, 1-hexyne and phenylacetylene are examples of alkynes that gave brown polymeric residues in this reaction.

In attempting to extend this reaction to derivatives of chromium and molybdenum, completely different results were obtained under comparable reaction conditions. Reaction of 3-hexyne with $(CH_3CN)_3Mo(CO)_3$ did not appear to give a compound corresponding to the tungsten acetylenic complex. Hexaethylbenzene isolated from the reaction mixture indicates that the molybdenum complex causes cyclotrimerization of the acetylenic molecule.

The reaction of the chromium analogue was complicated. Reaction of diphenylacetylene with $(CH_3CN)_3Cr(CO)_3$ gave a mixture of products. Thin layer chromatography showed a major component which was yellow but soon turned violet upon exposure to air. A chromatogram in the second dimension showed it to have an R_f value identical to authentic tetraphenylcyclopenta-dienone. Apparently an air sensitive cyclopentadienone chromium complex is formed by cyclodimerization of the diphenylacetylene with inclusion of a carbonyl group. This is consistent with results reported by others (2-6).

The molecular weights of [I], [II] and [III] indicate a single mononuclear species. When the acetylene complexes are refluxed in ethanolic I_2 solution, only one molar equivalent of CO is evolved. Furthermore, spectroscopic examination of the residues after carbonyl determination gives no evidence of carbonyl groups either organic or inorganic. From this one may conclude that a single carbonyl group exists in the molecule.

Hexyne-3 displays a typical A_2B_3 type NMR spectrum with the methylene group appearing as a quartet and the methyl group as a triplet. In contrast, the spectrum of [I] displays two quartets having equal intensity and two triplets (overlapping) having equal intensity. Also, there is no evidence for non-equivalence of protons within a methylene or methyl group. This suggests that all three acetylenic molecules are equivalent but that in each of the ligands one of the two ethyl groups is in a different environment than the other. Thus, we can conclude that the three ligand molecules in [I] are symmetrically arranged around the metal atom with 3 ethyl groups in the vicinity of the CO.

The NMR spectrum of [III] is consistent with those of [I] and [II] . This is the only compound in the series in which the ligand is not symmetrical about the triple bond and, therefore, it is possible to have geometric isomers within the same basic structure. The methyl resonance line in pure methyl-phenylacetylene appears at 2.01 ppm below TMS. It was observed with the hexyne that the resonance lines of the methylene groups in the environment of the CO are shifted downfield 1,24 ppm. and those of the ones extending in the opposite direction are shifted downfield 0,77 ppm. Assuming that approximately the same shifts are observed in this case, one would expect resonance lines for the methyl group to appear at approximately 2,78 and 3,25 ppm. Resonance lines are observed in these regions and this constitutes strong evidence that geometrical isomers are indeed present.

The IR spectra of the mulls of the complexes [I], [II] and [III] show one strong band for the A_1 metal carbonyl stretching mode assuming C3 or C3v symmetry. A weak, sharp band on the low frequency side of the metal carbonyl band is assigned as the ^{13}CO mode. The isotopic shift for the stretching frequency of the unique CO group (see Table I) is much larger than usually found for the isotope shift in other metal carbonyls (7). A broad, weak band in the region of 1700 cm^{-1} is assigned to the coordinated C≡C stretching vibration. The shift of over 400 cm^{-1} from the usual position for this vibration indicates that coordination to the metal is of a doubly π-bonded nature as in Pt(R$_3$P)$_2$ac (8) and ReCl(ac)$_2$ (9). Except for small shifts in position and slight changes in relative intensity, the ligand vibrations are essentially those of the free ligands.

TABLE I

INFRARED ABSORPTIONS OF (RC≡CR')$_3$W(CO) COMPLEXES (in cm^{-1})

Compound/Medium	ν CO	ν $_{13}$CO	ν C≡C
[I]/n-hexane	2034,3	1987,5	1702
[II]/CCl$_4$	2071,8	2022,3	1680
[III]/CCl$_4$	2057,0	2008,3	1731

Based on the evidence cited above, there are two structures that seem most probable (Fig. 1). Figure 1a shows a structure with C3v symmetry where the ligands may be considered arranged pseudotetrahedrally (to the center of the triple bond). This allows the symmetrical placing of the alkynes, and the different environment of one end of the alkyne by proximity to the carbonyl. By twisting the alkynes out of the plane of the metal—CO bond and considering the complex as a type of 7 coordination, one gets a C$_3$ symmetry that is also consistent with the NMR evidence (Fig. 1b).

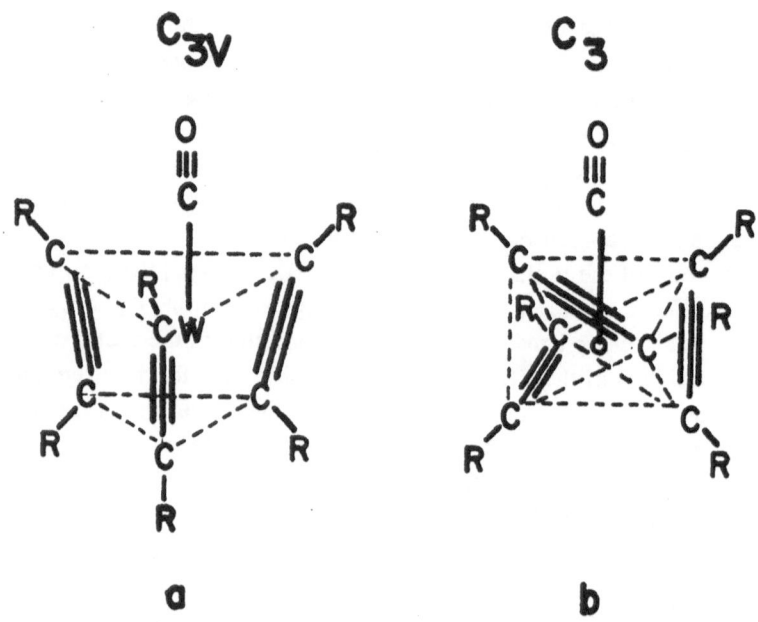

Figure 1
Proposed Tetrahedral and Distorted Octahedral
Configurations for (RC≡CR')W(CO).

The effective atomic number (EAN) for tungsten requires that only ten electrons should be doned by the acetylenic ligands. This means two of the acetylenic ligands donate both their π-electron pairs (similar to Pt(R$_3$P)$_2$ac (8) and ReCl(ac)$_2$ (9)) and the third only one. Since H^{1}MR, however, indicates that all acetylene ligands are bonded equivalently, we have to assume that the lone electron pair is resonated through the entire carbon system.

A simple M.O. treatment shows that the electron pair can be accommodated in a suitable non-bonding orbital in either C$_3$ or C$_{3v}$ symmetry. Further confirmation of structure must await X-ray crystallographic analysis which is currently underway (10).

Present address of D.B.TATE: The Firestone Tire & Rubber Co., Akron, Ohio
Present address of J.M. AUGL: Melpar, Inc., Falls Church, Virginia

References

1. TATE, D.P., and AUGL, J.M., J.Am.Chem.Soc., 85, 2174 (1963).
2. HÜBEL, W., and WEISS, E., Chem.Ind. (London), 1959, 703.

3. TSUTSUI, M., and ZEISS, H., J.Am.Chem.Soc., 82, 6255 (1960).

4. SCHRAUZER, G.N., Ber., 94, 1403 (1961).

5. LETO, J.R., and COTTON, F.A., J.Am.Chem.Soc., 81, 2970 (1959).

6. STROHMEIER, W., LAPORTE, H., and Von HOBE, D., Ber., 95, 445 (1962).

7. (a) SHUFLER, L., STERNBERG, H.W., and FRIEDEL, R.A., J.Am.Chem. Soc., 78, 2687 (1956).

   (b) FRIEDEL, R.A., WENDER, I., SHUFLER, S.L., and STERNBERG, H.W., J.Am.Chem.Soc., 77, 3951 (1955).

   (c) ROSS, B.L., GRASSELLI, J.G., RITCHEY, W.M., and KAESZ, H.D., Inorg.Chem., 2, 1023 (1963).

8. CHATT, J., ROWE, G.A., and WILLIAMS, A.A., Proc.Chem.Soc., 1957, 208.

9. COLTON, R., LEVITUS, R., and WILKINSON, S.G., Nature, 186, 233 (1960).

10. DAHL, L.F., University of Wisconsin, Madison, Wisconsin.

14 B

BONDING IN SOME NOVEL ORGANOMETALLIC COMPLEXES OF THE FIRST AND SECOND ROW TRANSITION METAL IONS.

M.R. Churchill and R. Mason
Department of Chemistry, University of Sheffield, England, U.K.

X-ray crystallographic analyses of a number of organometallic molecules have indicated that bonding of the cyclic ligands to transition metal ions differs considerably from that in such molecules as the bis π-cyclopentadienyl metal complexes. Thus in π-cyclopentadienyl-1-phenylcyclopentadiene-cobalt (1), cobalt (2,3) tetrakis(trifluoromethyl) cyclopentadienone iron tricarbonyl (3), π-cyclopentadienyl hexakis(trifluoromethyl) benzenerhodium (4) and tricarbonyloctafluorocyclohexa-1,3-diene iron (5), the essentially planar structure of the substituted diene, dienone or benzene ligand is considerably modified on bonding to the metal ion. The non-planarity of the ligand varies from some 20° in the case of the cyclopentadienone complex to approximately 47° in the benzene and hexa 1-3 diene structures.

The effect of such asymmetric bonding, which may be given a σ-π description or can, quite equivalently, be discussed in terms of a 'back-donation' model, is to lead to an asymmetry in the metal-cyclopentadienyl bond, the result of which is that the cyclopentadienyl ligand loses its five-fold symmetry

(6). Moreover, in the case of the tricarbonyl complexes the carbonyl groups are not arranged according to strict C_{3v} symmetry. In the hexadiene complex the (O)C-Fe C(O) angles of $88,8^O$, $97,6^O$ and $97,6^O$ show the same form of distortion as those reported for $C_8H_8Fe(CO)_3$ (7) ($92,7^O$, $100,7^O$, $100,7^O$), $(OC)_3FeC_8H_8Fe(CO)_3$ (7) (91°, 101°, 102° and 94°, $97,5^O$) (3). They are also entirely similar to the values in butadiene iron tricarbonyl (8) and in view of all the accumulated structural data, together with the nuclear magnetic resonance spectrum (9), it is felt that the early suggestion that this latter molecule is properly formulated as a simple π-complex must be re-examined.

All of this data is examined in the light of molecular orbital theory and its implications on the relative stabilities of olefin complexes of first-, second-, and third-row transition metal ions examined.

References

1. CHURCHILL and MASON, Proc.Chem.Soc., 1963, 112.
2. GERLOCH and MASON, Proc.Roy.Soc., in press.
3. BAILEY, GERLOCH and MASON, Nature, 201, 72 (1964).
4. CHURCHILL and MASON, Proc.Chem.Soc., 1963, 365.
5. CHURCHILL and MASON, Proc.Chem.Soc., in press.
6. BENNETT, CHURCHILL, GERLOCH and MASON, Nature, 201, 1318 (1964).
7. DICKENS and LIPSCOMB, J.Chem.Phys., 37, 2084 (1962).
8. MILLS and ROBINSON, Acta Cryst., 16, 758 (1963).
9. GREEN, PRATT and WILKINSON, J.Chem.Soc. (London), 1959, 3753.

4 B 2

DONATOR-AKZEPTOR-KOMPLEXE VON MONOSUBSTITUIERTEN BENZOLEN C_6H_5X MIT ORGANISCHEN DONATORVERBINDUNGEN.

M.G. Woronkow und A.J. Deitsch
Institut für organische Synthese der Akademie der Wissenschaften
der Lettischen SSR, Riga 6, UdSSR.

Mit Methoden der physikalisch-chemischen Analyse (Untersuchung der Isothermen: Zusammensetzung - Eigenschaft), der Kryoskopie, Kolorimetrie, der IR- und UV-Spektroskopie wurde gezeigt, daß viele polare monosubstituierte Benzolderivate vom Typ C_6H_5X (X ist ein elektronegativer Substituent, der die Dichte der π-Elektronenwolke des Benzolrings herabsetzt), die ein Dipolmoment nicht unter 1,4 D besitzen, imstande sind, in Lösungen labile Molekularkomplexe mit organischen und elementorganischen Lewis-Basen (D) zu bilden. Als solche können donatorartige Moleküle figurieren, die ein

Heteroatom mit einem freien Elektronenpaar, wie z.B. O, N u. a., enthalten.

Die refraktometrischen Daten zeigen, daß die größten Abweichungen der Isothermen $n(v)$ von der Geraden in Systemen auftreten, bei denen der Substituent X zur 2.Ordnung gehört (-I, -C-Effekt) und die benzoide Komponente C_6H_5X unter den Molekülen dieses Typs ein maximales Dipolmoment aufweist (X = NO_2, CN, NCS usw.).

Als nichtbenzoide Komponente in diesen Systemen dienen hauptsächlich Verbindungen mit klar ausgeprägtem Elektronendonator-Charakter (aliphatische Amine, Tetrahydrofuran, Aceton, Trimethyläthoxymethan). Systeme, in denen die aromatische Komponente ein niedriges Dipolmoment besitzt, was auf eine nur unbedeutende Verringerung der Elektronendichte bzw. auf eine Steigerung derselben hinweist (X = CH_3), sind fast ideal oder stehen einem Idealsystem nahe. Dasselbe gilt für Systeme, in denen die Donatorkomponente D nur schwach ausgeprägte basische Eigenschaften zeigt (D = $C_6H_5OCH_3$, $(C_6H_5)_2O$, n-C_4H_9OH, Pyrrol usw.).

Die Untersuchung der Dichte- und Viskositätsisothermen von vielen der genannten Systeme führt zu analogen Schlußfolgerungen.

In einer Reihe von Fällen (insbesondere, wenn ein Amin als Donatorkomponente dient) weist das Auftreten einer intensiven gelben, gelborangen oder roten Färbung beim Mischen der Bestandteile eindeutig auf die Bildung von Komplexen vom Typ C_6H_5X. D hin.

Die UV-Spektren einer Anzahl von Binärsystemen C_6H_5X-D werden durch eine Verschiebung des Absorptionsbandenrandes in das langwellige Gebiet gekennzeichnet. In inerten Lösungsmitteln (Hexan) erfolgt gewöhnlich nur eine Schwächung der Absorptionsintensität. Das deutet auf die Unbeständigkeit der Komplexe hin, die unter dem Einfluß des Lösungsmittels zerfallen.

Die Untersuchung der IR-Spektren zeigt, daß z.B. im System Nitrobenzol-Trimethyläthoxysilan eine Aufspaltung der Banden mit den Wellenzahlen 753 und 950 cm^{-1} erfolgt, die zu den Schwingungen Si-O bzw. C-O gehören. Hiebei tauchen neue intensive Banden bei 776 und 942 cm^{-1} auf. Ein ähnliches Bild ist im IR-Spektrum des Systems $C_6H_5NO_2$-$(CH_3)_2CO$ festzustellen. Hier spalten sich die äußerst intensiven Banden der Valenzschwingungen der Carbonyl- und Nitrogruppe bei 1228 bzw. 1525 cm^{-1} auf, wobei neue Absorptionsmaxima bei 1248 bzw. 1510 cm^{-1} entstehen. Außerdem wurden neue Banden im Gebiet 850 - 1010 und 1755 - 2805 cm^{-1} beobachtet. Dies ist vermutlich auf die Verminderung der Gesamtsymmetrie des Molekülsystems im Ergebnis der Komplexbildung zurückzuführen.

Die kolorimetrische Untersuchung des Systems $C_6H_5NO_2$-$(C_2H_5)_3N$ u. ä. zeigt, daß in derartigen Systemen 1:1-Molekularverbindungen vorliegen.

Die Bildung von Molekülkomplexen in den Systemen $C_6H_5NO_2\cdot(CH_3)_2CO$ und $C_6H_5NO_2\cdot(CH_3)_3SiOC_2H_5$ wurde kryoskopisch unter Beweis gestellt.

Die Komplexbildung in den Systemen C_6H_5X-D trägt einen Donator-Akzeptor-Charakter; aus diesem Grunde können die entstehenden Molekül-verbindungen der Komponenten als "umgekehrte π-Komplexe" bezeichnet werden. Diese Bezeichnung ist völlig berechtigt, da das Elektronensystem der beschriebenen Komplexe sich von den gewöhnlichen π-Komplexen nur durch die gegenläufige Richtung der π-Elektronenverschiebung unterschei-det. Das Additionszentrum des Addenden D ist nicht an den einzelnen Atomen des Benzolrings lokalisiert, sondern ist am ehesten auf die ganze aromatische π-Molekülbahn verteilt. Zunächst schien es wahrscheinlich, daß die umgekehrten aromatischen π-Komplexe durch Überlappung der π-Elektronenwolke des Benzolrings mit der Bahn des anteiligen Elektronenpaares am Sauerstoff bzw. Stickstoff des Donators zustande kommen, was schema-tisch folgendermaßen dargestellt werden kann:

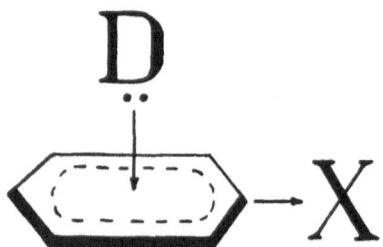

In diesem Falle müßte jedoch die Neigung zur Komplexbildung parallel mit der Fähigkeit des Donators, sein freies Elektronenpaar abzugeben, d.h. mit seiner Basizität steigen. In Wirklichkeit aber ist bei Sauerstoffverbindun-gen die Intensität der Wechselwirkung nur wenig von der Basizität abhängig. Mehr noch: Die größte Neigung zur Komplexbildung ist gerade bei Verbin-dungen mit herabgesetzten Basizitäten und Polaritäten (Donator-Eigenschaften) zu verzeichnen, z.B. bei den Äthoxysilanen, Vinylalkyläthern, bei Furan, Acetal usw. In diesen Verbindungen ist das freie Elektronenpaar des Sauer-stoffatoms delokalisiert: Bei den Äthoxysilanen durch d_π-p_π-Wechselwirkung mit dem benachbarten Siliciumatom, bei den Vinylalkyläthern und beim Furan durch p_π-p_π-Wechselwirkung mit der Doppelbindung und beim Acetal durch die Gemkonjugation mit dem in Gemstellung befindlichen Sauerstoff-atom.

Da die zur Herabsetzung der Basizität führende Delokalisierung des freien Elektronenpaares am Heteroatom eine Verringerung des Ionisationspotentials nach sich zieht, müssen die beschriebenen Molekularverbindungen als Ladungs-übertragungskomplexe ("charge-transfer complexes") betrachtet werden.

COORDINATION POLYMERS.

Maurice L. Huggins

Stanford Research Institute, Menlo Park, California, U.S.A.

Molecules or ions containing two or more atoms or groups capable of acting as ligands to appropriate metal atoms can form chelate complexes, if the geometrical relationships are suitable. Otherwise, each such molecule or ion would be expected to link together two or more metal atoms. If enough metal atoms are linked together in this way, the product can be called "a coordination polymer" or polyion.

In linear coordination polymers the metal atoms are connected together by nonmetallic molecules or ions into long chains. Many attempts have been made to prepare compounds of this sort which are of high molecular weight and stable with regard to chemical and thermal decomposition, and which have physical properties (strength, flexibility, etc.) comparable to those of common organic linear high polymers. Usually, metal ions of coordination number four have been connected through organic groups - ions of equal charge, to make the polymer chain neutral - each of these groups having two suitable placed pairs of coordinating atoms. Each metal atom is thus connected to two different organic moeities, filling all four sites in its coordination shell, and each organic moeity is bonded to two metal atoms. Although the synthesis of such polymers appears, at first sight, simple, most of the products which have been reported have either had relatively low molecular weights or have been branched or crosslinked. The probable causes of these undesirable results and possible ways of avoiding or overcoming the difficulties will be discussed.

Polyphosphinate polymers (1), in which (ideally) consecutive metal ions are believed to be connected through double bridges will be especially considered. Formulas proposed for some of these polymers are the following:

One might guess that organic molecules containing two carboxyl groups would be suitable for connecting metal atoms together to form linear chains,

but structures in which the two oxygen atoms of each carboxylate ion are connected to two different metal atoms are apparently preferred to those in which they both bond to a single metal atom. Because of this, three-dimensional network polymers are usually formed when dicarboxylic acids react with transition metal compounds to give anhydrous salts.

In some other classes of organic compounds, containing two pairs of oxygen atoms suitable for coordination, the two atoms in each pair do both bond to a single metal atom. The cupric salt of 2,5-dihydroxyquinone, for example, consists of rodlike polymer molecules (2-4):

Work now in progress in the writer's laboratory on the synthesis and study of other rodlike, sheetlike, and 3-dimensional network coordination polymers in which metal atoms are connected together through rigid organic groups will be reported.

Network coordination polymers with flexibility in the moeities connecting the metal atoms together, corresponding to thermoset and chemically cross-linked organic polymers, should have useful and interesting viscoelastic properties.

An important problem is that of synthesizing them in two stages, in such a way as to permit the production of desired shapes by procedures now used for organic plastics.

References

1. BLOCK, B.P., and coworkers, J.Am.Chem.Soc., <u>84</u>, 1749, 3200 (1962); <u>85</u>, 2018 (1963).
2. KANDA, S., and SAITO, Y., Bull.Chem.Soc.Japan, <u>30</u>, 192 (1957).
3. KANDA, S., Kogyo Kagaku Zasshi, <u>66</u>, 641 (1963), (in Japanese).
4. KOBAYASHI, H., HASEDA, T., KANDA, E., and KANDA, S., J.Phys. Soc. Japan, <u>18</u>, 349 (1963).

15 B 1

POLYNUCLEAR CARBONYLS OF MANGANESE, TECHNETIUM AND RHENIUM (1a).

W. Fellmann, D.K. Huggins (1b) and H.D. Kaesz
Department of Chemistry, University of California,
Los Angeles, California, U.S.A. (1c)

Recently, polynuclear tetracarbonyls of the two heavier metals in the manganese sub-group have been discovered. The first was for technetium (2) as a major by-product during attempts to prepare the pentacarbonyl hydride, and more recently, for rhenium (3) by an analogous route. The dimetal deca-carbonyl is treated first with sodium amalgam in tetrahydrofuran solution. After a time, the solvent is evaporated and cyclohexane and phosphoric acid are added and the heterogeneous mixture is refluxed; desired product is found in the cyclohexane layer. It was later discovered that the use of sodium boro-hydride in the first step instead of sodium amalgam led to the same products, but in greatly increased yields (3). For the derivative of technetium, an analysis, IR spectrum and molecular weight have been obtained. This and additional data are now available for the derivative of rhenium and these will be reviewed briefly.

We now wish to report the synthesis and properties of an analogous tetra-carbonyl of manganese. The preparation is similar to that of the tetracarbonyls of the other two metals in the sub-group; the use of sodium borohydride rather than sodium amalgam with the decacarbonyl in the first step also gives better yields of lower carbonyl. However, in the reaction sequence for manganese, we have isolated two principal products. The first is a yellow-orange trimer whose spectrum resembles that of the analogous derivatives of its congeners, and is shown in the attached figure. Analysis and magnetic and chemical properties will be discussed. Also, we will present chemical and spectroscopic data concerning the question of the presence of hydrogen in this derivative.

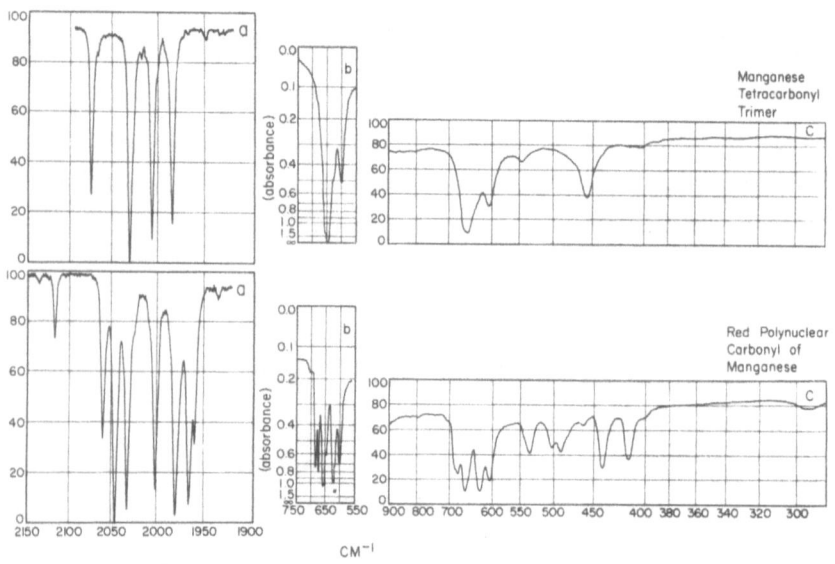

Legend: (a) Cyclohexane solution, LiF prism (b) KBr Pellet, grating instrument
(c) KBr Pellet, CsBr prism.

In addition to the yellow-orange trimer mentioned above, a large amount of a red derivative was also isolated, whose spectrum is shown in the lower part of the figure. Analysis, molecular weight and magnetic and chemical properties of this derivative will be reported. This second compound may be related to the red carbonyl hydride of manganese reported by HIEBER, BECK and ZEITLER (4), obtained by similar chemical transformations on the decacarbonyl. Chemical and spectroscopic evidence concerning the presence of hydrogen in the red derivative of manganese will be presented.

References

1. a) Supported by National Science Foundation Grant GP 1696, U.S.A.;
   b) present address: Research Laboratories, Allied Chemical Corp., Morristown, New Jersey, U.S.A.;
   c) Publication No. 1661.
2. KAESZ, H.D., and HUGGINS, D.K., Canad.J.Chem., <u>41</u>, 1250 (1963).
3. HUGGINS, D.K., FELLMANN, W., SMITH, J.M., and KAESZ, H.D., Abstracts of Papers Presented to the 147th Meeting of the American Chemical Society, April 1964, p. 11 L; manuscript in preparation.
4. HIEBER, W., BECK, W., and ZEITLER, G., Angew.Chem., <u>73</u>, 364 (1961).

15 B 2

UNUSUAL CARBON BRIDGING GROUPS IN BINUCLEAR COMPLEXES.

O.S. Mills
Department of Chemistry, University of Manchester,
Manchester 13, England, U.K.

A review of currently published structural investigations of binuclear transition metal complexes shows that where bridging groups are found there are at least two groups present, or alternatively one group with two atoms which are equidistant from the metals eg. $Co_2(CO)_6RC_2R'$. Metal-carbon distances and distortion of bond angles at these bridging carbon centres are discussed.

In addition to the published reports further values obtained from detailed analyses of $Ni_2(C_5H_5)_2RC_2R'$ [R = R' = phenyl] and two forms of $Co_2(CO)_9HC_2H$ will be available.

Further X-ray structural investigations have conclusively proved the existence of a bridging iso-nitrile group in the compound $Fe_2(C_5H_5)_2(CO)_3CNR$ [R = phenyl], a structure in which the cyclopentadienyl groups and terminal CO groups are arranged cis. The structure is compared to that of $Fe_2(C_5H_5)_2(CO)_4$.

The preparation of benzene derivatives containing ortho-substituted t-butyl groups via cobalt complexes of the formula type $Co_2(CO)_4[acetylene]_3$ has raised questions of the structural arrangements in these complexes. The analysis of $Co_2(CO)_4[Bu^tC_2H]_2[HC_2H]$, from which ortho-di t-butylbenzene is obtained, has been shown to involve six carbon atoms bonded bifunctionally to the two cobalts. Alternative descriptions of this arrangement involve (a) a heterocyclic eight-ring, twisted into the form of a figure 'eight', which contains two adjacent metal atoms or (b) an unusual six-carbon bridging group. In either description bonding of three carbon atoms to each cobalt requires either an allylic attachment (involving both σ - and π -types of metal attachment) or a combination of alkyl and olefinic bonds. This structure is discussed in detail.

Although the general rule for mixed cyclopentadienyl carbonyl complexes is that metals with odd atomic number form mononuclear complexes whilst those of even atomic number form binuclear complexes, it has been reported that mononuclear cyclopentadienyl rhodium dicarbonyl polymerises to a binuclear complex. The structure of a binuclear complex with an unusual carbonyl bridging group is reported.

5 B 3

ALKYNE-SUBSTITUTED IRON CARBONYLS OF FORMULA $RC_2R'Fe_x(CO)_y$.

C. Hoogzand and W. Hübel
Union Carbide European Research Associates, Brussels 18, Belgium.

Substitution of carbon monoxide ligands by one acetylene is the first step in the reaction of alkynes with metal carbonyls. This is well-established in the case of cobalt carbonyls by the isolation of substitution-type complexes of formula $RC_2R'Co_2(CO)_6$ (1) and $RC_2R'Co_4(CO)_{10}$ (2). Three different types of alkyne-substituted iron carbonyls have been isolated from iron carbonyls particularly with acetylenes having bulky substituents (3). They have recently been shown to be intermediates in the formation of complexes with more complicated organic ligands as well as of cyclic organic products.

1. Yellow complexes of formula $RC_2R'Fe(CO)_4$ have been obtained from $Me_3CC\equiv CCMe_3$, $Me_3SiC\equiv CSiMe_3$ and $PhC\equiv CSiMe_3$, and are regarded as alkyne substituted, pentacarbonyliron compounds for which a structure analogous to acrylonitrile $Fe(CO)_4$ (4) has been adopted. Their further reaction with other alkynes at room temperature leads to complexes of types $(RC_2R)_2COFe_2(CO)_6$, $(RC_2R)_2Fe_2(CO)_6$, and $(RC_2R)_3Fe_2(CO)_6$ and is discussed together with their role in the formation of quinones.

2. Compounds of composition $RC_2R'Fe_2(CO)_7$ have been isolated with diphenyl- and di-t-butylacetylene. These dark green complexes have only terminal CO ligands but correspond formally to a substitution type complex

of $Fe_2(CO)_9$. They react with alkynes at 20° to give primarily complexes of type $(RC_2R)_2COFe(CO)_6$ besides $(RC_2R)_2Fe_2(CO)_6$ and minor quantities of tri-carbonylquinoneirons or quinones.

3. Representatives of complexes of the class $RC_2R'Fe_3(CO)_9$ have been obtained from the reaction of $Fe_2(CO)_9$ with diphenylacetylene and from $Fe_3(CO)_{12}$ with methylphenylacetylene. These complexes have only terminal carbon monoxide groups and may be regarded as alkyne-substituted dodeca-carbonyltriirons. On refluxing $PhC_2PhFe_3(CO)_9$ with more diphenylacetylene in benzene/methanol for a short period violet $(PhC_2Ph)_2Fe_3(CO)_8$ is formed, which isomerized to the black form of $(PhC_2Ph)_2Fe_3(CO)_8$ on prolonged heating. Since the structure of both $(PhC_2Ph)_2Fe_3(CO)_8$ complexes has been ascertained by X-ray analysis (5) the following sequence demonstrates how the linkage of two alkynes can be achieved within the framework of organo-iron complexes:

violet   black

References

1. GREENFIELD, H., STERNBERG, H.W., FRIEDEL, R.A., WOTIZ, J.H., MARKBY, R., and WENDER, I., J.Am.Chem.Soc., 78, 120 (1956).
2. KRÜERKE, U., and HÜBEL, W., Chem.Ber., 94, 2829 (1961).
3. BRAYE, E.H., HOOGZAND, C., HÜBEL, W., KRÜERKE, U., MERÉNYI, R., and WEISS, E., in S.Kirschner, Ed., Advances in the Chemistry of the Coordination Compounds, The Macmillan Co., New York, 1961, p.190.
4. LUXMOORE, A.R., and TRUTER, M.R., Acta Cryst., 15, 1117 (1962).
5. DODGE, R.P., and SCHOMAKER, V., Ann.Meeting of Am.Cryst.Assoc., June 1962, Villanova University, Penn.

6 B 1

SULFUR-CONTAINING COBALT CARBONYL DERIVATIVES.

László Markó, György Bor and Egon Klumpp

Hungarian Oil and Gas Research Institute "M.Á.F.K.I.", Veszprém, Hungary.

Elemental sulfur and some organic sulfur compounds inhibit hydroformylation and hydrogenation catalyzed by cobalt carbonyls (1, 2). In order to elucidate the mechanism of this "poisoning" effect, several new, sulfur-containing cobalt carbonyl derivatives have been prepared.

TABLE

Cobalt carbonyl derivative		Products
Composition	S:Co ratio	
$Co_3(CO)_9S$	1:3	Chief products of the reaction
$[Co_2(CO)_5S]_n$	1:2	between $Co_2(CO)_8$ and S; by-products in all reactions of cobalt carbonyls with sulfur compounds
$Co_3(CO)_7S_2$	1:1,5	$Co_2(CO)_8$ and S
$Co_3(CO)_6(S)(SCH_2C_6H_5)$	1:1,5	$Co_2(CO)_8$ and $C_6H_5CH_2SH$
$Co_4(CO)_7(SC_2H_5)_3$	1:1,33	$Co_2(CO)_8$ and C_2H_5SH
$Co_4(CO)_5(SC_2H_5)_7$	1:0,57	or $C_2H_5SSC_2H_5$
$[Co_2(CO)(SC_2H_5)_2]_n$	1:0,5	$Co(SC_2H_5)_2$ and CO

The infrared spectra of the reaction mixtures indicated in some cases the presence of other complexes too, which, however, could not be prepared in crystalline form.

All these complexes are dark colored, more or less soluble in hydro-carbons, air sensitive, especially in solution. Their thermal decomposition, beginning usually at 100-140° yields carbon monoxide, hydrocarbons and a cobalt sulfide containing a solid residue, when sulfur is lost in form of volatile products, when the S:Co ratio in the complex exceeds one. The cobalt-sulfur bond seems therefore to be rather strong, irrespective the character of the bond as is seen from comparing $Co_3(CO)_9S$ and $Co_3(CO)_6(S)(SCH_2C_6H_5)$.

The structure of these complexes has not been determined. Based on infrared data structures for $Co_3(CO)_9S$, $Co_3(CO)_7S_2$ and $Co_3(CO)_6(S)(SR)$ are suggested. The "inert gas rule" seems not to be followed in most of these compounds and some of them are supposed to have rather complex structures. Both observations are thought to be due to the great variability of the cobalt-sulfur bonding.

$Co_3(CO)_9S$ and complexes of the type $Co_3(CO)_6(S)(SR)$ were found to be stable under hydroformylation conditions ($120-200°$, 100-200 at of carbon monoxide and hydrogen, respectively) and according to our observations play an important role in the desulfurization reactions observed under such conditions (3).

If iron pentacarbonyl was present in the hydroformylation reaction mixture too, a mixed complex, $Co_2Fe(CO)_9S$ was formed in good yield (4). The tendency of formation of this very stable complex is so strong, that even thiophene is easily desulfurised under such conditions, a reaction, which is not significant with cobalt carbonyl alone. Even the iron orginating from a small corrosion of the walls of the reaction vessel leads to the predominant formation of this complex.

The fact, that all three complexes, which are stable under hydroformylation conditions, contain the $(metal)_3S$ trigonal pyramid as the skeleton of the molecule, points to the stability of this configuration.

The (cobalt carbonyl)-(carbon disulfide) complexes prepared in this laboratory by KLUMPP and coworkers (5, 6) belong apparently to a different class of sulfur-containing cobalt carbonyl derivatives.

References

1. MACHO, V., Chem. Zvesti (Bratislava), 15, 181 (1961).
2. MARKÓ, L., paper to be presented on the Symposium on Coordination Chemistry in Tihany (Hungary), 15. Sept. 1964.
3. MARKÓ, L., KHATTAB, S.A., Chem. and Ind. in the press.
4. KHATTAB, S.A., MARKÓ, L., BOR, G., Organometallic Chem., 1, 373 (1964).
5. MARKÓ, L., BOR, G., KLUMPP, E., Angew. Chem., 75, 248 (1963).
6. KLUMPP, E., Diss., Univ. for Chem. Ind., Veszprém (1964).

THIOCYANATO METAL CARBONYLS.

Andrew Wojcicki and Michael F. Farona
The McPherson and Evans Chemical Laboratories, The Ohio State University,
Columbus 10, Ohio, U.S.A.

The first known simple thiocyanato metal carbonyl, $Mn(CO)_5SCN$, was prepared by the reaction of $NaMn(CO)_5$ with ClSCN in tetrahydrofuran (1). This golden yellow compound, soluble in chloroform, acetone and methanol, is diamagnetic and a nonelectrolyte in nitromethane. Its IR spectrum in the $1900 - 2200 \text{ cm}^{-1}$ region is consistent with a distorted C_{4v} symmetry. The distortion probably arises from nonlinearity of the Mn-S-CN moiety and suggests a sulfur-bonded structure (2). This mode of attachment is supported by the CS stretching frequency, which occurs at 676 cm^{-1} and is diagnostic of an S-bonded thiocyanate (3).

Reactions of $Mn(CO)_5SCN$ with a variety of neutral ligands in chloroform yield complexes which are analogous to those prepared from $Mn(CO)_5Br$ (4). The mode of the thiocyanato-manganese bonding in these compounds depends on the nature and position of the neutral ligand(s) present. Strong bases such as pyridine, p-toluidine (L) and 2,2'-bipyridine (L-L) give cis-disubstituted thiocyanato-N carbonyl complexes.

$$Mn(CO)_5SCN + 2 L \longrightarrow cis\text{-}[Mn(CO)_3L_2NCS] + 2 CO$$

$$Mn(CO)_5SCN + L\text{-}L \longrightarrow cis\text{-}[Mn(CO)_3(L\text{-}L)NCS] + 2 CO$$

The structures of these compounds have been elucidated by means of IR spectroscopy, both in the $1900 - 2200 \text{ cm}^{-1}$ (CO and CN) and $660 - 860 \text{ cm}^{-1}$ (CS) region. Unlike the amines, weaker bases such as $P(C_6H_5)_3$, $As(C_6H_5)_3$ and $Sb(C_6H_5)_3$ (L') react more slowly and yield a mixture of products, three of which have been separated by alumina chromatography. These are cis-$[Mn(CO)_4L'SCN]$, cis-$[Mn(CO)_3L_2'SCN]$ and trans-$[Mn(CO)_3L_2'NCS]$. The compounds cis-$[Mn(CO)_4L'SCN]$ and cis-$[Mn(CO)_3L_2'SCN]$ contain sulfur-bonded thiocyanate. The fact that the latter has not been isolated with $P(C_6H_5)_3$ may be due to the small size of phosphorus, relative to arsenic and antimony, which considerably crowds the phenyls around the metal. This explanation is supported by molecular models (Stewart) of the compounds. The different modes of Mn-CNS attachment in the cis- and trans-disubstituted complexes may be due to a greater steric hindrance in the former. Molecular models indicate that a bent S-bonded thiocyanate can avoid the bulky phenyl groups better than a normally linear N-bonded thiocyanate. However, an alternative explanation, in terms of a possible mechanism of formation of these complexes form the parent $Mn(CO)_5SCN$, is equally plausible at present.

Examination of the Mn-CNS bonding in the above-mentioned compounds reveals that, aside from those complexes where steric factors may be of prime importance, all of the disubstituted compounds contain an N-bonded thiocyanate. However, the parent and monosubstituted complexes are S-bonded. These observations suggest that with an increase in the negative charge at the metal, due to replacement of carbon monoxide by a more basic ligand, the bonding changes from M-SCN to M-NCS.

In order to unambiguously asses the effect of metal charge on the nature of thiocyanato-metal attachment, the anionic thiocyanatopentacarbonyl complexes of tungsten, molybdenum and chromium (M) were prepared, utilizing a reaction similar to that described previously (5),

$$M(CO)_6 + (CH_3)_4NSCN \longrightarrow (CH_3)_4N[M(CO)_5NCS] + CO \quad .$$

IR spectra (1800 - 2200 cm^{-1}) of these complexes are in agreement with the C_{4v} symmetry of the anions. Furthermore, the CS stretching frequencies in the 790 - 810 cm^{-1} region indicate that the thiocyanate is bonded through the nitrogen. Thus a decrease in charge on the metal from $Mn(CO)_5SCN$ to the isoelectronic $[Cr(CO)_5NCS]^-$ changes the bonding from M-SCN to M-NCS.

Preliminary kinetic studies of reactions of $Mn(CO)_5SCN$ with pyridine in chloroform revealed that there is a rapid Mn-SCN to Mn-NCS rearrangement prior to the replacement of carbon monoxide by the ligand. In view of this observation, infrared spectrum of the parent thiocyanato carbonyl was examined in several donor solvents which do not replace carbon monoxide under mild conditions. Of these, one solvent, acetonitrile, caused complete $Mn(CO)_5SCN$ to $Mn(CO)_5NCS$ isomerization within 5 min. at 25°,

$$(CO)_5Mn-SCN \xrightarrow{\text{CH}_3\text{CN}} (CO)_5Mn-NCS \quad .$$

Addition of petroleum ether to a concentrated solution of $Mn(CO)_5NCS$ in CH_3CN precipitates the N-bonded complex as a yellow powder. The IR spectrum of the solid supports the Mn-NCS bonding assignment. The compound $Mn(CO)_5NCS$, suspended in Nujol, rearranges back to the parent thiocyanate in approximately 5 hrs. at 25°,

$$(CO)_5Mn-NCS \xrightarrow{\text{Nujol}} (CO)_5Mn-SCN$$

This then is the first example of linkage isomers which can be interconverted. Furthermore, the facile interconversion of the two forms of $Mn(CO)_5CNS$ illustrates the importance of the solvent in determining the nature of the linkage isomer obtained.

References

1. WOJCICKI, A., and FARONA, M.F., Inorg.Chem., **3**, 151 (1964).
2. LEWIS, J., NYHOLM, R.S., and SMITH, P.W., J.Chem.Soc. (London), **1961**, 4590, summarize X-ray structural data on thiocyanato complexes.
3. TURCO, A., and PECILE, C., Nature, **191**, 66 (1961).
4. ANGELICI, R.J., and BASOLO, F., J.Am.Chem.Soc., **84**, 2495 (1962).
5. ABEL, E.W., BUTLER, I.S., and REID, J.G., J.Chem.Soc. (London), **1963**, 2068.

6 B 3

NEW ORGANOSULFUR DERIVATIVES OF METAL CARBONYLS.

R.B. King and M.B. Bisnette
Mellon Institute, Pittsburgh 13, Pennsylvania, U.S.A.

Numerous unusual organosulfur transition metal compounds have been obtained by reactions of various metal carbonyls and cyclopentadienyl-metal carbonyls with certain mercaptans, sulfides, disulfides, and dithietenes (1,2). Other types of organosulfur transition metal compounds have been obtained from metal sulfides and certain acetylene derivatives (3). This paper discusses a new and entirely different synthesis of new types of organosulfur derivatives of metal carbonyls with metal-sulfur bonds. These new compounds have been obtained by thermal or photochemical decarbonylation of compounds of general formula $CH_3S(CH_2)_nM(CO)_x(C_5H_5)_y$ (n = 1, 2 or 3; M = Fe [x = 2, y = 1], or Mn [x = 5, y = 0]) without metal-sulfur bonds. These sulfur-containing transition metal alkyl derivatives, none of which has been previously reported, may in turn be obtained from metal carbonyl anions and the chloroalkyl methyl sulfides $CH_3S(CH_2)_nCl$.

Thus, the sodium salt $NaMo(CO)_3C_5H_5$ (4) reacts with chloromethyl methyl sulfide in tetrahydrofuran solution at 25^0 to give the yellow crystalline alkyl tricarbonyl derivative $CH_3SCH_2Mo(CO)_3C_5H_5$, m.p. 66-67^0. On heating to $\sim 70^0$ or ultraviolet irradiation this tricarbonyl derivative forms in up to 60%

yield the likewise yellow crystalline dicarbonyl derivative
$CH_3SCH_2Mo(CO)_2C_5H_5$, m.p. 66-67°.

The infrared and proton n.m.r. spectra and close analogies in stoichiometry, physical properties, and general method of preparation to π-cyclo-
pentadienyl-π-allylmolybdenum dicarbonyl (5) lead us to propose structure I
for the dicarbonyl derivative $CH_3SCH_2Mo(CO)_2C_5H_5$. In structure I a carbon-
sulfur double bond is π-bonded to the molybdenum atom and the neutral
CH_3SCH_2 group donates three electrons (6) to the molybdenum atom in a
similar manner to the π-allyl ligand. The relationship between the metal-
ligand bonding in $CH_3SCH_2Mo(CO)_2C_5H_5$ (I) and the π-allyl derivative
$C_3H_5Mo(CO)_2C_5H_5$ is very similar to the relationship between the metal-
ligand bonding in benzene-chromium tricarbonyl and thiophene-chromium
tricarbonyl (7) and is a further demonstration of the well-known analogy
between the groupings $>C=C<$ and $>S\,\vdots\,$.

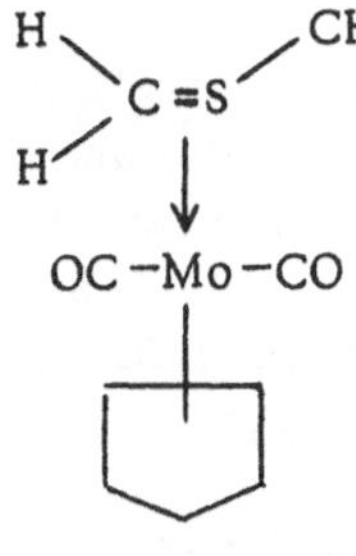

I

Treatment of $NaMn(CO)_5$ with 2-chloroethyl methyl sulfide in boiling
tetrahydrofuran gave a 26% yield of pale yellow diamagnetic crystalline
volatile $CH_3SCH_2CH_2COMn(CO)_4$, m.p. 56,5-58,5°. The infrared spectrum
of this new organomanganese compound besides very strong bands at 2071,
1996, and 1972 cm^{-1} due to the terminal metal carbonyl groups exhibits a
very strong band at 1622 cm^{-1} which may be assigned to an acyl carbonyl
group suggesting structure II.

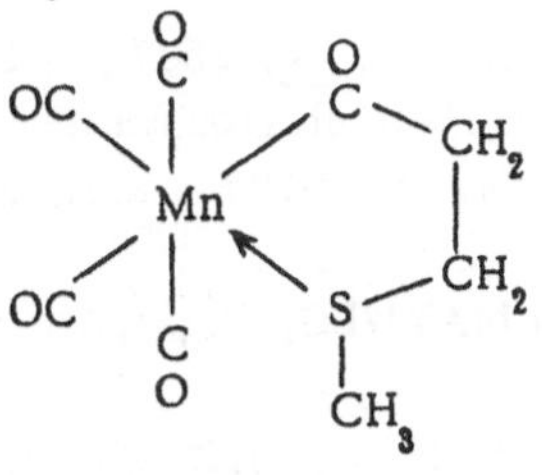

II

Reaction between $NaFe(CO)_2C_5H_5$ (4) and 2-chloroethyl methyl sulfide in tetrahydrofuran solution at room temperature yields a rather unstable orange liquid indicated by analysis and other properties to be $CH_3SCH_2CH_2Fe(CO)_2C_5H_5$. Ultraviolet irradiation of this material in benzene solution gives a complex mixture from which the following compounds can be isolated by chromatography on alumina in benzene solution followed by crystallization from pentane or ethanol of the fractions thus obtained: (a) Unchanged $CH_3SCH_2CH_2Fe(CO)_2C_5H_5$. (b) Brown crystalline $[CH_3SFeCOC_5H_5]_2$ identical to the compound of this composition previously isolated in low yield from the reaction between $[C_5H_5Fe(CO)_2]_2$ and dimethyl-disulfide (8). (c) $[C_5H_5Fe(CO)_2]_2$. (d) Brown crystalline $CH_3SFe(CO)_2C_5H_5$, m.p. 67-69°. (e) Red crystalline $CH_3SCH_2CH_2COFeCOC_5H_5$, m.p. 70°, exhibiting a single strong terminal metal carbonyl band at 1934 cm^{-1} and a single strong acyl carbonyl band at 1618 cm^{-1} in its infrared spectrum clearly indicating a structure analogous to the structure II proposed for $CH_3SCH_2CH_2COMn(CO)_4$.

The <u>monomeric</u> product $CH_3SFe(CO)_2C_5H_5$ is the first example of a methylthio derivative of a transition metal where the sulfur atom does not bridge between two metal atoms. This organosulfur derivative may also be obtained in 13% yield from $C_5H_5Fe(CO)_2H$ (9) and dimethyldisulfide. Similar reactions of dimethyldisulfide with $HMn(CO)_5$ and $C_5H_5Mo(CO)_3H$ have previously been shown to give the <u>dimeric</u> derivatives $[CH_3SMn(CO)_4]_2$ and $[CH_3SMo(CO)_2C_5H_5]_2$ (10).

On heating $CH_3SFe(CO)_2C_5H_5$ to 70° at atmospheric pressure, gas is evolved and the dimeric derivative $[CH_3SFeCOC_5H_5]_2$ may be isolated in 26% yield. Treatment of $CH_3SFe(CO)_2C_5H_5$ with excess methyl iodide forms in an exothermic reaction yellow solid ionic $[C_5H_5Fe(CO)_2S(CH_3)_2]I$, m.p. 104° (dec.).

The authors are indebted to the U.S. Air Force Office of Scientific Research for partial support of this work under GRANT AF-AFOSR-580-64.

References

1. KING, R.B., J.Am.Chem.Soc., <u>85</u>, 1587 (1963), and references to previous work from various research groups cited therein.
2. DAVISON, A., EDELSTEIN, N., HOLM, R.H. and MAKI, A.H., Inorganic Chemistry, in press.
3. SCHRAUZER, G.N., and MAYWEG, V., J.Am.Chem.Soc., <u>84</u>, 3221 (1962).
4. PIPER, T.S., and WILKINSON, G., J.Inorg.Nucl.Chem., <u>3</u>, 104 (1955).

5. COUSINS, M., and GREEN, M.L.H., J.Chem.Soc. (London), 1963, 889.
6. Complexes III and IV may possibly be regarded as derivatives of the unsaturated sulfonium ion $H_2C=S^+CH_3$ where the carbon-sulfur double bond is π-bonded to the transition metal like a carbon-carbon double bond.
7. FISCHER, E.O., and ÖFELE, K., Ber., 91, 2395 (1958).
8. KING, R.B., TREICHEL, P.M., and STONE, F.G.A., J.Am.Chem. Soc., 83, 3600 (1961).
9. GREEN, M.L.H., and NAGY, P.L.I., J.Organometal.Chem., 1, 58 (1963).
10. TREICHEL, P.M., MORRIS, J.H., and STONE, F.G.A., J.Chem.Soc. (London), 1963, 720.

1 C 1

ELECTRON TRANSFER REACTIONS BETWEEN Co(III), Rh(III) AND Ir(III) COMPLEXES AND Cr(II) THROUGH BRIDGING GROUPS.

R.T.M. Fraser (1), D.E. Peters and G.T. Takaki
Department of Chemistry, The University of Kansas,
Lawrence, Kansas, U.S.A.

In effects of the size of bridging groups and the nature of the coordinating atom on the rate of electron transfer have been investigated.

TABLE I

RATE CONSTANTS FOR Cr(II) REDUCTIONS OF
CARBOXYLATOCOBALT(III) COMPLEXES ($\mu = 1, 0$)

Complex	°C	k (M^{-1} sec^{-1})	Ref.
pentaammineacetato	25	0, 14-0, 17	2
pentaammineformato		0, 13	
pentaamminebutyrato		0, 08	3
pentaamminecyclopropane-carboxylato		0, 12$_5$	
pentaamminecyclobutane-carboxylato		0, 056	
pentaamminecyclopentane-carboxylato		0, 072	
pentaamminecyclohexane-carboxylato		0, 048	
pentaamminebenzoato		0, 14	4
tetraammineglycinato		0, 15	

Table I lists some rate constants at room temperature for the chromium(II) reduction of pentaammine cobalt (III) complexes containing monocarboxylic ligands. The tetraammineglycinate is a complex with a minimum of steric hindrance. The rate of reaction is very similar to that of the acetato complex, indicating that inductomeric effects are of little importance in these redox processes, so that the observed decreases in the reaction rates may be related to the steric hindrance of the ligand. As the bulk of the carbon ring increases the rate decreases, although this decrease is accompanied by an increase in activation entropy (4) (from -38 e.u. for the cyclopropane complex to -23 for the cyclohexane). These changes in rate constant occur in a series where the path for electron transfer is fixed, so that care must be taken in comparing experimental values with those predicted from theory (5, 6): examples are valid only if there are large differences in mobile bond orders or in the conducting bridge.

TABLE II

CHROMIUM(II) REDUCTIONS OF COBALT(III) COMPLEXES CONTAINING
COORDINATED OXYGEN, NITROGEN AND SULFUR ($\mu = 1$)

Complex	Formal charge	k ($M^{-1} \sec^{-1}$)	°C
pentaammine (formamide-N)	3+	3,3	20
pentaammine (urethane-N)	3+	3,3	25,3
pentaammine (glycine-N)	3+	0,01	24
pentaammine (methyl-glycinate-N)	3+	0,009	24
pentaammine (ethyl 4-amino-butyrate-N)	3+	0,002	24
hexaammine	3+	0,002	25
pentaammineformato	2+	0,13	25
pentaamminecarbamato	2+	1,7	25
pentaammineglycinato	2+	0,53	24
pentaamminesulfato	1+	23	25
pentaamminethiosulfato	1+	0,88	25
pentaammineselenato	1+	360	25

Table II lists rate constants for the Cr(II) reduction of some cobalt(III) complexes containing ligands bound to the metal through oxygen, nitrogen and sulfur atoms. The advantage of coordinated nitrogen over oxygen is that the ligands are substituted amines, the parent compound being ammonia with a rate constant of 0,002 $M^{-1} \sec^{-1}$ (7). It is thus possible to investigate the mediating ability of weakly conducting ligands where the rate constant for the chromium(II) reduction lies in the range 0,002 - 0,16 $\sec^{-1}$. The NH_2 group in the carbamato complex is conjugated with the carbonyl oxygen (in contrast to that in the tetraammine glycinato) and the rate of reduction shows a significant increase, probably due to an electromeric effect

$$(NH_3)_5Co-O-\overset{\overset{\textstyle O}{\|}}{C}-NH_2$$

which makes electrons available for reductant bonding without markedly affecting the transfer path. The acceleration noted with the pentaammine glycinato is due to another effect, reductant chelation.

Reductions through coordinated nitrogens proceed more rapidly than through the corresponding oxygen compounds, even though the nitrogen complexes bear a charge of 3+. Conjugation is still an important factor, since the introduction of even one $-CH_2-$ into the transfer path decreases the rate by a factor of 300. Introduction of more carbon atoms into the path decreases the conductivity to that of the coordinated NH_3 group, indicating that the mode of attack is probably the same - via an outersphere activated complex rather than by bridging.

The reduction of the thiosulfato is interesting. Recent work suggests that the group is bound to the cobalt through the sulfur atom (8) rather than an oxygen. The rate of reduction of the complex presumably proceeds by attack of the ligand at one of the oxygens followed by a subsequent rearrangement of the products first formed. The reduction is much slower than would be expected for electron transfer through a coordinated oxygen atom; it is twenty-five times slower than the rate of reduction of the corresponding sulfato complex and four hundred times slower than the reduction of the selenato.

TABLE III

CHROMIUM(II) REDUCTIONS OF RHODIUM(III) AND IRIDIUM(III) COMPLEXES
$(25^{o}, \mu = 1)$

Complex	k_2/k_1	k_{-1} sec^{-1}	k M^{-1} sec^{-1}	k_{bi} M^{-1} sec^{-1}
$[(NH_3)_5RhF]^{2+}$	-	-	-	$1.6 \cdot 10^{-2}$
$[(NH_3)_5RhCl]^{2+}$	0.6	$1.8 \cdot 10^{-2}$	0.89	-
$[(NH_3)_5RhBr]^{2+}$	-	-	-	$1.3 \cdot 10^{-2}$
$[(NH_3)_5RhI]^{2+}$	< 0.075	-	-	$7.0 \cdot 10^{-3}$
$[(NH_3)_5RhO_2CCH_3]^{2+}$	< 0.1	-	-	$2.8 \cdot 10^{-3}$
$[(NH_3)_5RhO_2CC_3H_7]^{2+}$	-	-	-	$1.9 \cdot 10^{-3}$
$[(NH_3)_5IrCl]^{2+}$	0.2	$2.6 \cdot 10^{-3}$	0.13	-
$[(NH_3)_5IrBr]^{2+}$	1.1	$3.2 \cdot 10^{-4}$	$1.6 \cdot 10^{-2}$	-
$[(NH_3)_5IrI]^{2+}$	0.5	$2.4 \cdot 10^{-4}$	$1.2 \cdot 10^{-2}$	-

Chromium(II) reactions with pentaammine rhodium(III) and iridium(III) complexes are much slower and involve more than one step (Table III). Final products in the rhodium(III) reductions are usually metallic rhodium and a chromium(III) complex except when the ligand is a carboxylic acid, when the ligand remains associated to a large extent with the rhodium.

This research supported by the National Science Foundation, the Petroleum Research Foundation and the University of Kansas General Research Fund.

References

1. SLOAN, A.P., Research Fellow, 1964 - 1966.
2. SEBERA, D.K., and TAUBE, H., J.Am.Chem.Soc., 83, 1785 (1961).
3. FRASER, R.T.M., and TAUBE, H., J.Am.Chem.Soc., 83, 2239 (1961).
4. FRASER, R.T.M., in "Advances in the Chemistry of the Coordination Compounds", S.Kirschner, ed., The MacMillan Company, New York 1961, p.287.
5. HALPERN, J., and ORGEL, L.E., Dis.Faraday Soc., 29, 32 (1960).
6. MANNING, P.V., JARNAGIN, R.C., and SILVER, M., J.Phys.Chem., 68, 265 (1964).
7. ZWICKEL, A., and TAUBE, H., J.Am.Chem.Soc., 81, 2915 (1959).
8. BABAEVA, A.V., BARANOVSKII, I.B., and KHARITONOV, Yu.Ya., Russ.J.Inorg.Chem., 8, 307 (1963).

1 C 2

ELECTRON TRANSFER REACTIONS OF PENTACYANOCOBALTATE(II).

Jack Halpern and Shuzo Nakamura
Department of Chemistry, University of Chicago, Chicago, Illinois, U.S.A.

The oxidation of $[Co(CN)_5]^{3-}$ to cobalt(III) is accompanied by an increase in the coordination number of the cobalt ion. The mechanism of electron transfer may be inner- or outer sphere, depending on whether the sixth ligand required to complete the octahedral coordination shell of the cobalt ion is derived directly from the oxidant or, alternatively, is an anion (e.g. CN^-) derived from the solution. The substitution-inertness of the product, and the different kinetic patterns of the two paths, permit this ligand to be identified and the mechanism of electron transfer to be established.

The $[Co(CN)_5]^{3-}$-catalyzed substitution of pentaaminecobalt(III) complexes by CN^- has been interpreted in terms of an electron transfer reaction between $[Co(CN)_5]^{3-}$ and the cobalt(III) complex, which may occur either through an inner- or outer-sphere mechanism (1,2). In the former case the stoichiometry of substitution corresponds to $[Co(III)(NH_3)_5X] + 5\,CN^- \longrightarrow [Co(III)(CN)_5X] + 5\,NH_3$ and the rate law is found to be $k_i \cdot c_{[Co(NH_3)_5X]} \cdot c_{[Co(CN)_5]^{3-}}$. This behaviour is characteristic of the cases $X = Br^-$, Cl^-, OH^-, N_3^-, NCS^-, NO_2^-, ONO^- and $S_2O_3^{2-}$ and is interpreted in terms of an inner sphere electron transfer between $[Co(CN)_5]^{3-}$ and $[Co(III)(NH_3)_5X]$ through the bridged intermediate $[(CN)_5Co(II)-X-Co(III)(NH_3)_5]$.

In other cases, notably where $X = PO_4^{3-}$, CO_3^{2-}, SO_4^{2-}, NH_3, OAc^- and other carboxylate ions, the substitution follows a different course, the stoichiometry being given by $[Co(III)(NH_3)_5X] + 6\,CN^- \longrightarrow [Co(CN)_6]^{3-} + 5\,NH_3 + X$, and the rate law by $k_o' \cdot c_{[Co(III)(NH_3)_5X]} \, c_{[Co(CN)_5]^{3-}} \cdot c_{[CN]}^-$. This behaviour is interpreted in terms of an outer-sphere electron transfer between $[Co(CN)_6]^{4-}$ (presumed to co-exist in equilibrium with $[Co(CN)_5]^{3-}$ and CN^-) and $[Co(III)(NH_3)_5X]$.

In this paper we extend our earlier (2) observations on these systems and report related measurements on the oxidation on $[Co(CN)_5]^{3-}$ by other cobalt(III) complexes, e.g. substituted cobalt(III) ethylenediaminetetra-acetates, which exhibit analogous behaviour.

Kinetic measurements on these reactions, made with a stopped-flow apparatus, are summarized in Tables I and II. In a few cases, notably with $[Co(NH_3)_5F]^{2+}$, $[Co(NH_3)_5NO_3]^{2+}$ and, possibly, $[Co(NH_3)_5SO_4]^+$ as oxidants, both inner- and outer-sphere electron transfer paths are simultaneously observed. In other cases only an upper limit can be assigned to the rate constant for one of the paths. The significance of the trends in Tables I and II will be discussed, and conclusions drawn concerning the factors which influence the choice of mechanism.

TABLE I

RATE CONSTANTS FOR THE OXIDATION OF $[Co(CN)_5]^{3-}$ BY VARIOUS COBALT(III) COMPLEXES AT 25^0, $\mu = 0.2$.

Oxidant	k_i ($M^{-1}\,sec^{-1}$)	k_o' ($M^{-2}\,sec^{-1}$)
RBr^{2+}	$> 2.10^9$	--
RCl^{2+}	$\sim 5.10^7$	--
$RS_2O_3^+$	$\sim 1.10^7$	--
RN_3^{2+}	$1.6.10^6$	$(< 8.10^5)^a$
$RNCS^{2+}$	$1.1.10^6$	$(< 5.10^5)^a$

TABLE (contd.)

$RONO^{2+}$	$4,2.10^5$ d	$(< 2.10^5)^a$
ROH^{2+}	$9,3.10^4$	$(< 5.10^4)^a$
RNO_2^{2+}	$3,4.10^4$ d	$(< 3.10^4)^a$
RNO_3^{2+}	$(\leqslant 1.10^4)^c$	$2,4.10^5$
RF^{2+}	$(1,8.10^3)^c$	$1,7.10^4$.
RNH_3^{3+}	$(< 4.10^2)^a$	9.10^4
$Co(en)_3^{3+}$	$(< 2.10^2)^a$	5.10^4
$ROAc^{2+}$ b	$(< 1.10^2)^a$	$1,1.10^4$
RSO_4^+	$(\leqslant 2.10^2)^c$	$3,6.10^4$
RCO_3^+	$(10)^a$	$1,5.10^3$
RPO_4	$(1)^a$	$5,2.10^2$
CoY^-	$(\leqslant 3.10)^c$	$3,8.10^2$
$CoYOH^{2-}$	$1,7.10$	(< 10)
$CoYCl^{2-}$	$2,7.10^5$	--
$CoYBr^{2-}$	$> 1.10^7$	
$CoYNO_2^{2-}$	9.10^2 d	$(< 5.10^2)$

[R = $Co(NH_3)_5$; Y = ethylenediaminetetraacetate]

[a] Upper limit based on absence of detectable contributions from this path.

[b] Similar rate constants found for R.Fumarato⁺, R.Maleato⁺, R.Oxalato⁺ and R.Succinato⁺; in none of these cases was there any detectable contribution from an inner sphere path.

[c] Estimate based on CN⁻-independent kinetic contribution, but product not identified.

[d] $RONO^{2+}$, RNO_2^+ and $CoYNO_2^{2-}$ appeared to yield the same product which was not identified but is believed to be $[Co(CN)_5NO_2]^{3-}$.

TABLE II
ACTIVATION PARAMETERS

Oxidant	Path	$\Delta H^{\ddagger}$ (kcal/mole)	$\Delta S^{\ddagger}$ (e.u.)
RNH_3^{3+}	Outer (k_o')	4,1	-22
$ROAc^{2+}$	Outer (k_o')	3,0	-31
RCO_3^+	Outer (k_o')	2,9	-34
RSO_4^+	Outer (k_o')	1,6	-32
RPO_4	Outer (k_o')	1,4	-41
ROH^{2+}	Inner (k_i)	6,9	-13
$RNCS^{2+}$	Inner (k_i)	6,3	-10
RN_3^{2+}	Inner (k_i)	3,7	-17

It is of interest that the product of outer sphere oxidation of $[Co(CN)_5]^{3-}$ appears in every case to be $[Co(CN)_6]^{3-}$, rather than $[Co(CN)_5OH_2]^{2-}$, despite the fact that the latter, once prepared, is a stable species. This observation is not readily reconciled with the formulation of the reductant as $[Co(CN)_5OH_2]^{3-}$ rather than $[Co(CN)_5]^{3-}$.

This work was supported by a grant from the National Science Foundation.

References

1. ADAMSON, A.W., J.Am.Chem.Soc., 78, 4260 (1956).
2. CANDLIN, J.P., HALPERN, J., and NAKAMURA, S., J.Am.Chem.Soc., 85, 2517 (1963).

1 C 3

METAL COORDINATION - A KEY TO OXIDATIVE STABILITY.

Horst G. Langer
Eastern Research Laboratory, The Dow Chemical Company
Framingham, Mass., U.S.A.

The oxidation of metal ions can be pH dependent because of kinetic effects as expressed in equation {1}

$$2 A^+ + H_2O + 1/2\, O_2 \;\rightleftharpoons\; 2\, A^{2+} + 2\, OH^- \qquad \{1\}$$

or might be based on a difference in stabilities of various metal complexes as indicated below {2}

$$[M(OH_2)_6]^{2+} \text{ is more stable than } [M(OH_2)_6]^{3+}$$

$$\{2\}$$

$$[M(OH)_6]^{3-} \text{ is more stable than } [M(OH)_6]^{4-} \quad.$$

This latter type of coordination chemistry confronted us after we had prepared three different tin (II)-EDTA chelates (1) and found unexpected differences in solubilities as shown in Table I and oxidative stabilities as shown in Table II.

TABLE I

Compound	Solubility mMoles/Liter	pH Sat. Solution
$Na_2SnY \cdot 2\,H_2O$	1000	7,0
$CaSnY \cdot 4\,H_2O$	50	5,5
$Sn_2Y \cdot 2\,H_2O$	>1	3,0
$(Y = C_{10}H_{12}N_2O_8{}^{4-})$		

TABLE II - OXIDATION

Compound	H_2O/Air 0,01 molar		H_2O/N_2 0,01 molar	
$Na_2SnY \cdot 2\,H_2O$	< 8[h]	open	pH 8	14[d]
			7,4	17[d]
			2	15[d]
$CaSnY \cdot 4\,H_2O$	~ 3[d]	bubbler	6,0	80[d]
$Sn_2Y \cdot 2\,H_2O$	~ 7[d]	bubbler	8	35[d]
			3,5	50[d]
			2,5	70[d]

These differences can not be explained as being due to different cations of one and the same anionic complex of tin. We could show, however, that 1) these three solid materials represent two types of EDTA complexes, namely a mononuclear tetra- or hexadentate complex represented by $Na_2SnY \cdot 2H_2O$ and two binuclear complexes represented by the remaining two compounds; 2) both types of complexes are observed in aqueous solution; 3) tin (II) in the mononuclear chelate is very readily oxidized while it is stable towards oxidative attack in both binuclear forms; 4) both forms undergo complicated rearrangements in the course of dissociation, hydrolysis, and oxidation.

Polarography has been helpful and almost the only tool to study these rearrangements in aqueous solution. The similarity between curves measured for both binuclear complexes (Fig. 1) is obvious and the absence of polaro-graphic waves in this area for the mononuclear complex indicates a high reduction potential (2). The appearance of two waves for the binuclear com-plexes is explained by dissociation of the water molecules attached to metal in each case, which is higher for the more basic solution of the mixed chelate and low for the more acidic solution of the distannous EDTA. Addition of basic or neutral electrolytes has no effect on the polarogram of disodium tin (II)-EDTA, however, acid produces a dramatic change as shown in Table III.

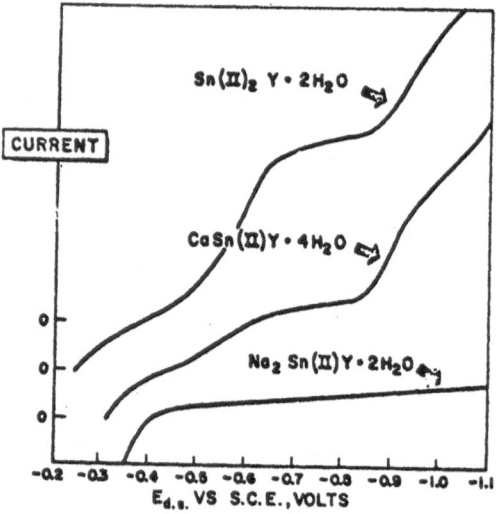

TABLE III - Na$_2$Sn(II)Y

[HClO$_4$] moles/1	E$_{1/2}$
0,0	-
0,01 ↓ 1,0	0,55 (0,65) ↓ 0,42
KCl/H$_2$O	-

The wave at -0,42 volts is due to free stannous ion (2) and the protonated form of the mononuclear complex represented by a wave at -0,55 volts is rapidly oxidized to a tin (IV)-EDTA chelate (3). For this oxidation no rearrangement nor breaking or forming of any bonds is necessary.

The chemistry of both binuclear complexes, however, involve rearrangement in almost every step and is summarized in Table IV.

TABLE IV

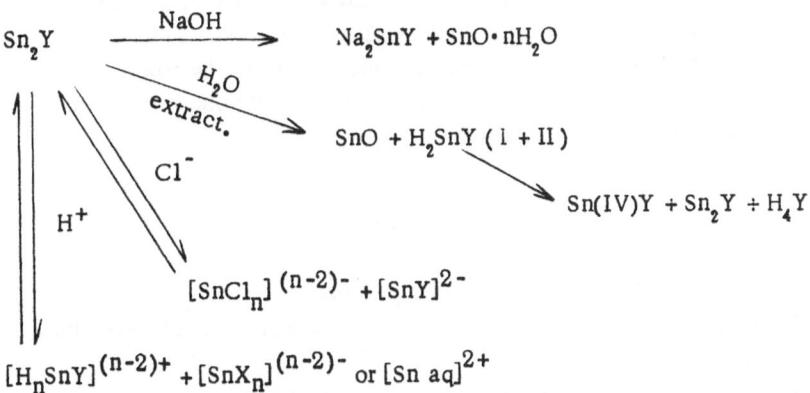

Addition of base produces soluble disodium tin (II)-EDTA and the rate of oxidation depends strictly on the presence of this compound. Precipitated stannous oxide does not react any further. Water extraction in a soxhlet apparatus produces stannous oxide in the thimble while two acid forms of 1:1 chelates are found in the extract. One of these two forms corresponds to the protonated 1:1 complex described above and is immediately oxidized while the second form rearranges and reforms the binuclear complex and one mole of free acid. In the presence of halides an equilibrium is formed between EDTA compounds and various halogen-containing complexes dominantly $[SnCl_3]^-$.

TABLE V - Sn(II) Y

$HClO_4$ moles/l	$E_{1/2}$	Species	
0,0	0,55; 0,9	H_2O-SnYSn-OH_2; HO-SnY-Sn-OH	H_3SnY
0,0 (+Cl⁻)	0,45 ; 0,55; 0,9	H_2O-SnYSn-OH_2; ⁻HO-SnY-Sn-OH⁻; $SnCl_3^+$; Sn_{aq}^{2+}	H_3SnY (II)
0,01	0,55	H_2O-SnYSn-OH_2	H_4SnY^+
0,03-0,3	0,45 ; 0,65	[(HOOC-CH_2)$_2$NH-(CH_2)$_2$-N-(CH_2-COO)$_2$Sn aq]; Sn_{aq}^{2+}	
0,01-0,3 (+Cl⁻)	(0,45)× 0,55; (0,65)	Sn_{aq}^{2+} ; $SnCl_3^+$; complex	H_5SnY^{2+}
		$H_5Y.SnCl_3.2 H_2O$ isolated	

Table V represents the relationship betwenn half wave potentials and acidity for the binuclear complexes with a completely different sequence of protonated compounds. All of these structures are different from type I described above and they cannot be oxidized to the tin (IV) chelate without bond breaking and bond forming. This explains the higher stability of the binuclear chelates in acidic solution since their rates of oxidation are determined by the rates of rearrangement to the octahedral or pseudo-octa-hedral configuration (type I) of the mononuclear compound. Additional proof for this unsymmetric protonation sequence has been the isolation of several compounds which are presently being studied and one of which has been identified as $H_5Y.SnCl_3.2 H_2O$.

References

1. LANGER, H.G., J.Inorg.Nucl.Chem. (in press).
2. MEITES, L., "Polarographic Techniques", Interscience Publishers, Inc., New York, p.286-87 (1955).
3. LANGER, H.G., J.Inorg.Nucl.Chem., 26, 59 (1964).

ETUDE POLAROGRAPHIQUE DES SOLUTIONS AQUEUSES
D'AQUOPENTACYANOFERRATE(II).

Guy Emschwiller, Loïk Viet et (Madame) Claude Friedrich
Laboratoire de Chimie physique, Ecole supérieure de Physique et de Chimie,
10 rue Vauquelin, Paris 5eme.

La complexité des solutions aqueuses d'aquopentacyanoferrates se manifest par un ensemble de propriétés singulières, en particulier des propriétés électrochimiques, ayant suscité des interprétations diverses. C'est ainsi que les anomalies de conductivité ont conduit à invoquer l'existence de deux formes en équilibre acidobasique (1) et l'allure inusuelle des courbes de potentiels d'oxydoréduction celle d'associations plurimoléculaires (2,3). L'étude cinétique de la réaction du nitrosobenzène a permis de reconnaître la présence de deux types au moins de constituants, dont les relations seraient celles de monomère à polymère, comme aussi de les doser (4); il a été reconnu effectivement que l'un est susceptible de dialyser et l'autre pas à travers une membrane de cellophane disposée entre la solution d'aquopentacyanoferrate(II) et de la méthyléthylcétone saturée d'eau (5). Nous nous sommes proposé de rechercher quelle contribution pourrait apporter une étude polarographique.

La solution d'aquopentacyanoferrate(II) de sodium ne donne pas de vague avec l'électrode à goutte de mercure, tout au moins quand on part d'un produit pur, car, si la purification a été insuffisante, on trouve les trois vagues caractéristiques du nitrosopentacyanoferrate(III) (6), utilisé comme matière première. Cette impureté peut d'ailleurs être ainsi non seulement reconnue mais dosée.

Il a donc fallu avoir recours à une électrode en fil de platine, déjà précédemment employée pour les solutions de ferrocyanure de potassium (7). Nous avons veillé à traiter l'électrode de telle manière qu'elle soit aussi sensible que possible et conduise à des résultats reproductibles; à cette fin, l'électrode a toujours été portée au rouge, après lavage, avant chaque prise de polarogramme. Nous avons opéré, à l'abri de l'oxygène et de la lumière, sur des solutions aqueuses maintenues à la température de 0°, le plus souvent en milieu tamponné (tampon de SÖRENSEN au borate, de pH 10 environ). Les vitesses de polarisation ont été généralement faibles, de 0,1 ou 0,2 V/mn. L'électrode de référence a toujours été l'électrode au calomel à solution de chlorure de potassium saturée et c'est à elle que sont rapportés les potentiels ci-dessous relatés, compte non tenu des potentiels de jonction.

Nous avons reconnu que les solutions d'aquopentacyanoferrate(II) présentent un comportement polarographique particulier qui les différencie nettement des autres complexes pentacyanés que nous avons examinés. En effet, nous avons trouvé que les solutions d'ammino- et d'arsénito-pentacyanoferrates(II) donnent des polarogrammes comparables à ceux des hexacyanoferrates(II) (ferrocyanures), comme il apparaît sur la figure 1; les potentiels de demi-vague ont été trouvés respectivement égaux à + 0,47 V pour le complexe arsénité, + 0,23 V pour le ferrocyanure, + 0,13 V pour le complexe amminé. Au contraire, avec l'aquo-pentacyanoferrate(II), on observe deux vagues distinctes, aussi bien en solution de chlorure de potassium décinormale qu'en milieu tamponné (voir fig. 1). Les hauteurs de chacune des deux vagues, sensiblement égales, sont proportionnelles à la concentration. Les potentiels de demi-vague sont quelque peu variables avec la concentration et avec la nature de l'électrolyte de base; le potentiel de demi-vague de la vague A varie approximativement de + 0,28 à + 0,40 V, celui de la vague B de - 0,05 à + 0,10 V.

Opérant avec une électrode tournante à disque de platine, nous avons observé que les hauteurs des vagues varient proportionnellement à la racine carrée de la fréquence de rotation (jusqu'à 1500 tours par minute), ce qui semble prouver, conformément aux conceptions de LEVICH (8), que les intensités de courant sont bien régies par la diffusion.

Les résultats qui précèdent sont relatifs aux solutions obtenues par dissolution du produit préparé suivant la méthode d'HOFMANN (9). Sa pureté a été contrôlée par voie polarographique, comme indiqué ci-dessus, et aussi par spectrométrie IR. La proportion du constituant réagissant rapidement sur le nitrosobenzène, soit donc de la "forme rapide" considérée comme monomère, a toujours été comprise entre 20 et 25% (10).

La question se pose de savoir quelle est l'origine de l'existence de deux vagues, alors qu'il n'y en a qu'une avec les autres complexes pentacyanés. Il est exclu qu'elles soient liées à la présence d'une forme acide et d'une forme basique, car nous n'avons décelé aucune influence appréciable du pH sur les hauteurs respectives des deux vagues dans le domaine de pH s'étendant entre 5,4 et 11,7. On est ainsi conduit à se demander si l'existence de deux vagues ne serait pas en relation avec la présence des deux types même de constituants révélée par les études cinétiques de la réaction sur le nitroso-benzène. A cette fin, il est apparu nécessaire d'essayer de séparer les deux types de constituants.

Il ne nous a pas été permis d'avoir recours à la méthode de séparation par dialyse précédemment signalée (5), car la présence de méthyléthylcétone contrarie les déterminations polarographiques. Nous avons fait appel à la technique de la chromatographie sur colonne en utilisant comme phase stationnaire le produit dénommé Sephadex, décrit pour la première fois par

PORATH et FLODIN (11). Cette substance agit, en quelque sorte, par un effet de filtration moléculaire et permet de séparer les constituants de degrés de polymérisation variables, les temps de passage étant d'autant plus grands que les masses moléculaires sont plus faibles. C'est la qualité dénommée "Sephadex G-25 medium" (vendue par la firme PHARMACIA de Uppsala, Suède) qui nous a donné les meilleurs résultats. La méthode colorimétrique au nitrosobenzène a permis de suivre l'efficacité du fractionnement. Si la séparation n'a pas été absolument totale, les fractions de tête n'en ont pas moins été considérablement enrichies en le constituant à réaction la plus lente sur le nitrosobenzène; il est ainsi confirmé qu'il diffère bien de l'autre constituant par une masse moléculaire plus grande, donc qu'il s'agit de polymère; la teneur en polymère a pu être élevée jusqu'à 95%. De même, les fractions de queue sont enrichies en monomère jusqu'à des teneurs dépassant 90%.

L'étude polarographique a porté sur les différentes fractions et nécessité l'emploi de cellules de petite capacité. Elle a révélé qu'il y a toujours eu oxydation partielle des solutions malgré les précautions prises au cours des passages sur colonne, effectués cependant à 0° et en atmosphère d'azote. Nous avons rassemblé sur la figure 2 les polarogrammes obtenus avec la solution de départ à 22% de monomère (courbe b), avec une fraction de tête à 8% de monomère (courbe a), avec une fraction de queue à 88% de monomère (courbe c).

De façon générale, les polarogrammes des solutions enrichies en polymère présentent toujours deux vagues, même pour les teneurs les plus faibles en monomère; les deux vagues demeurent d'importance comparable, bien qu'il semble que la vague A acquière une hauteur un peu supérieure à celle de la vague B; mais l'interprétation des polarogrammes est devenue délicate; en effet, au lieu que les deux vagues soient séparées par un palier ou un pseudo-palier, la courbe présente une bosse avec un maximum suivi d'un minimum très prononcé, ainsi qu'il apparaît nettement sur la courbe a de la figure 2. Par contre, pour les solutions enrichies en monomère, l'importance relative de la vague A décroît progressivement au fur et à mesure qu'augmente la proportion du monomère; dans les mélanges les plus riches en monomère, il ne subsiste guère que la vague B, la vague A n'étant plus qu'à peine marquée. On est donc conduit à penser que l'existence même de la vague A serait liée à la présence du polymère.

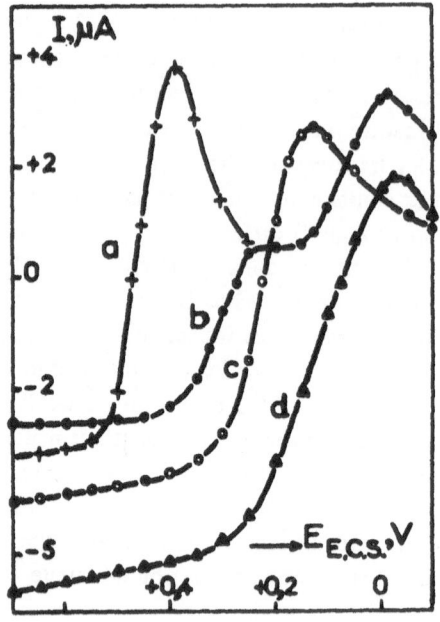

Figure 1

(Les intensités de courant sont rapportées à une concentration millimolaire et à une aire d'électrode de 10 mm²; température 0°; vitesse de polarisation 0,2 V/mn; solutions tamponnées de pH 10,1, sauf indication contraire).

a   arsenitopentacyanoferrate(II)
b   aquopentacyanoferrate(II) à 21% de monomère
c   ferrocyanure
d   amminopentacyanoferrate(II) (pH ≈ 11,1)

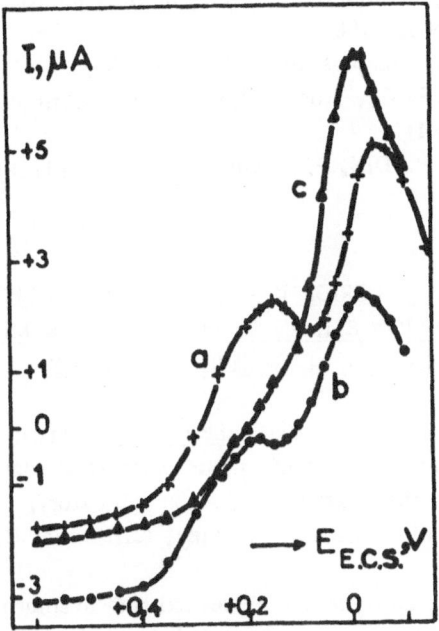

Figure 2

(Les intensités de courant sont rapportées à une concentration millimolaire et à une aire d'électrode de 10 mm²; température 0°; vitesse de polarisation 0,1 V/mn; solutions tamponnées de pH 10,1).

a aquopentacyanoferrate(II) à 8% de monomère
b aquopentacyanoferrate(II) à 22% de monomère (solution initiale)
c aquopentacyanoferrate(II) à 88% de monomère

References

1. HÖLZL, F., Monatshefte, 56, 79, 253 (1930).
2. DAVIDSON, D., J. Am. Chem. Soc., 50, 2622 (1928).
3. MICHAELIS, L., et SMYTHE, C.V., J. Biol. Chem., 94, 329 (1931) et Comptes rendus Trav. Lab. Carlsberg, série chim., 22, 347 (1938).
4. EMSCHWILLER, G., C.R. Acad. Sciences, 238, 341 (1954).
5. EMSCHWILLER, G., COHN, C., et LEGROS, J., C.R. Acad. Sciences, 239, 1213 (1954).
6. LANZA, P., et CORBELLINI, A., Contributi teoretici e sperimentali di Polarografia, Supplemento a "La Ricerca Scientifica", 22, 127 (1952).
7. GLASSTONE, S., et REYNOLDS, G.D., Trans. Faraday Soc., 29, 399 (1933).

8. LEVICH, B., Physicochemical Hydrodynamics, Prentice Hall Inc., Iglewood Cliffs N.J., 1962.

9. HOFMANN, K.A., Liebigs Ann., _312_, 1 (1900).

10. cf. EMSCHWILLER, G., Proceedings 7.ICCC - Abstracts, Stockholm and Uppsala, 1962, 244.

11. PORATH, J., et FLODIN, P., Nature, _183_, 1657 (1959).

2 C 1

THE LIFETIMES OF CHLORIDE ION AND WATER MOLECULES IN THE FIRST COORDINATION SPHERE OF FERRIC ION AS MEASURED BY NUCLEAR MAGNETIC RESONANCE,

Elaine Blatt and Robert E. Connick
Department of Chemistry and Inorganic Materials Research Division
of the Lawrence Radiation Laboratory,
University of California, Berkeley, California, U.S.A.

The several ferric chloride complexes existing in solutions containing chloride ion exchange both their coordinated waters and coordinated chloride ions with waters and chloride ions in the bulk of the solution. Under favorable conditions it is possible to ascertain the rates of such exchange reactions by NMR measurements.

Observed was the broadening of the resonances of the O^{17} and Cl^{35} nuclei in the bulk of the solution caused by the presence of ferric ions in the solution. When the nucleus undergoes transverse relaxation rapidly in the first coordination sphere of the paramagnetic ion relative to its lifetime in that environment, the slow and therefore controlling step for the broadening of the resonance of the nuclei in the bulk of the solution is the rate of chemical exchange. There is strong evidence that such a situation obtains in the present two cases as shown by the temperature dependence of the broadenings. Relaxation effects of the ferric ion outside its first coordination sphere are believed to be unimportant.

On the assumption that the broadening of the resonance was controlled by chemical exchange, rates of exchange were obtained for H_2O and Cl^- at 25° for concentrations of chloride ion ranging from 0 to 1 M and 0,1 to 5 M, respectively. The functional dependence of these apparent first order rate constants of the bulk species on chloride concentration was expressed as a power series in chloride ion. Each coefficient of the series should consist of a combination of rate and equilibrium constants corresponding to the rate of exchange for particular ferric chloride complexes.

An uncertainty in the interpretation arises from the values of the equilibrium quotients for the several complexes, which have not been measured for precisely our conditions and are inconsistent with each other (1-3). Values of 5, 2, and 1 liters per mole were chosen for the interpretation.

Assuming for simplicity that the stable complexes are 6 coordinated species, the following reactions lead to exchange of O^{17}:

$$[Fe(H_2O)_{6-n}Cl_n]^{3-n} + H_2\overset{*}{O} \rightleftharpoons [Fe(H_2O)_{5-n}(H_2\overset{*}{O})Cl_n]^{3-n} + H_2O$$

$$[Fe(H_2O)_{6-n}Cl_n]^{3-n} + H_2\overset{*}{O} \rightleftharpoons [Fe(H_2O)_{6-n}(H_2\overset{*}{O})Cl_{n-1}]^{4-n} + Cl^-$$

$$[Fe(H_2O)_{5-n}Cl_n]^{3-n} + H_2\overset{*}{O} \rightleftharpoons [Fe(H_2O)_{5-n}(H_2\overset{*}{O})Cl_n]^{3-n}$$

The first is a bimolecular or S_N2 exchange, the second the formation of the next higher complex and the reverse, and the last is the completion of an S_N1 substitution by reaction with a five coordinated intermediate. Writing $^{in}k^{out}_{complex}$ for the rate constants in the above equations, where "in" and "out" refer to the ligand entering and leaving, respectively, and the subscript designates the formula of the reacting complex, one obtains as the expression for the coefficient of the nth power of chloride:

$$k_n = \left\{ {}^{H_2O}k^{H_2O}_{6-n,n} + {}^{H_2O}k^{Cl}_{6-n,n} + \frac{{}^{O}k^{H_2O}_{6-n,n}}{[H_2O]} \right\} Q_1 Q_2 \cdots Q_n [Fe^{3+}]$$

The symbol Q_n represents the usual equilibrium quotient for the formation of the nth chloride complex from the n-1 complex. The values found for the part of k_n shown in braces above were approximately: $3, 1.10^2$ (4), 6.10^2, $1, 0.10^3$, and $1, 0.10^4$ M.sec^{-1} for n = 0 to 3, respectively. While it is not possible to distinguish easily between the contributions from the first and last rate constants, the second rate constant can be measured independently by studying the rate of complex formation. For n = 1 this value is $0, 034$ M^{-1} sec^{-1} (5) and therefore contributes inappreciably to the water exchange. It seems likely that it is also unimportant for higher values of n in the chloride range studied and that water exchange occurs predominantly through replacement of water by water in an S_N1 or S_N2 mechanism.

Comparison of the apparent first order exchange constants for chloride ion with those for water reveals a striking similarity for all except uncomplexed ferric ion. The simplest interpretation for the chloride complexes is that chloride as well as water exchange through replacement of water by an S_N1

mechanism, although an S_N2 mechanism is not ruled out definitely.

References

1. RABINOWITCH, E., and STOCKMAYER, W.F., J.Am.Chem.Soc., 64, 335 (1942).
2. OLERUP, H., Thesis, Lund (1944).
3. WOODS, M.J.M., GALLAGHER, P.K., and KING, E.L., Inorg.Chem., 1, 55 (1962).
4. GENSER, E., University of California Lawrence Radiation Laboratory Report UCRL-9846, January 1962.
5. CONNICK, R.E., and COPPEL, C.P., J.Am.Chem.Soc., 81, 6389 (1959).

2 C 2

SOME RECENT STUDIES OF RAPID REACTIONS OF METAL COMPLEXES IN AQUEOUS SOLUTION.

C.D. Hubbard, P. Moore and R.G. Wilkins
University of Sheffield, England, and
State University of New York at Buffalo, U.S.A.

There has been an increasing use of flow methods to investigate the redox and substitution reactions which certain complex ions undergo within the $10 - 10^{-3}$ second range. This paper describes two such studies.

Formation of transition metal-terpyridine and phenanthroline complexes

By using micromolar solutions of reagents and the stopped-flow spectral method (1) it was possible to measure the large second-order formation rate constants for the above complexes. Observations were made in the 3200 - 3400 A region where the complexes species (but not free ligand or metal ion) have high absorption coefficients. All experiments were carried out in the neutral pH region, thus obviating protonated species and possible difficulties in interpretation (2). Results obtained so far can be understood in terms of rate-determining coordination of one of the nitrogens, followed by rapid chelation. In the systems where intermediate complex species can be isolated and disproportionate only slowly in aqueous solution, i.e. $[Ni(phen)_2(H_2O)]^{2+}$, $[M(terpy)(H_2O)_3]^{2+}$ M = Fe, Co, Ni (3), it was possible to measure the rate of formation of the highest species. Data obtained so far for terpyridine complexes indicate a formation constant for the bis species some 200 times larger than for mono, suggesting marked enhanced lability of coordinated water when one terpyridine molecule is also attached to the metal ion. Results are shown in the Tables.

TABLE I - RATE DATA FOR THE FORMATION OF $[M(terpy)]^{2+}$ AT 25^0

M =	Mn*	Fe	Co	Ni	Cu	Zn*	Cd*
k_2 ($M^{-1} sec^{-1}$)	$1,7.10^5$	$5,6.10^4$	$2,3.10^4$	$\sim 9.10^2$	$\geq 10^7$	$1,1.10^6$	$3,2.10^6$
E_2 (kcal.mole^{-1})	11	9	10	-	-	6	7

* On basis that formation of mono was much slower than that of bis (further work in progress)

TABLE II - RATE DATA FOR THE FORMATION (k_2) AND DISSOCIATION (k_1) OF $[M(phen)]^{2+}$ AT 25^0

M	k_2* ($M^{-1} sec^{-1}$)	E_2* (kcal.mole^{-1})	$k_1^{\neq}$ (sec^{-1})	$E_1^{\neq}$ (kcal.mole^{-1})
Co	3.10^5	11,0	16.10^{-3}	19,4
Ni	4.10^3	13,7	10^{-5}	24,5

M	K_k (M^{-1})	K_t (M^{-1})	ΔH_k (kcal.mole^{-1})	ΔH_t (kcal.mole^{-1})
Co	2.10^7	$1,8.10^7$	8,4	9,1
Ni	4.10^8	6.10^8	10,8	11,2

* present studies
$\neq$ previous dissociative data (4)

The equilibrium constant ($K_k \sim k_2/k_1$) and the heat change ($\Delta H_k \sim E_1 - E_2$) were determined from the present formation kinetics and previous dissociative data (4). The agreement with the values obtained (5) from thermodynamic studies (K_t and ΔH_t) was excellent. First published data for reactions of silver(I) (with phenanthroline) was obtained - $k_2 \sim 3.10^5$ M^{-1} sec^{-1} at 0^0. All results were in agreement with current ideas on the dominance of the water exchange process in formation reactions.

Some rapid reactions of chromium(VI)oxy-anions

The hydrolysis of bichromate ion in alkaline solution ($Cr_2O_7^{2-}$ + 2 OH^- ⟶ 2 CrO_4^{2-} + H_2O) formerly considered first-order (6) was recently reported second order (7). These contradictory results were obtained by continuous flow methods using the thermal method. The reaction has been reinvestigated by stopped-flow using spectral differences between reactants and products at 4750 A. The second-order kinetics were confirmed over a range of reactant concentrations at ionic strength 0,1 M, units in moles.1^- and seconds. The rate law was

$$d[Cr_2O_7^{2-}]/dt = 10^{6,82} \exp[-5800/RT] [Cr_2O_7^{2-}] [OH^-].$$

The hydrolysis kinetics at low alkalinity were also measured, by using an anucleophilic buffer of lutidine/lutidinium ion at $p_H \sim 8$. The rate constant for this reaction ($Cr_2O_7^{2-}$ + H_2O ⇌ 2 $HCrO_4^-$) agreed well with that obtained recently by a relaxation method (8). Arrhenius parameters for this and for the ammonia-catalysed hydrolysed (7) are being obtained, and it is hoped to report on the acid-catalysed interconversion (9) of hydrogen chromate and bichromate ions.

The kinetics of formation of the blue species, believed CrO_5, ($HCrO_4^-$ + 2 H_2O_2 + H^+ ⟶ CrO_5 + 3 H_2O) were investigated by stopped-flow at 5800 A before subsequent decomposition (10). The reaction was governed by the rate law at ionic strength 0,1 M:

$$d[CrO_5]/dt = 10^{7,57} \exp[-4500/RT] [HCrO_4^-] [H_2O_2] [H^+]$$

The results apply to a limited range of reactant concentrations and it is hoped to report on further work on this reaction.

References

1. MELSON, G.A., and WILKINS, R.G., J.Chem.Soc. (London), 1962, 4208.
2. MARGERUM, D.W., BYSTROFF, R.I., and BANKS, C.V., J.Am. Chem.Soc., 78, 4211 (1956).
3. HOGG, R., and WILKINS, R.G., J.Chem.Soc. (London), 1962, 341.
4. ELLIS, P., and WILKINS, R.G., J.Chem.Soc. (London), 1959, 299; ELLIS, P., HOGG, R., and WILKINS, R.G., J.Chem.Soc. (London), 1959, 3308.

5. ANDEREGG, G., Helv.Chim.Acta, $\underline{46}$, 2813 (1963).
6. La MER, V.K., and READ, C.L., J. Am. Chem.Soc., $\underline{52}$, 3098 (1930).
7. LIFSHITZ, A., and PERLMUTTER-HAYMAN, B., J.Phys.Chem., $\underline{65}$, 2098 (1961).
8. SWINEHART, J.H., and CASTELLAN, G.W., Inorg.Chem. $\underline{3}$, 278 (1964).
9. SCHWARZENBACH, G., and MEIER, J., J.Inorg.Nuc.Chem. $\underline{8}$, 302 (1958).
10. See, for example, EVANS, D.F., J.Chem.Soc. (London), $\underline{1957}$, 4013.

2 C 3

EFFECT OF COORDINATED LIGANDS ON THE KINETICS OF SUBSTITUTION REACTIONS IN METAL COMPLEXES.

Dale W. Margerum and Manfred Eigen
Department of Chemistry, Purdue University, Lafayette, Indiana, U.S.A., and Max Planck-Institut für Physikalische Chemie, Göttingen, Germany.

In recent years considerable progress has been made in the study of the fundamental kinetic steps which occur in coordination substitution reactions. Many of the rate constants for the substitution of coordinated water in aquo-metal complexes now have been determined (1). For the general case of metal ions which are not hydrolyzed, the rate of water substitution is the rate limiting reaction step. This rate step is characteristic for each metal ion and is found in substitution reactions where the aquo-metal ion is one reactant. The next piece of fundamental information needed is the effect which other coordinated ligands have on subsequent rates of water substitution. Many substitution reactions of interest involve the replacement of water not from the aquo-metal ion but rather from a metal complex which has several co-ordination sites occupied by other ligands. Information is sparse on this subject for the divalent metal ions but it has been suggested that an increase in the substitution rate accompanies a decrease in the charge of the complex. Thus, the mono-glycine complexes of nickel and cobalt are more labile to the substitution of a second glycine than are the aquo-metal ions (2). The same is known also for some metal hydroxo species as compared to the corresponding aquo-metal ions, cf. (1). On the other hand, the neutral imidazole complexes of the same metals do not appear to change the rate of the water substitution steps (2).

In the present work a test is made of the hypothesis that the rate of water substitution increases as the metal complex charge decreases. Considered as a general rule the hypothesis fails. Some of the factors which are believed to influence such reactions are discussed.

Temperature-jump relaxation methods were used to study reactions of the type given in {1} where L is a chelate and M is a divalent metal ion.

$$ML(H_2O)_x + NH_3 \underset{k_{21}}{\overset{k_{12}}{\rightleftharpoons}} ML(H_2O)_{x-1} NH_3 + H_2O \qquad \{1\}$$

This reaction was coupled with much faster reactions {2} and {3} to give an indicator color change for a sensitive method of detecting the relaxation.

$$NH_3 + H_2O \rightleftharpoons NH_4^+ + OH^- \qquad \{2\}$$

$$OH^- + HInd \rightleftharpoons Ind^- + H_2O \qquad \{3\}$$

Conditions were selected to give a simple expression for the relaxation time, τ, of the form

$$1/\tau = k_{21} + k_{12} (\overline{C}_{ML} + \overline{C}_{NH_3}) \qquad \{4\}$$

Chelates are used in {1} because their rates of substitution are much slower than for monodentate ligands. In this way substitution by NH_3 with one to four coordinated water molecules can be studied without interference from multiple relaxations. The multidentate ligands used (see Table I) permitted variation of the number of amine nitrogens and of carboxylate groups and, of course, variation in the charge of the complex.

There are three important advantages in studying the rate of substitution of water with ammonia rather than other ligands.

1. The ammonia molecule is small so that effects due to steric hindrance in entering the inner coordination sphere are minimal.

2. Since the molecule is neutral the outer sphere association will be very weak.

3. What outer sphere association does exist between the complexes and ammonia will be approximately the same for the 2+ and 2- or for the 1+ and 1- complexes.

The stabilities of the mixed NH_3-EDTA complexes are known (3) and plots of equation {4} gave good agreement for the slope to intercept ratio. For the other complexes it was possible to estimate k_{12}, k_{21} and hence K for the mixed complex from the relaxation data itself. Work in progress will give independently the stability constants for the mixed NH_3 complexes (4).

TABLE I

M	L	Coordinate Groups	Charge of Complex	k_{12} $M^{-1} sec^{-1}$
Ni(II)	Glycinate	1 N, 1 COO⁻, 4 H₂O	+1	1.10^4
Ni(II)	Iminodiacetate	1 N, 2 COO⁻, 3 H₂O	0	6.10^3
Ni(II)	Nitrilotriacetate	1 N, 3 COO⁻, 2 H₂O	-1	6.10^3
Ni(II)	Hydroxyethyl-ethylenediamine-tetraacetate	2 N, 3 COO⁻, 1 H₂O	-1	$\sim 1.10^3$
Ni(II)	Ethylenediamine-tetraacetate	2 N, 3 COO⁻, 1 H₂O	-2	$1,5.10^3$
Cu(II)	Ethylenediamine-tetraacetate	2 N, 3 COO⁻, 1 H₂O	-2	$2,0.10^6$
Ni(II)	Diethylenetri-amine	3 N, 3 H₂O	+2	7.10^4
Ni(II)	Triethylene-tetramine	4 N, 2 H₂O	+2	6.10^3
Ni(II)	Tetraethylene-pentamine	5 N, 1 H₂O	+2	$< 10^3$

It is clear from Table I that there is no evidence that a 1- or 2- complex loses its coordinated water at a faster rate than a 1+ or 2+ complex. In fact, the [Nidien]$^{2+}$ complex is much more labile to NH_3 substitution than are the negatively charged complexes of nickel. It is also not true that the binding strength of a ligand in the inner coordination sphere of a metal ion depends necessarily on its overall charge (cf. SO_4^{2-} as compared to H_2O (1)).

With the exception of the dien complex there is a fair correlation between the number of water molecules coordinated to nickel and the rate constant. Such a statistical factor would be reasonable if the inner sphere can rotate rather freely relative to an ammonia molecule outside the inner sphere.

The outer sphere association between a complex such as Niglycine$^+$ and NH_3 must be less than unity so that the actual first order rate constant for the dissociation of water is greater than $1.10^4 sec^{-1}$. This is in agreement with the previous estimate that this value for the nickel glycine complex is $1,2.10^5 \ sec^{-1}$ (2) which is about ten times the rate of loss from the aquo nickel ion. However, the ten fold increase in rate cannot be attributed to the carboxylate portion of the glycinate complex, and the reduced charge of the complex, because similar or greater effects are observed with the amine complexes. Furthermore, additional charge reduction gives no evidence of increased lability. The data suggest that amine nitrogen groups accelerate water loss (in contrast to the weaker imidazole complexes) and that groups which are strongly coordinated (those which supply a greater electron density) will have the greatest influence on the rate of loss of remaining water molecules. However, the chelate ring itself may have an effect. The $[Nidien]^{2+}$ data serve as warnings against premature generalizations. In the dien case the complex is at least ten times more labile than any of the other nickel complexes studied. This cannot be correlated directly to the number of nitrogens bonded to nickel. The influence of chelation in coordination structure is demonstrated by the data for Cu- and Ni-EDTA. The ratio of rate constants for copper and nickel is $1,3.10^3$ as compared to a ratio of about 40.10^3 for the aquo-metal ions. In the case of Cu, chelation apparently hinders the fast communication between the non-equivalent axial and planar positions (5) and thus slows down the substitution.

It can be seen that several factors not tested as yet may affect the rate of the substitution of water in the divalent metal complexes. A knowledge of the individual influences of different ligands is most important for an understanding of the rate behavior of enzyme metal substrate complexes.

References

1. For a review see EIGEN, M., and deMAEYER, L., "Technique of Organic Chemistry" A. Weissberger, Ed., Vol. VIII, Part II, pp. 895-1054. Intersciences Publishers, New York, 1963.
2. HAMMES, G.G., and STEINFELD, J.I., J. Am. Chem. Soc., 84, 4639 (1962).
3. BHAT, T.R., and KRISHNAMURTHY, M., J. Inorg. Nucl. Chem., 25, 1147 (1963).
4. MARGERUM, D.W., and JACKOBS, N.E., work in progress.
5. EIGEN, M., Ber. Bunsenges. phys. Chem., 67, 753 (1963).

3 C 1

FURTHER STUDIES ON THE KINETICS AND MECHANISM OF HYDROLYSIS OF Co(III) DIACIDO CHELATE COMPLEXES.

Vincenzo C a g l i o t i and Gabriello I l l u m i n a t i
Istituto di Chimica dell'Università and Centro di Chimica Generale
del C.N.R., Roma, Italy.

In continuation of our studies on the hydrolysis of organic diacido co-ordination compounds, special attention was given to the position of bond-fission, to the characterization and kinetic behaviour of the intermediate complex ions of reactions {1} and, finally, to the role of the chelating ligand in the reaction mechanism.

$$[Co(AA)_2(OAc)_2]^+ \xrightarrow[k']{\text{hydrol.}} \begin{cases} [Co(AA)_2(OAc)_2(OH)] \\ \text{or} \\ [Co(AA)_2(OAc)(OH_2)]^{2+} \end{cases} \xrightarrow[k'']{\text{hydrol.}} \begin{cases} [Co(AA)_2(OH)_2]^+ \\ [Co(AA)_2(OH)(OH_2)]^{2+} \\ \text{or} \\ [Co(AA)_2(OH_2)_2]^{3+} \end{cases}$$

{1}

AA = en, dipy.

Position of bond-fission. Previous work had shown that in the complex ion series trans-$[Coen_2(OCOC_6H_4R)_2]^+$ cobalt-oxygen, rather that carbonyl-oxygen bond-fission is involved in base hydrolysis. This conclusion was based on the use of linear free-energy correlations with the structure of the substrate, as a diagnostic criterion; we wished to confirm it by O^{18} studies and to extend the work to the cis-isomers. In all tested cases, including substituents of widely varying polar type (trans-R=H; cis-R=H, p-MeO, m-NO$_2$, m-Cl, m-F), the substituted benzoic acid recovered from base hydrolysis in O^{18}-enriched solvent showed an isotopic composition practically identical to that of natural oxygen. The result was made particularly clean-cut by the fact that the benzoic acids chosen as ligands did not exchange under the conditions used in the O^{18} experiments with the complexes. Co-O bond-fission was then con-firmed and found to be generally occurring in the base hydrolysis of cis- and trans-$[Coen(OCOC_6H_4R)_2]^+$.

Intermediate Complex Ions. These compounds were isolated on hydrolysis at a pH dependent on the relative k' and k" values involved. Quenching was effected at a convenient reaction time in such a way as to keep any subsequent hydrolysis to a minimum.

In the case of cis- and trans-$[Coen_2(OAc\chi OH_2)]^{2+}$ the detection and isolation were made possible by ion-exchange chromatography. Trans-$[Coen_2(OAc\chi OH_2)]^{2+}$ was found to isomerize into the cis-form on standing at room temperature in acidic solution; this change is conveniently followed by spectrophotometry and shows a half-life of about 7 hours. In basic solution, no stereochemical change seems to occur prior to hydrolysis. Both cis and trans isomers undergo base hydrolysis at a rate lower than that of the first functional group in $[Coen_2(OAc)_2]^+$, viz., k' > k" (Table I).

TABLE I

KINETIC DATA FOR THE HYDROLYSIS OF SOME COMPLEX IONS OF THE SERIES $[Co(AA)_2 X_2]^+$ OR $[Co(AA)_2 XY]^{n+}$ AT DIVERSE INITIAL pH VALUES (a)

Complex Ion	pH			
	3	6-8	7-8	11-12
cis-$[Coen_2(OAc)_2]^+$			k_1', 3, 4, 10^{-5} (35°)	k_1', 2, 4, 10^{-2} (25°)
cis-$[Coen_2(OAc\chi OH)]^+$				k_2'', x, 2, 10^{-3} (25°)
trans-$[Coen_2(OAc)_2]^+$				k_1', 5, 0, 10^{-2} (25°)
trans-$[Coen_2(OAc\chi OH)]^+$				k_2'', 1, 1, 10^{-3} (25°)
cis-$[Codipy_2(OAc)_2]^+$		k_1', 1, 3 - 3, 10^{-4} (35°)		k_1', 2, 6, 10^{-4} (35°)
cis-$[Codipy_2(OAc\chi OH)]^+$				k_1'', ~8, 10^{-4} (35°)
cis-$[Codipy_2(OAc\chi OH_2)]^{2+}$	k_1'', ~10^{-7} (35°)			

(a) First-order rate constants (k_1' and k_1'') in sec^{-1}; second-order rate constants (k_2' and k_2'') in $l.m^{-1}.sec^{-1}$. Ionic strengths for the first-order reactions are in the range of 10^{-3} to 10^{-2}.

As to the bis-dipyridyl complexes, all attempts to prepare trans-isomers of the general type $[Codipy_2 X_2]^+$ have failed; in particular, only cis-$[Codipy_2(OAc)_2]^+$ and cis-$[Codipy_2(OAc\chi OH)]^+$ could be obtained and either step-wise or over-all base hydrolysis seem to yield cis-complexes in this series. The base hydrolysis of cis-$[Codipy_2(OAc\chi OH)]^+$ is markedly faster than that of the first functional group in cis-$[Codipy_2(OAc)_2]^+$, viz., k" ≫ k' (Table I).

Role of the Chelating Ligand. Bis-ethylenediaminodiacetato complexes hydrolyse in basic solution by a 2nd-order reaction. In contrast, bis-dipyridyl diacetate shows 1st-order kinetics and rate constants independent of the hydroxide ion concentration (Table I). According to these results, there appears to be no tendency to bimolecular attack by the hydroxide ion and either water effectively competes with the negatively-charged nucleophile as assisted by a concentration effect or Co-OAc bond ionization is rate-determining. The latter alternative case is a better explanation for the very low reactivity of the bis-dipyridyl complex. Comparison in 2nd-order k units shows a deactivation with respect to the bis-ethylenediamino complex by a factor of about half a million (cis-isomers). Such a great difference in reactivity can be related to the fact that the bis-dipyridyl complex is unable to produce a conjugate base.

The markedly higher rate of the intermediate $[Codipy_2(OAc)(OH)]^+$ ($k'' \gg k'$) can also be explained in terms of a structural effect. The effect of the hydroxo group is rate-enhancing both because of a greater electron-releasing power, as compared to the OAc group, and because either O-H ionization or hydrogen-bond interaction with the basic medium is possible. In acid solution (pH = 3), where the reactive species is $[Codipy_2(OAc)(OH_2)]^{2+}$, is $k' > k''$ (Table I). The rate-enhancing effect of the hydroxo group should be considerably less important in the bis-ethylenediamino series where large effects of similar kind due to the chelating ligand are preexisting in the starting diacetato ion and should remain predominant in the intermediate. This is in fact what has been found in these series ($k' > k''$). To sum up, the present data are better explained with a dissociative mechanism of hydrolysis than with a bimolecular reaction and are consistent with the intervention of the more reactive conjugate bases in the presence of acidic groups of the type metal-ZH, where Z may be a nitrogen- or oxygen-containing moiety.

The valuable collaboration and discussions by our research associates, Drs. F. APRILE, V. CARUNCHIO, F. MASPERO, F. MONACELLI and G. ORTAGGI, are gratefully acknowledged here.

3 C 2

KINETIC STUDY OF THE ISOTOPIC EXCHANGE OF EDTA COMPLEXES OF THE GALLIUM GROUP ELEMENTS.

Kazuo Saito and Michiko Tsuchimoto
Institute for Nuclear Study, the University of Tokyo,
Tanashi, Tokyo, Japan.

Little information is available concerning the isotopic exchange of co-ordination compounds of tervalent gallium group elements, which have the typical d^{10} structure. We have undertaken the kinetic study by use of EDTA complexes of these elements labelled with radioisotopes. Similar studies with EDTA-OH (N-2-hydroxyethylethylenediamine-N, N', N'-triacetate) complexes were also made for comparison.

The exchange of the EDTA complexes is very slow in neutral solutions and very fast in basic and strongly acid solutions. The rate is measurable in an acid and a weakly basic region as illustrated in the figure. When the rate is expressed by McKAY's formula, $R = -[2.303(a+b)/ab][\log(1 - F)]/t$, it is

independent of the concentration of uncomplexed metal, a (initially inactive)
in both acid and basic solutions although its dependence upon the pH, the
temperature and the concentration of other cosolutes differs remarkably in
these two regions and from metal to metal. Dependence of the rate constant
k which is expressed by R/b (b, the concentration of EDTA complex) upon.
the hydrogen ion concentration and the Arrhenius activation energy are
summarised in the figure. Plausible reaction mechanism has been reported
individually elsewhere (1-3), and general discussion and mutual comparison
will be made here.

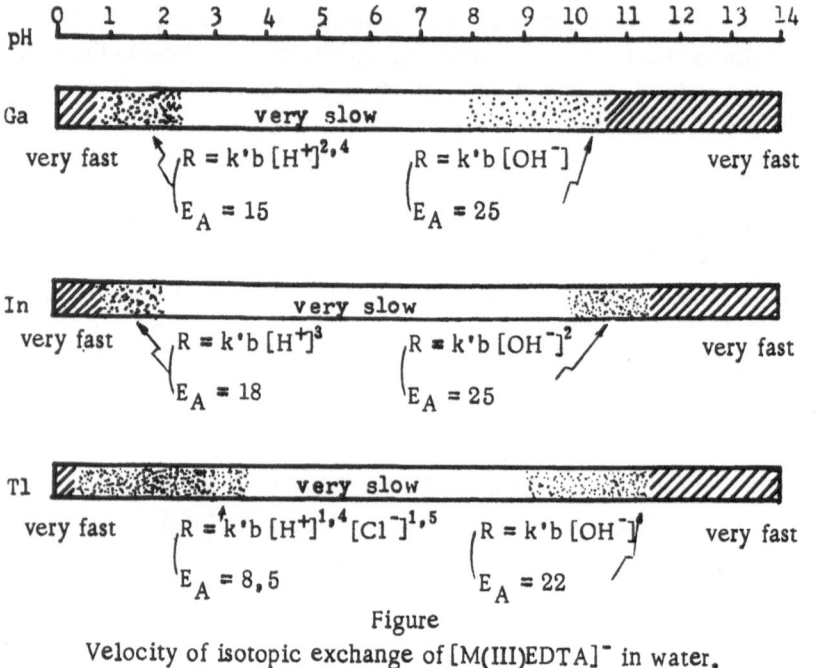

Figure

Velocity of isotopic exchange of [M(III)EDTA]⁻ in water.
(b, concentration of the complex; E_A in kcal/mole)

In acid solutions, the exchange seems to proceed through protonated
complexes such as [MYHOH₂] and [MYH₂OH₂]⁺ as intermediates, and via
substantial decomposition of the complexes. The non-integral exponents on
[H⁺] indicate that two or more reactions proceed side by side for gallium
and thallium. The high exponent for indium suggests that this complex is
the most inert among the three. The exchange is accelerated by acid radicals
such as acetate, citrate and chloride and this effect is most obvious for thallium,
which gives quite a complicated kinetic relationship with a and b in the
absence of chloride.

The kinetics in the basic media are summarised in the table. Since the rate is not affected by the buffer under due conditions, the reaction mechanism seems to involve essentially that of hydroxo complexes except for thallium, and the attack of hydroxide ions upon the complex appears to play an important role. The fact that the rate of the indium complex is proportional to the square of the hydroxide ion concentration tells that indium is the most inert among the three. Such an inertness of indium complexes may be reflected in the small value of its apparent dissociation constant of the coordinated water molecule, K_2.

In the presence of certain buffers, the rate is affected by their concentration, and so is the dependence of k upon the hydroxide concentration. The influence of ionic strength upon the rate is rather small. EDTA-OH complex gives similar kinetics for gallium, but not for indium (1, 4).

TABLE

COMPARISON OF THE KINETIC DATA IN BASIC MEDIA. *

	$[GaY.OH_2]^-$	$[InY.OH_2]^-$	$[TlY.OH_2]^-$
K_2	$1,2.10^{-6}$	$1,6.10^{-9}$	$0,32.10^{-6}$
k	$k'[OH^-]$	$k'[OH^-]$	$k'[OH^-][TEA]$ **
k'	11,3	52	3,4 ***
	(1/mole/sec)	$(l^2/mole^2/sec)$	(1/mole/sec)
E_A (kcal/mole)	25	25	22
k at pH 10 (sec^{-1})	$1,1.10^{-4}$	$1,2.10^{-6}$	$4,1.10^{-5}$ ***

* buffer: p-hydroxybenzoate (0,0125 M); at 25°; $\mu = 0,2$

** TEA = triethanolamine

*** TEA conc. ~ 0,12 M

$$K_2 = [H^+][MY.(OH)^{2-}] / [MY.OH_2^-]$$

References

1. SAITO, K., and TSUCHIMOTO, M., J.Inorg.Nucl.Chem., 23, 71 (1961).

2. SAITO, K., and TSUCHIMOTO, M., J. Inorg. Nucl. Chem., 25, 1245 (1963).
3. SAITO, K., and TSUCHIMOTO, M., unpublished.
4. SAITO, K., and TSUCHIMOTO, M., Bull. Chem. Soc. Japan, 35, 368 (1962); 36, 1073 (1963).

3 C 3

THE AQUATION AND BASE HYDROLYSIS OF CHLOROBIS(ETHYLENEDIAMINE)COBALT(III) CATIONS CONTAINING AMMONIA DERIVATIVES AS ORIENTING LIGANDS.

S.C. Chan and F. Leh
Department of Chemistry, University of Hong Kong, Hong Kong.

In the octahedral cobalt(III) system, the most widely studied reactions in aqueous solutions are the solvolytic aquation and the base hydrolysis. Both reactions have been shown to take place by a rearrangement, either S_N1 or S_N2, between the solvation shell and the coordination shell of the complexes (1, 2). The dependence of the rates of solvolytic aquation of complexes of the type $[Co\ en_2\ A\ Cl]^{n+}$ (where en = ethylenediamine and A = a variable orienting ligand) on the nature of A has led to the postulation of a duality of mechanism for the process (3, 4). Electron-repelling groups facilitate dissociation of the chlorine ligand and give rise to a unimolecular mechanism, while electron-attracting groups assist a bimolecular reaction. For the base hydrolysis of these cations, the assignment of mechanism has been controversial. BASOLO and PEARSON (5) suggested that an S_N1CB mechanism operates involving a pre-equilibrium in which the hydroxide ion removes a proton from one of the amine nitrogens forming an amido conjugate base which then undergoes a rate-determining dissociation of the chloride ion. BROWN, INGOLD and NYHOLM (6) however preferred an S_N2 mechanism in which the hydroxide ion directly replaces the chlorine ligand. The work of CHAN and TOBE (2) later indicated that neither the simple collision-activated bimolecular substitution nor the conjugate base mechanism occur, but that the reaction is a bimolecular rearrangement between the solvation shell and the coordination shell, involving deprotonation of the solvation shell by the hydroxide ion and Grotthus chain transfer to the metal. This mobility of the hydroxide ion about the complex may be indistinguishable from ion-pair formation, although evidence is still lacking. The work presented by us represents the beginning of a programme designed to give further clarification to the mechanism of octahedral substitutions.

The complexes under investigation are of the type $\underline{cis}$-[Co en$_2$ RNH$_2$ Cl]$^{z+}$ where R = H, CH$_3$, CH$_3$.CH$_2$ and OH. They were all obtained by the action of the appropriate amine on $\underline{trans}$-dichlorobis(ethylenediamine)cobalt(III) chloride.

$$\underline{trans}\text{-[Co en}_2 \text{ Cl}_2]\text{Cl} + \text{RNH}_2 = \underline{cis}\text{-[Co en}_2 \text{ RNH}_2 \text{ Cl]Cl}_2$$

The $\underline{cis}$ configurations of the complexes were confirmed by resolution into optical enantiomorphs and on the basis of infra-red observations using the CH$_2$-rocking region of the cobalt-ethylenediamine chelate ring (7). For the kinetic studies the chlorides of the complexes were converted into the corresponding nitrates by the action of 60% nitric acid. The case where R = H has been investigated by other workers (8) who measured the rates of solvolytic aquation of $\underline{cis}$-[Co en$_2$ NH$_3$ Cl]$^{2+}$ cations at 62^0 and the base hydrolysis at 0^0 at ionic strengths ranging from 1-4. 10^{-2} M. In order to compare with our data for the other complexes, we have extended their investigations. The solvolytic aquation obeys a first-order, and the base hydrolysis, a second-order rate law, the kinetics being followed by measuring the increase of free chloride ion concentration in solution by titration with silver nitrate solution (Volhard method). The anionic chloride was previously separated from the complex cations by means of a cation exchange resin (Amberlite IR 120; H$^+$ form). The first-order rate constants, k$_1$ for the aquation and the second-order rate constants, k$_2$ for the base hydrolysis were obtained from the appropriate expressions. The aquations did not go quite to their stoicheiometric completion, but it was found that when the complexes were supplied as the nitrates, the retrograde reaction had only a neglegible effect during the first half-life.

For R = OH, the first-order rate constant of the solvolytic aquation at 24,4^0 is 1,58 . 10^{-4} sec^{-1}, and the activation energy is 12,9 kcals/mole. The high rate of aquation compared with the case where R = H (k$_1$ = 2,30 . 10^{-4} sec^{-1} at 80,5^0; E$_a$ = 23,2 kcal / mole) is due, partly at least, to the specific hydrogen bonding of the hydroxylamine ligand bringing a water molecule into a highly favourable attacking position. This is supported by the relatively low activation energy. The second-order rate constant for the base hydrolysis of the chlorohydroxylamine complex at 0^0 and ionic strength of 0,1 M is 0,181 l.sec^{-1}.mole^{-1}, which is lower than the corresponding rate for the chloroammine complex (k$_2$ = 0,232 l.sec^{-1}.mole^{-1}). Hydroxylamine is a much weaker base than ammonia or ethylenediamine, and consequently an S$_N$1CB mechanism would predict a considerably higher rate for the chlorohydroxylamine complex. Our present data are therefore inconsistent with the conjugate base mechanism, but support the S$_N$2 mechanism of CHAN and TOBE (2). Qualitatively they indicate that ion-pair association is involved in the reaction because the concentration of the ion-pair in solution decreases as the size of ions increases. Quantitative con-. firmation, however, is not possible for two reasons. Firstly, for ions of the sizes concerned, ion-pair association constants can be directly measured only when $z_1 z_2 = 3$ or greater, and secondly, the reactivity of the cations under investigation with hydroxide ion renders any direct ion-pair measurements impossible.

The rates of solvolytic aquation decrease along the series R = H, CH_3, $CH_3.CH_2$ ($k_1 = 2,30, 2,03$ and $1,82 . 10^{-4}$ sec^{-1} respectively at $80,5^0$). Since the electron-attracting powers of the orienting groups RNH_2 also decrease along the series R = H, CH_3, $CH_3.CH_2$, the data support a bimolecular mechanism. The base hydrolysis rates for the complexes where R = CH_3 and $CH_3.CH_2$ are similar to (slightly less than) that for the chlorohydroxylamine complex. These rates decrease as R passes along the sequence H, OH, CH_3, $CH_3.CH_2$ ($0,232, 0,181, 0,173$ and $0,158$ l.sec^{-1}.mole^{-1} respectively at 0^0 and ionic strength of $0,1$ M). Some idea of the relative stability of the ion-pairs may be obtained by investigation of the association of the hydroxide ion with the corresponding hexammine-type cations, cis-[Co en$_2$ RNH$_2$ NH$_3$]$^{3+}$ which are much more stable towards bases and for which $-z_1 z_2 = 3$. For R = H and OH, these complexes were readily available, but corresponding complexes in which R = CH_3 and $CH_3.CH_2$ have not been obtained. As far as size is concerned, the [Coen$_3$]$^{3+}$ ion is comparable to the cis-[Co en$_2$ CH$_3$.CH$_2$NH$_3$ NH$_3$]$^{3+}$ complex and may be used as a substitute. The measurements depend on the decomposition of diacetone alcohol in solutions of the hexamminecobalt(III) hydroxides (9). The association equilibrium constants were found to decrease along the sequence R = H, OH and $CH_3.CH_2$, thus supporting an ion-pair mechanism.

The authors thank Professor J. MILLER for his interest and the Committee on Higher Degrees and Research Grants of the University of Hong Kong for the award of a research grant.

References

1. INGOLD, NYHOLM and TOBE, Nature, 187, 477 (1960).
2. CHAN and TOBE, J. Chem. Soc. (London), 1962, 4531.
3. ASPERGER and INGOLD, J. Chem. Soc. (London), 1956, 2862.
4. BASOLO and PEARSON, "Mechanism of Inorganic Reactions", John Wiley and Sons, Inc., New York, p. 166 (1958).
5. BASOLO and PEARSON, reference (4), p. 124.
6. BROWN, INGOLD and NYHOLM, J. Chem. Soc. (London), 1953, 2674.
7. BALDWIN, J. Chem. Soc. (London), 1960, 4369.
8. NYHOLM and TOBE, J. Chem. Soc. (London), 1956, 1707.
9. CATON and PRUE, J. Chem. Soc. (London), 1956, 671.

KINETICS AND MECHANISM OF THERMAL DECOMPOSITION OF Mn(III) COMPLEX OXALATES IN SOLUTION.

Ilie G. Murgulescu and Tatiana Oncescu

Physical Chemistry Laboratory, University of Bucarest, Roumania.

Thermal decomposition of $[Mn(C_2O_4)_2(H_2O)_2]K$ in solution. On dissolution in water a partial change takes place in hydroxo salt (1), according to

$$[Mn(C_2O_4)_2(OH_2)_2]^- \rightleftharpoons [Mn(C_2O_4)_2(OH)(OH_2)]^{2-} + H^+ \quad \{1\}$$

characterised by an increase of the extinction and a decrease of the pH value to 4. At pH = 4 reaction $\{1\}$ is completely on the left side with quantitative formation of the $[Mn(C_2O_4)_2(H_2O)_2]^-$ ion. The decomposition can be referred to the following consecutive reactions.

$$[Mn(C_2O_4)_2(H_2O)_2]^- \xrightarrow{k_1} MnC_2O_4 + [C_2O_4]^- + 2\,H_2O \quad \{2\}$$

$$[Mn(C_2O_4)_2(H_2O)_2]^- + [C_2O_4]^- \xrightarrow{k_2} MnC_2O_4 + [C_2O_4]^{2-} + 2\,CO_2 + 2\,H_2O \quad \{3\}$$

$$2\,[Mn(C_2O_4)_2(H_2O)_2]^- \longrightarrow 2\,MnC_2O_4 + [C_2O_4]^{2-} + 2\,CO_2 + 4\,H_2O$$

$$-\frac{d\,(a-x)}{dt} = k_1\,(a-x) + k_2\,(a-x)\,c_{ox}- \quad \{4\}$$

With the steady-state approximation $\dfrac{dc_{ox}-}{dt} = 0$, one obtains by integration:

$$k_1 = \frac{2,303}{2\,t}\,\log\frac{a}{a-x} \quad \{5\}$$

which is confirmed by the experimental results. In table I, the rate constants obtained at various temperatures are given.

From these data an activation energy equal to 20,2 Kcal/mol is obtained smaller than that found by DUKE (2) in the acid medium (25,0 kcal/mol). In the third column, the values of the rate constants calculated with the Arrhenius equation, have been noted. The frequency factor A is ten times smaller than

TABLE I

pH = 4

t	k_1, sec^{-1}		log A
	found	calc.	
21^0	0,000403	0,000418	11,599
30^0	0,00112	0,00111	11,622
38^0	0,00300	0,00274	11,675
49^0	0,00708	0,00792	11,564
			11,615

the theoretical one, for which we find at 35^0 after HERZFELD's (3) formula: log A = 12,801 and after MOELWYN-HUGHES (4): log A = 12,670. The equilibrium is therefore a so called "slow" reaction.

<u>Thermal decomposition of $[Mn(C_2O_4)_2(OH)(OH_2)]K_2$ in solution</u>. By dissolving the diaqua complex in KOH (pH = 10) equilibrium {1} is shifted to the right, a brown solution of hydroxo salt is formed accompanied with an increase of the extinction while the pH decreases to 4. Then the thermal decomposition of the hydroxo complex takes place, accompanied with a rise of the pH. The decomposition mechanism can be represented by the system of reactions:

$$[Mn(C_2O_4)_2(OH)(OH_2)]^{2-} \xrightarrow{k_2} Mn(C_2O_4) + [C_2O_4]^- + OH^- + H_2O \quad \{6\}$$

$$[Mn(C_2O_4)_2(OH)(OH_2)]^{2-} + [C_2O_4]^- \xrightarrow{k_3} Mn(C_2O_4) + [C_2O_4]^{2-} + OH^- + 2\,CO_2 + H_2O \quad \{7\}$$

$$2\,CO_2 + 2\,OH^- \xrightarrow{K} 2\,[HCO_3]^-$$

$$2\,[Mn(C_2O_4)_2(OH)(OH_2)]^{2-} \longrightarrow 2\,Mn(C_2O_4) + [C_2O_4]^{2-} + 2\,[HCO_3]^- + 2\,H_2O$$

The hydroxo salt concentration will thus evolve according to

$$\frac{dx}{dt} = k_2(a-x) + k_3(a-x)\,c_{ox^-} \quad \{8\}$$

In the steady-state $c_{ox^-} = \dfrac{k_2}{k_3}$, and the kinetic equation becomes by integration:

$$k_2 = \frac{2,303}{2t} \log \frac{a}{a-x} \qquad \{9\}$$

In table II, the values of rate constants obtained at various temperatures are given.

TABLE II

t	k_2, sec^{-1}		log A
	found	calc.	
30^0	0,000405	0,000321	7,962
49^0	0,00135	0,00150	7,816
59^0	0,00290	0,00314	7,826
67^0	0,00526	0,00551	7,841
			7,861

From these an activation energy of 15,74 kcal/mol is calculated. The frequency factor A, resulting from experimental data is approximately 10^5 times smaller than the theoretical one; the decomposition of hydroxo-complex is therefore an especially "slow" one.

Solution containing a mixture of $[Mn(C_2O_4)_2(H_2O)_2]K$ and $[Mn(C_2O_4)_2(OH)(OH_2)]K_2$. By the diaqua complex, a mixture of diaquo and hydroxo salts is produced, according to $\{1\}$. The two ions decompose in a reaction of the first order. The rate constants of the decomposition of the diaqua complex are approximately twice larger. The experimental data show that on this interval of temperature the equilibrium of reaction $\{1\}$ is noticeably displaced to the right with the formation of hydroxo-complex in a predominant proportion.

Thermal decomposition of $[Mn(C_2O_4)_3]K_3$ in solution. The decomposition was investigated in aqueous solution in the presence of $K_2C_2O_4$ at c = 10^{-2} m/l. By analogy with the cobalt(III) trioxalate (5), the following reaction mechanism can be adopted:

$$[Mn(C_2O_4)_3]^{3-} + H_2O \underset{k'}{\overset{k}{\rightleftharpoons}} [Mn(C_2O_4)_2OC_2O_3H_2O]^{3-} \qquad \{10\}$$

$$[Mn(C_2O_4)_2OC_2O_3H_2O]^{3-} \xrightarrow{k_1} [Mn(C_2O_4)_2]^{2-} + [C_2O_4]^- + H_2O \qquad \{11\}$$

$$[Mn(C_2O_4)_3]^{3-} + [C_2O_4]^- \xrightarrow{k_2} [Mn(C_2O_4)_2]^{2-} + [C_2O_4]^{2-} + 2\,CO_2 \qquad \{12\}$$

$$2\,[Mn(C_2O_4)_3]^{3-} \longrightarrow 2\,[Mn(C_2O_4)_2]^{2-} + [C_2O_4]^{2-} + 2\,CO_2 \qquad \{13\}$$

In this case, the kinetic equation is: $\dfrac{dx}{dt} = k\,(a-x)c_{H_2O} - k'c_A + k_2(a-x)c_{ox}$.

where c_A is the concentration of intermediate complex in which an oxalate ligand occupies only one coordinative position. By applying the steady-state approximation it results:

$$\frac{dx}{dt} = 2\,k_1\,\frac{k}{k'+k_1}\,(a-x)\,c_{H_2O} = 2\,k_3\,(a-x)\,c_{H_2O} \qquad \{14\}$$

which gives by integration: $k_0 = 2\,k_3\,c_{H_2O} = \dfrac{2,303}{t}\,\log\dfrac{a}{a-x}$ because

the water concentration c_{H_2O} remains practically constant during the reaction.

TABLE III

$a = 2.10^{-3}\ m/l$

t	k_0, min^{-1}	k_3 $mol^{-1}.cm^3.sec^{-1}$		log A
		found	calc.	
0^0	0,00120	0,000181	0,000190	14,235
10^0	0,00533	0,000834	0,000821	14,295
20^0	0,0231	0,00361	0,00321	14,337
30^0	0,0726	0,0113	0,0115	14,279
				14,286

The values of the rate constants are near to those obtained by FURLANI and CIANA (6) by the polarographic study of the thermal decomposition of $[Mn(C_2O_4)_3]Na_3$. From the data-presented in table III an activation energy of 22,49 kcal/mol is deduced, a value much higher than that estimated by GOSH and KAPPANA (7) from two rate constants (13,0 kcal/mol) but near to that found by TAUBE (8) for the decomposition of the reaction product from $KMnO_4$ and $C_2O_4H_2$ (22,2 kcal/mol).

References

1. CARTLEDGE, G.H., and ERICKS, W.P., J.Am.Chem.Soc., 58, 2061 (1936).
2. DUKE, F.R., J.Am.Chem.Soc., 69, 2885 (1947).

3. HERZFELD, K., Ann.Phys., 59, 635 (1919).
4. MOELWIN-HUGHES, F.A., "The Kinetics of Reactions in Solution", Oxford, p. 10, 238 (1950).
5. MURGULESCU, I.G., and ONCESCU, T., Z.Physik.Chem., 214, 238 (1960).
6. FURLANI, C., and CIANA, A., Ann.chim. (Roma), 47, 1081 (1957).
7. GOSH, J.C., and KAPPANA, A.N., Chem.Zentralblatt II, 2144 (1926).
8. TAUBE, H., J.Am.Chem.Soc., 70, 1216-20 (1948).

4 C 1

METAL ION AND METAL CHELATE CATALYSIS OF SALICYL PHOSPHATE HYDROLYSIS.

Arthur E. Martell
Illinois Institute of Technology, Chicago, Illinois, U.S.A.

New kinetic and equilibrium data (1-4) on the hydrolysis of salicyl phosphate and its analogs in the presence of metal ions and metal chelate compounds reveals many patterns of behavior which determine: (a) whether catalysis does or does not take place, and (b) the degree of catalytic activity achieved.

The spontaneous hydrolysis of salicyl phosphate occurs most likely through nucleophilic attack on the negative carboxylate group on the adjacent phosphorus atom of [II] and [III] to give a penta coordinated activated intermediate, which then decomposes directly to hydrolysis products, as suggested by CHANLEY (5) and by MARTELL et al (6). Hydrolysis of [II] via this mechanism would require the shift of a proton from the carboxyl group to the phosphate oxygen to give a tautomeric form. Estimation of the concentration of this form leads to a specific rate constant higher than that of [III], in accordance with what would be expected as the result of the influence of an additional proton on the phosphate group. The mechanism proposed by BENDER (7) involving transfer of a proton from the carboxyl group to the phenolate ester oxygen is not in accord with the lower rate constant observed experimentally for [II], since this mechanism would predict the rate of hydrolysis of [II] to be greater than that of [III].

The data in Table I indicate that the rates of hydrolysis of the ionized species of 1,3-dicarboxy-2-phenylphosphate [IV-VII] are higher than the analogous forms of salicyl phosphate [I-IV], as would be expected from the presence of the second negative carboxylate group adjacent to the phosphate

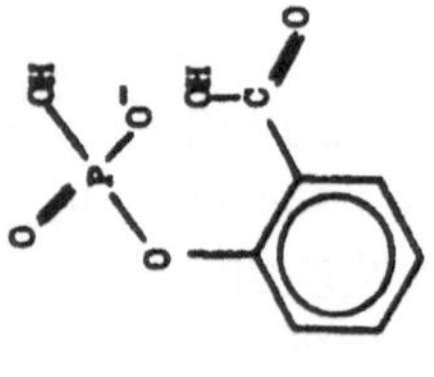

I

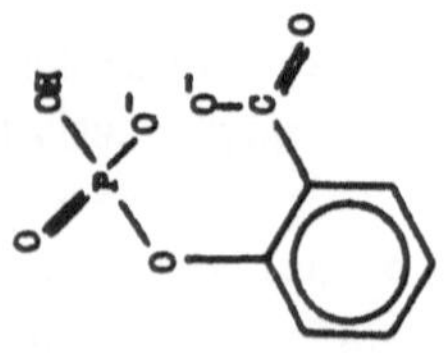

II

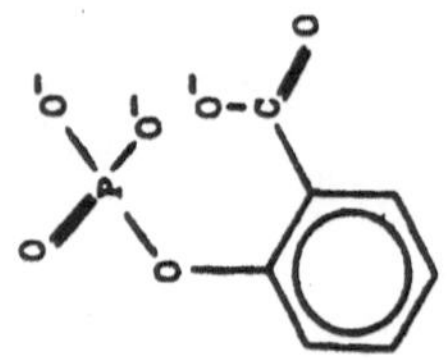

III

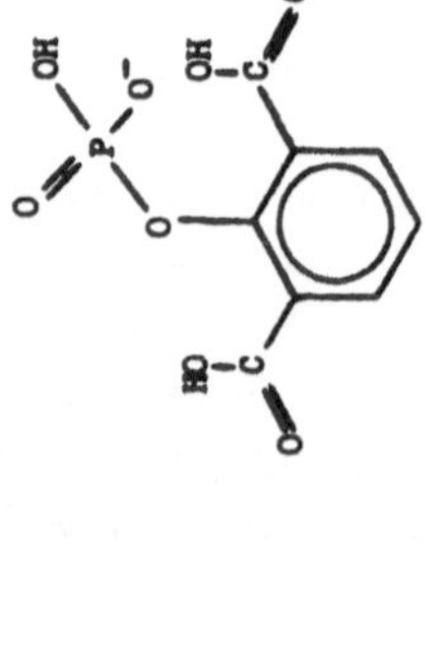

IV

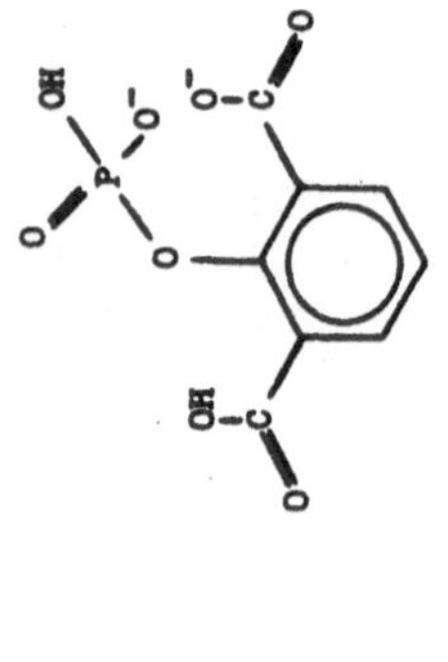

V

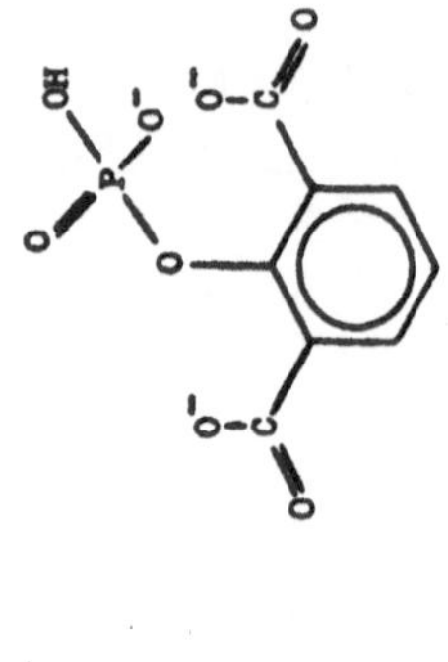

VI

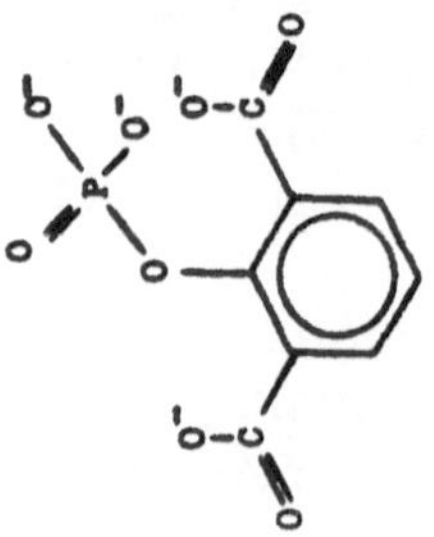

VII

ester group. The greatest increase (from [II] to [IV]) is interpreted as being due to the concerted influence of nucleophilic attack by an adjacent negative carboxylate group, and hydrogen bonding by an adjacent carboxyl group.

The fact that divalent metal ions, the vanadyl ion, and their chelates, do not catalyze the hydrolysis forms [I] and [IV] to an appreciable extent is in accord with the proposed mechanism, since coordination of the phosphate group by the metal ion would be prevented or greatly reduced by the presence of two protons. Accordingly metal ion catalysis by Cu^{2+} and VO^{2+} ions increases in effectiveness as the number of protons on the substrate is successively reduced. Such behavior would not be expected if transfer of a hydrogen-bonded proton from the carboxyl group to the phenolic ester oxygen were the only pathway for the reaction. Metal ion catalysis of the hydrolysis of [V], [VI], and [VII] was not measured because of the formation of a solid phase in the presence of Cu^{2+} and VO^{2+} ions.

The 1:1 dipyridyl-Cu(II) chelate was found to have no catalytic activity toward the hydrolysis of [II] and [III]. This result is interpreted as being due to the formation of a mixed ligand chelate compound in which the carboxylate group is bound to the metal ion and is thus prevented from attacking the phosphate group. A similar interpretation is offered for the failure of the 1:1 Cu(II)-N-hydroxyethylethylenediamine chelate to show catalytic activity. However, the 1:1 vanadyl complexes of 3,5-disulfopyrocatechol, 5-sulfo- -8-hydroxyquinoline, and 5-sulfosalicylic acid which do not form stable mixed ligand chelates with salicyl phosphate, were found to have considerable catalytic activity.

The catalytic activity observed for the 1:1 Cu(II)-dipyridyl chelate on the hydrolysis of [V] is in accord with the interpretation given for its lack of activity with respect to the hydrolysis of [II]. Combination of Cu(II)-dipyridyl with [V] to form a mixed ligand chelate would not prevent, and should accelerate, the attack of the phosphorus atom by the free carboxylate group. Further, the catalytic effect is seen to increase as the number of protons bound to the substrate is reduced ([VI] and [VII], Table I).

These results indicate that the catalytic activity of a metal chelate is very dependent on the number and arrangement of the electron donor groups of the substrate, since combination with the substrate can take place in such a manner as to interfere with the freedom of the functional groups of the ligand to form a reactive intermediate or a transition state complex. This type of interaction could also occur with metal ions, but would lead to the formation of a complex that would itself have catalytic activity because of the presence of unsubstituted positions in the coordination sphere of the metal ion.

TABLE I

SECOND ORDER RATE CONSTANTS FOR METAL ION AND METAL CHELATE CATALYSIS
OF SALICYL PHOSPHATE AND 1,3-DICARBOXYPHENYL-2-PHOSPHATE

Catalyst	Specific Rate Constants, $M^{-1} sec^{-1}$						
	I	II	III	IV	V	VI	VII
None[a]	$1.3 \cdot 10^{-6}$	$7.6 \cdot 10^{-6}$	~0	$3.1 \cdot 10^{-6}$	$2.3 \cdot 10^{-9}$	$1.0 \cdot 10^{-4}$	~0
Cu[2+]	~0	$4.0 \cdot 10^{-2}$	~3.7				
VO[2+]	~0	$2.1 \cdot 10^{-1}$					
Cu-Dipy[b]	0	~0	~0	~0	active	14[c]	11[c]
Cu-Hen[d]	0	~0	~0				
VO-Tiron[e]	~0	$7.0 \cdot 10^{-2}$					
VO-SSA[f]	~0	$6.0 \cdot 10^{-2}$					
VO-Hqs[g]	~0	$6.1 \cdot 10^{-2}$					

[a] First order constants, sec^{-1}, reported for ligand alone; [b] Cu-Dipy is the 1:1 Cu(II)-dipyridyl chelate; [c] catalyst is Cu(OH)Dipy$^+$; [d] Cu-Hen is the 1:1 Cu(II) chelate of N-hydroxyethylethylenediamine; [e] VO-Tiron is the 1:1 VO(IV) chelate of 3,5-disulfopyrocatechol; [f] VO-SSA is the 1:1 VO(IV) chelate of 5-sulfosalicylic acid; [g] VO-Hqs is the 1:1 VO(IV) chelate of 5 sulfo-8-hydroxyquinoline.

References

1. MARTELL, A.E., "Advances in Chemistry Series", No.37, pp.161-173 (1963).
2. MURAKAMI, Y., and MARTELL, A.E., J.Phys.Chem., 67, 582 (1963).
3. MURAKAMI, Y., and MARTELL, A.E., J.Am.Chem.Soc., 86, in press.
4. MONT, G., and MARTELL, A.E., J.Am.Chem.Soc., to be submitted.
5. CHANLEY, J.D., footnote 24a in BENDER, M.L., and LAWLER, J.M., J.Am.Chem.Soc., 85, 3010 (1963).
6. MARTELL, A.E., MURAKAMI, Y., and MONT, G., J.Am.Chem.Soc., in press.
7. BENDER, M.L., and LAWLER, J.M., J.Am.Chem.Soc., 85, 3010 (1963).

C 2

CATALYTIC HYDROGENATION BY PENTACYANOCOBALTATE(II):
π- AND σ- ALLYLIC COMPLEX INTERMEDIATES.

Jack Kwiatek and Jay K. Seyler
U.S. Industrial Chemicals Co., Division of National Distillers
and Chemical Corp., Cincinnati, Ohio, U.S.A.

The catalytid hydrogenation of a variety of organic substrates may be effected by pentacyanocobaltate(II) anion in aqueous solution (1). A general requirement for the reduction of the carbon-carbon double bond is the presence of conjugation; thus, conjugated dienes are reduced to the monoolefin stage. A further requirement appears to be the ability or the diene to attain a "cisoid" conformation (2). For example, 2,5-dimethyl-2,4-hexadiene, in which the cisoid conformation is sterically hindered, is not reduced, while bicyclo-[2,2,1]heptadiene, in which a cisoid conformation is fixed, is reduced, although its resonance stabilization in the ground state is considered to be negligible (3).

The isomer distribution of the butene product obtained from butadiene is dependent on the cyanide-to-cobalt ratio employed in formation of the catalyst (4). Thus, at low ratios, as much as 86% trans-butene-2 has been obtained, while at high ratios, 87% butene-1 has been found, the greatest change in isomer distribution occurring between cyanide-to-cobalt ratios of 5,5 and 6,0. This effect has been ascribed (4) to the possible intermediacy of a 1-methyl-π-allyl cyanocobaltate complex in reductions carried out at low cyanide-to-cobalt ratios, and a sigma-bonded methylallyl cyanocobaltate complex at high ratios. Evidence for such intermediates, as well as the relationship of diene and allylic halide reductions, is now presented.

Mechanism of Butadiene Reduction

The catalytic hydrogenation of butadiene apparently occurs in the following stepwise manner (4):

$$2\,[Co(CN)_5]^{3-} + H_2 \;\rightleftharpoons\; 2\,[Co(CN)_5H]^{3-} \qquad \{1\}$$

$$[Co(CN)_5H]^{3-} + C_4H_6 \;\rightleftharpoons\; [Co(CN)_5(C_4H_7)]^{3-} \qquad \{2\}$$

$$[Co(CN)_5H]^{3-} + [Co(CN)_5(C_4H_7)]^{3-} \longrightarrow 2\,[Co(CN)_5]^{3-} + C_4H_8 \;\{3\}$$

Evidence for the formation of the hydrido complex has been presented by other workers (5,6), although a NMR study (7) revealed the presence of only a minor percentage of metal-bonded hydrogen. Our data support the intermediacy of the hydrido complex in the hydrogenation of dienes.

Absorption of butadiene by a hydrogenated solution of pentycyano-cobaltate(II) is rapid, one mole equivalent being taken up for each atom equivalent of hydrogen preabsorbed. The solution so formed gradually evolves an approximately equimolar mixture of butadiene and isomeric butenes when places in an inert atmosphere. Reversibility of {2} is further demonstrated by adding deuterium to butadiene and employing deuterium oxide as the solvent. The predominant product, butene-1, was 66% di-

deuterated and 15% each mono- and trideuterated, the remainder being
divided between non-, tetra- and pentadeuterated species. Since the pure
butene isomers were shown not to exchange under the conditions of the
reaction, reversibility of {2} is established.

The relatively slow rates of this reverse reaction and the final cleavage
step {3} allow isolation of the solid butenylcyanocobaltate salt. The presence
of the butenyl group in this salt is indicated by its PMR spectrum (D_2O solvent),
butene-1 evolution upon acid treatment, mixed butene (composition of the
butenes is dependent on the cyanide-to-cobalt ratio employed as previously
noted) evolution upon treatment with hydrido complex and spontaneous decomposition of aqueous solutions of the salt to yield approximately equal portions
of butadiene and butenes.

Mechanism of Allylic Halide Reductions

Hydrogenolysis of allylic compounds was rather general, including
halides, acetates, alcohols, ethers and amines.

A pathway for the catalytic reduction of allyl halides is formulated as
involving homolytic cleavage of the substrate by pentacyanocobaltate(II),
with subsequent cleavage of the allyl complex so formed by hybrido complex:

$$2\,[Co(CN)_5]^{3-} + CH_2{=}CH{-}CH_2X \longrightarrow [Co(CN)_5(C_3H_5)]^{3-} + [Co(CN)_5X]^{3-} \qquad \{4\}$$

$$[Co(CN)_5H]^{3-} + [Co(CN)_5(C_3H_5)]^{3-} \longrightarrow 2\,[Co(CN)_5]^{3-} + C_3H_6 \qquad \{5\}$$

The release of one molar equivalent of propylene for each two equivalents
of pentacyanocobaltate(II) present, upon acidification of a reaction mixture
containing the complex ion and an excess of allyl iodide, provides evidence
for the stoichiometry shown in {4}. The PMR spectrum of the isolated mixture
of complexes (D_2O solvent) clearly shows the presence of a σ-bonded allyl
group which gradually rearranges to a π-allyl group. The σ-bonded group
is immediately reformed upon addition of cyanide ion, and propylene is
rapidly evolved upon addition of hydrido complex.

$$[(CH_2{=}CH{-}CH_2)Co(CN)_5]^{3-} \rightleftharpoons [(CH_2{\cdots}CH{=\!=}CH_2)Co(CN)_4]^{2-} + CN^- \qquad \{6\}$$

The reduction of vinylacetyl chloride also yields propylene, presumably
via decarbonylation of a butenoyl complex (8).

There is evidence for the reduction of butenyl halides in a similar manner.

The complex isolated from the reaction of crotyl bromide with pentacyano-
cobaltate(II) undergoes the same reactions with hydrido complex and acid as
that isolated from butadiene {2}. Reduction of α- and γ-methylallyl chlorides
yields the same distribution of butene isomers as obtained from butadiene.
Treatment of these isomeric chlorides with pentacyanocobaltate(II) in the
absence of hydrogen yields equal quantities of butadiene and butenes.

At high cyanide concentrations, the various σ-bonded butenyl complexes
formed might be expected to equilibrate rapidly via internal ion-pair complexes, while at low concentrations equilibration may take place via π-bonded
complexes.

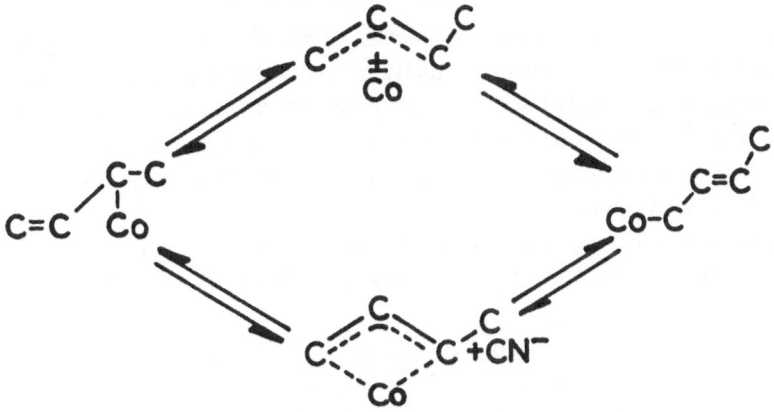

Fig. 1 - Intermediates in the reduction of butadiene
and methylallyl halides.

Nature of the Intermediate Complexes

The PMR spectra of the σ- and π-allyl complexes correspond very well
with the spectra of the corresponding manganese and cobalt carbonyl complexes (9). Although the exact location of the π-allyl group with respect to
the metal is not known, the reaction with cyanide ion indicates that the
π-allyl group may be considered to be bidentate, a conclusion in full accord
with the displacement of carbon monoxide in the conversion of σ-to-π-allyl
cobalt and manganese carbonyls (9), and with the coordination of dimethyl-
sulfoxide in the conversion of π-to-σ-allyl palladium chloride (10). Structure(I)
is tentatively proposed for the π-allyl cyanocobaltate complex.

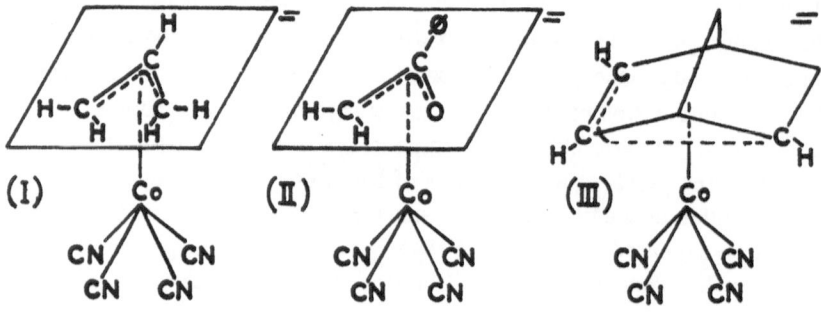

Fig.2 - Types of π-allylic cyanocobaltates.

The complex obtained from the reaction of phenacyl bromide with penta-cyanocobaltate(II) is assigned structure(II), a π-oxaallyl type (11), on the basis of absence of carbonyl absorption (IR) and vinylidene hydrogen (NMR); acidification releases acetophenone. Evidence is being sought for the possible intermediacy of a π-homoallyl complex (structure III) in the reduction of bicyclo[2,2,1]heptadiene.

This research was supported by the Directorate of Chemical Sciences, Air Force Office of Scientific Research, under Contract Nr. AF 49(638)-1214.

References

1. KWIATEK, J., MADOR, I.L., and SEYLER, J.K., J.Am.Chem.Soc., 84, 304 (1962).
2. To be published.
3. TURNER, R.B., MEADOR, W.R., and WINKLER, R.E., J.Am.Chem. Soc., 79, 4116 (1957).
4. KWIATEK, J., MADOR, I.L., and SEYLER, J.K., Advances in Chemistry Series, No.37, pp.201-15, ACS (1963).
5. KING, N.K., and WINFIELD, M.E., J.Am.Chem.Soc., 83, 3366 (1961).
6. deVRIES, B., Koninkl.Ned.Akad.Wetenschap., Proc.Sect.B, 63, 443 (1960); J.Catalysis, 1, 489 (1962).
7. GRIFFITH, W.P., and WILKINSON, G., J.Chem.Soc. (London), 1959, 2757.
8. HECK, R.F., and BRESLOW, D.S., J.Am.Chem.Soc., 82, 750 (1960).
9. McCLELLAN, W.R., HOEHN, H.H., CRIPPS, H.N., MUETTERTIES, E.L., and HOWK, B.W., J.Am.Chem.Soc., 83, 1601 (1961).
10. CHIEN, J.C.W., and DEHM, H.C., Chem.and Ind., 1961, 745.
11. GOETZ, R.W., and ORCHIN, M., J.Am.Chem.Soc., 85, 2782 (1963).

KINETIC STUDIES OF SOME Rh(III) COMPLEXES.

E.J. Bounsall and A.J. Poë
Imperial College, London, U.K.

(i) The trans-effect.

The rates of substitution reactions of trans-$[Rhen_2Cl_2]^+$ are first-order in complex concentration and independent of the nature and concentration of the substituting group. They also proceed without stereochemical rearrangement (1). These reactions appear, therefore, to be free of the difficulties which complicate studies of Co(III) and Pt(IV) complexes and were accordingly chosen as suitable for systematic investigation of the trans-effect in octahedral complexes.

Interchange reactions of one halide ligand by another are, in general, easily followed by spectrophotometry and, unlike many acid hydrolyses, they have the advantage of going to completion. Studies of the kinetics of the reactions {1} give a relative measure of the trans-effect of X on the lability of the Rh(III)-Y bond.

$$\text{trans-}[Rhen_2XY]^+ + Z^- \longrightarrow \text{trans-}[Rhen_2XZ]^+ + Y^- \quad \{1\}$$

Data which show the effect of changing X and Y are given in Table I.

TABLE I

KINETIC DATA FOR {1}.

X :	Cl[a]	Br[b]	OH[c]	NO$_2$[c]	I[a]	Cl[b]	Br[a]	I[a]	OH[c]	I[a]
Y :	Cl	Cl	Cl	Cl	Cl	Br	Br	Br	I	I
$10^5 k_{70}^0$ (sec^{-1}) :	1,4	4,2	50		295	0,83	6,1	193	80	100
$10^5 k_{80}^0$ (sec^{-1}) :	4,2			410						
$\Delta H^{\ddagger}$ (kcal/mole) :	23,8	21,3			14,0	18,9	20,0	16,1	25,8	26,2

a) BASOLO, BOUNSALL and POË, Proc.Chem.Soc., 1963, 366;
b) BOTT and POË, unpublished observations;
c) BASOLO and KLABUNDE, personal communication.

When Rh(III)-Cl is the bond affected, the trans-effect of Br$^-$ and I$^-$ relative to Cl$^-$ is shown by an increase in $k_{70}0$ by factors of 6 and 420, respectively, the statistical effect having been allowed for. $\Delta H^{\ddagger}$ is reduced by 2,5 and 9,8 kcal/mole, respecitively. When Rh(III)-Br is being broken, trans-Br$^-$ and trans-I$^-$ increase $k_{70}0$ by factors of 4 and 230, and decrease

$\Delta H^{\ddagger}$ by -1,3 and 2,6 kcal/mole, respectively, compared with trans-Cl$^-$. Thus, when $\Delta H^{\ddagger}$ values are considered, the trans-effect on Rh(III)-Br is qualitatively, as well as quantitatively, different from that on Rh(III)-Cl. When measured by values of k_{700}, the trans-effect of I$^-$ relative to OH$^-$ is much greater on Rh(III)-Cl than on Rh(III)-I. It appears, therefore, that the trans-effect decreases, and may even change order, as the strength of the affected bond decreases. The fact that the relative trans-effect of the halides on Pt(IV)-Cl is much smaller (2) than that on Rh(III)-Cl, in spite of the presumed greater strength of the Pt(IV)-Cl bond, shows that the effect of oxidation state (and therefore, probably, of polarisability) of the metal ion plays an even greater part in determining the trans-effect than does the bond strength.

The data also show that the Rh(III)-Y bond becomes relatively easier to break along the series I < Br < Cl when an iodide is in the trans-position. This suggests that the effect of incipient solvation of the leaving ligand outweighs the bond-strength effects and these complexes can be considered as kinetically of class (b) type (3). When a chloride is in the trans-position $\Delta H^{\ddagger}$ values show bromide to be more easily removed than chloride so bond-strength effects are here more important than solvation effects and the complexes are kinetically of class (a).

(ii) Nucleophilic character of halides towards Rh(III).

Since the rates of the replacement reactions studied are independent of the nucleophilic ligands they must proceed via a rate-determining step to a reactive intermediate. This may be the five-coordinate [Rhen$_2$X]$^{2+}$ (S_N1 dissociative mechanism) or the six-coordinate [Rhen$_2$XH$_2$O]$^{2+}$ (S_N2 solvolytic mechanism). Since the reactions are reversible, they can be written as in {2} where $k_{+Z} \gg k_{-Y}$ and $k_{+Y} \gg k_{-Z}$.

$$[\text{Rhen}_2XY]^+ + Z^- \underset{k_{+Y}}{\overset{k_{-Y}}{\rightleftharpoons}} [\text{Rhen}_2X]^{2+}$$

or

$$[\text{Rhen}_2XH_2O]^{2+} + Z^- + Y^- \underset{k_{-Z}}{\overset{k_{+Z}}{\rightleftharpoons}} [\text{Rhen}_2XZ]^+ + Y^- \qquad \{2\}$$

The equilibrium constant for the interchange equilibrium is given by $K = k_{-Y} k_{+Z}/k_{-Z} k_{+Y}$. Independent measurements of K and k_{-Y}/k_{-Z} allow calculation of k_{+Z}/k_{+Y}, i.e. the relative rate constants for nucleophilic attack by X^- and Y^- on the reactive intermediate. Such measurements have been made for $X = I^-$, $Y = Cl^-$ or Br^-, and $Z = I^-$ and the results are given in Table II. Although the relative nucleophilic characters of the three halides are similar when based on rate constants, $\Delta H^{\ddagger}$ values are in the order $I^- \ll Br^- \sim Cl^-$.

TABLE II

KINETIC AND THERMODYNAMIC DATA FOR THE EQUILIBRIUM:

$$[Rhen_2IY]^+ + I^- \rightleftharpoons [Rhen_2I_2]^+ + Y^-$$

Y	K_{50^0}	k_{-Y}/k_{-I}	k_{+I}/k_{+Y}	ΔH^0	$\Delta H^{\ddagger}_{-Y} - \Delta H^{\ddagger}_{-I}$	$\Delta H^{\ddagger}_{+I} - \Delta H^{\ddagger}_{+Y}$
Cl	5,72	5,9	0,97	-6,2	-12,2	6,0
Br	3,22	3,6	0,90	-3,8	-10,1	6,3

Rate constant data for relative nucleophilic character towards five-coordinate $[Co(CN)_5]^{2-}$ (4) and $[Co(NH_3)_5]^{3+}$ (5), and towards four-coordinate $[PtdienCl]^+$ (6), have shown $I^- > Br^- \gtrsim Cl^-$ but $\Delta H^{\ddagger}$ values have not been obtained. The fact that, in our systems, the nucleophilic character of I^- is relatively less on the basis of $\Delta H^{\ddagger}$ values shows that bond-making must play a major part in the process since, if solvation effects were paramount, the relative nucleophilic character should be greater in terms of $\Delta H^{\ddagger}$. Systematic studies of relative nucleophilic character towards a variety of reactive intermediates might well give information of mechanistic significance.

References

1. BASOLO, JOHNSON and PEARSON, J.Am.Chem.Soc., 85, 1741 (1963).
2. ZVYAGINTSEV and SHUBOCHKINA, Russ.J.Inorg.Chem., 8, 300 (1963).
3. AHRLAND, CHATT and DAVIES, Quart.Rev., 12, 265 (1958).
4. GRASSI and WILMARTH, Proc. 7 ICCC, p.242.
5. HAIM and TAUBE, Inorg.Chem., 2, 1199 (1963).
6. GRAY, J.Am.Chem.Soc., 84, 1548 (1962).

5 C 1

HARD AND SOFT ACIDS AND BASES.

Ralph G. Pearson

Department of Chemistry, Northwestern University, Evanston, Illinois, U.S.A.

Recently (1) the rate data for the generalized nucleophilic displacement reaction were reviewed.

$$N + S\text{-}X \longrightarrow N\text{-}S + X \qquad \{1\}$$

Here N is a nucleophilic reagent (ligand, Lewis base) and S-X is a substrate containing a replaceable group X (also a base) and an electrophilic atom (Lewis acid) S. It was found that rates for certain substrates, S-X, were influenced chiefly by the basicity (toward the proton) of N, and other substrates had rates which depended chiefly on the polarizability (reducing power, degree of unsaturation) of N.

In this paper the equilibrium constants of {1} will be considered, instead of the rates.

$$N \text{ (base)} + S\text{-}X \text{ (acid-base complex)} \rightleftharpoons N\text{-}S \text{ (acid base complex)} + X \text{ (base)}$$

$$\{2\}$$

Thus the relative strengths of a series of bases, N, will be compared for various acids, S. The nature of N-S may be that of a stable organic or inorganic molecule, a complex ion, or a charge transfer complex.

In terms of equilibria, rather than rates, it again turns out that various substrate acids fall into two categories: those that bind strongly to bases which bind strongly to the proton, that is, basic in the usual sense; those that bind strongly to highly polarizable or unsaturated bases, which often have negligible proton basicity. It will be convenient to divide bases into two categories, those that are polarizable, or "soft", and those that are nonpolarizable, or "hard". Now it is possible for a base to be both soft and strongly binding toward the proton, for example, sulfide ion. Still it will be true that hardness is associated with good proton binding.

For the special case of metal ions as acids, AHRLAND, CHATT and DAVIES (2) made a very important and useful classification. All metal ions were divided into two classes depending on whether they formed their most stable complexes with the first ligand atom of each group, class (a), or whether they formed their most stable complexes with the second or a subsequent member of each group, class (b).

Table I contains a listing of all generalized acids for which sufficient information could be found in the literature to enable a choice between class (a) or class (b) to be made (3). In classifying Lewis acids, the criterion of AHRLAND, CHATT and DAVIES was used whenever possible. When this was not possible, two other criteria were used. One is that class (b) acids will complex readily with a variety of soft bases that are of negligible proton basicity. These include CO, olefins, aromatic hydrocarbons, and the like.

The other auxiliary criterion is that if a given acid depends strorgly on basicity and little on polarizability as fas as rates of nucleophilic displacements are concerned, then it will depend even less on polarizability as far as equilibrium binding to bases is concerned. Such an acid will therefore be in class (a).

TABLE I

CLASSIFICATION OF LEWIS ACIDS

Class (a) or hard

H^+, Li^+, Na^+, K^+
Be^{2+}, Mg^{2+}, Ca^{2+}, Sr^{2+}, Sn^{2+}
Al^{3+}, Sc^{3+}, Ga^{3+}, In^{3+}, La^{3+}
Cr^{3+}, Co^{3+}, Fe^{3+}, As^{3+}, Ir^{3+}
Si^{4+}, Ti^{4+}, Zr^{4+}, Th^{4+}, Pu^{4+},
  VO^{2+}
UO_2^{2+}, $(CH_3)_2Sn^{2+}$
$BeMe_2$, BF_3, BCl_3, $B(OR)_3$
$Al(CH_3)_3$, $Ga(CH_3)_3$,
  $In(CH_3)_3$
RPO_2^+, $ROPO_2^+$
RSO_2^+, $ROSO_2^+$, SO_3
I^{7+}, I^{5+}, Cl^{7+}
R_3C^+, RCO^+, CO_2, NC^+
HX (hydrogen bonding molecules)

Class (b) or soft

Cu^+, Ag^+, Au^+, Tl^+, Hg^+,
  Cs^+
Pd^{2+}, Cd^{2+}, Pt^{2+}, Hg^{2+},
  CH_3Hg^+
Tl^{3+}, $Tl(CH_3)_3$, BH_3
RS^+, RSe^+, RTe^+
I^+, Br^+, HO^+, RO^+
I_2, Br_2, ICN, etc.
Trinitrobenzene, etc.
Chloranil, quinones, etc.
Tetracyanoethylene, etc.
O, Cl, Br, I, R_3C (?)
M^0 (metal atoms)
Bulk metals

Borderline

Fe^{2+}, Co^{2+}, Ni^{2+}, Cu^{2+},
  Zn^{2+}, Pb^{2+},
$B(CH_3)_3$, SO_2, NO^+

The features which bring out class (a) behavior are small size and high positive oxidation state. Class (b) behavior is associated with a low or zero oxidation state and/or with large size. Both metals and nonmetals can be either (a) oder (b) type acids depending on their charge and size. Since the features which promote class (a) behavior are those which lead to low polarizability, and those which create type (b) behavior lead to high polarizibility, it is convenient to call class (a) acids "hard" acids and class (b) acids "soft" acids. We then have the useful generalization that hard acids prefer to associate with hard bases, and soft acids prefer soft bases.

Polarizability is simply a convenient property to use as a classification. It may well be that other properties which are roughly proportional to polarizability are more responsible for the typical behavior of the two classes of acids. For example, a low ionization potential is usually linked to a large polarizability and a high ionization potential to a low polarizability. Hence ionization potential, or the related electronegativity, might be the important property. Unsaturation, with the possibility of acceptor π-bonding in the acid-base complex, and ease of reduction, favoring strong electron transfer to the acid, are also associated with high polarizability. Different investigators, looking at different aspects, have come up with several explanations for the data of Table I (3).

1. The Ionic-Covalent Theory. - The class (a) acids are assumed to bind bases with primarily ionic forces and the class (b) acids hold bases by covalent bonds. High positive charge and small size would favor strong ionic bonding and bases of large negative charge and small size would be held most strongly. According to MULLIKEN bonding will be strong if the electron affinity of the acid is large and the ionization potential of the base is low. Softness in both the acid and base means that the repulsive part of the potential energy curve rises less sharply than for hard acids and bases. Thus closer approach is possible and better overlap of the wave functions used in covalent bonding.

In the theory of covalent bonding the coulomb integrals on both bonded atoms should be similar, and the sizes of the bonding atomic orbitals should be similar to get good overlap. These considerations show that hard acids will prefer hard bases even when considerable covalency exists. Soft bases will mismatch with hard acids for good covalency, and ionic bonding will also be weak because of the small charge or large size of the base.

2. The π-Bonding Theory. - According to CHATT the important feature of class (b) acids is considered to be the presence of loosely held outer d-orbital electrons which can form π-bonds by donation to suitable ligands. Such ligands would be those in which empty d-orbitals are available on the basic atom, such as P, As, S, I. Also unsaturated ligands such as CO and isonitriles would be able to accept metal electrons by the use of empty, but not too unstable, molecular orbitals. Class (a) acids would have tightly held outer electrons, but also there would be empty orbitals available, not too high in energy, on the metal ion. Basic atoms such as O and F particularly could form π-bonds in the opposite sense, by donating electrons from the ligand to the empty orbitals of the metal. With class (b) acids, there would be a repulsive interaction between the two sets of filled orbitals on metal and O and F ligands.

3. Electron Correlation Effects. - PITZER has suggested that London, or van der Waals, dispersion forces between atoms or groups in the same molecule may lead to an appreciable stabilization of the molecule. Such London forces are large when both groups are highly polarizable. It seems plausible to generalize and state that additional stability due to London forces will always exist in a complex formed between a polarizable acid and a polarizable base. In this way the affinity of soft acids for soft bases can be accounted for.

MULLIKEN has given a different explanation for the extra stability of the bonds between large atoms; for example, two bromine atoms. It as assumed that d_π-p_π orbital hybridization occurs so that both the π_u bonding molecular orbitals and π_g antibonding orbitals contain some admixed d-character. This has the twofold effect of strengthening the bonding orbital by increasing overlap and weakening the antibonding orbital by decreasing overlap.

There is considerable similarity to the proposals of London forces and of orbital hybridization. They both represent electron correlation phenomena. The basic cause is different in the two cases, however. London correlation occurs because of the electrostatic repulsion of electrons for each other. The proposed π-orbital hybridization occurs largely because of non-bonded repulsion effects arising from the Pauli exclusion principle. It would appear that the latter would be more important for interactions between bonded atoms and the former for more remote interactions.

MULLIKEN's theory is the same as CHATT's π-bonding theory as far as the π_u bonding orbital is concerned. The new feature is the stabilization due to the π_g molecular orbital. As MULLIKEN points out this effect can be more important than the more usual π-bonding. The reason is that the antibonding orbital is more antibonding than the bonding orbital is bonding, if overlap is included. For soft-soft systems, where there is considerable mutual penetration of charge clouds, this amelioration of repulsion due to the Pauli principle would be great.

4. The Solvation Theory. - Several workers have recently stressed the effect of solvents on reducing the basicity of small anions and hence causing large anions to appear abnormally strong. The implication that such solvation is the common explanation for the strong binding, or high rates of reaction, of polarizable bases seems to be incorrect. For one thing, differences in solvation energies between neutral molecules such as ROH and RSH are very much smaller than the differences for the corresponding anions. What solvation does do is to generally destroy class (a) character and enhance class (b) character. It does not explain why some acids have a strong preference for hard bases, and others for weak bases.

References

1. EDWARDS, J.O., and PEARSON, R.G., J.Am.Chem.Soc.,84, 16 (1962).
2. AHRLAND, S., CHATT, J., and DAVIES, N.R., Quart.Rev. (London), 12, 265 (1958).
3. PEARSON, R.G., J.Am.Chem.Soc.,85, 3533 (1963).

5 C 2

INTERACTION OF LEWIS ACIDS WITH TRANSITION METAL COMPLEXES.

D.F. Shriver, A. Luntz and J.J. Rupp
Department of Chemistry, Northwestern University,
Evanston, Illinois, U.S.A.

Previous work has shown that metal complexes in the presence of Lewis acids may lead to: (i) simple addition of the Lewis acid to unshared electron pairs of coordinated ligands (3), (ii) Lewis acid addition to unshared electron pairs of a metal (2, 4) and (iii) more complex reactions (1). This report covers some of our recent experiments designed to explore (i) - simple ligand basicity.

Bridge Addition Compounds of $Fe(phen)_2(CN)_2$.

Dicyano-bis(1,10-phenanthrene)iron(II) displays a visible charge transfer spectrum which is strongly dependent upon the hydrogen bonding propensity of the solvent in which it is dissolved (5, 6). This interesting spectral behavior and solubility of the complex in convenient solvents prompted us to investigate its interaction with a variety of Lewis acids. Direct reaction of solid $Fe(phen)_2(CN)_2$ with the appropriate gaseous Lewis acid led to formation of a series of adducts with the general formula $Fephen_2(CNBX_3)_2$ where X = H, CH_3, F, Cl, and Br (7). The composition of each was established by P, V, T measurement of initial and unused Lewis acid and by weight gain of the solid phase. These gas solid reactions do not occur in a stepwise manner and resulting addition compounds do not display appreciable Lewis acid dissociation pressures. However, the possibility of forming an intermediate mono-adduct, $Fe(phen)_2(CN)(CNBF_3)$, was demonstrated by a "continuous variations" study in s-tetrachloroethane.

Addition of BX_3 to $Fe(phen)_2(CN)_2$ results in an increase in the CN stretching frequencies. These frequency changes are symptomatic of CN bridges (3), expected in the formation of $Fe(phen)_2(CNBX_3)_2$. Hydrolysis of the bridge adducts (where X $\neq CH_3$) regenerates dicyano-bis(1,10-phenanthrene)iron(II). Colors of the adducts range from dark red when X = H to a bright orange-yellow when X = Cl or Br. A pair of fairly strong ($6\,000 \lesssim \epsilon \lesssim 10\,000$) absorption bands are responsible for the color and the hypsochromic shift for $Fe(phen)_2(CNBX_3)_2$ follows the series $BH_3 < B(CH_3)_3 < BF_3 < BCl_3 = BBr_3$. Except for the first two members this trend parallels the order of acidity toward nitrogen containing bases. A parallel also exists between the dipole moments of trimethylamino-BX_3 and the corresponding $Fe(phen)_2(CNBX_3)_2$ visible absorption.

Group four acids also add to $Fe(phen)_2(CN)_2$ and the compounds $Fe(phen)_2(CN)_2 \cdot MCl_4$ (M = Si, Ge or Sn) have been isolated. Infrared spectra indicate both CN ligands are in bridging environments so it is probable that the group four element is six coordinated in these adducts. A blue shift is again observed and follows the order Si $\sim$ Ge < Sn which roughly parallels the generally recognized order of acidity.

Addition of SO_2, HCl and HBr to $Fe(phen)_2(CN)_2$ has also been investigated. The latter pair of acids lead to $Fe(phen)_2(CN)_2(HX)_2$ and dissociation pressures are greatest when X = Cl. In harmony with our previous findings the hypsochromic shift is least when X = Cl. Gaseous SO_2 is absorbed by solid $Fe(phen)_2(CN)_2$ to the extent of ca. 0,6 SO_2 per complex. Infrared spectra, visible spectra and entropy of dissociation all indicate that the SO_2 is probably not bound by a CN bridge, as was the case for other acids.

New Bridging Groups.

Attempts to add gaseous BF_3 to solid $Fe_2(CO)_9$, $Fe(CO)_3[P(C_6H_5)_3]_2$ and $K_2Fe(CN)_5NO$ give no indication of $Fe-C\equiv O-BF_3$ or $Fe-N=O-BF_3$ bridges. These observations are not surprising since there is little precident for the basicity of oxygen in the parent compounds. Reaction of BF_3 with the complexes $K_3Cr(NCS)_6$ and $Pd(CNS)_2[As(C_6H_5)_3]_2$ leads to metathesis reactions and formation of $B(NCS)_3$. One simple thiocyanate bridge adduct was prepared, $Pd(bipy)(CNS)_2(BF_3)_2$. Boron trifluoride and the nitro compound $Co(NH_3)_3(NO_2)_3$ undergo a complex reaction. It is probable that this reaction is similar to the degradation of NO_2^- by BF_3 (8):

$$3\ NaNO_2 + BF_3 = 2\ NO + NaNO_3 + (Na_2OBF_3)$$

In summary, cyanide containing complexes appear to form bridge adducts with a variety of Lewis acids. In the case of $Fe(phen)_2(CN)_2$ bridge adducts a correlation exists between metal-ligand charge transfer spectra and the acidity of the acceptor molecule. Addition of BF_3 to other potentially basic complexes ($-NCS^-$, $-SCN^-$, and $-NO_2^-$ complexes) frequently leads to abstraction or distruction of the ligand, but one simple thiocyanate bridge adduct has been prepared (9).

References

1. SCHRAUZER, G.N., Chem.Ber. 95, 1438 (1962); THIERIG, D., and UMLAND, F., Angew.Chem. 74, 388 (1962).
2. PARSHALL, G.W., J.Am.Chem.Soc. 86, 361 (1964).
3. SHRIVER, D.F., J.Am.Chem.Soc. 84, 4610 (1962); ibid. 85, 1405 (1963).
4. SHRIVER, D.F., J.Am.Chem.Soc. 85, 3509 (1963).
5. BJERRUM, J., ADAMSON, A.W., and BOLSTRUP, O., Acta.Chem. Scand. 10, 329 (1956).
6. SCHILT, A.A., J.Am.Chem.Soc. 82, 3000 (1960).
7. The compounds in which X = H or F have been reported previously in ref. 3b.
8. We have not yet characterized the compound in parenthesis.
9. This work was supported by the N.S.F..and by the Advanced Research Projects Agency of the Department of Defense through the Northwestern Materials Research Center.

5 C 3

THE NATURE OF SOLUTIONS OF TRANSITION METAL SALTS OF VERY STRONG ACIDS IN NON-AQUEOUS SOLVENTS.

M.J. Baillie, D.H. Brown, K.C. Moss and D.W.A. Sharp
Chemistry Department, The Royal College of Science and
Technology, Glasgow, C.1., U.K.

Transition metal salts of strong acids have been prepared in solution in diethyl ether, nitromethane, and benzene by the reactions:

$$n \, AgR + MX_n \longrightarrow MR_n + n \, AgX\downarrow$$

and

$$n \, AgR + M \longrightarrow MR_n + n \, Ag\downarrow$$

R is the anion of the very strong acid: $[ClO_4]^-$, $[CF_3COO]^-$, $[BF_4]^-$, and $[SbF_6]^-$, X is Cl or Br, and M is the transition metal. Solutions have been prepared in the above solvents as follows.

$M[ClO_4]_n$ for M = Ti^{3+}, V^{3+}, Cr^{3+}, Mn^{2+}, Fe^{3+}, Fe^{2+}, Co^{2+}, Ni^{2+}, Cu^+, Cu^{2+}, Zn^{2+}.

$M[OOCCF_3]_n$ for M = Ti^{3+}, V^{3+}, Cr^{3+}, Mn^{2+}, Fe^{3+}, Fe^{2+}, Co^{2+}, Ni^{2+}, Cu^{2+}, Zn^{2+}.

$M[BF_4]_n$ for M $=$ Cr^{3+}, Fe^{2+}, Co^{2+}, Ni^{2+}, Cu^+, Cu^{2+}, Zn^{2+}.

$M[SbF_6]_n$ for M $=$ Co^{2+}, Cu^+, and Cu^{2+}.

Solutions of $M[BF_4]_3$ salts could not be prepared for M $=$ Ti^{3+}, V^{3+}, and Fe^{3+} and it is believed that, if formed, the salts decompose to the metal trifluorides and boron trifluoride. Apart from the limitations of this type of decomposition it appears that transition metal salts of any very strong acid can be prepared in solution in weakly basic organic solvents and solutions containing salts of the $[TiF_6]^{2-}$, $[VF_6]^-$, $[NbF_6]^-$, and $[SiF_6]^{2-}$ anions have been prepared as examples.

Solid trifluoroacetates have been prepared by removal of nitromethane and $Cu[ClO_4]_2$ and $Fe[ClO_4]_3$ were also prepared by this method. The two latter compounds are volatile in vacuo and are similar to the previously described substances. Solutions of metal fluoroborates decomposed on removal of solvent.

The parent silver salts $AgBF_4$, $AgClO_4$ (high temperature form), $AgSbF_6$, $AgNbF_6$, $AgTaF_6$, $AgVF_6$, Ag_2TiF_6, and Ag_2SiF_6 are isomorphous with corresponding potassium salts and are ionic. $AgOOCCF_3$ is similar to $AgOOCC_3F_7$ and is covalent. Infrared data on other solid transition metal trifluoroacetates are inconclusive. The Ag^+ ion (radius 1,21 Å) is much less polarising than other transition metal ions of high oxidation state (e.g. V^{3+} 0,64; Mn^{2+} 0,78, Zn^{2+} 0,72 Å).

Magnetic susceptibility measurements (mainly on diethyl ether solutions of metal perchlorates) are in agreement with octahedrally coordinated spin-free states for the transition metal ions. Where d-d transitions can be studied, the spectra show a close dependence on the anion.

TABLE I - POSITION OF MAXIMUM ABSORPTION FOR CUPRIC SALTS
$(cm^{-1}. 10^{-3})$

anion	$(C_2H_5)_2O$	C_6H_6	CH_3NO_2
$[ClO_4]^-$	12,6	12,4	13,7
$[CF_3COO]^-$	12,5	12,5	13,8
$[BF_4]^-$	10,9	-	14,3
$[SbF_6]^-$	12,9	-	-
solid $Cu[CF_3COO]_2$	12,9	$[Cu(H_2O)_6]^{2+}$	12,6

The position of absorption for the tetrafluoroborate in ether corresponds to coordination by the fluoroborate, the ligand-field exerted by the fluoro-borate plus solvent being stronger than that of the oxy-anions plus solvent.

TABLE II - LIGAND FIELD SPLITTING Δ (cm^{-1}. 10^{-3})

Ni^{2+}	$(C_2H_5)_2O$	C_6H_6	CH_3NO_2
$[ClO_4]^-$	8,32	-	8,5
$[CF_3COO]^-$	8,15	-	9,6
$[BF_4]^-$	9,0	-	10,1
$[Ni(OH_2)_6]^{2+}$		8,5	
Cr^{3+}			
$[ClO_4]^-$	16,0	-	17,1
$[CF_3COO]^-$	16,3	15,9	17,1
$[BF_4]^-$	-	16,7	-
$[Cr(OH_2)_6]^{3+}$		17,4	

Again there are strong anionic effects.

The order of ligand-field effect for the various anions is complex and is probably not obtainable from the type of data presented above as it is not known whether the anions are mono- or bi-dentate. The nephelauxetic effect as measured in these complexes is again complex. However, the results show unequivocally that the anions $[ClO_4]^-$, $[CF_3COO]^-$, and $[BF_4]^-$ derived from very strong acids interact with transition metal cations in non-aqueous solvents of relatively low basicity.

Spectroscopic results will be reported for solutions of other metals and results will be given for the effect of the addition of excess of anion to metal salts of strong acids.

This work has been supported by the U.S. Office of Naval Research.

AUTO-COORDINATION IN SOLID PHOSPHORYLCHLORIDE.

E.W. Wartenberg

Laboratorium für anorganische Chemie der Technischen Hochschule,
Stuttgart, Germany.

IR and Raman investigations on solid $POCl_3$ at temperatures between -40 to $-100°$ have shown a negative frequency shift of the PO-stretching frequency of 15 cm^{-1} on passing from the liquid to the solid state. In the IR spectrum one sharp peak was observed, whereas the Raman frequency of the liquid (1295 cm^{-1}) is split into two frequencies 1300 (w) and 1280 (st). Gaseous $POCl_3$ absorbs at 1324 cm^{-1}. The frequency shift gas/solid amounts to $\Delta \nu$ -45 cm^{-1}. Undercooled liquid $POCl_3$ ($-30°$) shows no frequency shift.

This is a considerable shift compared to shifts shown by $C=O$ frequencies in carbonyl compounds, which are rarely exceeding more than 25 cm^{-1} (1) on passing from the vapour to the solid state. That amounts to 1,5% compared to 3,4% in $POCl_3$. Even less are the effects in nonpolar molecules as CCl_4 and ethylene, where the shifts are less than 1%. Similar results as on $POCl_3$ have been obtained by GOUBEAU and BERGER on $(CH_3)_3PO$ (2). They found $\Delta \nu$ gas/solid to -68 cm^{-1}, 5,4%. The stronger effect can easily be explained by the smaller electronegativity of the methyl group in comparison to chlorine, which makes the $P=O$ bond more polar, increasing the donator strength. The shift has been explained by dipole-dipole interaction.

Since $POCl_3$ is used as a non protonic solvent for donor-acceptor reactions we will discuss the spectroscopical results from the viewpoint of a donor-acceptor reaction. This seems justified as a donor-acceptor interaction has been defined by LINDQVIST (3) as the "interaction between donor and acceptor molecules leading to the formation of an adduct molecule, which exhibits an increased coordination, and an associated gain in energy".

The coordination of the acceptor molecule can be increased either by different or like molecules as it is the case with Al_2Cl_6. One can therefore - basing on the spectroscopical results and the above argument - assume that solid $POCl_3$ consists of a molecular crystal, formed by a mechanism in which $POCl_3$ is donor and acceptor at the same time. The donor atom being oxygen, the acceptor atom phosphorus.

$$n\ Cl_3P=O \rightleftharpoons Cl_3P \overset{O}{\diagdown} PCl_3 \overset{O}{\diagdown} PCl_3 \overset{O}{\diagdown} PCl_3 \overset{O}{\diagdown} \cdots\cdots$$

Although a ring structure might also be conceivable, the splitting of ν P$=$O in 1300/1280 cm^{-1} is in favour of a chain structure where the rather weak Raman line at 1300 cm^{-1} could stem from an uncoordinated P$=$O end group.

It follows that the coordination number of phosphorus is changed from 4 to 5.

We must bear in mind that the molecular complexity increases from the vapour phase to the more condensed phases so that symmetry arguments become more complicated and less useful. Therefore we will rather base our arguments on shifts of group frequencies. Frequency shifts occur through an alteration of the electron densities around a given atom. In a wider sense one may regard donor-acceptor reactions as electron density changes as well on the donor as on the acceptor atom or molecule. Consequently frequency shifts are most useful in studying such reactions. Let us consider the donor-acceptor bond P$=\bar{O} \rightarrow$ P . We have firstly to be aware of two opposed effects on the P=O bond: a) electron withdrawal from the P$=$O bond to form the P$=\bar{O} \rightarrow$ P bridging bond leads to an increase of the relative electropositive character of the P atom. This increases the s-character of the P$=$O bond and therefore the force constant f_{PO}. Consequently the frequency should rise, an increasing frequency should furthermore result from coupling effects. b) Since the P$=$O bond is highly polar, the same electron withdrawal leads to a higher heteropolarity of the bond and therefore to a lower force constant, consequently the P$=$O frequency should fall. The latter is actually observed, and shows that the last effect is the more important one.

The next consideration will be given to the P-Cl bond. This bond is less polar than the P$=$O bond and in this case a bond strengthening should follow the electron withdrawal, which would result in positive shifts of the P-Cl valence frequencies. Whereas this has been observed in all MeCl$_3$.POCl$_3$ adducts (4) of the group III elements, no such shift was observed in solid POCl$_3$. This is understandable, because the molecule is a donor and acceptor at the same time, compensating the electron withdrawal from the P atom through the electrons from the donor oxygen of another POCl$_3$ molecule. In addition to the one PO valence frequency in POCl$_3$ a second one should be observed after bridging. Because the donor-acceptor bond is not very strong this frequency must be found in the far infrared region.

A resulting donor-acceptor bond must not always be accompanied by large frequency shifts. An example is the addition compound $TlCl_3(POCl_3)_{2\frac{1}{2}}$ (5). In this case $\Delta\nu$ P=O has the same value as in solid $POCl_3$, namely 15 cm^{-1}. It is a crystalline compound which melts slightly above room temperature, looses $POCl_3$ in CCl_4 or in a vacuum, where pure $TlCl_3$ is obtained. The reported addition compound $AlCl_3(POCl_3)_6$ (6) which is of similar instability might be formulated as $AlCl_3[(POCl_3)_2]_3$ with auto-coordinated phosphorylchloride molecules.

From the comparison $POCl_3$ gas/liquid $\Delta\nu$ P=O = -30 cm^{-1} it is evident that to a certain extent already in the liquid state $P=\bar{O}\rightarrow P$ bridging is present, changing the coordination sphere from 4 to 5. The importance of coordination processes in solvents has already been pointed out by GUTMANN (7), he states that "... für den Aufbau der Lösungen die Koordinationsfähigkeit der Solvens-Dipole eine ausschlaggebende Rolle spielt".

A comparison between the striking similarities between H_2O and $POCl_3$ in their physical constants and their solvent behaviour has been made by VAN WAZER (8). But the remarks that: "In spite of the similarities between H_2O and $POCl_3$ it must be remembered that there is a large amount of hydrogen bonding in pure water and in many aqueous solutions, whereas a similar internal bonding mechanism is not readily conceivable for phosphoryl-chloride". The similarities can now be better understood, if one compares hydrogenbonding H-O...H in water with phosphorylbridging P=O...P in phosphorylchloride.

References

1. BELLAMY, L.J., "The Infrared Spectra of Complex Molecules", J.Wiley, S.380 (1958).
2. GOUBEAU, J., and BERGER, W., Z.anorg.allg.Chem., 304, 148 (1960).
3. LINDQVIST, I., "Inorganic Adduct Molecules of Oxo-Compounds", Springer, S.87 (1963).
4. WARTENBERG, E.W., Proc., 7 ICCC, 214 (1962).
5. WARTENBERG, E.W., and GOUBEAU, J., Z.anorg.allg.Chem., im Druck.
6. GROENEVELD, W.J., and ZUUR, A.P., Rec.Trav.Chim. Pays-Bas, 76, 1005 (1957).
7. GUTMANN, V., XVIIth International Congress of Pure and Applied Chem., Verlag Chemie, Weinheim/Bergstr., p.106 (1959).

8. VAN WAZER, J.R., "Phosphorus and its Compounds", Interscience, S. 254 (1958).

6 C 2

THE CRYSTAL STRUCTURES OF SOME SOLVATES OF SELENINYLDICHLORIDE.

Yngve Hermodsson

Institute of Chemistry, University of Uppsala, Uppsala, Sweden.

Studies of the coordination in non-aqueous ionizing solvents have depended largely upon electrochemical techniques. However, a knowledge of the structures of the solid solvates, the compounds formed between solute and solvent, can also indicate the spatial arrangement likely to be adopted in solution. Most of the solvents investigated have been halides and oxy-halides, and of these seleninyldichloride, $SeOCl_2$, is one of the most strongly ionizing. The dielectric constant is 46 at 20°.

Of the known solvates of $SeOCl_2$ the following four have been examined using X-ray diffraction techniques: $SbCl_5.SeOCl_2$, $SnCl_4(SeOCl_2)_2$, $(CH_3)_4NCl(SeOCl_2)_5$ and $SeOCl_2(C_5H_5N)_2$. The last mentioned compound has been studied by LINDQVIST and NAHRINGBAUER (1) and the remainder have been investigated by the present author. The structure determinations have been made from three-dimensional single crystal data.

From electrochemical data it may be concluded that the solvates are composed of complex ions formed by halide ion transfer; thus the system $SbCl_5$-$SeOCl_2$ gives rise to $SeOCl^+$ and $SbCl_6^-$ ions. Polynuclear complexes are also possible. However, a structure determination has shown that crystals of the solvates $SbCl_5.SeOCl_2$ comprise discrete $SbCl_5.SeOCl_2$ molecules with an octahedral configuration in which five chlorine atoms and one oxygen atom coordinate the antimony atom. The coordination of the selenium atom is approximately pyramidal. The Sb-O distance corresponds to a single bond distance indicating that the interaction between the solute and the solvent molecule is strong.

The same type of structure occurs in $SnCl_4(SeOCl_2)_2$ (2). The tin atom has a distorted octahedral coordination in which the ligands are two oxygen atoms in the cis-position and four chlorine atoms.

In $SeOCl_2(C_5H_5N)_2$ the arrangement is much the same, although in this case $SeOCl_2$ acts as a Lewis acid. As a result the selenium atom accepts the electron pair of the nitrogen atom in each of the pyridine rings, thereby increasing its coordination from three to five. One of the Se–Cl bond lengths is significantly larger than the other (2,57 and 2,39 Å resp.), a difference which may be interpreted in terms of a tendency towards chloride ion transfer.

In the system $(CH_3)_4NCl-SeOCl_2$ the situation is quite different since the solute is an ionic compound. Depending on the strength of interaction between the solute and the solvent the solvates may exhibit the following main types of configuration in the solid state.

a) The interaction between $SeOCl_2$ and the two ions is very slight and structures consisting of $SeOCl_2$ molecules and $(CH_3)_4N^+$ and Cl^- ions are formed.

b) The interaction between $SeOCl_2$ and the two ions is somewhat stronger leading to structures composed of ions coordinated to $SeOCl_2$ molecules ("solvated ions"), and possibly together with more loosely bound $SeOCl_2$ molecules.

c) There is a strong interaction between $SeOCl_2$ and the chloride ions forming structures consisting of complex ions $(Se_mO_mCl_{2m+n})^{n-}$, e.g. $SeOCl_3^-$, and $(CH_3)_4N^+$ ions. These ions can be coordinated to further $SeOCl_2$ molecules.

The structure determination of the first solvate studied, $(CH_3)_4NCl(SeOCl_2)_5$, provides an example of type b-interaction. The structure consists of $(CH_3)_4N^+$ ions together with Cl^- ions each of which lies at the center of a distorted octahedron formed by the selenium atoms from six $SeOCl_2$ molecules. Two of these selenium atoms are shared by an adjacent octahedron. One of the six Cl^--Se distances (4,03 Å) is much greater than the other (between 2,94 and 3,08 Å). This may be explained by steric effects and by an interaction between the oxygen atom bound to this selenium atom and a neighbouring selenium atom, the intermolecular Se–O distance being 2,84 Å. Similar interactions are also found in the other solvates described. The shortest distances from a methyl group to a chlorine and to an oxygen atom are 3,66 and 3,23 Å respectively, a result which indicates that the tetramethyl-ammonium ion has no tendency to coordinate $SeOCl_2$ molecules.

The structure of $(CH_3)_4NCl(SeOCl_2)_5$ illustrates that even larger neutral molecules than water and ammonia can be incorporated into an ionic crystal structure. Furthermore the arrangement around the anion (Cl^-) can be compared with that of cations in those hydrates for which the oxygen coordination number of the cations is greater than the available number of water molecules.

The picture of negative ions coordinated to polar inorganic molecules is also to be found in the structure of $KF(SbF_3)_4$ (3). Here the fluoride ion is surrounded by four SbF_3 molecules. In the system $KF-SbF_3$, however, there are also complexes which exhibit stronger interaction between F^- and SbF_3. Thus the structure of $KF.SbF_3$ (4) contains tetranuclear $[Sb_4F_{16}]^{4-}$ ions while that of $(KF)_2SbF_3$ (5) contains $[SbF_5]^{2-}$ ions. Ionic complexes similar to these may therefore occur among the lower solvates in the system $(CH_3)_4NCl-SeOCl_2$.

References

1. LINDQVIST, I., NAHRINGBAUER, G., Acta Cryst., 12, 638 (1959).
2. HERMODSSON, Y., Acta Cryst., 13, 656 (1960).
3. BYSTRÖM, A., WILHELMI, K-A., Arkiv Kemi, 3, 17 (1951).
4. BYSTRÖM, A., BÄCKLUND, S., WILHELMI, K-A., Arkiv Kemi, 4, 175 (1952).
5. BYSTRÖM, A., WILHELMI, K-A., Arkiv Kemi, 3, 461 (1951).

6 C 3

COORDINATION PROBLEMS OF NITROGEN-IODINE-COMPOUNDS IN LIQUID AMMONIA.

Jochen Jander and Udo Engelhardt (1)
Anorganisch-Chemisches Laboratorium der
Technischen Hochschule, München, Germany.

By the reaction of iodine and small amounts of liquid ammonia at -75° a black brown solid is formed. A tensimetric deammonation shows that this solid is an addition compound of iodine and ammonia with 2 moles of ammonia per mole of iodine. Further deammonation effects formation of a compound $I_2.NH_3$; finally, I_2 remains. Ammonolysis of iodine takes place only to a small extent. These brown addition compounds of iodine correspond to the well known brown addition compounds of iodine and ether, alcohol, pyridine or liquid hydrogen sulfide respectively and are looked upon as charge-transfer complexes with iodine as electron acceptor and the solvents as electron donors. This view was proved by the measurement of the charge transfer band of iodine-ammonia in heptane at 229 mμ (2).

If one treats iodine with more liquid ammonia at -60°, the brown addition compound is not stable and by a solvolysis reaction forms soluble ammonium iodide and green $NI_3(NH_3)_3$, which in liquid ammonia is only slightly soluble with yellow colour. By tensimetric measurements and x-ray patterns, taken at -30° to -25°, it is shown, that the addition compounds $NI_3(NH_3)_{12}$ (3) and $NI_3(NH_3)_2$ (4), which by several authors were claimed to exist, do not exist; only the green $NI_3(NH_3)_3$ and the reddish brown $NI_3.NH_3$ do exist.

The yellow solutions of $NI_3(NH_3)_3$ in liquid ammonia do not contain this compound; the solutions become deep red at liquid air temperature. By a special tensimetric method a dark red compound may be isolated. The same compound is formed as an insoluble precipitate besides the soluble ammonium iodide, if iodine is ammonolysed with excess of liquid ammonia at -75°. Under these conditions the distribution of reaction heat is more effective than in less ammonia, and unstable compounds may therefore exist. Nevertheless the red compound changes in liquid ammonia slowly into the green compound via brown intermediates. The red compound shows its own x-ray diagramm. The tensimetric analysis leads to a nitrogen-iodine-ratio 2:1. Regarding the mechanism of the formation of nitrogen triiodide from iodine and liquid ammonia, it seems probable, that the formula of the red compound is $NH_2I.NH_3$ or $NHI_2.(NH_3)_3$. The formula $NH_2I.NH_3$ was proved by chemical evidence and IR-spectra: a) by pumping off ammonia at -90°, a black compound with a nitrogen-iodine ratio of 1:1 (NH_2I) is obtained. b) the ammonolysis of ICl, IBr, $INO_3.py$ and $IClO_4.py$ in liquid ammonia at -75° also leads to the red compound. c) the substance in liquid ammonia introduces iodine and the NH -group into organic compounds, for instance according to the scheme:

d) the IR-spectrum of the solid red substance at -90° shows bands typical for compounds containing NH_3 or NH_2 at 3000 (N-H stretching), 1600 (H-N-H bending) and 1100 cm^{-1}. The black substance (nitrogen-iodine ratio 1:1) in the N-H stretching area shows 2 bands expected for NH_2I. Further removal of ammonia at a somewhat higher temperature leads to a black substance with only 1 band in the N-H stretching area. The bands at 1600 and 1100 cm^{-1} show decreased intensity. These changes in IR-spectrum are expected for the transition $NH_2I \rightarrow NHI_2$. Further pumping finally leads to a black substance showing no IR-absorption in the 3 areas mentioned. This substance is mainly NI_3 which may be decomposed in a vacuum at room temperature.

References

1. ENGELHARDT, U., Thesis, Freiburg/Br. -München 1964.

2. JADA, H., TANAKA, J., and NAGAKURA, S., Bull.Chem.Soc.Japan, 33, 1660 (1960).

3. RUFF, O., Ber.dtsch.chem.Ges., 33, 3025 (1900).
   WATT, G.W., FOERSTER, D.R., J.Inorg.Nucl.Chem., 13, 313 (1960).

4. HUGOT, C., C.R.hebd.Séances Acad.Sci. 130, 507 (1900); Ann.chim.
   phys. [7] 21, 26 (1900).

7 C 1

CALORIMETRIC TITRATIONS FOR THE STUDY OF STEPWISE EQUILIBRIA IN SOLUTIONS.

Ingmar Grenthe and Ido Leden
Department of Inorganic and Physical Chemistry, Lund University,
Lund, Sweden.

During the last twentyfive years a large number of investigations of thermodynamic properties of various metal complexes have been published. In most of these investigations only the change in free energy for the various complex formation reactions was determined. The purpose of these measure- ments was in most cases only to decide, via the stability constants, which species actually existed in solution. During recent years there has been an increased interest in the magnitude of the enthalpy and entropy changes for complex formation reactions.

The best method of obtaining accurate thermodynamic functions seems to be a potentiometric determination of the various stability constants followed by a direct calorimetric determination of the corresponding enthalpy change. In this communication we will describe a reaction calorimeter developed by GERDING et al (1) and also give an example of its use for the study of the composition of the thallium(III)chloride and bromide complexes.

In a titration the calorimeter (1) was filled with 100 ml of a solution S, containing the metal ion to be studied. A known amount of the ligand solution T, was introduced from a syringe into a pipette with a ground seal. The system was equilibrated with the pipette in the calorimeter solution. After equilibrium had been attained the pipette was emptied and the temperature and enthalpy changes were determined. The composition of the liquid phase could be calculated from known values of the stability constants, the concentration and the amounts of calorimeter solution and added titrant. The enthalpy change could thus be referred to defined initial and final states.

From experimental values of the heat of reaction Q_{corr}, i.e. the heat of reaction for addition of a certain amount of ligand, corrected for heats of dilution, the total molar heat change Δh_v was computed. If the total volume of the solution in the calorimeter vessel is V ml and if v ml of titrant has been added we have:

$$\Delta h_v = \frac{1000}{c_M \cdot V} \cdot \sum_V Q_{corr} \qquad \{1\}$$

The various enthalpy changes can be computed from the relation:

$$\Delta h_v = \sum_1^N \Delta H_j^0 \cdot \alpha_j \qquad \{2\}$$

where ΔH_j^0 is the overall enthalpy change and α_j the degree of formation for the j:th complex.

In a mononuclear system Δh_v is a function only of the concentration of free ligand. We can thus get information, if polynuclear complexes are formed in a certain system by varying c_M in the calorimetric titrations. An example of a titration of this kind is given in Fig. 1 where Δh_v is plotted versus $\bar{n}$ for

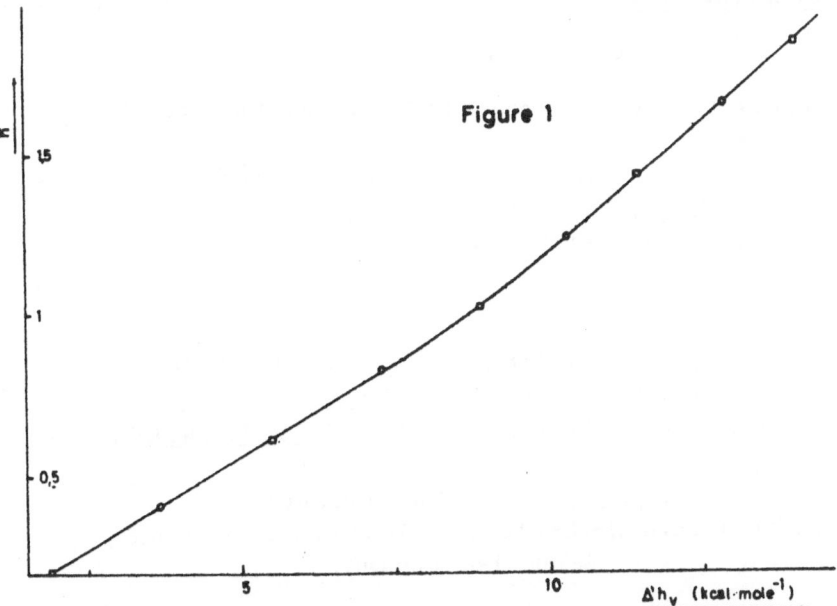

Fig. 1 - The molar enthalpy change Δh for the reaction between thallium(III)-ion and bromide as a function of the average ligand number $\bar{n}$ for two values of the thallium(III) concentration, viz. 9,75 mM (O) and 1,95 mM (□).

two values of c_M in the Tl^{3+}-Br^- system. From the figure it is obvious that Δh_v is independent of c_M and that only mononuclear complexes are formed. This piece of information could not be obtained in the potentiometric determination of the stability constants (2) where mononuclearity had to be assumed in order to integrate the BODLÄNDER equation. The calorimetric titration has given proof that this procedure was justified. In the Tl^{3+}-Br^- system a point of equivalence was obtained at $c_A = 4c_M$ as practically no heat was evolved after this point. This indicates strongly that no more than four bromide ions can be bound by a thallium(III)-ion. This fact is also in agreement with the result obtained by the potentiometric method (2).

If the standard deviations of the various enthalpy values are of the magnitude expected from the accuracy of the calorimeter, the composition of the solution in the calorimeter must be the one given by the stability constants. The calorimetric titration will thus give an independent check of the stability constants. In this context it can be mentioned that a least-square program "Leta Grop Kalle" has been developed by SILLÉN et al (3) for the simultaneous determination of stability constants and enthalpy changes from calorimetric data.

References

1. GERDING, P., LEDEN, I., and SUNNER, S., Acta Chem.Scand., <u>17</u>, 2190 (1963).
2. AHRLAND, S., GRENTHE, I., JOHANSSON, L., and NORÉN, B., Acta Chem.Scand., <u>17</u>, 1567 (1963).
3. SILLÉN, L.G., et al, to be published.

7 C 2

BESTIMMUNG DER BILDUNGSKONSTANTEN UND REAKTIONSENTHALPIEN MEHRSTUFIGER KOMPLEXBILDUNGS-GLEICHGEWICHTE DURCH THERMOMETRISCHE TITRATION.

E. Becker und H.M. Lüschow
Institut für Physikalische Chemie der Universität des Saarlandes,
Saarbrücken, Germany.

Die in Saarbrücken ausgeführten Untersuchungen über die Grundlagen der thermometrischen Titration (1 - 4) haben ergeben, daß diese Methode ein allgemein anwendbares kalorimetrisches Verfahren zur Untersuchung rasch eingestellter chemischer oder physikalischer Gleichgewichte unter

dem Einfluß einer kontinuierlichen Veränderung der Konzentration einer
der beteiligten Komponenten ist.

Wir verwenden eine Titrationsapparatur (2), bei der die Reagenslösung
aus einer thermostatisierten Bürette mit konstanter Geschwindigkeit in das
in einen Präzisionsthermostaten eingebaute Reaktionsgefäß fließt. Die Temperatur wird mittels einer Thermistorbrücke von einem Kompensationsschrei-
ber kontinuierlich als Funktion der Zeit registriert, wobei die Ansprechemp-
findlichkeit besser als 10^{-4} ist.

Für die Auswertung der Registrierkurven stehen zwei graphische Verfahren
zur Verfügung (1, 2): Das Tangentenverfahren liefert die thermische Leistung
der Reaktion, $w_R(t)$, als Funktion der Zeit, das Abschnittsverfahren den inte-
gralen Wärmeeffekt der Reaktion, $Q_R(t)$, vom Beginn der Reagenszugabe
($t_0 = 0$) bis zum Zeitpunkt t. Von besonderem Interesse ist die Anfangsleistung
der Reaktion, $w_R(t_0)$, die man aus der Steigung der Tangente an die Regi-
strierkurve bei $t_0 = 0$ erhält. Im Falle eines N-Stufen-Gleichgewichtes

$$MA_{i-1} + A \rightleftharpoons MA_i \quad (i = 1, 2, \ldots, N) \qquad \{1\}$$

gilt, wenn M (Konzentration m [Mol/Liter]) im Reaktionsgefäß vorgelegt
wird und A (Konzentration a) aus der Bürette zuläuft,

$$w_R(t_0) = -av'\Delta H_{MA} \frac{mK_1}{1 + mK_1} \qquad , \qquad \{2\}$$

wobei v' die Zulaufgeschwindigkeit [Liter/Min], K_1 die Bildungskonstante
[Liter/Mol] und ΔH_{MA} die Bildungsenthalpie [kcal/Mol] der 1. Stufe (Gl. {1},
i = 1) bedeutet. Die Messung der Anfangsleistung $w_R(t_0)$ bei zwei verschiedenen Werten von m und konstantem a erlaubt eine Berechnung von K_1 und
ΔH_{MA}, unabhängig von der Zahl der weiteren Gleichgewichtsstufen.

Legt man umgekehrt A im Reaktionsgefäß vor (Konzentration a^*) und
läßt M (Konzentration m^*) aus der Bürette zulaufen, so lautet die Anfangsleistung $w_R^*(t_0)$

$$w_R^*(t_0) = -m^* v' \frac{\sum_{i=1}^{N} \Delta H_{MA_i} K_i a^{*i}}{1 + \sum_{i=1}^{N} K_i a^{*i}} \qquad ; \qquad \{3\}$$

sie hängt also von den Bruttobildungskonstanten K_i und den Reaktionsenthalpien ΔH_{MA_i} sämtlicher Gleichgewichtsstufen ab.

Bei einem Einstufengleichgewicht müssen die aus jeweils zwei Werten von $w_R(t_0)$ und $w_R^*(t_0)$ berechneten Größen K_1 und ΔH_{MA} übereinstimmen. Diese Tatsache stellt gleichzeitig ein Kriterium für das Vorliegen einer einzigen Gleichgewichtsstufe dar. Im Falle eines Zweistufengleichgewichtes kann man aus jeweils zwei Werten von $w_R(t_0)$ und $w_R^*(t_0)$ alle 4 Unbekannten K_1, ΔH_{MA}, K_2, ΔH_{MA_2} berechnen. Grundsätzlich läßt sich dieses Verfahren auch auf Mehrstufengleichgewichte (N > 2) anwenden; ist N = 4, so sind z.B. zwei Werte von $w_R(t_0)$ und sechs Werte von $w_R^*(t_0)$ erforderlich. Für eine praktische Auswertung des Gleichungssystems {3} reicht aber die experimentell erzielbare Genauigkeit der $w_R^*(t_0)$-Werte (Unsicherheit mindestens ± 1%) dann nicht mehr aus.

Wendet man das Abschnittsverfahren auf ein Einstufengleichgewicht an, so lassen sich aus der Beziehung

$$-Q_R(t) \;=\; n_{MA}\Delta H_{MA} \qquad\qquad \{4\}$$

($n_{MA} = n_{MA}(t)$ ist die zur Zeit t in der Lösung vorhandene Molzahl der Verbindung MA) Gleichungen herleiten (vgl.(2)), mit deren Hilfe man bei Kenntnis von $Q_R(t)$ für zwei verschiedene Zeitpunkte K_1 und ΔH_{MA} errechnen kann. Im Falle eines Zweistufengleichgewichtes kann man den Ausdruck

$$-Q_R(t) \;=\; n_{MA}\Delta H_{MA} + n_{MA_2}\Delta H_{MA_2} \qquad\qquad \{5\}$$

in eine lineare Beziehung

$$F_1 \;=\; \Delta H_{MA} + F_2 \cdot \Delta H_{MA_2} \qquad\qquad \{6\}$$

umformen, wobei

$$F_1(Q_R, t, K_1, K_2) \;\equiv\; -Q_R(t)/n_{MA}$$
$$F_2(K_1, K_2, t) \;\equiv\; n_{MA_2}/n_{MA} \qquad\qquad \{7\}$$

ist. Man setzt zunächst Näherungswerte für K_1 und K_2 ein und variiert diese so lange, bis die Funktion $F_1 = f(F_2)$ im gesamten Zeitintervall linear wird (5). Dann gibt die Steigung der erhaltenen Geraden ΔH_{MA_2} und der Ordinatenabschnitt ΔH_{MA}.

Unter günstigen Bedingungen ist bei einem Vierstufengleichgewicht die Bestimmung aller 8 Unbekannten K_i und ΔH_{MA_i} (i = 1, 2, 3, 4) durch thermometrische Titrationen bei einer Temperatur möglich, wenn man das Tangentenverfahren und das Abschnittsverfahren in geeigneter Weise miteinander kombiniert.

Man beginnt mit der Bestimmung der Anfangsleistungen $w_R(t_0)$ für 2 Konzentrationen m und errechnet daraus K_1 und ΔH_{MA}. Durch geeignete Wahl von m und a ist es möglich, das Abschnittsverfahren unter Bedingungen durchzuführen, die eine Vernachlässigung der vierten Gleichgewichtsstufe erlauben. Dann läßt sich eine lineare Beziehung von der Form

$$F_1 = (\Delta H_{MA_2} - \Delta H_{MA}) + (\Delta H_{MA_3} - \Delta H_{MA_2}) F_2 \qquad \{8\}$$

mit

$$F_1(Q_R, t, K_2, K_3) \equiv [-Q_R(t) - \Delta H_{MN}(n_{MA} + n_{MA_2} + n_{MA_3})]/(n_{MA_2} + n_{MA_3})$$

$$F(t, K_2, K_3) \equiv n_{MA_3}/(n_{MA_2} + n_{MA_3}) \qquad \{9\}$$

aufstellen. Das richtige Wertepaar K_2, K_3 erkennt man daran, daß die Auftragung $F_1 = f(F_2)$ eine Gerade ergibt. Die erforderlichen Rechnungen lassen sich am schnellsten mit einer elektronischen Rechenmaschine durchführen. K_4 und ΔH_{MA_4} bestimmt man aus den Anfangsleistungen $w_R^*(t_0)$ für zwei Konzentrationen a^* nach Gl. $\{3\}$ unter Verwendung der bereits ermittelten Werte für K_1, K_2, K_3 und ΔH_{MA}, ΔH_{MA_2} und ΔH_{MA_3}.

Als Anwendungsbeispiel wurde das in der Literatur (6) als Vierstufengleichgewicht beschriebene System

$$Cd^{2+} + 4I^- \rightleftharpoons [CdI_4]^{2-} \qquad \{10\}$$

gewählt. Sämtliche Messungen wurden bei 25°C unter Verwendung von $Cd(ClO_4)_2$ - und NaI-Lösungen durchgeführt, die durch Zusatz von 0,1 M

Tab. 1: Thermodynamische Daten des Vierstufengleichgewichtes
$Cd^{2+} + 4I^- \rightleftharpoons [CdI_4]^{2-}$ bei 25° und I = 2.

Gleich-gewichts-stufe	k_{298} bzw. K_{298}	$-\Delta H_{298}^{\circ}$ kcal/Mol	$-\Delta G_{298}^{\circ}$ kcal/Mol	ΔS_{298}° cal/Mol. Grad
1	130 ± 15	2,32 ± 0,03	2,88	+ 1,88
2	4 ± 2 ⎫	-3,45 ± 1,50 ⎫	0,82	+14,32
3	90 ± 5 ⎬ 360 ± 20	7,65 ± 1,50 ⎬ 4,20 ± 0,10	2,67	-16,70
4	2,5 ± 0,25	1,43 ± 0,10	0,54	- 2,99
1	130 ± 15	2,32 ± 0,03	2,88	+ 1,88
1 - 2	520 ± 260	-1,13 ± 1,50	3,70	+16,20
1 - 3	(4,68± 0,60).10^4	6,52 ± 0,15	6,37	- 0,50
1 - 4	(1,17± 0,12).10^5	7,95 ± 0,10	6,91	- 3,49

HClO$_4$ und von NaClO$_4$ auf eine Ionenstärke I = 2 [Mol/Liter] gebracht waren.
Die erhaltenen Resultate sind in Tabelle 1 zusammengestellt. Die Fehlergren-
zen sind, bedingt durch das Auswerteverfahren, für die zweite individuelle
Gleichgewichtsstufe am größten; sie verringern sich jedoch sehr, wenn man
die zweite und dritte Stufe als Bruttoprozeß betrachtet.

Literatur

1. BARTHEL, J., BECKER, F., und SCHMAHL, N.G., Z.physik.Chem.
N.F. 29, 58 (1961).
2. BECKER, F., BARTHEL, J., und SCHMAHL, N.G., Z.physik.Chem.
N.F. 37, 33 (1963).
3. BECKER, F., BARTHEL, J., SCHMAHL, N.G., LÜSCHOW, H.M.,
Z.physik.Chem. N.F. 37, 52 (1963).
4. BECKER, F., SCHMAHL, N.G., und PFLUG, H.D., Z.physik.Chem.
N.F. 39, 306 (1963).
5. BJÖRKMAN, M., und SILLÉN, L.G., Kungl.Tekn.Högskolans Handl.
Stockholm, No. 199, 1-19 (1963).
6. BJERRUM, J., SCHWARZENBACH, G., und SILLÉN, L.G., "Stability
Constants of Metal-ion Complexes", London 1957, S. 120.

7 C 3

CONFIGURATIONAL EQUILIBRIA OF SOME TRANSITION METAL COMPLEXES IN SOLUTION.

H.C.A. King, E. Körös, S.M. Nelson and T.M. Shepherd
Department of Chemistry, Queen's University of Belfast, N.Ireland, U.K.

A series of complexes of the types ML$_2$X$_2$ and ML$_4$X$_2$ have been examined
both in the solid state and in solution in organic solvents; M = Co(II), Ni(II)
or Zn(II); L = pyridine, monosubstituted pyridine, quinoline or isoquinoline;
X = halide or pseudohalide ion. Solid state structures have been assigned on
the basis of magnetic moments and electronic spectra and (in the case of zinc
complexes) X-ray powder crystallography and IR spectra. The bisamine complexes, ML$_2$X$_2$, were found, for the most part, to have either 4-coordinate
(tetrahedral) or 6-coordinate (bridged octahedral) structures; a few of the
Ni(II) complexes are planar. Many of the ML$_2$X$_2$ complexes dissolve in
chloroform, with configurational rearrangement to the tetrahedral form
where necessary, and also exist in equilibrium, in the presence of excess of
the appropriate amine, with the corresponding mononuclear octahedral complex.

$$ML_2X_2 + 2L \rightleftharpoons ML_4X_2$$

(tetrahedral)   (octahedral)

It is clear that the position of this equilibrium is a sensitive indicator of the preferred coordination number of the metal for a given type of ligand environment ant that it provides a useful background to a discussion of solid state stereochemistry.

Association constants were determined spectrophotometrically in chloroform at $20°$ by utilising the different spectra of the two stereochemical forms. Enthalpies of association were determined calorimetrically, where possible, and the corresponding entropy changes calculated. The results refer, for the most part, to complexes of Co(II) though thermodynamic data are also available for a few Ni(II) systems.

The association constant K is very dependent on the nature of both L and X, and both steric and electronic factors are important. With L = 2-methylpyridine or quinoline, for example, K is immeasurably small and it was not found possible to isolate the corresponding ML_4X_2 complexes in the solid state. With complexes containing heterocyclic amines in which there is no substitution alpha to the nitrogen atom the variations in the thermodynamic functions are interpreted in terms of electronic factors. For a series of systems with a given ligand L it was found that the strong dependence of K on X is due more to entropy than to enthalpy factors. For the case of L = pyridine an interpretation in terms of a variable degree of back-coordination of electrons from metal to amine has been proposed (1). The more polarisable the anionic ligand X the more charge is transferred to the metal and this, in turn, enhances the release of non-bonding d electrons to anti-bonding amine π-orbitals. The results indicate that the effectiveness of the anionic ligand in promoting $M \rightarrow L$. π-bonding is I > Br > Cl > NCO > NCS.

A similar pattern of results was obtained for complexes containing a variety of substituted pyridines as ligands. However, a number of significant differences have been found which seem to support the above interpretation. For example, for L = isoquinoline the results are consistent with the view that isoquinoline forms weaker σ-bonds to the metal than does pyridine but that it is better able to accept back-coordinated electrons in a π-bond. Similarly, marked differences in the data obtained for complexes with 3-alkylsubstituted and 4-alkylsubstituted pyridines suggest that the former amines are poorer π-acceptors, a conclusion in accord with published information on the mesomeric effects of 3- and 4-substituents and the calculated electronic charge densities at different ring positions (2).

A comparison of thermodynamic functions for the Ni(II)/pyridine/iodide system with the corresponding Co(II) system confirms the expected effect on $-\Delta H$ of the greater crystal field stabilisation energy of 6-coordinate Ni(II). However, a comparison of the difference in the observed enthalpies of association of the two systems with the difference in the calculated gains in CFSE of the two metal ions on passing from the tetrahedral to the octahedral state indicates that the stability advantage of octahedral Ni(II) over tetrahedral Ni(II), when corrected for CFSE, is substantially less than for the corresponding Co(II) case. This confirms a suggestion previously made by GILL and NYHOLM (3) that there is a convergence in the slopes of the CFSE - corrected curves representing the heats of formation of tetrahedral and octahedral complexes on passing from Mn(II) to Zn(II). Further support for this proposal comes from the observation that all of the ZnL_2X_2 complexes studied have tetrahedral configurations. In contrast, tetrahedral complexes of Mn(II) with the same ligands are unknown.

Correlations are drawn between the magnitude of the thermodynamic functions for the tetrahedral-octahedral equilibria in chloroform solution and the configuration (tetrahedral or polymeric octahedral) of the bisamine complex in the solid state. While the correlation is good for many of the systems studied certain anomalies suggest that other factors such as the bridging power of X and packing conditions in the crystal can play an important role in deciding solid state stereochemistry.

References

1. KING, KÖRÖS and NELSON, J.Chem.Soc. (London), 1963, 5449.
2. CARTIER and SANDORFY, Canad.J.Chem., 41, 2759 (1963).
3. GILL and NYHOLM, J.Chem.Soc., (London), 1959, 3997.

7 C 4

THE RELATIVE STABILITY OF OCTAHEDRAL AND TETRAHEDRAL COMPLEXES OF TRANSITION METAL IONS.

Irmina Uruska and Włodzimierz Libuś
Department of Physical Chemistry, Technical University of Gdańsk,
Gdańsk, Poland.

It is well known that the relative stability of octahedral and tetrahedral complexes in solution depends on the properties of ligands in addition to being dependent on the central metal ion. For a discussion of the latter effect it is desirable to possess data concerning the complexes of different metals with the same ligands. With the object of obtaining equilibrium constants fulfilling this requirement an investigation of equilibria of the type:

$$MCl_2py_2 + 2\,py = MCl_2py_4 \qquad \{1\}$$

M - being a divalent transition metal ion - has been undertaken in the present work. In the case of cobalt(II) this equilibrium was investigated by KATZIN (1) and quite recently by KING et al. (2). This work is a continuation of earlier investigations performed by one of us (3).

A method of determination of the equilibrium constant of $\{1\}$ based on measurements of solubilities has been divesed. So far equilibria involving Co(II), Ni(II), and Zn(II) were investigated, chlorobenzene being used as an inert solvent. The determined solubility isotherms for chlorobenzene-pyridine mixtures are shown in Fig. 1, the compositions of solid phases in equilibrium being indicated. It is seen that solid complexes are very slightly soluble in pure chlorobenzene but their solubility increases more or less rapidly on increasing pyridine concentration. This effect is a result of the formation of

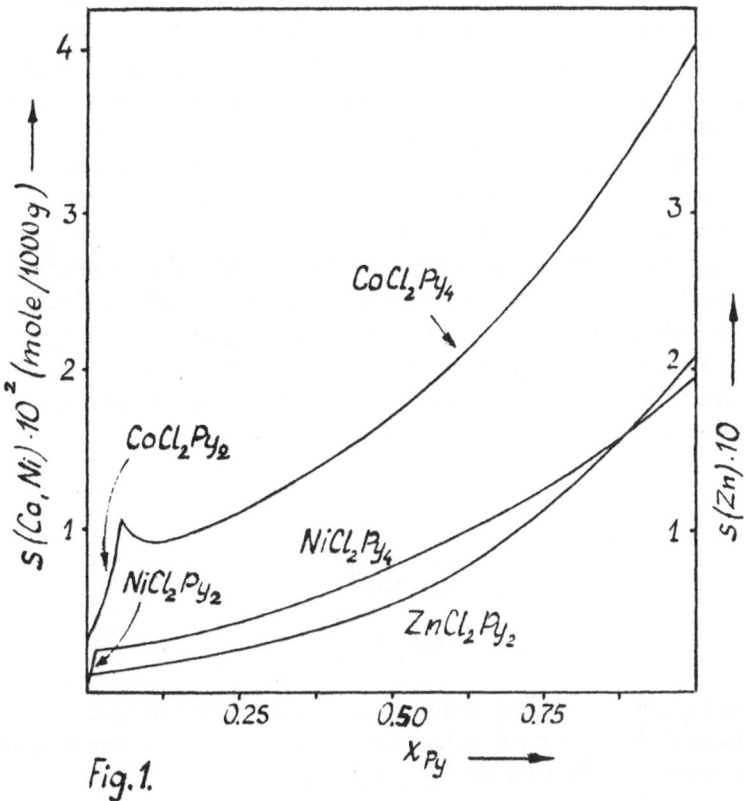

Fig. 1.

octahedral complexes MCl_2py_4 in solution according to $\{1\}$, in addition to a presumably smaller and nonspecific effect of solvent composition on the activity coefficients γ_0 of individual complexes. For an evaluation of equilibrium constants from solubility data the second effect must first be considered. For the octahedral complex $NiCl_2py_4$ the dependence of γ_0 on solvent composition may be found directly from solubility data, owing to the fact that, except for smallest pyridine concentrations, practically all nickel(II) in solution is in the form of $NiCl_2py_4$. A similar dependence on γ_0 on solvent composition for $CoCl_2py_4$ may be calculated from both solubility and optical density measurements the latter enabling for a direct determination of concentrations of $CoCl_2py_2$ and $CoCl_2py_4$ complexes in solution. As a final result of these calculations it has been found that γ_0 is the same function of solvent composition (chlorobenzene + pyridine mixtures) for both $NiCl_2py_4$ and $CoCl_2py_4$ octahedral complexes. In a similar way the values of γ_0 for the tetra-hedral complex $CoCl_2py_2$ were calculated for lower pyridine concentrations making use of solubility and spectral data. In turn, the same dependence of γ_0 on solvent composition has been found for $ZnCl_2py_2$ from solubility data assuming it to be the only zink complex in solution. From the above results two conclusions may be drawn; 1. The dependence of γ_0 on the composition of the solvent (chlorobenzene + pyridine mixtures) is the same for complexes of the same constitution regardless of the central metal ion. 2. In solutions of $ZnCl_2py_2$ in chlorobenzene-pyridine mixtures practically no octahedral complexes are formed. In other words the increase of solubility of $ZnCl_2py_2$ on increasing concentration of pyridine is due entirely to solvation effects.

Assuming that in a saturated solution of MCl_2py_2 (M = Ni(II) or Co(II)) the only complex species present are MCl_2py_2 (tetrahedral) and MCl_2py_4 (octa-hedral) the solubility data together with the activity coefficients, found in the above discussed way, are sufficient for a calculation of equilibrium constants of $\{1\}$. It was defined as

$$K = \frac{a_{MCl_2py_4}}{a_{MCl_2py_2} \, a^2_{py}} \qquad ,$$

the activities being normalized assuming ideal solutions of unit molar concentrations of the reagents in chlorobenzene as the corresponding standard states. The resulting values for 20° are given below:

	Co(II)	Ni(II)	Zn(II)
solubility method:	6,5	$1,2.10^4$	0,0
spectrophotometric:	8,3	-	-

In the case of cobalt(II) the results obtained by the solubility method could be verified by a spectrophotometric determination of the equilibrium constant. The value obtained in this way is given above, the agreement of the two methods being satisfactory. The advantage of the solubility method lies in its applicability in cases in which the spectrophotometric method cannot be used.

The work is in progress, equilibria involving other divalent transition metals being investigated.

References

1. KATZIN, L., J.Chem.Phys., 35, 467 (1961).
2. KING, H.C.A., KÖRÖS, E., NELSON, S.M., J.Chem.Soc. (London), 1963, 5449.
3. LIBUŚ, W., Roczniki Chem., 33, 931 (1959); 33, 951 (1959); 34, 29 (1960); 35, 411 (1961); 36, 999 (1962); Theory and Structure of Complex Compounds, Proc.Symp.Wrocław, 1963, p.537-550.

8 C 1

FORMATION AND STABILITY OF MIXED AND POLYNUCLEAR COMPLEXES IN NON-AQUEOUS SOLUTIONS.

M.T. Beck and F. Gaizer
Reaction Kinetical Research Group of the Hungarian Academy
of Sciences and Institute of Inorganic and Analytical Chemistry,
The University, Szeged, Hungary.

The study of coordination phenomena in non-aqueous solutions has at least twofold importance. On the one hand the comparison of results obtained in different solvents may give valuable information even about the chemistry of aqueous solutions, on the other hand the non-aqueous solvent makes possible the study of particular reactions which cannot occur in aqueous solution due to the limited solubility and to disturbing side reactions. The subject matter of the present paper illustrates both the above mentioned features.

1. <u>Stability of mercury(II) cyanide-halide mixed complexes in dioxane.</u>
The equilibrium constant of the following disproportionation reaction

$$2 \ HgCNX = Hg(CN)_2 + HgX_2$$

has been determined. If X = Cl or Br the value is greater, if X = I it is
smaller than in aqueous system. On the basis of these findings it can be argued
that the polar medium stabilizes the mixed complex.

At our surprise, it was found that in dioxane the formation of the mixed
complex from the parent complexes takes place in a fairly slow reaction
(velocity order: Cl < Br < I), while it is known that the substitution reactions
of mercury(II) in water are extraordinarily fast. The reaction is markedly
catalyzed by water and by traces of unknown impurities of dioxane. Due to
this latter fact careful purification of dioxane is necessary. The formation
of the mixed complexes is evidently preceeded by formation of a transition
complex of the following structure:

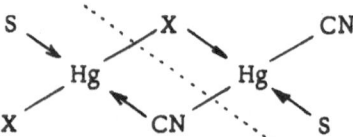

S means the solvent molecule. The mixed complex is formed by rupture
along the dotted line.

2. <u>Formation of mixed complexes in the system BiCl$_3$ + BiI$_3$ in dioxane.</u>
In comparison to the $MA_2 + MB_2$ system, $MA_3 + MB_3$ is more complicated and
can be characterized by the followint equilibria:

$$MA_3 + MB_3 = MA_2B + MAB_2 \qquad \{K_{11}\}$$

$$MA_2B + MB_3 = 2 \ MAB_2 \qquad \{K_{12}\}$$

$$MAB_2 + MA_3 = 2 \ MA_2B \qquad \{K_{21}\}$$

But $K_{11} = K_{12} \cdot K_{21}$. The study of such equilibria in aqueous solutions is hindered
by the slight solubility of the compounds and the tendency of tervalent cations
to hydrolyze. A general spectrophotometric method is elaborated for determination of K_{12} and K_{21} and the system BiCl$_3$ + BiI$_3$ was studied. The equilibria
set in instantaneously.

3. <u>Interaction of iodides on non-transitional metals in dimethylformamide.</u>
The iodides of the bivalent and tervalent ions of non-transitional elements

- except TlI$_3$ - are very well soluble in dimethylformamide while the solubility of the iodides of monovalent ions is small. BiI$_3$ and PbI$_2$ form stable crystalline adducts with dimethylformamide. The interaction of different iodides was observed and in several particular cases quantitatively studied. BiI$_3$, CdI$_2$ and PbI$_2$ react with HgI$_2$ in a 1:2 ratio, TlI and AgI in a 2:1 ratio. The latter compounds form the wellknown iodomercurates and with the former compounds heteropolynuclear complexes are formed. In concentrated solutions (over 10^{-2} M) BiI$_3$ reacts with HgI$_2$ in a 1:3 ratio, and a compound Hg$_3$BiI$_9$(DMF)$_8$ could be prepared. In most of the cases oily substances were obtained. To determine the molecular weight of complexes formed, sedimentation experiments were performed in collaboration with Dr. A. CZUPPON.

HgI$_2$ dissolved in dimethylformamide reacts with triphenylphosphine and, according to the spectrophotometric experiments, at least three different complexes are formed. Similarly complex formation was observed with o-phenantroline, dipyridyl, 8-hydroxyquinoline and dithiooxamide. The quantitative study of these reactions is in progress.

According to the preliminary experiments the same behaviour can be observed in dimethylsulphoxide.

8 C 2

THE FORMATION OF METAL COMPLEXES IN SOLUTION.

George H. Nancollas

Chemistry Department, The University, Glasgow, W.2., Scotland, U.K.

One of the more recent developments in the field of coordination chemistry has been the determination of the heats of formation of metal complexes and ion-pairs in solution. It has been recognised that the measurement of the association constants at only one temperature is of very limited value since it is preferable to regard the free energy change as a consequence of changes in the heat and entropy of association. An effectively constant ΔG for a series of complex formation reactions can mask important changes in ΔH and ΔS.

In Table I, thermodynamic properties are given for a number of complex formation reactions

$$M^{2+} + A^{n-} \rightleftharpoons MA^{(2-n)+}$$
$$MA^{(2-n)+} + A^{n-} \rightleftharpoons MA_2^{2(1-n)+}$$

involving divalent transition metal ions. Thermodynamic association constants for the dicarboxylates and glycinates were calculated from the emf's of cells (1-4) of the types

$$H_2, \ Pt/H_2A, \ NaOH, \ MCl_2/AgCl/Ag.$$

The ΔH values in Table I have been obtained both from the temperature coefficients of the association constants, and from direct calorimetric measurements; the agreement between the two methods is very good (5, 6).

TABLE I

Reaction	ΔG (kcal. mole^{-1})	ΔH (kcal. mole^{-1})	ΔS (cal.deg^{-1} mole^{-1})	ΔS hyd (MA) (cal.deg^{-1} mole^{-1})	Ref.
Glycinate					
$Mn^{2+} + A^-$	-4,32	-0,29	13,5	-29,6	4
$Co^{2+} + A^-$	-6,29	-2,82	13,7	-33,6	4
$Ni^{2+} + A^-$	-8,43	-4,09	14,5	-33,8	4
$CoA^+ + A^-$	-5,42	-3,55	6,3	-4,5	4
$NiA^+ + A^-$	-6,75	-4,69	6,9	-4,8	4
Ethylenediamine					
$Mn^{2+} + en$	-3,75	-2,80	3,0	-	7
$Co^{2+} + en$	-8,10	-6,90	4,0	-	7
$Ni^{2+} + en$	-10,50	-8,90	5,5	-	7
Oxalate					
$Mn^{2+} + A^{2-}$	-5,41	1,42	22,9	-54,8	1
$Co^{2+} + A^{2-}$	-6,54	0,59	23,9	-55,8	1
$Ni^{2+} + A^{2-}$	-7,05	0,15	24,2	-56,1	1
Malonate					
$Mn^{2+} + A^{2-}$	-4,48	3,68	27,4	-57,3	5
$Co^{2+} + A^{2-}$	-5,13	2,90	26,9	-60,0	5
$Ni^{2+} + A^{2-}$	-5,60	1,77	24,8	-63,0	2

The ethylenediamine data are those obtained by CIAMPOLINI, PAOLETTI and SACCONI from calorimetric measurements in 1M KCl (7). It can be seen from Table I that the heat of reaction is more important than the entropy in determining the association constant. The endothermic heat changes are to be expected for the essentially electrostatic interactions involving dicarboxylate ions. The exothermic heat changes accompanying the formation of glycinate complexes reflect the greater covalency of the metal-nitrogen bond. The effect is even more marked when the 5 membered chelate ring is formed through two nitrogen atoms as in the ethylenediamine complexes. With the glycinate complexes, the ΔH values for successive steps increase very slightly and this has been observed with a number of uncharged ligands (8). It seems that the bonding of the second glycinate molecule is not appreciably affected by the presence of the first even though different electrostatic conditions are involved.

Plots of the heat changes against atomic number of the transition metal have been made by a number of workers (e.g. 8). The exothermicity of the reaction with ethylenediamine increases from Mn(II) to Cu(II) and then falls at Zn(II) paralleling the trend observed in the heats of hydration of these metal cations (7). With the glycinate complexes crystal-field effects will account for the considerable increase in exothermicity in going from Mn(II) to Ni(II).

A more detailed discussion of the mechanism of complex formation may be made by combining these thermodynamic properties with kinetic data. There are now a number of techniques available for the study of fast reactions, and the rate of formation of nickel monooxalate complex has been measured spectrophotometrically at a number of temperatures using a flow technique (9). For this system, all the equilibrium thermodynamic properties have been measured (1), and the kinetic data are consistent with the assumption that the reaction of Ni(II) with oxalate ion in neutral solution occurs in the following steps (9)

$$[Ni(H_2O)_6]^{2+} + A^{2-} \rightleftharpoons (H_2O)_5Ni(H_2O)A \qquad K_0 \qquad \{1\}$$

$$(H_2O)_5Ni(H_2O)A \underset{k_{01}}{\overset{k_{10}}{\rightleftharpoons}} (H_2O)_5NiA + H_2O \qquad K_{10} \qquad \{2\}$$

$$(H_2O)_5NiA \underset{k_{02}}{\overset{k_{20}}{\rightleftharpoons}} (H_2O)_4NiA + H_2O \qquad K_{20} \qquad \{3\}$$

where $(H_2O)_5NiA$ represents a half-bonded intermediate. If equilibrium $\{1\}$ is rapidly attained

$$d[(H_2O)_4NiA]/dt = k_{20}[(H_2O)_5NiA] - k_{02}[(H_2O)_4NiA]$$

When $k_{20} \gg k_{01}$, a steady state assumption for the concentration of $(H_2O)_5 NiA$ gives $k_{1f} = k_{10} K_0$. Under these conditions, the formation of $(H_2O)_4 NiA$ is determined primarily by the loss of a coordinated water molecule from the Ni^{2+} ion. Using the Debye-Hückel interaction potential, K_0 has been estimated and the derived k_{10} values are approximately constant for a wide variety of ligands. Some values of the kinetic parameters are given in Table II and it can be seen that the heats and entropies of activation for reaction {2} are close to those for the exchange of water molecules between the inner coordination shell of a Ni^{2+} ion and the bulk of the solution (10).

DAVIES and SMITH have used a stopped flow technique to study the reaction between nickel and thiocyanate ions

$$Ni^{2+} + SCN^- \underset{k_{1d}}{\overset{k_{1f}}{\rightleftharpoons}} NiSCN^+$$

at an ionic strength of 0,5M (11). They also give values for the association constants, K_1, obtained from spectrophotometric measurements at a number of temperatures. The heat of formation, -4,7 kcal.mole^{-1}, obtained from a plot of $\log K_1$ against T^{-1}, appears to be too exothermic. Because of the difficulty of interpreting temperature coefficients of spectral measurements, we have determined ΔH by a direct calorimetric method at an ionic strength of 0,5M, obtaining a value of $\Delta H = -0,6$ kcal.mole^{-1} (12). Since k_{1d} values were not given by DAVIES and SMITH, they have been recalculated from K_1 and k_{1f} at each temperature ($k_{1d} = k_{1f}/K_1$). The activation energy for the decomposition reaction, $E_d = 15$ kcal.mole^{-1}, was obtained from a plot of $\log k_{1d}$ against T^{-1}, and the calculated k_{10}, $\Delta H_{10}^{\ddagger}$ and $\Delta S_{10}^{\ddagger}$ values are given in Table II. It can be seen that these values are similar to oxalate and water exchange data, providing additional evidence for the view that the rates of complex formation are controlled by the rate of loss of a coordinated water molecule from the nickel ion.

TABLE II

KINETIC PARAMETERS FOR THE FORMATION OF NICKEL COMPLEXES

	k_{10} (sec^{-1})	$\Delta H_{10}^{\ddagger}$ (kcal.mole^{-1})	$\Delta S_{10}^{\ddagger}$ (cal.deg^{-1}mole^{-1})
Oxalate	6.10^3	11,2	-3,5
Bioxalate	3.10^3	12,4	-0,8
Thiocyanate	4.10^3	13,0	+2,4
Water	$2,7.10^4$	11,6	+0,6

References

1. McAULEY, A., and NANCOLLAS, G.H., J.Chem.Soc. (London), 1961, 2215.
2. NAIR, V.S.K., and NANCOLLAS, G.H., J.Chem.Soc. (London), 1961, 4367.
3. McAULEY, A., and NANCOLLAS, G.H., J.Chem.Soc. (London), 1961, 4458.
4. BRANNAN, J.R., DUNSMORE, H.S., and NANCOLLAS, G.H., J.Chem. Soc. (London), 1964, 304.
5. McAULEY, A., and NANCOLLAS, G.H., J.Chem.Soc. (London), 1963, 989.
6. BOYD, S., and NANCOLLAS, G.H., Unpublished results.
7. CIAMPOLINI, M., PAOLETTI, P., and SACCONI, L., J.Chem.Soc. (London), 1960, 4553.
8. CIAMPOLINI, M., PAOLETTI, P., and SACCONI, L., "Advances in the Chemistry of the Coordination Compounds", Ed.Kirschner, MacMillan, 1961.
9. NANCOLLAS, G.H., and SUTIN, N., Inorg.Chem., 3, 360 (1964).
10. SWIFT, T.J., and CONNICK, R.E., J.Chem.Phys., 37, 307 (1962).
11. DAVIES, A.G., and SMITH, W.MacF., Proc.Chem.Soc., 1961, 380.
12. NANCOLLAS, G.H., and TORRANCE, K., Unpublished results.

8 C 3

IRON(III) AND PROTON ASSOCIATIONS WITH SOME SINGLY SUBSTITUTED PHENOLATE IONS IN AQUEOUS SOLUTION.

Kaizer E. Jabalpurwala and Ronald M. Milburn
Department of Chemistry, Boston University, Boston, Mass., U.S.A.

In an earlier study (1) in which the stabilities of 1:1 iron(III)-phenolate complexes were measured in aqueous solution for phenol and several singly substituted para or meta derivatives, an approximately linear relation was observed to exist between the free energies for the association reactions:

$$Fe^{3+} + OC_6H_4X^- \rightleftharpoons FeOC_6H_4X^{2+} \qquad \{1\}$$

$$H^+ + OC_6H_4X^- \rightleftharpoons HOC_6H_4X \qquad \{2\}$$

The approximateness of the relation was recognized; deviations from linearity being small but significantly greater than could be attributed to experimental error. It was therefore suggested that the two sets of reactions might have non-parallel entropy dependencies and/or that there might be double bond character to the iron(III)-phenolate bonds (1). In considering the limitations of the above and similar linear free energy relations, WILLIAMS et al (2) have emphasized the importance of double bond character, and have pointed out that conjugated π donor substituents should stabilize iron(III) complexes compared to corresponding proton complexes, while conjugated π acceptor substituents should have the reverse effect.

The present paper describes additional investigations, covering a series of closely related phenols and including temperature dependent measurements, which have been made to specifiy more precisely the factors determining the relative stabilities of iron(III) and proton phenolate complexes. Requirements are outlined for establishing the relative importance of entropy, metal-ligand π bonding, and other contributions to deviations from the strictly linear free energy relation.

Potentiometric titration methods have been used to measure equilibrium quotients at 25° for reactions of type $\{1\}$ and $\{2\}$, with X equal to ortho, meta and para nitro-, fluoro-, chloro-, bromo- and iodo-phenol. The medium was in all cases aqueous 0,10 M perchlorate ($NaClO_4/HClO_4$).

The equilibrium quotients for $\{2\}$ were determined with use of glass and calomel electrodes. Resulting $-\log K_H$ ($= pK_a$) values are: o-NO_2, 7,15; m-NO_2, 8,14; p-NO_2, 7,13; o-F, 8,60; m-F, 8,92; p-F, 9,57; o-Cl, 8,44; m-Cl, 8,87; p-Cl, 9,21; o-Br, 8,33; m-Br, 8,86; p-Br, 9,17; o-I, 8,43; m-I, 8,85; p-I, 9,15; H, 10,01. Uncertainties are of the order $\pm 0,02$ units.

Equilibrium quotients for reactions of type $\{1\}$ were determined by measurements of $[Fe^{3+}]$ and $[H^+]$ with the cell systems: Cal.electr./Satd. KCl/Fe^{2+}, Fe^{3+}, H^+, phenol/Pt or Glass electr. Iron(II)-phenol complexes were absent over the relevant pH range. By maintaining $[Fe^{3+}]_t \leqslant 10^{-3}$ M, $[H^+] \geqslant 2.10^{-3}$ M and $[ClO_4^-]_t = 0,1$ M, polynuclear iron(III) species could be neglected and it was necessary to consider only Fe^{3+}, $FeOH^{2+}$, $Fe(OH)_2^+$ and $FeOC_6H_4X^{2+}$. While activity coefficients and the liquid junction potential are constant, $\log([Fe^{3+}]_t / [Fe^{3+}]) = (E_0' - E) / (0,05916)$, where E and E_0' refer respectively to the measured redox potentials and to the potential the cell would have in the absence of hydrolyzed and phenolyzed iron(III). Examination of the hydrolysis in the absence of phenol demonstrated that E_0' could not be obtained satisfactorily by extrapolation of E to large acidities. The difficulty is attributed to changes in the liquid junction potential as the acidity is significantly increased. Therefore, successive approximations in E_0' followed by least squares refinement of the two hydrolysis quotients were carried out until input and output E_0' values agreed. The resulting first hydrolysis quotient ($2,84.10^{-3}$) is in excellent

agreement with the value appropriate to the medium as obtained earlier by
spectrophotometry (3). The $E_0^!$ from this analysis was used together with the
two hydrolysis quotients to obtain equilibrium quotients for {1} (K_X) by a least
squares procedure. Good agreement with values determined by spectrophotometry is obtained. Resulting $-\log K_X$ values are: o-NO$_2$, 5.93; m-NO$_2$, 6.91;
p-NO$_2$, 5.66; o-F, 7.23; m-F, 7.96; p-F, 8.62; o-Cl, 7.20; m-Cl, 7.49;
p-Cl, 7.89; o-Br, 6.94; m-Br, 7.62; p-Br, 7.85; o-I, 7.03; m-I, 7.53;
p-I, 8.22. While the absolute uncertainty in values is of the order ± 0.20
units, the values within the set are internally consistent to about ±0.05 units.

Fig. 1 shows a plot of $-\log K_X$ against $-\log K_H$. Included are points also
for phenol and p-cresol where the K_X values were obtained by spectrophotometry. Minor corrections have been applied to the latter values, so that these
and the potentiometrically obtained values would constitute a set of internally
consistent data.

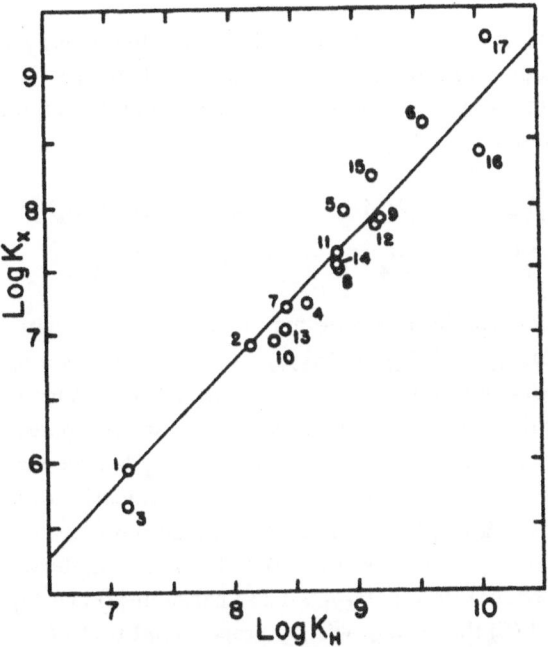

Fig. 1 - Comparison of Equilibrium Quotients for Reactions of types
{1} and {2}. Substituent X: 1, o-NO$_2$; 2, m-NO$_2$; 3, p-NO$_2$;
4, o-F; 5, m-F; 6, p-F; 7, o-Cl; 8, m-Cl; 9, p-Cl; 10, o-Br;
11, m-Br; 12, p-Br; 13, o-I; 14, m-I; 15, p-I; 16, H;
17, p-CH$_3$. The line is of unit slope.

The distribution of points for the larger group of phenols is summarized rather well by a single line of unit slope. The absence of serious scatter for ortho substituted phenols is noteworthy, and indicates that any special inter- actions of the iron with the chosen ortho substituents are effectively balanced by proton-substituent interactions in the free phenols. In ÅGREN's studies on iron(III)-phenolate chelates, on the other hand, correlation of the logarithms of the proton and iron(III) association constants required three separate lines (4). The scatter in figure 1 for certain of the phenols is nevertheless somewhat larger than can be reasonably attributed to experimental error.

In considering the distribution of points for phenol and its meta and para substituted derivatives, arguments on metal-ligand π bonding can be applied (2) to explain some of the scatter. In this way the stabilization of the iron complexes for the π electron donor p-CH_3, p-F and p-I and the destabilization of the complex for the π electron acceptor p-NO_2 could be accomodated. But the positions of p-Cl, p-Br and m-F would still leave some questions.

However, for deviations in the linearity of the free energy plot to reflect bonding differences within the iron(III) and proton phenolate complexes requires special circumstances, which can be specified by reference to the exchange reactions:

$$FeOC_6H_5^{2+} + OC_6H_4X^- \rightleftharpoons OC_6H_5^- + FeOC_6H_4X^{2+} \qquad \{3\}$$

$$HOC_6H_5 + OC_6H_4X^- \rightleftharpoons OC_6H_5^- + HOC_6H_4X \qquad \{4\}$$

For any reaction of type $\{3\}$ or $\{4\}$ we may write, in accord with HEPLER (5), $\Delta F = \Delta H_{int} + \Delta H_{ext} - T(\Delta S_{int} + \Delta S_{ext})$, where subscripts refer to internal contributions arising within the complex and to external contributions associated with solute-solvent interaction. Now, linearity between ΔF values for $\{1\}$ and $\{2\}$ is equivalent to direct proportionality between ΔF values for $\{3\}$ and $\{4\}$. To be able to equate the deviations in linearity of the ΔF plots (for $\{1\}$ & $\{2\}$, or $\{3\}$ & $\{4\}$) to the differences which substituents X cause in the internal bond energies for the iron(III) and proton complexes, there must be a direct proportionality of $(\Delta H_{ext} - T\Delta S)$ between $\{3\}$ and $\{4\}$. However, for $\{4\}$ it is argued (5) that ΔH_{ext} will be proportional to ΔS (= ΔS_{ext}, since $\Delta S_{int} \sim 0$), and on similar grounds there should be a proportionality between ΔH_{ext} and ΔS for $\{3\}$. The requirement of proportionality of $(\Delta H_{ext} - T\Delta S)$ between the reactions sets thus reduces to proportionality between the ΔS values.

Values of ΔH and ΔS for $\{4\}$ with X equal to o-NO_2, m-NO_2, p-NO_2, o-Cl and p-Cl are available through calorimetric studies (6). For four of these five substituents the entropy contribution to the free energy at 25° exceeds the heat contribution; also, there is no general proportionality between ΔH and ΔS. We have made measurements to determine ΔH and ΔS for reactions of type {3} with a similar group of substituents. The necessity of working with very dilute iron(III) solutions (to avoid polynuclear species) places special demands on the precision required for calorimetric measurement of the heats for these reactions. Measurements to determine ΔH and ΔS have therefore been made through study of the temperature dependence of the equilibria $Fe^{3+} + HOC_6H_4X \rightleftharpoons FeOC_6H_4X^{2+} + H^+$, where equilibrium quotients have been determined spectrophotometrically with correction for the hydrolysis from previous work (7). The relationships between the heats and entropies for reactions of types {3} and {4} will be discussed.

References

1. MILBURN, R.M., J.Am.Chem.Soc., 77, 2064 (1955).
2. JONES, J.G., POOLE, J.B., TOMKINSON, J.C., and WILLIAMS, R.J.P., J.Chem.Soc. (London), 1958, 2001.
3. MILBURN, R.M., and VOSBURGH, W.C., J.Am.Chem.Soc., 77, 1352 (1955).
4. ÅGREN, A., Svensk.Kem.Tidskr., 68, 189 (1956).
5. HEPLER, L.G., J.Am.Chem.Soc., 85, 3089 (1963).
6. FERNANDEZ, L.P., and HEPLER, L.G., J.Am.Chem.Soc., 81, 1783 (1959).
7. MILBURN, R.M., J.Am.Chem.Soc., 79, 537 (1957).

9 C 1

COMPLEXE EQUILIBRIA OF UO_2^{2+} IN AQUEOUS SOLUTIONS WITH LIGANDS HAVING OXYGEN DONOR ATOMS.

L. Sommer and M. Bartušek
Department of Analytical Chemistry of the University of Brno, Czechoslovakia.

The strong affinity of UO_2^{2+} to oxygen donor atoms causes a anomal tendency of this ion and its complexes to hydrolyze in aqueous solution. This fact often influences the data of complex equilibria and their constants and many differences could be found in the literature.

Having proved the hydrolysis of UO_2^{2+} in perchlorate and nitrate medium there follows the only main equilibrium:

$$2 UO_2^{2+} + 2 H_2O \rightleftharpoons [UO_2(OH)_2UO_2]^{2+} + 2 H^+$$

in diluted solution with $c_M = 2.10^{-4} - 4.10^{-3}$ gat U/1 from pH 4,0 - 5,5.
This dimer is present also in the medium of 30 Vol% or 50 Vol% ethanol being
formed already in more acidic solutions. At the beginning of hydrolysis a certain
amount of dimeric $[(UO_2)OH(UO_2)]^{3+}$ is formed at pH 3 - 4 and $c_M < 3.10^{-3}$
gat U/1. At the conditions mentioned above no monomer $[UO_2OH]^+$ or higher
polymers of the core+ links type could be found. At pH ≤ 3 the hydrated
monomeric UO_2^{2+}-ion is the only species in the solution (1).

From the ligands having oxygen donor atoms the monohydric and polyhydric phenoles, o-phenolecarboxylic acids, hydroxy-4-pyrones are of
particular interest for binding UO_2^{2+} in complexes and its spectrophotometric
determination in aqueous solution. The monodentate phenolic hydroxyl is
already bound giving UO_2^{2+} complexes stable against hydrolysis up to pH 5.
If bidentate oxygen atom groups particularly o- and peri-diphenolic groups
are present in the ligand molecule, the stability of the UO_2^{2+} complexes
increases strongly.

In this case some monomeric complex species can be formed in steps:
$MR \rightarrow MR_2 \rightarrow MR_3$ according to the increasing pH of the solution. Depending
on pH and the ligand concentration in excess the hydrolysis of the MR complex
or chelate into dimeric species is taking part simultaneously. This hydrolysis
predominates for $c_R/c_M \leq 10$ at pH ≥ 4 and in the case of monohydric and
some sulfonated polyhydric o- and peri-diphenols also in solutions with a large
ligand excess.

The occuring UO_2^{2+}-complex equilibria with some ligands studied are
given shortly here (UO_2^{2+} as M, charges of complexes ommited).

Complexes or chelates with M:R = 1:1 are formed in the range up to
pH 4 - 4,5 following equilibria with Kojic acid (5-Hydroxy-2-hydroxymethyl-
-4-pyron):

$$M + RH \rightleftharpoons MR + H^+ \qquad \{1\}$$

$$M + RH \rightleftharpoons (MR)_{chelate} + H^+ \qquad \{2\}$$

with meconic acid (5-Hydroxy-2,6-dicarboxy-4-pyrone):

$$M + RH_3 \rightleftharpoons M(RH_2) + H^+ \qquad \{3\}$$

$$M + RH_2 \rightleftharpoons M(RH) + H^+ \qquad \{4\}$$

$$M(RH) \rightleftharpoons MR + H^+ \qquad \{5\}$$

with catechol, catechol-3-sulfonic, catechol-3,5-disulfonic, 2,3-dihydroxy-
naphtalene-6-sulfonic, chromotropic acids (their sodium salts):

$$M + RH_2 \rightleftharpoons MRH + H^+ \qquad \{6\}$$

$$MRH \rightleftharpoons MR + H^+ \qquad \{7\}$$

$$M + RH_2 \rightleftharpoons MR + 2 H^+ \qquad \{8\}$$

The ranges of protonized complex forms, with kojic acid the monodentate complex in contrary to the chelate, are often well distinguished in the presence of large ligand excess at pH $\leq 2,5$.

The second step is formed as follows with 5-Hydroxy-4-pyrones:

$$MR + RH \rightleftharpoons MR_2 + H^+ \qquad \{9\}$$

at pH 3,8 - 6 with meconic acid in 30 - 50 Vol% ethanol, at pH 4,9 - 6,3 with kojic acid; and with catechol and 2,3-dihydroxynaphtalene-6-sulfonic acid (DHNS) following $\{10\}$ at pH 4,8 - 6,8:

$$MR + RH_2 \rightleftharpoons MR(RH) + H^+ \qquad \{10\}$$

The third step is occuring with 5-Hydroxy-4-pyrones following:

$$MR_2 + RH \rightleftharpoons MR_3 + H^+ \qquad \{11\}$$

at pH 5,5 - 6,3 and with catechol and DHNS only at pH 6 - 7,9 in dependance on ligand excess:

$$MR(RH) + RH_2 \rightleftharpoons MR(RH)_2 + H^+ \qquad \{12\}$$

The formation of dimeric species is usually following the first step in the UO_2-chelate systems with catechol, catechol-3-sulfonic and catechol--3,5-disulfonic acids (Tiron) at pH 4 - 5 and probably accompanies other equilibria in the system of chromotropic acid at pH 3,9 - 4,2:

$$2 MR + H_2O \rightleftharpoons M_2R_2(OH) + H^+ \qquad \{13\}$$

In solutions with small ligand excess ($c_R/c_M = 20$) the formation of M_2R and $M_2R(OH)$ complexes resp. with tiron was observed already from pH > 2,5. With tiron the equilibration at pH > 5 is slow according to the monomerisation of the dimeric species taking part and binding of further molecules of ligand. Trimer formulated by GUSTAVSON et al. (2) could not be found in our experimental conditions in solutions with $c_R/c_M > 10$. In the system with

chromotropic acid it could not yet be distinguished between the following
equilibria at pH 4,8 - 6,9:

$$MR + H_2O \rightleftharpoons MR(OH) + H^+ \qquad \{14\}$$

$$M_2R_2(OH) + H_2O \rightleftharpoons M_2R_2(OH)_2 + H^+ \qquad \{15\}$$

$$MR + RH_2^{2-} \rightleftharpoons MR_2 + 2\,H^+ \qquad \{16\}$$

Depending on the c_R/c_M one of them prevails.

Pyrogallol reacts with UO_2^{2+} forming a M_2R-complex including solutions
with ligand excess at pH 4 - 5 as follows:

$$2\,M + RH_3 \rightleftharpoons M_2R + 3\,H^+ \qquad \{17\}$$

$$2\,M + RH_3 + H_2O \rightleftharpoons M_2(RH)(OH) + 3\,H^+ \qquad \{18\}$$

This dimer monomerized again at pH 5 - 7 probably binding further ligand
molecules to MR and M(RH)R chelates.

With phenol and resorcinol only MR complex is found following $\{1\}$ at
pH 3,2 - 4 with phenol and pH 3,3 - 4 with resorcinol, which is easily
hydrolyzed in $M_2R_2(OH)$ and $M_2(OH)_2$ species at pH 4,2 - 5,0.

The sufficient bond stability of UO_2^{2+} with monodentate phenolic hydroxyl
in aqueous solution allows the previous monodentate complex formation and
following chelation giving species of different optical properties [see also (3)].
The coordination of the third ligand molecule demands previous weakening
of the chelating ring of the second ligand molecule and only simple complex
bond of the third one is possible. This can be imagined for the catechol-UO_2^{2+}
complexes as follows:

To solve these complicated complex equilibria some graphical and numerical analysis of spectrophotometric curves Extinction $= f(pH)_{c_R, c_M}$

and Extinction $= f(c_R)_{c_M, pH}$ curves and potentiometric $Z = (c_{OH} + [H] - [OH])/c_M = f(pH)_{c_M, c_R}$ curves have been carried out.

Stability and equilibrium constants and optical characteristics of the complexes and chelates formed will be given at the conference.

References

1. BARTUŠEK, M., and SOMMER, L., Z.phys.Chem. (Leipzig), in press.
2. GUSTAVSON, R.L., RICHARD, C., and MARTELL, A.E., J.Am.Chem. Soc., 82, 1526 (1960).
3. ŠNAJDERMAN, S.Ja., and GALINKER, E.V., Z.neorg.chim., 7, 279 (1962).

9 C 2

STABILITY OF SOME METAL COMPLEXES CONTAINING OXYGEN, SULFUR, OR SELENIUM AS DONOR ATOMS.

Kazuo Yamasaki and Keinosuke Suzuki
Nagoya University, Chikusa, Nagoya, Japan.

Studies on the relative affinities of various donor atoms for acceptor atoms indicate that there are two classes of the latter (1):

(a) those which form their most stable complexes with the first donor atoms of each group of the periodic table, i.e., with N, O, and F; and

(b) those which form their most stable complexes with the second or a subsequent donor atom of each group.

In addition to these two classes there are borderline elements which can show both characters. Since their character depends on the donor atoms, it may be interesting to study the stability of the complexes of the borderline elements and look for factors which effect the strength of metal-donor bonds. From this point of view the stability constants of some metal complexes formed with the ligands containing O, S, or Se as the donor atom were studied and their affinities towards bivalent metals were compared (2).

As metals Cu, Ni, Zn, Cd and Pb and as ligands the following compounds were used:

(a)	(b)	(c)	(d)

(e)	(f)

These ligands have analogous structures with oxygen, sulfur, or selenium as the donor atoms. Glutaric acid, the ligand (a), was chosen for comparison.

The solutions containing metal perchlorate and ligand were prepared and they were titrated with standard alkali at 25°. The ionic strength was kept at 0,1 by adding $NaClO_4$. The acid dissociation constants of the ligands and the stability constants of the 1:1 complexes formed are listed in Table I and the ratio of these two constants in Table II.

TABLE I

ligand	acid dissociation constant		stability constant $(\log k_1)$				
	pK_1	pK_2	Cu	Ni	Zn	Cd	Pb
a	4,14	5,01	2,4	1,6	1,6	2,0	2,8
b	2,77	3,92	3,9	2,6	3,6	3,3	4,4
c	3,15	4,13	4,3	4,0	2,9	2,6	3,6
d	3,17	4,31	3,6	2,7	2,1	2,3	3,2
e	2,51	4,36,	3,1	2,0	2,6	2,0	~2,6 (ppt.)
f	2,93	4,01	3,0	1,9	1,8	2,0	~2,5 (ppt.)

TABLE II

ligand	$\log k_1/(pK_1 + pK_2)$				
	Cu	Ni	Zn	Cd	Pb
a	0,26	0,18	0,18	0,22	0,30
b	0,58	0,39	0,54	0,49	0,66
c	0,59	0,55	0,40	0,36	0,49
d	0,48	0,36	0,28	0,31	0,43
e	0,45	0,29	0,39	0,29	(0,38)
f	0,43	0,28	0,26	0,29	(0,36)

For the ligands (e) and (f) zinc alone shows an appreciable difference of stability, while other metals show very little differences. From the results obtained above, the five metals studied are divided into two classes:

A. larger affinity towards O than towards S and Se: Zn, Cd, Pb

B. larger affinity towards S than towards O and Se: Cu, Ni.

This classification holds except for the complexes of copper and nickel with the ligands (e) and (f).

The relative affinities of oxygen, sulfur and selenium seem to be

$$Pb > Cu > Zn > Cd > Ni$$
$$Cu > Ni > Pb > Zn > Cd$$

and
$$Cu > Pb > Ni > Cd > Zn \ .$$

The stability order for charged sulfur is different from the above and a larger affinity of zinc than nickel was reported by several authors (3). The largest stability of lead for ligands containing oxygen as donor atoms has already been reported (2, 4).

References

1. AHRLAND, S., CHATT, J., and DAVIES, N.R., Quart.Reviews, 12, 265 (1958), AHRLAND, S., et al., J.Chem.Soc. (London), 1958, 264, 1403.
2. YASUDA, M., YAMASAKI, K., and OHTAKI, H., Bull.Chem.Soc. Japan, 33, 1067 (1960); SUZUKI, K., and YAMASAKI, K., J.Inorg. Nucl.Chem., 24, 1093 (1962).
3. For example, LEUSSING, D.L., J.Am.Chem.Soc., 80, 4180 (1958); SCHWARZENBACH, G., et al., Helv.Chim.Acta, 40, 1147 (1955).

4. YASUDA, M., Z.physik.Chem. N.F., **29**, 377 (1961).

9 C 3

STABILITÄTS- UND DISSOZIATIONSKONSTANTEN EINIGER o-DIPHENOL-GERMANIUM-VERBINDUNGEN.

N. Konopik und P. Svejda
Institut für Physikalische Chemie der Universität Wien, Austria.
(to be presented in English)

Nitrobrenzcatechine bilden, analog dem Brenzcatechin, mit Germanium-säure mittelstarke 2-basige komplexe Säuren der Zusammensetzung

$$H_2\left[Ge\left(O\text{-}\bigcirc\text{-}NO_2\right)_3\right] \quad \text{resp.} \quad H_2\left[Ge\left(O\text{-}\bigcirc\text{-}NO_2\right)_3\right].$$ Auf Grund der charak-

teristischen Absorptionsspektren ihrer Anionen wurde eine neue optische Bestim-mungsmethode für Germanium ausgearbeitet (1). Im Rahmen dieser Studien sollten auch die Stabilitätskonstanten der komplexen Säuren sowie deren Dis-soziationskonstanten ermittelt werden.

Spektrophotometrische Untersuchungen ergaben, daß neben den oben angeführten Komplexen, in denen das Verhältnis Ge : o-Diphenol 1 : 3 be-trägt, und ihren 2fach negativ geladenen Ionen Ge-reichere Verbindungen in Lösung vorhanden sind.

Zur Aufklärung der in Lösung vorliegenden Gleichgewichte werden spektrophotometrische und potentiometrische Messungen bei konstanter Tem-peratur (25°) und konstanter Ionenstärke ($I = 1,0$ M $NaClO_4$) herangezogen. Ein geeignetes Meß- und Auswerteverfahren erlaubt, aus den ermittelten Po-tentialwerten Wasserstoffionenkonzentrationen zu berechnen.

Bezeichnet man die Germaniumsäure $Ge(OH)_4$ mit B, das komplexbild-dende o-Diphenol mit A, und mit b bzw. a die Konzentrationen an freiem B bzw. freiem A, so können die in Lösung vorliegenden Gleichgewichte durch folgende Reaktionen beschrieben werden:

$$B + A = AB \qquad k_1 = \frac{[AB]}{a.b} \qquad \{1\}$$

$$B + 2A = A_2B \qquad k_2 = \frac{[A_2B]}{a^2.b} \qquad \{2\}$$

$$B + 3A = A_3B \qquad k_3 = \frac{[A_3B]}{a^3.b} \qquad \{3\}$$

Bei dem Komplex A_3B handelt es sich, wie eingangs erwähnt, um eine relativ starke 2-basige Säure. Man hat somit ein weiteres Gleichgewicht, nämlich die Dissoziation von A_3B nach

$$A_3B = 2\,H^+ + A_3B^{2-} \qquad K_3 = \frac{[H^+]^2 \cdot [A_3B^{2-}]}{[A_3B]}$$

zu berücksichtigen.

Der Komplex A_2B war nur in fester Phase bekannt (2); in Lösung konnte er bisher nicht nachgewiesen werden. Nach ANTIKAINEN (3) handelt es sich bei A_2B um ein Phenolat mit der Struktur

das keine sauren Eigenschaften zeigt.

Aus den bei konstanter Wasserstoffionenkonzentration gemessenen Extinktionen E wird eine Funktion ϵ' nach

$$\epsilon' = \frac{E - \epsilon_A A_0 d}{B_0 d}$$

E ...... Extinktion
ϵ_i ..... Extinktionskoeffizient von i
A_0 ..... Gesamtkonzentration an o-Diphenol
B_0 ..... Gesamtkonzentration an Ge
d ...... Schichtdicke

berechnet, die mit den Gleichgewichtskonstanten und den Konzentrationen a und b auf folgende Weise verknüpft ist:

$$\epsilon' = \frac{\epsilon_B b + (\epsilon_{AB} - \epsilon_A)k_1 ab + (\epsilon_{A_2B} - 2\epsilon_A)k_2 a^2 b + (\epsilon_{A_3B} - 3\epsilon_A)k_3 a^3 b + (\epsilon_{A_3B^{2-}} - 3\epsilon_A)k_3 \frac{K_3}{[H^+]^2} a^3 b}{b + k_1 ab + k_2 a^2 b + k_3 a^3 b + k_3 \frac{K_3}{[H^+]^2} a^3 b}$$

Mit Hilfe der für verschiedene Gesamtkonzentrationen A_0 und B_0 ermittelten Funktionen ϵ' lassen sich nach der Methode der korrespondierenden Lösungen (4) die Konzentrationen an freiem o-Diphenol a sowie die mittleren Ligandenzahlen pro Ge-Atom $\bar{n}$ nach

$$\bar{n} = \frac{k_1 a + 2\,k_2 a^2 + 3\,k_3 a^3 + 3\,k_3 \dfrac{K_3}{[H^+]^2}\,a^3}{1 + k_1 a + k_2 a^2 + k_3 a^3 + k_3 \dfrac{K_3}{[H^+]^2}\,a^3}$$

berechnen.

Führt man die Messungen bei verschiedenen Wasserstoffionenkonzentrationen durch, so erhält man ein System von Gleichungen, die man nach den unbekannten Gleichgewichtskonstanten k_1, k_2, k_3 und K_3 auflösen kann.

Literatur

1. KONOPIK, N., und WIMMER, G., Mh.Chem., 93, 1404 (1962).
2. BÉVILLARD, P., Bull.Soc.Chim.France, 1954, 304.
3. ANTIKAINEN, P.J., Suomen Kemistilehti, 32B, 211 (1959).
4. BJERRUM, J., Kgl.Danske Videnskab.Selskab Mat. -fys.Medd., 21, Nr.4 (1944). FRONAEUS, S., Acta Chem.Scand., 5, 139 (1951). Vgl. auch ROSSOTTI, F.J.C., und ROSSOTTI, H., "The Determination of Stability Constants", McGraw-Hill Book Comp., Inc., New York, Toronto, London 1961.

0 C 1

STEPWISE FORMATION OF METAL COMPLEXES IN NON-AQUEOUS SOLVENTS.

David Dyrssen

Dept. of Analytical Chemistry, University of Gothenburg, Göteborg, Sweden.

Stepwise complex formation in non-aqueous solvents is a much less frequent phenomena than in aqueous solutions, where the stepwise substitution of water molecules by mono-dentate ligands is a common process with hydrated metal ions. Experience from work with aqueous solutions demonstrates two essential factors for a study of stepwise complex formation: a) effective solvation, which dissociates the metal salt MX_m into $M(H_2O)_n^{m+}$ and mX^-, b) the possibility to vary the central ion and the ligand concentration over a wide range. These two conditions are not so easy to realize in non-aqueous solvents, and this may be one reason why most stability constants have been determined in aqueous solutions. However, the knowledge of a metal-ligand system will be more complete if it is studied in several different media (states). The stability of a complex in water, may, for example, be due to a mixed ligand stabilization of a complex $ML_4(H_2O)_2$, which is built into the water lattice. Two examples are taken here to illustrate this point and to demonstrate some techniques of studying stepwise complex formation in non-aqueous solvents. Further material will be delivered at the conference.

A study (1) of the complex formation between Bi^{3+} and Cl^- in 4M $H(Cl, ClO_4)$ has given the stability constants and spectra of five $BiCl_n$ complexes. An X-ray structure determination (2) of $BiCl_{1,167}$ shows that this solid is composed of one Bi_9^{5+} nonagon, one-half $Bi_2Cl_8^{2-}$, and two $BiCl_5^{2-}$ groups. If a solution of trilaurylammoniumchloride (R_3NHCl) in xylene is saturated with $BiCl_3$, it can be shown (3) that the xylene phase contains one $BiCl_3$ per R_3NHCl. The UV-spectrum of the solution is very similar to the spectrum found for $BiCl_4^-$ in water (1, 4). If more R_3NHCl is added, the spectrum is changed to a spectrum very similar to the one found for $BiCl_6^{3-}$ in water (4), or $BiCl_5^{2-}$ (1). Furthermore, only one R_3NHCl per $R_3NHBiCl_4$ is needed to shift the UV-spectrum. This demonstrates that the step

$$BiCl_4^- + Cl^- \rightleftharpoons BiCl_5^{2-} \qquad \{1\}$$

has a very large formation constant in xylene. This also explains the relation-ship

$$\log D_{Bi} = 2 \log [R_3NHCl]_{xylene} + const. \qquad \{2\}$$

found by a study (3) of the distribution of radioactive $^{207}Bi^{3+}$ between solutions of R_3NHCl in xylene and 3 M $H(Cl, ClO_4)$ with constant $[Cl^-]$. $BiCl_6^{3-}$ is thus not found in xylene, cf. ref. (4, 5).

For the study of adduct reactions it may be an advantage to have the acceptor molecule in an inert non-solvating medium (e.g. carbontetrachloride or hexane), and then increase the concentration of the donor (e.g. dibutyl-sulfoxide, tributylphosphate) in the system. Very many metal salts are good acceptors, but in most cases they are not soluble in inert solvents. However, metal chelates can often be sufficiently dissolved in inert solvents or chloro-form to study the change of spectrum (6, 7), heat of reaction (7) or increase of distribution (8) on the addition of donor molecules. By application of the law of mass action it is thus possible to determine the constants β_n for the stepwise adduct formation in non-aqueous solvents

$$MA_m + nB \rightleftharpoons MA_mB_n \qquad \{3\}$$

The treatment of a system using this formula will (just as for aqueous systems) prove the existence of the complexes MA_mB_n. For many adducts (or solvates), e.g. $Eu(NO_3)_3(TBP)_3$, the possibility of finding intermediate complexes has been overlooked, but it has been shown (9) that the β-diketone complex $Eu(TTA)_3$ can form both $Eu(TTA)_3TBP$ and $Eu(TTA)_3(TBP)_2$ depending on the TBP concentration in the carbontetrachloride diluent. It may be guessed that $Zn(TTA)_2$ would also form two solvates, but distribution experiments (8) show that only $Zn(TTA)_2TBP$ is formed.

Variation of the concentrations of the reactions, measuring some quantity related to the reaction, and treating the experimental data by the law of mass action will probably be a more used technique to establish the composition of complexes in non-aqueous solvents, and give more stability constants to tabulate in the future.

References

1. HUME, D.N., and NEWMAN, L., J.Am.Chem.Soc., 79, 4576 (1957).
2. HERSHAFT, A., and CORBETT, J.D., J.Chem.Phys., 36, 551 (1962).
3. DYRSSEN, D., EKBERG, S., and SEKINE, T., to be published.
4. HAIGHT, Jr., G.P., SPRINGER, C.H., and HEILMANN, O.J., Inorg. Chem., 3, 195 (1964).
5. AHRLAND, S., and GRENTHE, I., Acta Chem.Scand., 11, 1111 (1957), 15, 932 (1961).
6. GRADDON, D.P., and WATTON, E.C., J.Inorg.Nucl.Chem., 21, 49 (1961).
7. MAY, W.R., and JONES, M.M., J.Inorg.Nucl.Chem., 25, 507 (1963).
8. SEKINE, T., and DYRSSEN, D., J.Inorg.Nucl.Chem., in press.
9. SEKINE, T., Proceedings 7ICCC, Abstract 7J4, 1962.

0 C 2

METHOXIDE COMPLEXES OF METAL IONS IN METHANOL.

Rudolf Gut

Eidgenössische Technische Hochschule, Zürich, Switzerland.

The formation of methoxide complexes was studied by measuring the pH in titrations, with methylate, of the chlorides and methylates of metals, in absolute methanol. The pH determinations were carried out with Pt/H_2 or Pd/H_2 electrodes, which were found to respond rapidly, reproducibly and strictly proportional to the logarithms of the concentrations of solvated protons. The problem of changing activity coefficients and ion-pair formation were avoided by the use of high concentrations (1 F) of a supporting electrolyte [N(CH$_3$)$_4$Cl or LiCl]. The latter effect may be serious because the dielectric constant of the solvent is about one third that of water. Because of the high concentration of supporting electrolyte the cations (H$^+$ and cationic methoxide complexes) will form ion pairs predominantly with Cl$^-$, with concentrations proportional to that of the free cations. Similar considerations apply to the anions (OR$^-$ and anionic methoxide complexes). Simple mass-action expressions can therefore be used, provided that symbols like M$^{\partial +}$ are under-stood to include solvated and ion-pair species that contain M$^{\partial +}$. The kinds of species involved here are, of course, not known.

All results apply to 1 F N(CH$_3$)$_4$Cl and 20^0 unless otherwise stated. The titration of methanolic hydrochloric acid with lithiummethylate shows the expected Nernst relation between potential of the hydrogen electrode and total concentrations of H$^+$ in all pertinent forms and that of methylate, respectively: pK$_M$ = -log [H$^+$][OR$^-$] = 16,60; pK$_M$ = 16,05, 1 F LiCl. This permits the definition of pH-scales based on molar concentrations. The titration curves or proton acids like acetic acid (pK = 8,9$_5$), acetylacetone (pK = 11,8), ammoniumion (pK = 11,2), oxalic acid (pK$_1$ = 8,4, pK$_2$ = 5,1$_5$) and pyrocatechol (pK$_1 \sim$ 15,4, pK$_2$ = 13,2) were as expected and permitted to check the measuring techniques.

The following results are of interest. Addition of chlorides to the solvent often results in extensive solvolysis under the conditions chosen (c $\sim$ 10^{-2} F).

Without the addition of methylate BCl$_3$ and SiCl$_4$ form quantitatively B(OR)$_3$ and Si(OR)$_4$ and an equivalent amount of solvated protons. Si(OR)$_4$ does not coordinate additional OR$^-$. B(OR)$_3$ adds one more OR$^-$, with the development of a normal buffer region: log [B(OR)$_4^-$] / [B(OR)$_3$][OR$^-$] = 5,6$_5$. PCl$_3$ forms dimethylphosphite (70%) according to PCl$_3$ + 3 ROH $\longrightarrow$ (RO)$_2$PHO + 2 RCl + 2 HCl and 30% P(OR)$_3$. The proton bound to the P atom in dimethylphosphite cannot be titrated, because of its weak acidity. GeCl$_4$ undergoes some solvolysis (2,6 OR$^-$ per Ge). Assuming mononuclear complexes only, the constant for the reaction Ge(OR)$_3^+$ + OR$^- \rightleftharpoons$ Ge(OR)$_4$ can be evaluated: log K = 13,6$_5$. The addition of 4 OR$^-$ is followed by a pH jump. More OR$^-$ is taken up in the alkaline region, but quantitative data can not be given because of poisoning of the hydrogen electrode.

NbCl$_5$ and TaCl$_5$ show some solvolysis (2,6 OR$^-$ per M). The constant for the addition of the third OR$^-$ can be calculated for Nb : Nb(OR)$_3^{2+}$ + OR$^- \rightleftharpoons$ Nb(OR)$_4^+$: log K = 14,4. Classical buffer regions have been found for the reactions M(OR)$_4^+$ + OR$^- \rightleftharpoons$ M(OR)$_5$ (log K = 10,4$_5$ für Nb, 11,4$_5$ for Ta) and M(OR)$_5$ + OR$^- \rightleftharpoons$ M(OR)$_6^-$ (log K = 5,4 for Nb, 6,7 for Ta in

1 F $N(CH_3)_4Cl$ and 6, 4_5 in 1 F LiCl).

$TiCl_4$ reacts according to $Ti^{4+} \longrightarrow Ti(OR)_2^{2+} + 2 H^+$. The two protons can be titrated like hydrochloric acid. Further addition of OR^- produces $Ti_2(OR)_8$, probably with the formation of polynuclear intermediates. $Ti_2(OR)_8$ adds one more OR^- to form $Ti_2(OR)_9^-$.

$AlCl_3$ undergoes practically no solvolysis. Addition of OR^- results first in the formation of $Al_2(OR)_3^{3+}$; more OR^- precipitates $Al(OR)_3$, which on further addition of OR^- goes into solution as $Al(OR)_4^-$. $GaCl_3$ shows no solvolysis and reacts with OR^- forming polynuclear species; more OR^- precipitates $Ga(OR)_3$, which dissolves as $Ga(OR)_4^-$, adding a fourth OR^-. $InCl_3$, investigated in 1 F LiCl, shows no solvolysis. The titration curve shows the ready addition of two OR^-; a buffer region reveals the coordination of a third and last OR^-, no precipitate is formed. Methanol-insoluble $Zn(OR)_2$ dissolves in 1 F $N(CH_3)_4Cl$ losing half a OR^- per Zn and forming $Zn_2(OR)_3^+$. The same and only species is found in the titration of $ZnCl_2$ in 1 F LiCl. Te, Sn(IV) and Sb(III) could not be investigated because of poisoning effects of the hydrogen electrode. $U(OR)_5$ is not soluble enough to be titrated.

It is exceedingly important that the water content of the solvent be very small compared to the amount of substance titrated. The influence of traces of water was investigated in titrations of $Nb(OR)_5$, where water was added intentionally. In such solutions the buffer region $Nb(OR)_5 + OR^- \rightleftharpoons Nb(OR)_6^-$ disappears. In its place a steep pH rise is found, indicating the formation of polynuclear complexes containing hydroxide. Similar effects are found in titrations of $NbOCl_3$ and $WOCl_4$. $NbOCl_3$ shows an uptake of two OR^- in the acidic region and a third OR^- is coordinated after a small pH change. Further addition of OR^- results in the steep pH rise mentioned. $WOCl_4$ produces in a first step three protons that can be titrated like hydrochloric acid; a fourth OR^- is added in a normal buffer region and a fifth OR^- is taken up during another steep pH rise. The importance of working in the strict absence of oxygen was confirmed and the presence of even trace amounts avoided by elaborate experimental precautions.

O C 3

COORDINATION CHEMISTRY OF SULPHUR TRIOXIDE WITH OXYGEN BASES.

Ram Chand Paul and M.S. Bains
Department of Chemistry, Panjab University, Chandigarh-3, India.

Octahedral configuration for the sulphur atoms is not very frequent.

Examples are sulphur hexafluoride, sulphur monochloro pentafluoride and di-
sulphur decafluoride. JANDER (1) from potentiometric titrations of sulphur
trioxide against anhydrous hydrocyanic acid has postulated the existence of
tribasic adducts with the proposed structure $(HO)_3S(CN)_3$ having octahedral
configuration.

The chemistry of sulphur trioxide, to any large extent has not been
studied in non-aqueous media. Compounds of sulphur trioxide with a few
electron donors such as pyridine, dimethylamine, dimethylaniline, dioxane,
acetic acid and butyric acid have been utilised for synthetic purposes without
much attention having been paid to their structure and their solution chemistry.
Based upon conductivity work the formation of compounds of sulphur trioxide
with ethers (2) and fatty acids (3) has been briefly reported. These and other
physicochemical studies such as potentiometric titrations, viscosity, density,
dipole moment and molecular weight determinations have revealed in our
work the formation of definite compounds of sulphur trioxide with alcohols,
ethers, esters and fatty acids.

In the case of sulphur trioxide-alcohol systems the conductivity-compo-
sition curve reveals a maximum at 1:4-5 sulphur trioxide/alcohol molar ratio
showing that the adducts formed are highly solvated. There is no break in the
curve at a molar ratio 1:3. There are, however, breaks at 1:2 and 1:1 sulphur
trioxide/alcohol molar ratios. The potentiometric, conductometric and visual
titrations show that sulphur trioxide solutions in these alcohols behave as
monobasic acids against nitrogen bases and indicate that the stable species
in alcohol solutions are the tetracoordinated alkyl hydrogen sulphates which
are monobasic ionizable acids.

In studies with esters 1:2 and 1:1 compounds are indicated. With higher
esters the curves show only one break at a 1:1 molar ratio. The behaviour of
ethers with sulphur trioxide is analogous.

With acetic acid the first maximum is at a 1:5 sulphur trioxide/acid
molar ratio. In this case no break is detected at a 1:3 ratio but there is a
break at 1:2 stoichiometry. With all other acids breaks in the conductivity-
composition curves at 1:3 or/and 1:2 sulphur trioxide/acid molar ratio are
observed. With acids 1:1 molar ratio region could not be studied as the
system became very viscous.

Molar volumes for mono adducts of esters and di-adducts of acids with
sulphur trioxide have shown that the average decrease in molar volume is
18,5 cc and 21,0 cc respectively. This decrease in volume seems due to substitution at the sulphur oxygen double bond. The density of the mixtures of
acids and sulphur trioxide varies linearly with mole percent of sulphur trioxide

showing that the chemical interaction is of the same nature i.e. addition at the S =O double bond. Moreover the increase in viscosity of the sulphur trioxide-acid systems is abnormal; a mixture of sulphur trioxide/acetic acid at a molar ratio of 0,325 has a viscosity of 97,5 c.p. at 26° whereas that of pure acetic acid is only 1,4 c.p. at the same temperature and that for sulphur trioxide is 1,82 c.p. at 25°.

The determined dipole moments are given against the most probable compound existing in solution of the parent acid or the parent ester.

$SO_3(CH_3CH_2COOH)_3$	9,50 D	$SO_3(C_2H_5COOC_2H_5)$	16,35 D
$SO_3(CH_3CH_3CH_2COOH)_3$	6,54 D	$SO_3(C_3H_7COOC_2H_5)$	9,20 D
$SO_3[(CH_3)_2CHCH_2COOH]_2$	7,72 D	$SO_3(C_4H_9COOC_2H_5)$	12,32 D
		$SO_3(CH_3COOC_4H_9)$	8,20 D

These values suggest a highly dipolar character for the molecular species produced in solution.

Some species were isolated from an excess of the oxygen base under vacuum or from carbon tetrachloride solution. The stable adduct separated as a definite layer, which was removed and put to vacuum. In the case of esters, ethers and alcohols 1:1 adducts were obtained. Carboxylic acids yielded the following most stable compounds under the experimental conditions.

$$SO_3(CH_3COOH)_3; \quad SO_3(C_2H_5COOH)_3; \quad SO_3(C_3H_7COOH)_3;$$
$$SO_3[(CH_3)_2CHCOOH]_2; \quad SO_3(CH_3)_2CH_2CH_2COOH.$$

By adding calculated quantities of the acids stable adducts with 1:1 and 1:2 molar ratios could be prepared.

The adducts are liquids which decompose on heating; they are susceptible to moisture.

The cryoscopic determinations of the molecular weights of some of the isolated acid adducts have provided interesting results.

Adduct	Solvent	Molecular weight		Complexity
		Found	Calculated	
$SO_3(CH_3COOH)_3$	Acetic acid	259,0	260	1,00
$SO_3(C_2H_5COOH)_3$	Cyclohexane	306,0	302	1,01
$SO_3[(CH_3)_2CHCH_2COOH]_2$	Cyclohexane	279,0	284	0,98
$SO_3(CH_3COOH)_3$	Dioxane	85,1	260	0,33
$SO_3(C_2H_5COOH)_3$	Dioxane	89,0	302	0,295
$SO_3(CH_3COOH)_2$	Dioxane	104,3	200	0,52
$SO_3(CH_3COOH \cdot C_4H_8O_2)_2$	Dioxane	142,0	284	0,50

The acetic acid compound with sulphur trioxide is insoluble in cyclohexane. Its molecular weight was therefore, determined in the parent acid. The propionic acid adduct is monomeric in cyclohexane. An octahedral configuration (Structure I) is proposed for these tri-adducts. The monomeric di-adduct of isobutyric acid may have an sp^3d or sp^2d^2 hybridisation (structures II and III).

(I)   (II)   (III)

References

1. JANDER, G., and SCHOLZ, G., Z. physik. Chem., 192, 163 (1943).
2. PAUL, R.C., and NARULA, S.P., J. Sci. Ind. Res., 20B, 184 (1961).
3. PAUL, R.C., NARULA, S.P., and MEYER Prabha, J. Ind. Chem. Soc., 39, 297 (1962).

1 C 1

METAL COMPLEXES OF THIO-DERIVATIVES OF β-DIKETONES.

Scot H.H. Chaston and Stanley E. Livingstone
School of Chemistry, University of New South Wales,
Kensington, Sydney, Australia.

In 1901 WERNER (1) reported the first metal chelate of acetylacetone and since then the study of metal complexes of β-diketones has been actively pursued. We have prepared thioderivatives RC(SH=CHCOR' (where R = Me, R' = Me, Ph, OEt; where R = Ph; R' = Ph, OEt) of β-diketones in order to study their metal complexes.

Ethyl thioacetoacetate $CH_3C(SH)=CHCOOEt$ {1} was prepared by MITRA (2) by passing hydrogen sulphide for 6 hours into an alcoholic solution of ethyl acetoacetate, saturated with dry hydrogen chloride. Ethyl thiobenzoylacetate {2} PhC(SH)=CHCOOEt was prepared by a modification of MITRA's method by REYES and SILVERSTEIN (3). Using REYES and SILVERSTEIN's method we have prepared the thio-derivative of acetylacetone, 4-mercaptopent-3-en-2-one {3}. However, this method gives little reaction with benzoylacetone and di-benzoylmethane. The thio-derivatives of these β-diketones, 3-mercapto--1-phenylbut-2-en-1-one {4} and 3-mercapto-1,3-diphenylprop-2-en-1-one {5}, were prepared by dissolving the diketone in alcohol saturated with hydrogen sulphide at -10° and passing dry hydrogen chloride into the solution for 5 minutes. This quicker method can also be used to prepare the compounds {1},{2} and {3}.

Mass spectrographic investigation (4) of {4} has shown that it has the structure MeC(SH)=CHCOPh and not PhC(SH)=CHCOMe.

The infrared spectra of these five thio-derivatives show no SH band at ca 2570 cm^{-1}; this indicates strong chelation as in acetylacetone, which shows no OH absorption above 3000 cm^{-1} but a weak band near 2700 cm^{-1} (5). All the compounds display three characteristic strong absorption bands which have been assigned as follows:

Infrared Bands of Thio-derivatives of β-Diketones:

1640 - 1587 cm^{-1}	C===O stretch
1565 - 1530 cm^{-1}	C===O stretch
1240 - 1190 cm^{-1}	C===S stretch

Whereas β-diketones react with nickel salts only under alkaline conditions to yield pale green dihydrates which are octahedral and paramagnetic, the monothio-derivatives {1-5} react with nickel acetate solution to give brown crystalline anhydrous complexes which are diamagnetic and presumably squareplanar and readily soluble in organic solvents.

The infrared spectra of metal acetylacetonates show three characteristic strong bands, which have been assigned by NAKAMOTO and coworkers (6):

Infrared Bands of Metal Acetylacetonates (6):

$1600 - 1570 \text{ cm}^{-1}$	$C = C$ stretch
$1590 - 1530 \text{ cm}^{-1}$	$C = O$ stretch
$490 - 422 \text{ cm}^{-1}$	$M—O$ stretch

The spectra of the nickel chelates of the thio-derivatives {1-4} show five characteristic strong bands.

Infrared Bands of Ni(II) Chelates of Thio-derivatives
of β-Diketones:

1588 cm^{-1}	$C = C$ stretch
$1567 - 1514 \text{ cm}^{-1}$	$C = O$ stretch
$1258 - 1218 \text{ cm}^{-1}$	$C = S$ stretch
$818 - 800 \text{ cm}^{-1}$	$C—S$ stretch + $C—H$ deformation
$430 - 420 \text{ cm}^{-1}$	$Ni—O$ stretch

The assignments for the $C = C$ and $C = O$ stretching frequencies are made on the basis of NAKAMOTO's assignments (6) of the bands in metal acetylacetonates.

The $C = S$ stretching mode has been reported in a number of compounds as an intense band occurring over a wide range from $1400 - 1025 \text{ cm}^{-1}$ (7). In thiofenchone, a thioketone where there is unlikely to be coupling with other modes, the band occurs at 1180 cm^{-1} (8). Accordingly, it appears that the strong band at $1240 - 1190 \text{ cm}^{-1}$ in the thio-derivatives and the intense band occurring in the range $1258 - 1218 \text{ cm}^{-1}$ in the nickel chelates is the $C = S$

stretching mode, possibly coupled with the C===C or the C==O stretching mode. It is significant that the frequency of this band alters little on co-ordination.

The C—S vibration usually appears as a weak absorption within the range $700 - 600$ cm^{-1} but in 1:2-dimercaptoethane it occurs at 735 cm^{-1}. In the platinum(II)chelate of 3-ethylthiopropane-1-thiol, $EtSCH_2CH_2CH_2SH$ (S—SH), there is a band of medium intensity at 830 cm^{-1} and the μ-thiolo-bridged compound $Pt_2(S-S)_2I_2$ shows a strong band at 856 cm^{-1} and a medium to strong band at 818 cm^{-1}. Consequently, the band of medium intensity at $837 - 805$ cm^{-1} in the thio-derivatives and the strong band at $818 - 800$ cm^{-1} in the nickel chelates are tentatively assigned to the C—S stretch coupled with another mode, possibly the C—H deformation.

The strong band at $430 - 420$ cm^{-1} is regarded as the Ni—O stretching mode since this is within the range of the M—O stretching frequency in metal acetylacetonates.

The nickel chelates of the β-thioketo-esters {1} and {2} readily take up two molecules of pyridine to form the crystalline complexes, Ni(thioketo-ester)$_2$(py)$_2$. The pyridine derivative of the nickel complex of {1} is green and of {2} is yellow; both are paramagnetic with moments of 3.1 Bohr magnetons.

References

1. WERNER, A., Ber., 34, 2584 (1901).

2. MITRA, S.K., J.Ind.Chem.Soc., 10, 71 (1933).

3. REYES, A., and SILVERSTEIN, R.M., J.Am.Chem.Soc., 80, 6367, 6373 (1958).

4. LIVINGSTON, S.E., and SHANNON, J.S., unpublished work.

5. RASMUSSEN, R.S., TUNNICLIFF, D.D., and BRATTAIN, R.R., J.Am. Chem.Soc., 71, 1068 (1949); MECKE, R., and FUNCK, E., Zeitschrift für Elektrochem., 60, 1124 (1956); COTTON, F.A., in "Modern Co-ordination Chemistry" (Lewis and Wilkins, eds.), Interscience, New York, 1960, p.379.

6. NAKAMOTO, K., and MARTELL, A.E., J.Chem.Phys., 32, 588 (1960); NAKAMOTO, K., McCARTHY, P.J., RUBY, A., and MARTELL, A.E., J.Am.Chem.Soc., 83, 1066, 1272 (1961); NAKAMOTO, A., "Infrared Spectra of Inorganic and Coordination Compounds", John Wiley, New York, 1963, p.216.

7. BELLAMY, L.J., "Infrared Spectra of Complex Molecules", Methuen, London, 2nd edn., 1958, p.355; BELLAMY, L.J., in "Organic Sulphur Compounds" (Kharasch, N., ed.), Pergamon, Oxford, 1961, Chap.6.

8. RAO, C.N.R., and VENKATARAGHAVAN, R., Spectrochim.Acta, 18, 541 (1962).

11 C 2

CO-ORDINATION CHEMISTRY OF METAL COMPLEXES OF SALICYLIDENEIMINES AND β-KETOIMINES.

Shoichiro Yamada, Hiroaki Nishikawa and Emiko Yoshida
Department of Chemistry, College of General Education,
Osaka University, Toyonaka, Osaka, Japan.

Co(II), Ni(II), Cu(II), Pd(II) and Pt(II) complexes of N-substituted salicylideneimines [I] and β-ketoimines [II], including a number of new compounds, have been prepared and their configuration examined mostly on the basis of their electronic spectra. The main results of the examination, which are summarized as follows, are understood in terms of the nature of the metal-ligand bond and of the electronic structure of the metal ions.

The complexes [I] with R being α-branched alkyls are considered to take configuration more or less distorted from the square-plane owing to the steric condition, even with those metal ions for which the tetrahedral configuration is not generally stable. In the present work, these complexes of Ni(II), Cu(II), Pd(II) and Pt(II) were prepared with R = cyclohexyl, isopropyl, sec-butyl and t-butyl groups. The distortion from the square-plane in Ni(II) and Cu(II) complexes increases in the above order of the substituents.

In general, the acetylacetoneimine complexes [II] of 3d-transition elements can hardly be obtained. On the contrary, a number of Pd(II) complexes [II] were prepared in fine crystals with R = methyl, ethyl, n-propyl and n-butyl groups. This is interpreted as representing stabilization of the β-ketoimines through bonding with the Pd(II) ion, which is generally expected to form stronger bond than the bivalent ions of the 3d transition elements.

The Co(II), Ni(II) and Cu(II) complexes of [I] with R = o,o'-dimethyl-, o,p,o'-trichloro- and o,p,o'-tribromo-derivatives were prepared. It is found that the Ni(II) as well as Cu(II) and Pd(II) complexes in non-donor solvents are squareplanar. It is to be noted that the benzene rings of the substituents in these planar complexes must be tilted from the $[M(O-)_2(N=)_2]$-plane to a considerable extent.

A number of N-aryl-β-ketoimines form the complexes with Co(II), Ni(II) and Cu(II), R being phenyl, p-tolyl, p-chlorophenyl, p-anisyl, α- and β-naphthyl and o,o'-xylidyl. All the cobalt(II) complexes examined are considered to be tetrahedral in non-donor solvents.

The configuration of the complexes prepared was examined in pyridine to see whether the pyridine molecules are co-ordinated with the metal ion already satisfying square-planar quadri-co-ordination. As shown in Table I, a marked difference is observed between Ni(II) salicylideneiminate and β-ketoiminate, indicating that the field produced by the former is weaker than that by the latter. A drastic change is also found between the ethylene-diimine(en) and trimethylenediimine(tn) derivatives of Ni(II), showing that the length of the polymethylene chain affects the bond strength considerably. The dendency of the square-planar complex to take additional ligands is greater for Cu(II) than for Ni(II) as well as for Pd(II).

TABLE I - CONFIGURATION OF Ni(II) COMPLEXES OF
N-SUBSTITUTED IMINES OF ACETYLACETONE(ACACIM) AND
SALICYLALDEHYDE(SALIM) IN PYRIDINE

R	acacim	salim
H	planar	6-co-ord.
aryl	6-co-ord.	6-co-ord.
en	planar	planar
tn	6-co-ord.	6-co-ord.
phn*		6-co-ord.

* phn = o,o'-phenylenediimine

Even the o-substituted complexes as well as the N-m- or N-p-aryl-complexes of [I] and [II] in pyridine form quinque- or sexa-co-ordinate complexes with one of two molecules of pyridine when the metal ions are Co(II), Ni(II) and Cu(II). The fact that the o,o'-xylidyl-derivatives of Ni(II) and Cu(II) in pyridine keep the square-planar configuration is ascribed to the steric condition. The corresponding Pd(II) complexes show very slight tendency to combine an additional molecule of pyridine.

A similar comparison was carried out with the N-isopropyl- and N-t-butyl-salicylideneiminato-complexes which were concluded to take a configuration more or less distorted from the square-planar configuration. Only the isopropyl-substituted complex of Ni(II) take sexa-co-ordinate configuration in pyridine, while the other keep the same configuration as is present in non-donor solvents. The stability of the distorted configuration against the octahedral one seems to be remarkably lower for Ni(II) than for Cu(II) as well as Co(II).

The compound, $[Co(II)(salen)(py)_2]$, has not been well characterized, while a series of $[Co(II)(N.R.sal)_2(py)_2]$ complexes have been obtained, R being n-alkyl and aryl groups, where py and salen denote a pyridine molecule and a bis(salicylaldehyde) ethylenediiminate anion, respectively. A number of new complexes, $[Co(III)(salen)X_2]^+$, have been synthesized, X being NH_3, n-alkyl amines, pyridine, H_2O, NCS^-, NO_2^- and CN^-. It seems that the Co(III) complexes of this type are obtained as crystals when X denotes one of those ligands which produce ligand field which is stronger than a certain value. It is to be noted that both $[Co(II)(N.CH_3.sal)_2(py)_2]$ and $[Co(III)(N.CH_3.sal)_2(py)_2]^+$ were isolated in crystals.

11 C 3

GLYCINE DERIVATIVES OF THE QUADRIVALENT PLATINUM.

A.A. Grinberg

Technological Institute of Lensowiet, Leningrad, U.S.S.R.

The most effective way to synthesize the glycine compounds of Pt(IV) is the oxidation of the corresponding compounds of Pt(II).

When acting with H_2O_2 on compounds of Pt(II), containing glycine-rests only in form of rings, two hydroxyls are added on the ends of the new co-ordinate

$$\begin{array}{c} \text{OCO} \quad \underset{Pt}{} \quad \text{NH}_2\text{CH}_2 \\ | \qquad \qquad | \\ \text{CH}_2\text{NH}_2 \quad \text{OCO} \end{array} + \text{H}_2\text{O}_2 \longrightarrow \begin{array}{c} \text{OH} \\ \text{OCO} \quad \underset{Pt}{} \quad \text{NH}_2\text{CH}_2 \\ | \qquad \qquad | \\ \text{CH}_2\text{NH}_2 \quad \text{OCO} \\ \text{OH} \end{array}$$

The action of H_2O_2 on compounds of Pt(II), containing ligands GlH is accompanied by an "innersphere" neutralisation of GlH and OH, followed by the formation of glycine rings on Pt(IV). We could synthesize geometrical-isomeric compounds of the composition $Pt(NH_3)_2Gl_2Cl_2$ and to prove their configuration by the resolution in the optical antipodes (in collaboration with O. N. ADRIANOVA).

$$\left[\begin{array}{cc} \text{NH}_3 & \text{NH}_2\text{CH}_2\text{COOH} \\ & Pt \\ \text{NH}_3 & \text{NH}_2\text{CH}_2\text{COOH} \end{array} \right] Cl_2 + H_2O_2 \longrightarrow \left[\begin{array}{cc} & \text{OH} \\ \text{NH}_3 & \text{NH}_2\text{CH}_2\text{COOH} \\ & Pt \\ \text{NH}_3 & \text{NH}_2\text{CH}_2\text{COOH} \\ & \text{OH} \end{array} \right] Cl_2 \longrightarrow$$

$$\longrightarrow \left[\begin{array}{cc} & \text{O}-\text{CO} \\ \text{NH}_3 & \text{NH}_2\text{CH}_2 \\ & Pt \\ \text{NH}_3 & \text{NH}_2\text{CH}_2 \\ & \text{O}-\text{CO} \end{array} \right] Cl_2 + 2 H_2O$$

$$\left[\begin{array}{cc} \text{NH}_3 & \text{NH}_2\text{CH}_2\text{COOH} \\ & Pt \\ \text{HOOCCH}_2\text{NH}_2 & \text{NH}_3 \end{array} \right] Cl_2 + H_2O_2 \longrightarrow$$

$$\longrightarrow \left[\begin{array}{cc} & \text{OH} \\ \text{NH}_3 & \text{NH}_2\text{CH}_2\text{COOH} \\ & Pt \\ \text{HOOCCH}_2\text{NH}_2 & \text{NH}_3 \\ & \text{OH} \end{array} \right] Cl_2 \longrightarrow$$

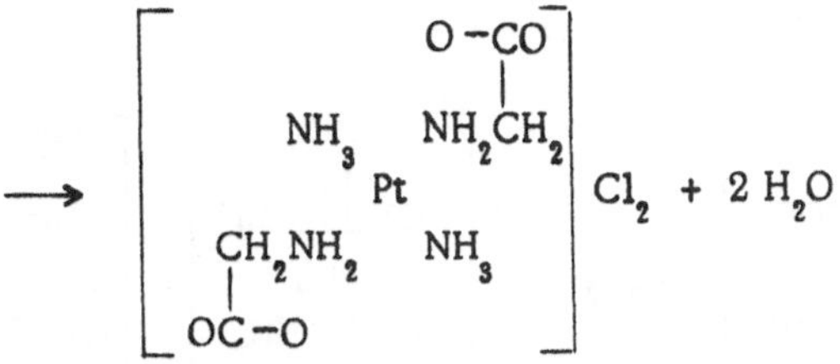

$$\longrightarrow \left[\begin{array}{c} \overset{O-CO}{\underset{|}{}} \\ NH_3 \quad NH_2CH_2 \\ Pt \\ CH_2NH_2 \quad NH_3 \\ \underset{|}{} \\ OC-O \end{array} \right] Cl_2 + 2\,H_2O$$

Cis-$[Pt(NH_3)_2Gl_2]Cl_2$ is characterised by a high value of optical rotation. We could prepare two new compounds $[PtGl_2Cl_2]$, containing two Cl-atoms in cis-position. Thus, we synthesized four isomers of the composition $[PtGl_2Cl_2]$ (out of five possible isomers).

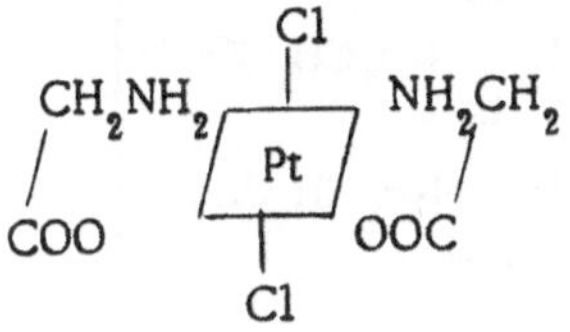

I

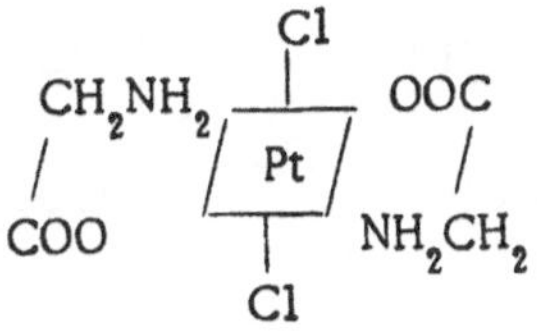

II

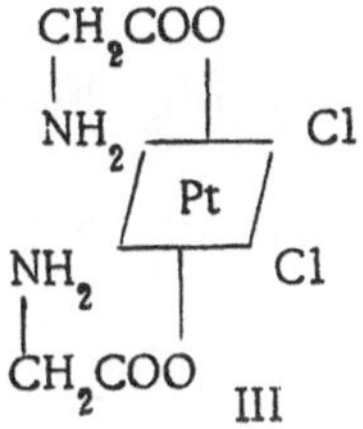

III

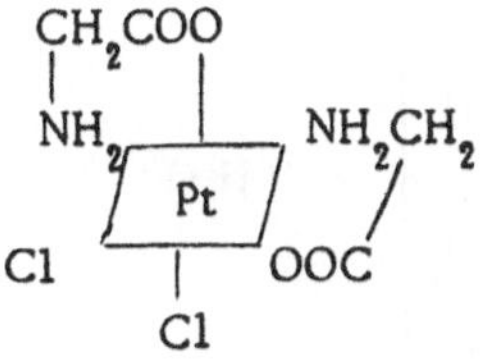

IV

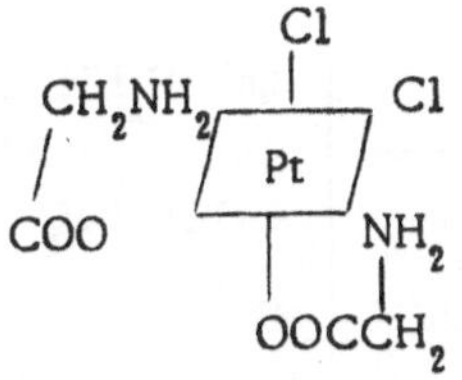

V

There are the ways used to prepare the mentioned isomers:

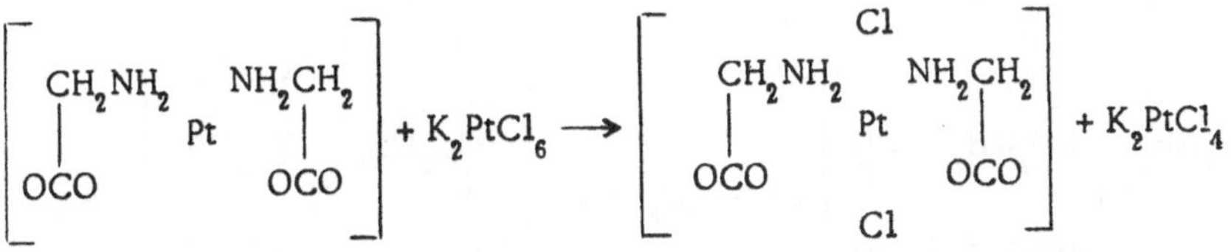

$$\left[\begin{array}{c} CH_2NH_2 \quad NH_2CH_2 \\ | \qquad\quad | \\ Pt \\ OCO \qquad\quad OCO \end{array} \right] + K_2PtCl_6 \longrightarrow \left[\begin{array}{c} Cl \\ CH_2NH_2 \quad NH_2CH_2 \\ | \qquad\quad | \\ Pt \\ OCO \qquad\quad OCO \\ Cl \end{array} \right] + K_2PtCl_4$$

I

$$\left[\begin{array}{c} CH_2NH_2 \quad OCO \\ | \qquad\quad | \\ Pt \\ OCO \qquad\quad NH_2CH_2 \end{array} \right] + K_2PtCl_6 \longrightarrow \left[\begin{array}{c} Cl \\ CH_2NH_2 \quad OCO \\ | \qquad\quad | \\ Pt \\ OCO \qquad\quad NH_2CH_2 \\ Cl \end{array} \right] + K_2PtCl_4$$

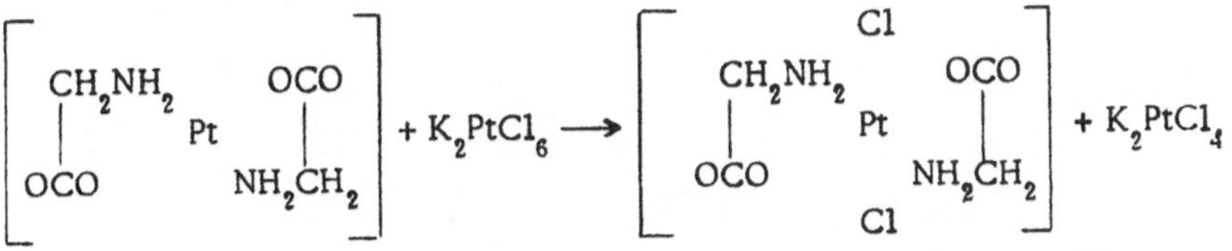

II

$$\left[\begin{array}{c} Cl \diagdown \quad NH_2CH_2COOH \\ \quad Pt \\ Cl \diagup \quad NH_2CH_2COOH \end{array}\right] + H_2O_2 \longrightarrow \left[\begin{array}{c} OOCCH_2 \\ Cl \big| \quad NH_2 \\ \quad Pt \\ Cl \big| \quad NH_2 \\ OOCCH_2 \end{array}\right] + \left[\begin{array}{c} CH_2COO \\ NH_2 \big| \quad NH_2CH_2 \\ \quad Pt \\ Cl \quad OOC \\ Cl \end{array}\right]$$

$$\qquad\qquad\qquad\qquad\qquad\qquad\qquad\qquad\qquad III \qquad\qquad\qquad IV$$

By the action of H_2O_2 on H_2PtGl_4 we have prepared the compound $PtGl_4$.

$$\left[\begin{array}{c} OOCCH_2NH_2 \quad NH_2CH_2COOH \\ Pt \\ OOCCH_2NH_2 \quad NH_2CH_2COOH \end{array}\right] + H_2O_2 \longrightarrow \left[\begin{array}{c} OH \\ OOCCH_2NH_2 \quad NH_2CH_2COOH \\ Pt \\ OOCCH_2NH_2 \quad NH_2CH_2COOH \\ OH \end{array}\right] \longrightarrow$$

$$\longrightarrow \left[\begin{array}{c} O-CO \\ OOCH_2NH_2 \quad NH_2CH_2 \\ Pt \\ OOCH_2NH_2 \quad NH_2CH_2 \\ O-CO \end{array}\right] + 2\,H_2O$$

We could also synthesize Pt(IV) derivatives containing three glycine-rings:

$$\left[\begin{array}{c} CH_2NH_2 \quad NH_2CH_2COOH \\ Pt \\ OCO \quad NH_2CH_2COOH \end{array}\right] Cl + H_2O_2 \longrightarrow \left[\begin{array}{c} O-CO \\ CH_2NH_2 \quad NH_2CH_2 \\ Pt \\ OCO \quad NH_2CH_2 \\ O-CO \end{array}\right] Cl$$

These compounds display an "imido"-reaction via the scheme:

$$\begin{bmatrix} & & \text{O-CO} & \\ \text{CH}_2\text{NH}_2 & & \text{NH}_2\text{CH}_2 & \\ & \text{Pt} & & \\ \text{OCO} & & \text{NH}_2\text{CH}_2 & \\ & & \text{O.-CO} & \end{bmatrix}^+ \rightleftarrows \begin{bmatrix} & & \text{O-CO} & \\ \text{CH}_2\text{NH} & & \text{NH}_2\text{CH}_2 & \\ & \text{Pt} & & \\ \text{OCO} & & \text{NH}_2\text{CH}_2 & \\ & & \text{O-CO} & \end{bmatrix}^0 + \text{H}^+$$

$$(K \cong 5.10^{-8})$$

Thus is shown the ability of $\text{NH}_2\text{CH}_2\text{COOH}$ to function as a dibasic acid splitting off protons from -COOH and -NH_2. The Pt(IV)-compound of the composition $[\text{PtGl}_2(\text{Gl-H})]^0$ is isolated. It is shown that the rest (Gl-H) gradually reduces Pt(IV) to Pt(II). To identify the new isomers we used infra-red spectroscopy. The work was realised in collaboration with IUAN KANG and I.S.WARSZAVSKI. With E.N.INKOVA and I.S.WARSZAVSKI we have shown that cis-PtGl$_2$ exists in two different forms.

11 C 4

SELENIUM AND SULFUR ANALOGS OF 8-QUINOLINOL AND THEIR METAL CHELATES.

Eiichi Sekido, Quintus Fernando and Henry Freiser
Department of Chemistry, University of Arizona, Tucson, Arizona, U.S.A.

A comparison of the behavior of a series of ligands in which one of the bonding atoms is replaced by another in the same group of the periodic table, helps in the understanding of the properties of these ligands and their metal chelates. For these reasons we have made a comparative study of the acid dissociation and certain metal chelate formation equilibria of 8-quino-linol, quinoline-8-thiol and quinoline-8-selenol by spectrophotometric and potentiometric techniques (Tables I and II).

TABLE I. TAUTOMERIC CONSTANTS AND ACID DISSOCIATION
CONSTANTS IN 50% v/v AQUEOUS DIOXANE AND IN WATER.

Compound	Medium	K_t	pK_{a_A}	pK_{a_B}	pK_{a_C}	pK_{a_D}
Quinoline-8-selenol	Water	2187	-0,08	3,26	8,75	5,41
Quinoline-8-selenol	50% Aqueous dioxane	59	0,13	1,90	8,50	6,72
Quinoline-8-thiol	Water	27 (1)	2,07	3,50	8,27	6,84
8-Quinolinol (2)	Water	0,04	6,60	5,14	8,42	9,98

TABLE II. log k_1 METAL CHELATE FORMATION CONSTANTS IN
50% v/v AQUEOUS DIOXANE.

Metal Ion	8-Quinolinol	Quinoline-8-Thiol (2)	Quinoline-8-Selenol
Zn^{++}	9,96	11,04	10,17
Ni^{++}	11,44	10,95	---
Cd^{++}	9,43	---	10,7
Pb^{++}	10,61	11,85	10,4
Mn^{++}	8,28	6,74	---

The acid dissociation phenomena observed for these three ligands require
the following explanatory scheme.

The relationships between the various equilibrium constants in this scheme and macroscopic acid dissociation constants, K_{a_1} and K_{a_2} are as follows:

$$K_{a_1} = K_{a_A} + K_{a_B} \qquad \{1\}$$

$$\frac{1}{K_{a_2}} = \frac{1}{K_{a_C}} + \frac{1}{K_{a_D}} \qquad \{2\}$$

$$K_t = \frac{K_{a_A}}{K_{a_B}} = \frac{K_{a_C}}{K_{a_D}} \qquad \{3\}$$

The values of pK_{a_D} are expected to be sensitive to the electronegativity of X and vary considerably from O to Se. Similarly, values of pK_{a_A}, which correspond to the loss of a proton from the group $-XH$ in the cationic species, are lower than those of pK_{a_D} by virtue of the presence of the positive charge, but change in the same manner as pK_{a_D}. The relationship between pK_{a_A} and pK_{a_D} can be expressed as

$$pK_{a_A} = 1,5 \, pK_{a_D} - 8,4 \qquad \{4\}$$

This relationship shows two consequences of the presence of the positive charge. First, its presence causes a general lowering of the pK_{a_A} and second, this lowering varies linearly with X, being greatest with quinoline-8-selenol.

The effect of changing X on the values of pK_{a_B} and pK_{a_C} would not be expected to be appreciable since the X atom is not directly involved in the dissociation processes. This is seen to be the case in the pK_{a_C} values observed, and were it not for 8-quinolinol, in the pK_{a_B} values as well. The deviation of the value of pK_{a_B} for 8-quinolinol may reflect the stabilization of the neutral species through hydrogen bonding.

Both quinoline-8-selenol and quinoline-8-thiol exist almost entirely in their zwitter ion forms in water. In 50% v/v dioxane-water (dielectric constant = 32), quinoline-8-selenol is still present entirely as the zwitter ion form, but quinoline-8-thiol, with its significantly lower K_t in water, is probably largely in its neutral form in the 50% aqueous dioxane medium. The concentration of the zwitter ion form in 8-quinolinol, even in aqueous solution, is negligible.

With a large number of chelating ligands of the N–O type the stability order observed is Ni> Zn> Pb. With N–S ligands however, the reverse order is characteristic. This is seen to be true for both the thiol and the selenol ligands. The metal stability sequence depends on such factors as crystal field stabilization energy, metal ion electronegativity, bonding atom polarizability, steric hindrance and, in the case of nickel, energy involved in spin-pairing. The results shown in Table II will be discussed in terms of such factors.

References

1. ALBERT, A., and BARLIN, G.B., J.Chem.Soc. (London), 2384 (1959), MASON, S.F., J.Chem.Soc. (London), 678 (1958).
2. CORSINI, A., FERNANDO, Q., and FREISER, H., Anal.Chem. 34, 1424 (1963).

12 C 1

AN INVESTIGATION ON SOME SULPHATO COMPLEXES IN AQUEOUS SOLUTION WITH INFRARED SPECTROSCOPY.

Sture Fronaeus and Ragnar Larsson
Department of Inorganic Chemistry, Institute of Chemistry,
University of Lund; Lund, Sweden.

Infrared spectra in the range 1400 - 850 K have been recorded for several metal sulphates in aqueous solution. For this purpose 8,8 μ thick cells with "IRTRAN - 2" windows have been used. In the frequency region mentioned the free sulphate ion absorbs radiation at 1104 K corresponding to the asymmetric S-O vibration (ν_3). From solid state investigations (1) it is known that in the spectrum of sulphato complexes this absorption peak is split into two or three new peaks depending upon whether the symmetry of the sulphate ion is decreased to C_{3v} or C_{2v}. This corresponds to a mononuclear, unidentate complex and a polynuclear or possibly a mononuclear bidentate complex, respectively. Furthermore, the symmetric S-O vibration (ν_1), which is infra-red inactive for the free ion, becomes apparent in the spectra of sulphato complexes.

In aqueous solution, these effects have been clearly observed in the case of the chromium (III) and indium systems. As the indium system has been the object of Raman investigations (2, 3) with conflicting results, it is of special interest that we find three new peaks at 1135 K, 1050 K and 975 K, i.e. corresponding to a sulphato complex of C_{3v} symmetry. The 1050 K absorption cannot be caused by HSO_4^- as this ion has an absorption band at 1200 K too, which was not observed. For solutions of other systems of such a composition that the first complex should dominate only a broadening of the ν_3-band was observed, together with a very feeble ν_1 absorption. These systems included the sodium and the cerium ones.

The measurements discussed in this paper have been primarily designed to determine the ratio between inner-sphere and outer-sphere complexes. It has been assumed that for sulphate ions bonded in an outer-sphere complex, i.e. a complex with one or two layers of water molecules interposed between the metal ion and the ligand, the disturbance of the symmetry should be so small that its spectrum would be identical with that of the free sulphate ion.

The observed increase of the halfwidth of the ν_3 band has consequently been interpreted as resulting from a superposition of spectra of inner-sphere complexes on spectra of outer-sphere ones. Then one may expect that the greater the ratio between inner- and outer-sphere complexes, the larger the halfwidth.

From this criterion one finds that the tendency towards inner-sphere complexity is greater for indium than for cerium and that for the divalent transition metals ($0,3$ M MSO_4 solutions) it increases in the order $Ni < Mn < Co < Zn < Cu < Cd$.

On the other hand, the integrated absorption of the ν_1-band is also a measure of the degree of inner-sphere complexity. This quantity has been measured (1 M MSO_4 solutions) for the same metals. From the results the series of increasing inner-sphere complexity was found to be $Mn < Zn \approx Co < Ni < Cu \approx Cd$. As the differences in halfwidth were not very great, the latter series is probably the more correct one.

We have also measured the ν_1 intensity for a series of solutions (M = Cu, Cd, Ni) with a constant metal concentration but increasing concentration of sulphate. The result of these measurements verify the existence of the second complex $M(SO_4)_2$ (4).

Furthermore, there is some evidence that the degree of inner-sphere complexity is greater for the second complex than for the first one and that this effect increases in the order $Cd < Ni < Cu$.

As it is impossible to determine the integrated absorption of the ν_1 peak for a pure inner-sphere complex, one cannot obtain an absolute determination of the degree of inner-sphere complexity. However, from a comparison of the integrated absorption of the ν_1 peak and the corresponding quantities of the three peaks emerging from the ν_3 absorption in spectra of solid, inner-sphere complexes, we have tried to estimate the integrated absorption, such as it would be for a pure inner-sphere complex. From measurements on $CuSO_4 . 5 H_2O$ and $[Co(NH_3)_4OH_2SO_4]Cl$ in KBr pellets we obtained β_1 in$/\beta_1 \approx 0,1$ for the divalent transition metals. The cerium and indium systems gave β_1 in$/\beta_1 = 0,12$ and $0,5$, respectively. These values are in general agreement with those given by EIGEN and TAMM (5).

References

1. NAKAMOTO, K., FUJITA, J., TANAKA, S., and KOBAYASHI, M., J.Am.Chem.Soc., 79, 4905 (1957).
2. HESTER, R.E., PLANE, R.A., and WALRAFEN, G.E., J.Chem.Phys., 38, 249 (1963).
3. McCARROLL, B., and LIETZKE, M.H., J.Chem.Phys., 35, 1276 (1960).
4. FRONAEUS, S., Acta Chem.Scand., 8, 1174 (1954).
5. EIGEN, M., and TAMM, K., Z.Elektrochem., 66, 107 (1962).

12 C 2

THE USE OF NMR FOR THE STUDY OF METAL CHELATES IN AQUEOUS SOLUTION.

Donald T. Sawyer and Richard J. Kula
Department of Chemistry, University of California,
Riverside, California, U.S.A.

Nuclear magnetic spectroscopy has been used to study a number of chelating agents and their metal chelates as a function of solution pH. This has permitted to determine when protonation of the ligand occurs and at what donor group the protons are attached. Such conclusions are made possible by observing the change in the chemical shift of the organic protons within the ligand adjacent to the donor group. The effect of varying the pH upon the NMR spectra of metal chelates permits determination of when metal-ligand bonding occurs in terms of pH and also permits establishment of which donor group of the ligand actually participates in the bonding. In many cases the chemical shifts for the free ligand are different from those for the coordinated ligand. When this is true NMR spectroscopy can be used to determine the correct formula for a specific metal chelate. Studies of chemical shifts as a function of pH also permit determination of stability constants which are related to a specific type of chelate bonding.

The relative rate of exchange for various metal-ligand bonds has been established by varying solution pH, metal-ligand concentration, and the temperature of the solution and then determining the effect of these upon the NMR spectra and the resonance line width. When the metal-ligand bonds are exchanging rapidly only a single averaged proton resonance is observed. However, when the exchange rate for the bond decreases line broadening is first observed and as the bonding becomes non-labile separate resonances are observed for the free ligand as well as the coordinated ligand. When non-labile bonds are present additional structural information can be obtained by observing spin-spin interactions (A-B type and A-B-X type) of the non equivalent protons in the ligand molecule.

These generalized approaches have been applied specifically to the NMR spectra of EDTA, DTPA, MIDA, and NTA and the metal chelates formed by these ligands. Included are the chelates of the ions of the alkali metals, the alkaline earths, zinc, mercury, lead, molybdenum, zirconium, and some of the platinum metals. For the four chelating agents studied the nitrogen atoms within the ligands are definitely protonated before the carboxylate groups as their solutions are acidified. The results for a study of the molybdenum-EDTA system have established that there are definitely two molybdenum ions per EDTA molecule and that there is non-labile bonding with both the nitrogen and oxygen donor groups of the ligand.

By using manganese (II) to cause line broadening of the metal chelate NMR spectra, the relative exchange rates for the metal-donor bonds for a series of metal-EDTA chelates have been established. This approach has permitted the determination of the relative stability for the metal-nitrogen versus the metal-oxygen bonds in a specific chelate and also has permitted a relative comparison of the exchange rates for a large group of metal-donor bonds.

Additional NMR studies of the metal chelates formed with the hydroxy acids are discussed. The general approach has been similar to that discussed in the preceding paragraphs.

2 C 3

NUCLEAR RESONANCE STUDIES OF STRUCTURAL EQUILIBRIA IN SOLUTIONS OF NICKEL(II) COMPLEXES.

R.H. Holm, A. Chakravorty, G.W. Everett, Jr.
Department of Chemistry, Harvard University, Cambridge, Mass., U.S.A.

Recent work (1-7) has demonstrated that in solutions of non-coordinating solvents certain general types of four-coordinate nickel(II) complexes undergo a rapid dynamic structural equilibrium between planar and tetrahedral configurations. This existence of this equilibrium has already been demonstrated in the cases of bis(N,N-disubstituted-aminotroponeimine) (5), bis(N-substituted-salicylaldimine) (1,2,6), and bis(N-substituted-o-hydroxynaphthaldimine) (3,7) complexes. For these general types of complexes it has been shown (2-5) that an analysis of the isotropic proton hyperfine contact shifts (8) manifested in the nuclear magnetic resonance spectra leads to an evaluation of the thermodynamic parameters (ΔF, ΔH, ΔS) of the structural change, and to a determination of the spin density distribution in the π-system of the coordinated ligand.

From the results of magnetic susceptivility, optical spectral, and proton resonance studies it is demonstrated that the planar-tetrahedral structural equilibrium also occurs in solutions of bis(β-ketoamine)Ni(II) of the type $Ni[CH_3C(O)CHC(NR)CH_3]_2$. The synthesis of these complexes is discussed, and it is shown that these complexes manifest proton contact shifts which increase in the order R = CH$_3$, aryl < C$_2$H$_5$ < i-C$_3$H$_7$ at a given temperature. This is the order of increasing proportions of tetrahedral forms in solution. The proton resonance data are shown to be interpretable using a simple valence bond model placing positive spin density at the central carbon and negative spin density at the other two carbons of the chelate ring. Thermodynamic parameters for the planar $\longrightarrow$ tetrahedral change are reported for a number of complexes with systematically varied R substituents. A rationalization of the recently reported (9) partial optical resolution of these complexes with the presence of the structural interconversion is given, and it is suggested that the diamagnetic ("planar") forms may not possess a strictly coplanar arrangement of chelate rings.

The proton resonance spectra of the three structurally isomeric groups of bis(o-hydroxynaphthaldimine)Ni(II) are presented and the assignments discussed. A satisfactory qualitative explanation of the spin density distributions in the three isomeric groups is given in terms of a valence bond model of spin delocalization. The spin density distributions for the 2-hydroxy-1- or -3-naphthaldimine complexes and 1-hydroxy-2-naphthaldimine complexes are similar in sign to those observed for β- and α-naphthyl groups, respectively.

Ligand exchange reactions have been found to occur with those bis(R-N-salicylaldimine)Ni(II) complexes which undergo the structural change. The exchange is very rapid and nearly statistical, and leads to mixed complexes of the type RR'. The proton resonance spectra of these complexes are discussed and a procedure for making unambiguous assignments is outlined. The spin density distributions in these unsymmetrical complexes are evaluated, and the relative spin densities at a given position in the two chelate rings are discussed in relation to a recent model proposed by LIN and ORGEL (10).

Complexes of the bis(salicylaldimine) series have been prepared in which the nitrogen substituent R carries a functional group such as -OCH$_3$, -OH, -N(CH$_3$)$_2$. These complexes are shown to exist as octahedral complexes in the crystalline state, but in solution are involved in observably temperature dependent structural equilibria amongst octahedral, tetrahedral, and planar forms with the proportion of tetrahedral species increasing with increasing temperature. The magnetic susceptibilities, optical spectra, and proton resonance contact shifts are analyzed in terms of these equilibria.

References

1. HOLM, R.H., and SWAMINATHAN, K., Inorg.Chem., 2, 181 (1963).
2. HOLM, R.H., CHAKRAVORTY, A., DUDEK, G.O., J.Am.Chem.Soc., 86, 387 (1964).

3. CHAKRAVORTY, A., and HOLM, R.H., Inorg.Chem., in press.

4. EVERETT, G.W., Jr., and HOLM, R.H., J.Am.Chem.Soc., 86, in press.

5. EATON, D.R., PHILLIPS, W.D., CALDWELL, D.J., J.Am.Chem.Soc., 85, 397 (1963).

6. SACCONI, L., PAOLETTI, P., and CIAMPOLINI, M., J.Am.Chem. Soc., 85, 411 (1963).

7. SACCONI, L., CIAMPOLINI, M., and NARDI, N., J.Am.Chem.Soc., 86, 819 (1964).

8. EATON, D.R., JOSEY, A.D., PHILLIPS, W.D., and BENSON, R.E., J.Chem.Phys., 32, 347 (1962).

9. HSEU, T.-M., MARTIN, D.F., MOELLER, T., Inorg.Chem., 2, 587 (1963).

10. LIN, W.C., and ORGEL, L.E., Mol.Phys., 7, 131 (1963).

3 C 1

THE NATURE OF MOLYBDIC ACID IN PERCHLORID ACID MEDIUM

A.K. Banerjee and S.Y. Tyree
University of North Carolina, Chapel Hill, N.C., U.S.A.

The nature of molybdic acid in water and in acidified solutions has been studied by relatively few workers. JONES (1) proposed the existence of isopoly-cations of molybdenum to explain the solubility of molybdic acid in acid solutions. More recently, AVESTON, ANACKER, and JOHNSON (2) report ultracentrifuge studies of molybdic acid in hydrochloric acid medium. They find a degree of polymerization of about 3 in 0.8 M HCl, but note that equilibrium is not attained. On the other hand they find no aggregation and apparent equilibrium in 6 M HCl. No quantitative results were reported of experiments in $HClO_4$ medium.

We wish to report the results of a light scattering study of molybdic acid in varying concentrations of aqueous $HClO_4$. Stock solutions of molybdic acid, approximately 0.2 M in MoO_3, were prepared from solutions of ammonium molybdate, by passing the latter through columns of Dowex-50 cation exchange resin in the H^+ form. By dilution, experimental solutions were prepared varying in total molybdenum molarity from 0.01 to 0.1. When the diluent was aqueous $HClO_4$ (molarity range, 0.01 to 1.1) the resulting solutions appeared to attain equilibrium. When the diluent was aqueous $NaClO_4$ the resulting solutions exhibited irreproducible variations in turbidity, but of the order of 5 times that of the corresponding stable $HClO_4$ solutions.

Turbidities and refractive index gradients at room temperature were measured on all solutions in the manner described previously (3).

The value of $\Delta n/M$ for the molybdic acid-water solutions was taken to be 0,0272 1/mole, the average of 4 determinations, with a scatter from 0,0264 to 0,0281. The value of $\Delta n/M$ for molybdic acid in molar $HClO_4$ was taken as 0,0290 1/mole, the average of 4 determinations, with a scatter from 0,0269 to 0,0297.

In accordance with our customary treatment of light scattering data, we might assume a polymeric component:

$$[MoO_4H_p(ClO_4)_q^Z + z\ ClO_4^- + z/2\ (H^+ClO_4^-)]_N$$

in which z can assume either plus or minus values, corresponding to isopoly-cation or isopolyanion formation. Ordinarily p would be held constant and the usual deviation plot of $\dfrac{1}{N_z} = \dfrac{HcM'}{\tau} - \dfrac{10^3 z^2 c}{2 m_3 M'}$ would be made.

Figure 1

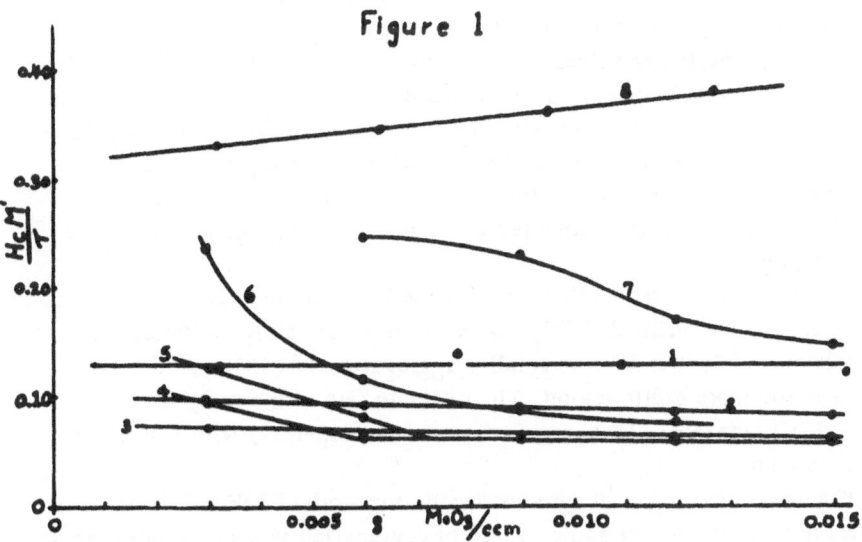

Fig. 1. $\dfrac{HcM'}{\tau}$ for Molybdic acid in various solvents.

Curve (1)   pure water
Curves (2-8)   0,017, 0,044, 0,088, 0,132, 0,22, 0,44,
       and 1,1 molar, respectively, $HClO_4$.

However, as JOHNSON and coworkers (2) have pointed out it is not possible to assure a constancy of p. Consequently we have made the simplifying assumptions that z = 0, p = 2, and q = 0. In this limit, $\dfrac{1}{N_z} = \dfrac{H \, c \, M'}{\tau}$, where M' = assumed molecular weight of molybdic acid, H_2MoO_4 or MoO_3.

Using concentration units of grams of MoO_3 per cc., the calculations yield the results shown in Fig. 1. One argument may be presented a posteriori in partial justification of the simple treatment of data. pH measurements were made on all solutions. Over the range of 0.01 to 0.10 molar total Mo concentration the maximum variation in pH was 0.45 pH units, and this was for the MoO_3-H_2O solution series. In 0.09 molar $HClO_4$ the effect of varying MoO_3 from 0.01 to 0.10 M was less than 0.1 pH unit. In all acid solutions we feel that p does not differ much from 2; although when $HClO_4/MoO_3$ becomes large, z seems to become positive, corresponding to p increasing above 2.

The intercept of the curve for pure water solvent corresponds to N = 7.5 - 8.5, octamolybdic acid. It would be most pleasant to conclude that the solute is dispersed as octamers. However we think such a conclusion is premature. The larger values of N read from the 0.017 and 0.044 M $HClO_4$ solutions may be indicative of an increased degree of aggregation, but it can equally well be the result of activity coefficient effects (4). Thus MoO_3.aq may be octameric in pure water and undergo further aggregation as $HClO_4$ is first added. Alternatively MoO_3.aq may be more highly aggregated in pure water and we can only get a measure of its degree of aggregation as we add supporting electrolyte.

The sharp increases in 1/N values for the high dilution points in the 0.088, 0.132, and 0.22 M $HClO_4$ curves are most certainly an indication of breakdown of "octamer" into smaller aggregates. The $HClO_4/MoO_3$ is highest in the more dilute region. The ultimate fate of the molybdic acid in 1.1 M $HClO_4$ appears to be about a trimeric solute species, with z assuming a positive sign.

The above results are in agreement with the work of JOHNSON and co-workers in that we find the same degree of aggregation in approximately molar acid, despite the different acids used. In contrast to their findings we find apparent equilibrium over the entire range of acid concentrations covered. Solutions of molybdic acid in $HClO_4$ supporting electrolyte are stable for periods up to one month (so far). Similar solutions in pure water begin to yield precipitates in 7 - 10 days.

We wish to acknowledge a grant from the U.S. Army Research Office (Durham).

References

1. JONES, M.M., J.Am.Chem.Soc., 76, 4233 (1954).
2. AVESTON, J., ANACKER, E.W., and JOHNSON, J.S., Preprint kindly furnished by Dr.JOHNSON, Oak Ridge, National Laboratory.
3. TOBIAS, R.S., and TYREE, S.Y., J.Am.Chem.Soc., 82, 3244 (1960).
4. KRONMAN, M.J., and TIMASHEFF, S.N., J.Phys.Chem., 63, 629 (1959).

13 C 2

STUDY OF THE COMPLEX FORMATION WITH RARE METALS BY THE HIGH FREQUENCY METHOD.

D.I. Ryabchikov and V.A. Zarinsky
V.I.Vertadsky Institute of Geochemistry and Analytical Chemistry,
Academy of Sciences, Moscow, U.S.S.R.

The high frequency titration (h.f.t.) (1,2) as a method for complexing study in solution was used by R.HARA and P.WEST (3) for the first time. This method is based on the determination of the molar ratio of the metal to the addend from an inflection point in the titration curve due to a change in conductivity in the moment of completion of complex formation, especially with complexing agent, which will liberate hydrogen ions, such as EDTA.

The h.f.t. as well as other physicochemical methods was used for the determination of the composition and, in some cases, the structure of com - plex compounds of Th(IV) (7-9), In(III) and several rare earths ions primarily for using its properties for the development of new analytical procedures (4-6).

Fig. 1 shows the h.f.t. curves for the titration of 2.10^{-3} M thorium chloride solutions with di-Na salts of EDTA and cyclohexanediaminetetraacetic acids. The molar ratio of Th(IV) to the complexon for stable compounds is equal to 1 : 1, which is in agreement with data obtained by other methods.

By means of the above method the Th complex composition with a number of aminopolycarboxylic acids was studies: uranyldiacetic, nitrilotriacetic, cyclopentane-, cyclobutane- diaminecyclohexanetetraacetic acids. The results closely agree with those obtained by other methods (7).

The influence of steric factors was investigated in the course of interaction of thorium(IV) with hexamethylendiaminetetraacetic acid (HMDTA). The reaction followed the scheme:

$$Th^{4+} + H_2A^{2-} \longrightarrow ThA + 2H^+ \qquad ,$$

where A = anion of the complexon and was accompanied by the precipitation of the ThA.n H$_2$O; its composition was checked by gravimetric and thermo-gravimetric methods (8).

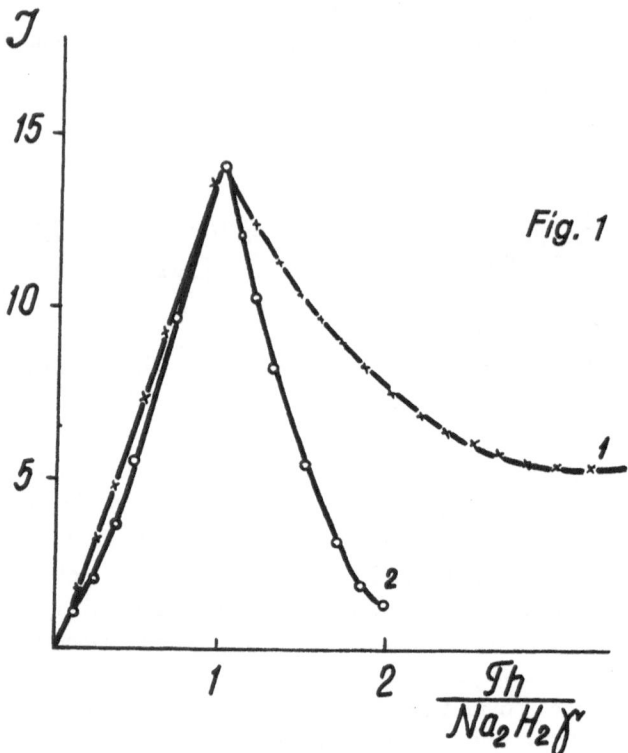

Fig. 1

On the basis of results obtained the possibility of thorium determination in the presence of five-fold excess of rare earths was stated by means of HMDTA titration and h.f.t. detection of the equivalent point.

In the ammonium or alkali metal carbonate media (9) Th(IV) was found to form a complex with the molar ratio of Th(IV): CO$_3^{2-}$ = 1 : 4. In this case the complex anion was believed to have the formula [Th(CO$_3$)$_4$]$^{4-}$ with the coordination number of eight. The detailed thermogravimetric and X-ray studies permit to suggest the following crystallographic formula of the compound:

$$Me_4[Th(CO_3)_4] \cdot MeCO_3 \cdot 12 \; H_2O \qquad ,$$

which shows that the single CO$_3^{2-}$-group is not connected with thorium.

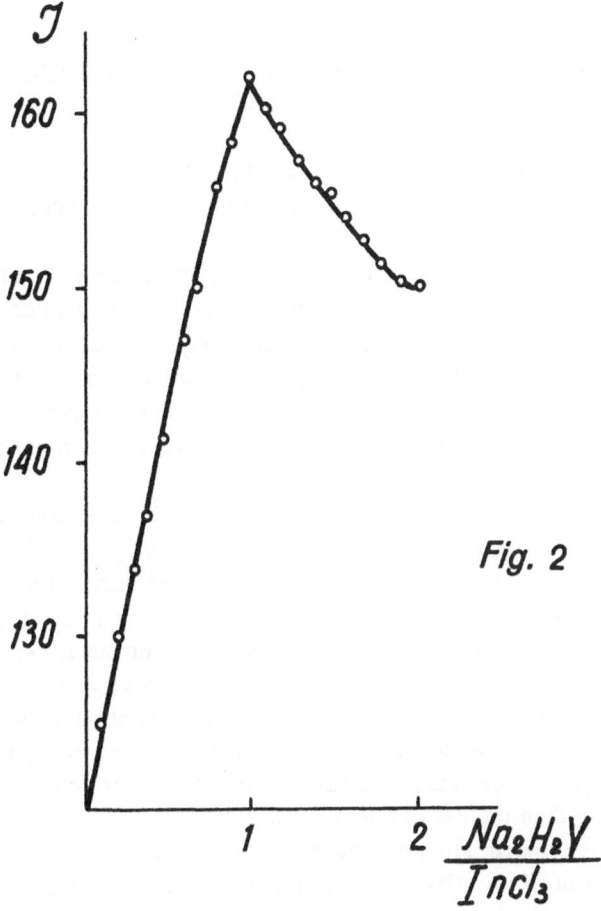

Fig. 2

A study was carried out of the formation of In(III) complex with citric acid and some aminopolycarboxylic acids (10, 11). In acid neutral media indium forms a 1 : 1 complex with citric acid as well as with EDTA, di-aminocyclohexanetetraacetic and HMDTA acids, but it forms complexes with the molar ratio of 1 : 1 and 1 : 2 with nitrilotriacetic and oxyethyl-iminoacetic acids.

3 C 3

ISOMERISM AND REDUCTION PRODUCTS OF THE MOLYBDIC HETEROPOLYACIDS.

P. Souchay, R. Massart, M. Biquard
Laboratoire de Chimie, Université de Paris, France.

The first example of isomerism in the molybdic heteropolyacids was described by STRICKLAND for silicomolybdic acid. In order to compare the chemical properties of the isomers by polarography, the use of a dropping mercury electrode appeared inadequate, since mercury reacts with these compounds. A platinum electrode gives reproducible polarograms displaying perfect reversibility which makes possible the study and characterization of these compounds and of their reduction-products.

The β-isomer is notably more oxidizing than the stable α-isomer, a fact which allows their quantitative determination; we have proved that the different preparative methods of the β-isomer in solution yield in fact $\beta+\alpha$ mixtures. The SKERLAK's method which is carried out in a solvent with a high amount of alcohol is the only one which leads to a pure β-compound. We have worked out a preparative method of the β-compound crystallized and free from α.

The reduction of the pure compounds yields, as is already known, compounds of a deep blue colour in which a proportion of the Mo(VI) is replaced by Mo(V); this reduction was performed electrolytically which provides a more progressive way than the use of chemical reducing agents.

Spectrophotometric and potentiometric studies indicate that the α-compound, in acid medium gives 2 reduced compounds containing 2 and 4 Mo(V) atoms out of 12, which are distinct on the polarograms. The reduced acids have been obtained through precipitation in their concentrated solutions with concentrated hydrochloric and perchloric acids.

Potentiometric and conductimetric titration curves indicate the existence of 6 acid H atoms in the compound with 2 e$^-$ and 7 in the compound with 4 e$^-$, which in fact does contain 8 but the last acidity is very weak (the seventh has already a pk of 9); on the other hand, in strong alkaline medium, an octo-potassium salt is obtained.

One may compare this octo-acidity with the tetra-acidity of the non-reduced acid; the stability of the anion to the alkaline media is remarkable if one bears in mind that the non-reduced acid undergoes destruction at pH > 5.

The reduction process may be written as follows:

$$\left[SiMo_{12}O_{40} \right]H_4 \xrightarrow{+2\,e^- + 2\,H^+} \left[Si\genfrac{}{}{0pt}{}{Mo_{10}}{Mo(V)_2}O_{40} \right]H_6$$

$$\xrightarrow{+2\,e^- + 2\,H^+} \left[Si\genfrac{}{}{0pt}{}{Mo_8}{Mo(V)_4}O_{40} \right]H_8$$

On the other hand, the anion with 2 e$^-$ undergoes disproportionation in alkaline medium and gives the anion with 4 e$^-$; as for the non-reduced anion which should form simultaneously, it is destructed into silica and molybdates.

Similar reactions are observed for the β acid, but whereas the non-reduced β-compound converts itself quite fast into the α-form (except in alcoholic medium), the reduced β-products do not transform into the α-form, but the β-compound with 2 e$^-$ disproportionates in acid medium into the β-compound with 4 e$^-$ and the non-reduced β-acid which in its turn changes into the α-form.

One must emphasize that when the isolation of the reduction products is attempted from the reactants in calc. proportions, the β-products are formed almost exclusively. This fact is paralleled by formation in high ratio of the β-isomer, which afterwards transforms into the α-form, when the non-reduced ion is prepared in an identical way. Since the polarograms involve 3 reduction steps with 2 e$^-$ each we have prepared the reduced compound with 6 e$^-$.

In acid medium, a reduction of the α-isomer carried over the 4 e$^-$ step results in a destruction of the structure of the heteropolyacid, as is proved through the formation of Mo(V) (which is not extractable in isoamylic acid, in opposition to these compounds).

However a certain quantity of Mo is extracted in this alcohol which corresponds to the β-compound with 4 e$^-$; its oxidation in acid medium gives the β-compound with 2 e$^-$ which should disproportionate after a time (see above) and then the β non-reduced acid.

At pH $\sim$ 4,5 the reduction process gives the α and β-products with 6 e$^-$; below pH $\sim$ 4,5 they undergo disproportionation into compounds with 4 e$^-$ and pentavalent molybdenum.

The possibility of an exact knowledge of the behaviour of the α and β-isomers has led us to the investigation of the formation of the β-isomer in the other molybdic heteropolyacids where it certainly shows a lower stability.

We have actually defined the conditions of existence of these isomers not only for the P and Ge compounds but also for those of As, for which even the α-isomer had not been obtained with certainty, the arsenio-9-molybdic anion forming itself preferably with other series.

We have also investigated the conditions of the existence of the reduced compounds.

4 C 1

THE NATURE OF THE HALIDE COMPLEXES OF Fe(III), Co(II), Ga(III) AND In(III) EXTRACTED FROM AQUEOUS CHLORIDE MEDIA BY HIGH MOLECULAR WEIGHT SUBSTITUTED ALKYL AMMONIUM COMPOUNDS.

Mary L. Good and Suresh C. Srivastava
Louisiana State University, New Orleans, Louisiana, U.S.A.

The use of high molecular weight amines, amine salts and quaternary salts as extracting reagents for various anionic metal species has been exten- sively investigated in recent years. However, most of the studies have not included definitive experimentation on the nature of the extracted metal entity. In an attempt to obtain such specific information on several extracted species we have undertaken a series of studies on the organic extracts obtained when aqueous halide solutions of Fe(III), Co(II), Ga(III), or In(III) are extracted with various organic solutions of amine salts of quaternary compounds. The com- position of the organic extract (including the concentration of the metal ion, the titratable hydrogen ion and the halide ion) and the infrared spectra of the extracted solutions have been determined. In addition, the usual extraction data including plots of log extractant concentration vs. log of the distribution coefficient and the determination of extraction isotherms have been obtained. All of this information has been correlated with other previously reported data and specific formulations of the extracted species have been postulated. Com- plete data were obtained only for the metal chloride systems extracted with tri-n-hexyl amine hydrochloride (TNH.HCl) or tricapryl-monomethylammo- nium chloride (Aliquat 336) in toluene; therefore only these systems will be reported on here.

The extraction isotherms determined for the metal ions at various chloride and acid concentrations were consistent and indicated the following extractant to metal ratios for macro concentrations of metal ion under loaded extractant conditions (references given in brackets):

Metal Ion	Extractant	
	TNH.HCl	Aliquat 336
Fe(III)	1:1 (1)	1:1 (this work)
Co(II)	2:1 (2)	2:1 (this work)
In(III)	2:1 (3)	1:1 (3)
Ga(III)	1:1 (3)	1:1 (3)

Note that in all cases, the extractant to metal ratio is the same for the amine salt and the quaternary salt except for In(III).

The plots of log extractant concentration vs. log distribution coefficient where tracer concentrations ($< 10^{-4}$ M) of metal ions are employed, gave variable results in some cases, depending on the chloride of hydrogen ion concentration of the aqueous solution. Furthermore, the slopes were not constant over a reasonable range of extractant concentrations. However, in all cases, the slopes observed for the extraction of the metal ion from a given system were less for the quaternary salt than for the amine salt, usually by a factor of 1,5 to 2,0. This would imply that the extracted species obtained at low metal concentrations for amine salt extractants are different than those obtained for quaternary extractants. Comparing these results to those give above in connection with the extraction isotherms, we find that only for the case of In(III) are the data consistent for both macro- and tracer concentrations of metal ion.

The results of studies on the composition of the extracted species are outlined below:

Metal Ion	Extractant	[Cl⁻] / Metal Ion	Titratable [H⁺]
Fe(III)	TNH.HCl	4:1	= [TNH.H⁺]
Co(II)	TNH.HCl	4,3:1	> [TNH.H⁺]
Ga(III)	TNH.HCl	4,5:1	> [TNH.H⁺]
In(III)	TNH.HCl	5:1	= [TNH.H⁺]
In(III)	Aliquat 336	4:1	None

Note that in the case of the TNH.HCl extraction of Co(II) and Ga(III), excess acid is present in the organic extract.

The far infrared spectra are given below with previously reported Raman and infrared data on the spectra of the metal tetrachloride ions:

Metal Ions	Fundamental frequencies in cm^{-1} for tetrachloride ion				Ref.	Present work	
	ν_1	ν_2	Infrared Active			Extract Media	Abs. max. cm^{-1}
			ν_3	ν_4			
In(III)	321	89	337	112	(4)	TNH.HCl	332
						Aliquat 336	335
Ga(III)	346	114	386	149	(4)	TNH.HCl	382
						Aliquat 336	384
Fe(III)	330	106	385	133	(4)	TNH.HCl	383
						Aliquat 336	383
Co(II)				310	(5)	TNH.HCl	305

Note that in all cases reported, the spectrum obtained is that of the tetra-halide metal ion.

A consideration of all the preceding data leads one to predict the following formulations for the extracted metal species:

Metal Ion	Extractant	Probable Extracted Metal Species
In(III)	TNH.HCl	$R_3NH^+InCl_4^-$, $R_3NH^+Cl^-$
In(III)	Aliquat 336	$R_4N^+InCl_4^-$
Ga(III)	TNH.HCl	$R_3NH^+GaCl_4^-$ with some $R_3NH^+Cl^-$. $HGaCl_4$
Fe(III)	TNH.HCl	$R_3NH^+FeCl_4^-$
Fe(III)	Aliquat 336	$R_4N^+FeCl_4^-$
Co(II)	TNH.HCl	$(R_3NH^+)_2CoCl_4^{2-}$ with some $R_3NH^+HCoCl_4^-$

Thus it would appear that the metal pentachloride ion (i.e. $InCl_5^{2-}$) does not form in the organic extract for any of these metals; at least not where macro concentrations of metal ion are involved. This work also points out some of the possible pitfalls involved when the coordination number of the metal ion is deduced from the usual extraction data.

References

1. GOOD, M.L., and BRYAN, S.E., J.Am.Chem.Soc., 82, 5636 (1960).
2. GOOD, M.L., and BRYAN, S.E., J.Inorg.Nucl.Chem., 20, 140 (1961).
3. GOOD, M.L., and HOLLAND, F.F., Jr., J.Inorg.Nucl.Chem., 26, 321 (1964).
4. WOODWARD, L.A., and TAYLOR, M.J., J.Chem.Soc. (London), 1960, 4473.
5. CLARK, R.J.H., and DUNN, J.M., J.Chem.Soc. (London), 1963, 1198.

14 C 2

DI-COORDINATE ANION COMPLEXES OF HYDROGEN.

J.D. Cotton, J.A. Salthouse and T.C. Waddington
University Chemical Laboratories, Cambridge, England, U.K.

The two unequivocal and well documented cases of symmetrical hydrogen bond occur in the hydrogen maleate ion (1) and in the hydrogen difluoride ion (2). In both cases neutron diffraction studies have established that the hydrogen atom is centrally situated (3). Recently salts of the higher hydrogen dihalides have been prepared, $MHCl_2$ (4), $MHBr_2$ (5) and MHI_2 (6, 7); and also, though they are not strictly examples of symmetrical hydrogen bonds, MHClI and MHBrCl. Here we report on the IR and NMR spectra, conditions of formation and stability of these compounds. Following the characterization of the difluoride ion attention was focussed on tetra-alkylammonium halides containing an extra formula unit of the hydrogen halide. WADDINGTON prepared tetramethyl-ammonium dichloride, $(CH_3)_4NHCl_2$ and calculated its lattice parameters from X-ray powder photography. The IR spectrum of the dichloride ion was similar to that of the difluoride ion (8) (see Table). Crystal entropy measurements on the dichloride ion confirm that it is linear and symmetrical (9). In the corresponding dibromide and diiodide ions the IR spectra strongly suggest that the hydrogen bond is linear and symmetrical. The IR spectra of the two mixed hydrogen dihalide anions indicate that these also are very strongly hydrogen bonded. Crystalline samples of the tetra-n-butylammonium salts of the hydrogen bromide chloride and hydrogen chloride iodide anions, $[HBrCl]^-$ and $[HClI]^-$ respectively have been isolated.

TABLE

INFRA-RED FREQUENCIES OF HYDROGEN DIHALIDE ANIONS.

Anion	cm^{-1} (Assym. Stretch)	cm^{-1} (Bend)	Sample	Ref.
$[HF_2]^-$	1450	1223	KHF_2	(8)
$[HCl_2]^-$	1565	1180	Me_4NHCl_2	(4)
	1565	1160	Me_4NHCl_2	(6)
	1540	1150	$n-Bu_4NHCl_2$	(6)
	1625	1150	$Si(acac)_3HCl_2$	(6)
$[HBr_2]^-$	1700		Et_4NHBr_2	(5a)
	1670	1170	Et_4NHBr_2	(6)
	1690	1170	$n-Bu_4NHBr_2$	(6)
$[HI_2]^-$	1650	1165	$n-Bu_4NHI_2$	(6)
$[HBrCl]^-$	1650	1100	$n-Bu_4NHCII$	
$[HClI]^-$	2000	1000	$n-Bu_4NHCII$	

End points, at a molar ratio of 1:1 in conductometric titrations of the hydrogen halides against tetraalkyl ammonium halides in liquid hydrogen sulphide, provide evidence for the formation of $[HBrCl]^-$ and $[HClI]^-$, but not for $[HBrI]^-$ (fig. 1). Similar conductometric titrations have demonstrated the formation of the species $[HCl_2]^-$, $[HBr_2]^-$ and $[HI_2]^-$.

The proton NMR of the hydrogen dichloride ion has been studied in chloroform solution, and a limiting τ value of -2.7 observed, relative to tetramethyl silane as 10.0. A similar value has been obtained in nitromethane solution. The concentration dependence of the shift, in chloroform, is shown in fig. 2. The chemical shift for the dichloride ion is far to low field of the values for hydrogen chloride in chloroform, and for liquid hydrogen chloride. We therefore conclude that the shielding of the proton is much reduced in the dichloride ion.

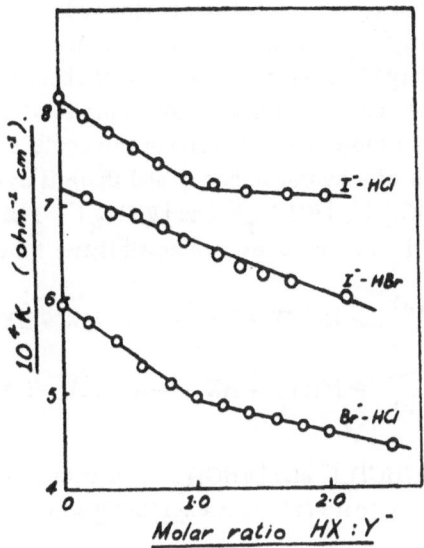

Figure 1.

Conductometric titration curves of hydrogen halides
against tetraalkylammonium halides in liquid
hydrogen sulphide.

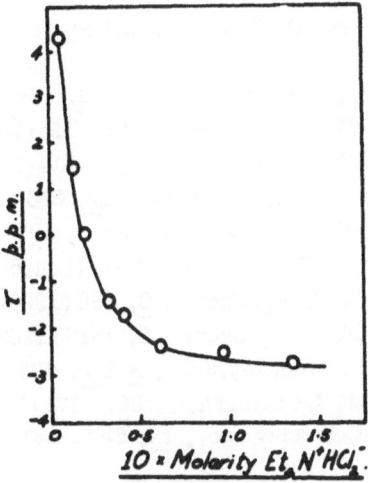

Figure 2.

Concentration dependence of the chemical shift of the
dichloride anion in chloroform.

The stability of the hydrogen dihalide ion is especially interesting. As would be expected the $[HX_2]^-$ species decompose to form HX and X^-. The difluoride ion, having a large hydrogen bond energy of about 58 kcals per mole (10) is the most stable and the di-iodide ion is the least. At low temperatures it appears that more than one solvent molecule is coordinated to the halide ion by means of a hydrogen bond, and there is evidence for the ions $[Cl(HCl)_2]^-$, $[Br(HCl)_2]^-$, $[I(HCl)_2]^-$ and $[I(HBr)_2]^-$. The mixed species, $MX(HY)_2$ are not very stable and there are two possibilities for decomposition.

$$MX(HY)_2 \begin{array}{l} \xrightarrow{(a)} MXHY + HY \longrightarrow MX + HY \\ \xrightarrow[(b)]{} MHY_2 + HX \longrightarrow MY + HY \end{array}$$

It appears that the $[Br(HCl)_2]^-$ and $[I(HCl)_2]^-$ ions decompose by route (a) but $[I(HBr)_2]^-$ decomposes by route (b), thus explaining why $[HBrI]^-$ is not formed.

References

1. CARDWELL, DUNITZ and ORGEL, J.Chem.Soc. (London), 1953, 764, 3740.
2. WESTRUM and PITZER, J.Am.Chem.Soc., 71, 1940 (1949).
3. (a) PETERSON and LEVY, J.Chem.Phys., 29, 948 (1958).
   (b) PETERSON and LEVY, J.Chem.Phys., 20, 704 (1952).
4. WADDINGTON, J.Chem.Soc. (London), 1958, 343, 1708.
5. (a) TUCK and WOODHOUSE, Proc.Chem.Soc., 1963, 53.
   (b) WADDINGTON and WHITE, J.Chem.Soc. (London), 1963, 502, 2701.
6. McDANIEL and VALLEE, Inorg.Chem., 2, 996 (1963).
7. HARMON and GEBAUER, Inorg.Chem., 2, 1319 (1963).
8. COTÉ and THOMPSON, Proc.Roy.Soc., A 210, 206 (1952).
9. CHANG and WESTRUM, J.Chem.Phys., 36, 2571 (1962).
10. WADDINGTON, Trans.Faraday Soc., 54, 25 (1958).

HALIDE COMPLEXES OF IONS WITH ELECTRONIC STRUCTURES $4d^{10}5s^2$ AND $5d^{10}6s^2$.

G.P. Haight, Jr.

Chemistry Department, Swarthmore College, Swarthmore, Pa., U.S.A.

Weak halide complexes of Sn^{2+}, Sb^{3+}, Pb^{2+}, and Bi^{3+} are being studied by solubility (1) and spectrophotometric techniques in aqueous solutions of high, constant, ion normality in an effort to determine the maximum ligand numbers and the relative importance of various intermediate complexes. The work so far tends to give more information on complexes of highest and next highest ligand number than on those of low ligand number.

Patterns of Complex Formation. Results of present work and some selected results of previous work are summarized in Table I. Present work indicates the formation of $[BiCl_6]^{3-}$ and $[SbCl_6]^{3-}$ directly from tetrachloro complexes with negligible formation of pentachloro complexes at equilibrium. Our work and SCOTT's (2,3) indicate that ML_3 complexes are relatively unimportant for Tl^+, Pb^{2+}, and Bi^{3+} complexes. $K_2 > K_1$ for $[PbCl_n]^{2-n}$. There is thus an inference that the more symmetrical complexes of even n are the more stable. $[BiBr_5]^{2-}$ appears to form, though it is less important than $[BiBr_4]^-$ or $[BiBr_6]^{3-}$. $PbBr_2$ is much more soluble in HBr than is $PbCl_2$ in HCl. Thus Bi^{3+} and Pb^{2+} appear to be class b ions with respect to formation of higher complexes. However, the lower complexes which have overall (+) charge seem to be as strong or stronger for Cl^- as for Br^-, characteristic of class a metal ions.

Spectrophotometric Analysis of Equilibria Between Two Complexes of Highest and Next Highest n.

It is often possible to observe at high ligand concentration a region where spectra show isosbestic points indicating the presence of only the two species of highest and next highest n. It may be difficult to obtain limiting spectra characteristic of only one or the other species. In such cases it is necessary to make assumptions concerning Δn (the difference in n between the two species), and the absorbance due to one or the other species in order to analyze the data (5). Solubility studies (1, 7) have been interpreted in terms of a single equilibrium between $[BiCl_4]^-$ and $[BiCl_6]^{3-}$ and a similar equilibrium between corresponding Sb(III) complexes. A spectrophotometric study (5) has proposed $[BiCl_4]^-$ and $[BiCl_5]^{2-}$ as the two species represented in the range, $0.5 \text{ M} \leq (\text{Cl}^-) \leq 4.0 \text{ M}$, used in both studies. For the case where C_M is constant and ligand conc., L, is varied

$$A = \frac{A_1 + A_2 KL^{\Delta n}}{1 + KL^{\Delta n}} \qquad \{1\}$$

where A is measured absorbance at a given wave length, A_1 is the absorbance if all the metal ion is present in complex of lower n, A_2 is the absorbance if all the metal ion is present in complex of higher n. K is the formation constant for the complex of higher n being formed by the addition of Δn ligands to the complex of lower n. If A_1, A_2, and K are all inaccessible by direct measurement equation $\{2\}$ may be used to analyze data.

$$\frac{(L_1^{\Delta n} - L^{\Delta n})}{(A' - A)} = K'(L - KL^{\Delta n}) \qquad \{2\}$$

K' is a collection of constants. L_1 and A' are from one point used as a reference point. Given data on $[\text{BiCl}_n]^{3-n}$ spectra, the left hand side of equation $\{2\}$ plotted against $L^{\Delta n}$ gives a straight line and proper intercept for $n = 1, 2,$ or 3. Thus one cannot determine n unambiguously unless A_1 or A_2 is known. In the case cited A_2 was determined from the spectrum in 6-10 M HCl making possible the use of the equation

$$1 + KL^{\Delta n} = \frac{(A_2 - A_1)}{(A_2 - A)} \qquad \{3\}$$

L^2 vs. $1/(A_2 - A_1)$ gave a straight line while L vs. $1/(A_2 - A_1)$ was curved in agreement with the solubility study.

Spectra of Pb(II) in HCl show no isosbestic point until $(\text{HCl}) > 6.0 \text{ M}$. Solubility studies show $\bar{n} = 4.0$ in 4 M HCl. This means that maximum n for $[\text{PbCl}_n]^{2-n} > 4$ and, by analogy with the bismuth study, probably reaches 6 in conc. HCl.

Difficulties with Solubility Studies. A simple analysis of the solubility of $[(\text{CH}_3)_4\text{N}]_3\text{Bi}_2\text{Br}_9$ vs. (Br^-) indicates $\bar{n}$ changing from 6 to 7 over a range of (Br^-) where the spectrum is invariant. Tests for polynuclear complexes were negative. DIAMOND's proposal of ion pairs between alkyl ammonium ions and either bromide or iodide ions may explain this inconsistency. Previous results in bromide media (1) may have to be reevaluated.

Changes in the solid phase. The solubility of $(CH_3)_4NHgCl_3$ vs. $[(CH_3)_4N]^+$ shows a sharp break, becoming virtually constant at high conc. of the alkyl ammonium ion, probably due to the formation of $[(CH_3)_4N]_2HgCl_4$. A similar effect was noted in the study of the solubility of $[(CH_3)_4N]_2SnI_4$ suggesting that it may be possible to make salts of $[SnI_6]^{4-}$.

TABLE I

FORMATION CONSTANTS.

Complex	K_1	K_2	K_3	K_4	K_5	K_6	Medium	Technique	Ref.
$[TlCl_n]^{1-n}$	4	0,5					v. KCl	sol.	2
$[TlBr_n]^{1-n}$	4,2	3,3	--0,06--[a]				v. KBr	sol. sp.	3
$[PbCl_n]^{2-n}$	10	44	≤ 0,5	2,6			4M(HCl + HClO$_4$)	sol.	
$[PbBr_n]^{2-n}$	12	2	7,5?				1M NaClO$_4$	sol.	4
$[BiCl_n]^{3-n}$	270	200	23	2,7	3,0		4M Na$^+$ 1M H$^+$	sol. sp.	5
	229	14	71	5,6	41,0	0,7	2M Na$^+$ 1M H$^+$	sol. Bi(Hg)	6
			--380--[a]		--6,0--[a]		4M H$^+$		
$[BiBr_n]^{3-n}$	180	100	83	25	38	1,25	2M Na$^+$ 1M H$^+$	sol. Bi(Hg)	6
$[SnCl_n]^{2-n}$	28	8	1,3	0,7			4M H$^+$	sol.	8
$[SnBr_n]^{2-n}$	7,9?	6,7	2,5	0,34	2,1?		4M H$^+$	sol.	1
$[SbCl_n]^{3-n}$					--0,17--[a]		4M H$^+$	sol.	1

[a] $K = (ML_{n+2})/(ML_n)(L)^2$

More data an $[BiBr_n]^{3-n}$ and $[PbBr_n]^{2-n}$ are expected by September, 1964.

References

1. HAIGHT, Jr., G.P., Proc. VIIth ICCC, Stockholm (1962), p.318.
2. HU, K.H., and SCOTT, A.B., J.Am.Chem.Soc., 77, 1380 (1955).
3. SCOTT, A.B., DARTAU, R.G., and SAPSOONTHORN, S., Inorg. Chem., 1, 313 (1962).
4. KIVALO, P., Suomen.Kem., 29B, 8 (1956).
5. HUME, D.N., and NEWMAN, L., J.Am.Chem.Soc., 79, 4576 (1957).
6. AHRLAND, S., and GRENTHE, I., Acta.Chem.Scand., 11, 1111 (1957).
7. HAIGHT, Jr., G.P., SPRINGER, C.H., and HEILMANN, O.J., Inorg. Chem., 3, 195 (1964).
8. HAIGHT, Jr., G.P., ZOLTEWICZ, J., and EVANS, W., Acta Chem. Scand., 16, 311 (1962).
9. DIAMOND, R.M., J.Phys.Chem., 67, 2513 (1963).

5 C 1

COMPLEX FORMATION IN TERTIARY AMINE EXTRACTION
OF TRIVALENT METALS. [*]

W. Müller [**], J. Fuger [***] and G. Duyckaerts
Université de Liège, Laboratoire de Chimie Nucléaire, Belgium.

Toluene solutions of trilaurylammoniumchloride in equilibrium with dilute hydrochloric acid contain an amount of water which corresponds to amine salt monohydrate. In contact with hydrochloric acid of increasing concentration, the organic phase extracts increasing amounts of excess acid, whereas the water content decreases. From distribution studies with acidic concentrated chloride solutions, it can be concluded that the extraction of excess acid is due mainly to amine dihydrochloride formation. The water content is independent of excess acid present in the organic phase, and is proportional to the water activity. Physicochemical investigations of the organic phases (infrared absorption, conductivity, cryoscopic measurements) confirm the interpretation of analytical data as to amine dihydrochloride formation and to amine salt hydration.

Extractions of trivalent iron in macroscopic amounts were carried out from 1 or 10 M hydrochlorid acid, the preequilibrated organic phases containing approximately amine hydrochloride monohydrate or dihydrochloride respectively.

Two different complexes between trilaurylammoniumchloride and ferric chloride have been observed in the organic phase after extraction from 1 M HCl. A complex containing 2 molecules of amine salt per molecule of ferric chloride (2:1 complex) exists when the equilibrium concentration of iron in the aqueous phase is small. A 1:1 complex, however, is extracted either from 1 M HCl containing up to 3 M/l $FeCl_3$, or from 10 M HCl which in the last case is accompanied by a decrease in excess acid content in the organic phase. The formation of these different complexes and the decrease of hydration water content, as followed by chemical analysis of the equilibrium solutions, are confirmed by infrared spectra of the organic phases. In both cases, formation of 2:1 or 1:1 complexes, visible spectra indicate the existence of tetrachloroferrate ion in the organic phase.

[*]   Work performed under Euratom Contract No. 003-61-2 TPU B with University of Liège, Belgium.

[**]   Euratom

[***] Institut Interuniversitaire des Sciences Nucléaires, Belgium.

Tracer amounts of ferric (Fe-59) chloride have been extracted from HCl and from acidic LiCl solutions. The variation of the extraction coefficient E_a^0 (E_a^0 = ratio of total metal concentrations in the organic and in the aqueous phase) with the total amine concentration C in the organic phase supports the results obtained with macroscopic amounts of metal: For small concentrations of HCl or LiCl, the slope of the curve $\log E_a^0 = f(\log C)$ equals 2, indicating the 2:1 complex. The slope decreases with increasing aqueous acid or chloride concentrations and tends to 1 at HCl or LiCl molarities of the order of 10. Extraction of tracer iron decreases with increasing acidity in HCl-LiCl mixtures of constant total chloride concentration (10 M). This decrease can be explained by assuming two competing reactions involving the amine salt: the ferric chloride is extracted by the ammoniumchloride, whereas excess acid forms dihydrochloride in the organic phase.

The extraction coefficients of trivalent actinides, extracted from con-centrated chloride solutions, vary with the second power of total amine con-centration and decrease with increasing aqueous acidity. A complex containing 2 molecules of trilaurylammoniumchloride per actinide is formed, which is in contrast to the 1:1 complex observed with iron under comparable conditions. Again, the decrease of the extraction coefficient as a function of acidity can be explained by competition between metal extraction and amine dihydro-chloride formation. Actinide extraction decreases with temperature. Scandium and trivalent lanthanide extraction show approximately the same dependence on amine salt concentration and on acidity as the actinide extraction. The order of extractibility is

$$Cf > E \gg Am > Cm \gg Sc \gg lanthanides.$$

Efficient separation of trivalent actinides from lanthanides is possible.

15 C 2

EXTRACTION SPECIES OF NIOBIUM AND TANTALUM WITH SOME NOVEL EXTRACTING AGENTS.

C. Djordjević
Institute "Rudjer Bošković", Zagreb, Yugoslavia.

Niobium and tantalum have virtually no cationic chemistry, but form several kinds of anionic species. Such behaviour is in particular important in extraction mechanism of all types of extracting agents. The tendency of both metals to form polynuclear species, dependent on metal concentration, pH, ligand type and electronegativity, is well known. However, even in systems such as oxalic and tartaric acid solutions, the characterisation of anionic metal complexes has not been fully established (1).

In the present investigation the extraction of niobium(V) and tantalum(V) from acid oxalic and tartaric solutions with dioctylmethylenebisphosphonic acid, $CH_2[P(O)(OC_8H_{17})OH]_2$, tri-n-octylamine and dioctylethanolamine is studied, with special regard to extraction species formed. $CH_2[P(O)(OC_8H_{17})OH]_2$ and $(C_8H_{17})_2N(C_2H_5OH)$ can act as bidentate ligands, penetrating into the metal coordination sphere, whereas in the case of tri-n-octylamine anionic metal complexes present in the solution are expected to follow the ammonium type of extracting agent cation into the organic phase. To study the extraction species organic phase was examined in solution and by evaporating the solvent, under controlled conditions, solid substances were isolated from it. They were analysed and examined by some physico-chemical methods. Metal-ligand ratio found in solution from extraction data was generally the same as in the isolated solid substances. IR spectra prooved to be very useful in many respects, even indispensable for some conclusions.

Dicotylmethylenebisphosphonic acid was prepared (2) and used in our laboratories as extractant for a variety of metal ions. In some cases metal complexes were isolated (3) and their infrared spectra examined (4). This reagent is extracting niobium(V) and tantalum(V) from acid oxalic and tartaric solutions in petrolether very efficiently (5). Metal-reagent ratio found in solutions from extraction data is in some cases dependent upon normality of the acid used and can be correlated with the composition of the solids isolated from the corresponding organic phase. For niobium crystalline substances were isolated from petrolether, as well as solids forming the third phase, which appeared under certain conditions. For tantalum gelatinous products were obtained. The IR spectra in all the examined cases confirmed the chemical evidence on absence of oxalato and tartarato groups in the extraction species. The analyses for metal, carbon, hydrogen and phosphorus are in agreement with a polynuclear species built from two or more units, probably of the type $\{CH_2[P(O)(OC_8H_{17})O]_2M(V)(OH)_3\}_n$, with oxygen bridges between niobium and tantalum, respectively. No free OH groups were found in the spectra of all the extraction species isolated, with only one absorption occuring at about $1600\ cm^{-1}$, indicating strong hydrogen bonding. Another characteristic band in this kind of phosphonic derivatives belongs to the vibrational mode of the phosphoryl group, $P = O$ (6, 7). In the free acid it appears at $1210\ cm^{-1}$. In niobium and tantalum derivatives this absorption is shifted to $1180 - 1190\ cm^{-1}$, indicating some participation of phosphoryl group in the hydrogen bonding. In the low region spectra (down to 400 cm^{-1}) no bands were found which could be assigned to a metal-oxygen vibrational mode. These modes are therefore of a lower energy, giving rise to frequencies below 400 cm^{-1}, the metal-oxygen bond being hence of a corresponding low bond order.

The extraction of niobium(V) and tantalum(V) with tri-n-octylamine from oxalic acid solutions shows different behaviour for the two metals and also an interesting irregular dependence of distribution coefficient upon the normality of the acid in the water phase. This is to be expected for niobium and tantalum owing to different polynuclear anionic complexes formation, dependent upon the solution conditions. Isolation of some species from organic phase formed in different acidity range was performed and characterisation of the solids attempted. The composition of these substances can be correlated with the metal reagent ratio found in solution. Infrared spectra are very useful in this case, in studying the extractant change as well as the metal-ligand situation in the anionic part of the complex. Oxalates are present in the extraction species, analysis indicating three oxalato groups per metal atom.

Dioctylethanolamine was prepared and its extracting ability for niobium(V) and tantalum(V) from acid oxalic solutions of the metals was examined. In a certain acidity range satisfactory extraction of the metals into organic phase, chloroform, occured. The extraction species were isolated and studied as described briefly for other two extractants employed. Tantalum reagent ratio in the solution, for example, was again found to be dependent upon the acidity. The IR spectra of the reagent, as well as those of the solids isolated from the organic phase were studied. As with tri-n-octylamine, the extraction species contain oxalate.

References

1. FAIRBROTHER, F., and TAYLOR, J.B., J.Chem.Soc. (London), 1956, 4946; NABIVANEC, B.I., Zhur. neorg. khim., 8, 2302 (1963).
2. GORIČAN, H., and GRDENIĆ, D., Proc.Chem.Soc., 1960, 228.
3. GORIČAN, H., and GRDENIĆ, D., J.Chem.Soc. (London), 1964, 513.
4. DJORDJEVIĆ, C., and GORIČAN, H., to be published.
5. DJORDJEVIĆ, C., GORIČAN, H., LOVRECEK, D., and TAN, S.L., to be published.
6. MOEDRITHER, K., and IRANI, R.R., J.Inorg.Nucl.Chem., 22, 297 (1961).
7. PEPPARD, D.F., FERRARO, J.R., and MASON, G.W., J.Inorg.Nucl. Chem., 12, 60 (1959).
8. DJORDJEVIĆ, C., GORIČAN, H., and TAN, S.L., to be published.
9. DJORDJEVIĆ, C., GORIČAN, H., and SEVDIĆ, D., to be published.

5 C 3

THE COORDINATION BEHAVIOR OF Co, Ni, Cu AND Zn
IN A CHELATING ION-EXCHANGE RESIN (1).

Chaim Eger (2) and Jacob A. Marinsky
State University of New York at Buffalo, Buffalo, New York, U.S.A.

Recently JAM showed that the thermodynamic properties of ion-exchange resins could be evaluated by consideration of macro- and small-ion interaction in the gel phase (3-5). Good agreement was obtained between the pH values of the basic monomer units and the intrinsic pK values calculated for cross-linked polymethacrylic acid and the acid-form of the cross-linked Dowex Chelating A-1 resin, a polystyrene, divinyl benzene copolymer containing iminodiacetic acid functional groups attached to the hydrocarbon matrix. In the calculation it was assumed, without detriment, that interaction between neighboring functional groups was negligible and that the cross-linked polymer was sufficiently homogeneous to support symmetry approximations.

On the basis of the successful correlation that was obtained it was predicted that the thermodynamic properties of cross-linked polymer systems could, in general, be anticipated from the monomer properties of the repeating functional group. The coordination behavior of Co, Ni, Cu and Zn in the Dowex Chelating A-1 resin was consequently investigated in an attempt to substantiate further this estimate. A comparison of the dissociation constant of chelate forms of this resin with those of the monomer unit, available from data due to CHABEREK and MARTELL (6), was believed to be worthwhile for this purpose.

It was expected that the unfavorable geometry of the cross-linked system would preclude the possibility of multiple ligand attachment to the metal as is observed with the monomer analogue, thereby simplifying the analysis of the data.

Potentiometric titration and batch equilibration methods were employed to examine the above estimate of the chelating behavior of Dowex A-1 resin. A typical set of potentiometric titration curves is presented in fig. 1 for the H_2R, CuR, Na_2SO_4 system, where the pH is plotted versus the moles of base added per mole of metal present. Each point represents a measurement made after the allowance of at least a one week period for equilibration following the addition of each increment of standard alkali. The 4 curves represent experiments in which the initial molar ratio of Cu^{++} to H_2R was 0; 0.25; 0.5; and 1.0 respectively. Similar curves were obtained for the other systems which were studied in exactly the same manner.

The shapes of the curves that are obtained indicate quite conclusively that, contrary to expectation, the sequence of complexes, MR, MHR_2 and MR_2, forms exactly as in solutions of the monomer analogues (6).

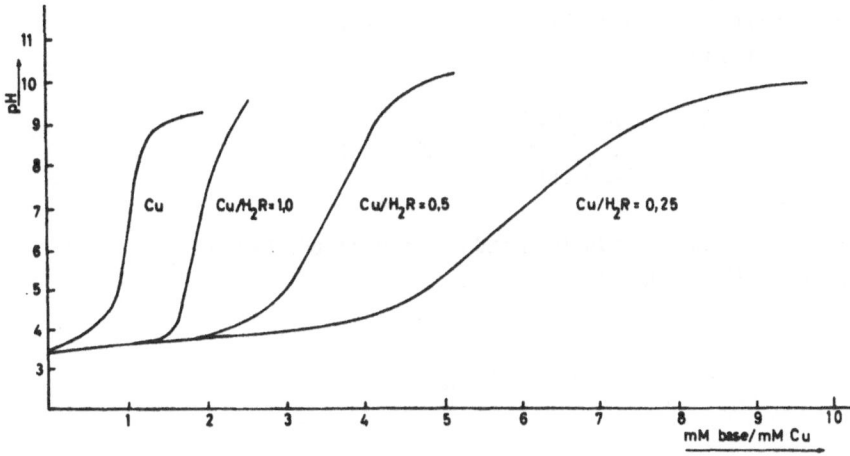

Analysis of these data to provide a quantitative estimate of the stepwise complex constants has not been completed. The problem inherent in the treatment of a heterogeneous system are further complicated by the need to develop a proper consideration of macro- and small-ion interaction which is encountered in the multiple ligand species.

The batch equilibration experiments were conducted at 25^0 in 10^{-3} M HNO_3 (Co-, Ni-, Zn-R) and 10^{-2} M HNO_3 (Cu-R) with resin samples already partially converted from the hydrogen form to the various metal forms. At these high acidity values the stability of the multiple ligand forms are not of sufficient magnitude to achieve the dissociation of the 2 moles of H_2R necessary for their formation and the complex species, MR, should dominate the system. As a consequence these data were expected to allow a quantitative estimate of the dissociation constant of the 1:1, bidentate metal complex. The expression given below was employed for the computation:

$$K_{MR} = \frac{K_{H_2R} [H_2R] [M^{2+}]}{[MR] [H^+]^2} \qquad \{1\}$$

$\{1\}$ derives from the following consideration of the ion-exchange reaction:

$$2\,\overline{H^+} + M^{2+} = \overline{M^{2+}} + 2\,H^+ \qquad \{2\}$$

By defining the standard chemical potential of the ion species in the resin (designated by a bar placed above an ion-participant in the reaction) and solution phases exactly the same, the distribution of ions can be represented by the mass-action expression given below: Brackets are used to designate the activity of each ion species.

$$\frac{[\overline{M^{2+}}]\,[H^+]^2}{[\overline{H^+}]^2\,[M^{2+}]} = K = 1 \qquad \{3\}$$

The following expressions for $[\overline{H^+}]$ and $[\overline{M^{2+}}]$ in the equilibrium system, when substituted in $\{3\}$, yield $\{1\}$.

$$[\overline{H^+}]^2 = K_{H_2R}\frac{[H_2R]}{[R^{2-}]} \qquad \{4\}$$

$$[\overline{M^{2+}}] = K_{MR}\frac{[MR]}{[R^{2-}]} \qquad \{5\}$$

In $\{1\}$ the published dissociation constant of iminodiacetic acid is used since reasonable agreement between the two has already been demonstrated (5). The experimentally determined composition of the resin at equilibrium is used directly in the equation to represent the activity of acid $[H_2R]$ and metal chelate $[MR]$ forms of the resin at equilibrium. Since ionization is negligible in the metal and acid forms these numbers are an accurate estimate of their concentration values. For this reason, also, there is little interaction between them to disturb their ideal behavior and the measured concentration ratio is not drastically different from their thermodynamic activity ratio.

The activity of the ions, $[H^+]$ and $[M^{2+}]$, respectively, in the solution phase at equilibrium, as used in $\{1\}$, are estimated with reasonable accuracy from the measured pH value and the published mean activity coefficient of the metal salt at the experimental ionic strength. In the series of experiments that are reported herein the total ionic strength was kept sufficiently low to allow direct use of the concentration of the metal in $\{1\}$ without the introduction of significant error.

The results of the equilibration experiments with the Dowex Chelating A-1 resin are presented in Table I together with the dissociation constants calculated from these data. The dissociation constants that have been published for the analogous 1:1 metal chelate complexes of iminodiacetic acid are also included in the table for comparison.

The discrepancy between the dissociation constant values observed in the 1:1 metal chelate forms of the resin and their monomer analogues increases from Cu to Ni to Co to Zn. Only for the Cu system can this discrepancy be almost completely attributed to neglect of the lowering of the activity coefficient of the metal ion at the higher ionic strength employed in the study of the monomer systems. The lack of agreement in the case of the other chelate systems may be due to differences which arise from macro-ion characteristics that have been neglected. The most likely source of complication, however, may be the occurrence of a small quantity of the multiple ligand ion species, MHR_2^-. Its existence was indicated in the potentiometric study and the formation of this species is not energetically excluded. It may very well be that its incipient formation is a complicating factor which becomes more significant when the dissociation of the 1:1 complex becomes relatively greater.

TABLE I

THE DISSOCIATION OF THE 1:1 RESIN CHELATE COMPLEXES OF Co, Ni, Cu AND Zn

Sample	$(H_2R)/(MR)$	K_{MR}	$K_{MR_{ave}}$	K_{MIDA} [6] *
		CoR		
10	19,35	$7,26.10^{-9}$	$8,5.10^{-9}$	$9,8.10^{-8}$
20	12,53	7,74		
B	11,00	8,35		
A	8,00	8,57		
40	3,26	8,32		
60	2,06	8,67		
80	1,43	10,5		

TABLE (contd.)

		NiR		
10	11,83	$1,03.10^{-9}$	$1,2.10^{-9}$	$5,0.10^{-9}$
B	10,23	1,07		
A	9,96	1,34		
20	7,49	1,38		
40	2,98	1,41		
60	1,92	1,09		
80	1,44	0,89		
		CuR		
10	4,29	$0,66.10^{-11}$	$1,2.10^{-11}$	$2,81.10^{-11}$
A	3,84	1,10		
B	3,67	0,90		
20	2,15	1,12		
40	1,60	1,35		
60	1,14	1,55		
80	0,79	1,70		
		ZnR		
10	16,0	$4,77.10^{-9}$	$5,1.10^{-9}$	$9,3.10^{-8}$
20	10,47	4,69		
B	8,09	4,84		
A	6,37	5,19		
40	3,31	5,39		
60	2,10	5,61		
80	1,34	5,48		

* $t = 30^{\circ}$, $\mu = 0,1$ (KCl)

Financial support through Contract No. At (30-1)·2269 with the U.S. Atomic Energy Commission is gratefully acknowledged.

References

1. Dowex Chelating Resin A-1, a product of the Dow Chemical Company, Midland, Michigan.
2. Post-Doctoral Research Fellow, 1961-1963.
3. CHATTERJEE, A., and MARINSKY, J.A., J.Phys.Chem., 67, 41 (1963).
4. MARINSKY, J.A., and CHATTERJEE, A., J.Phys.Chem., 67, 47 (1963).

5. KRASNER, J., and MARINSKY, J.A., J.Phys.Chem., 67, 2559 (1963).

6. CHABEREK, Jr., S., and MARTELL, A. E., J.Am.Chem.Soc., 74, 5052 (1952).

16 C 1

<u>ZUR MAGNETOCHEMIE DES SYSTEMS Fe/PORPHYRIN/PROTEIN.</u>

W. Haberditzl

Physikalisch Chemisches Institut der Humboldt-Universität Berlin, Germany.

Temperatur-Gleichgewichte zwischen "low-spin"- und "high-spin"-Zuständen von d^5-Ionen-Komplexen wurden vom Vortragenden 1956 für das System Fe/Porphyrin/Protein vorausgesagt (1, 1a). Sie sind inzwischen gefunden worden (2, 3). Die Voraussage solcher Gleichgewichte, die für relativ kleine Temperaturintervalle im Raumtemperatur-Bereich nur in ganz seltenen Fällen auftreten, beruhte auf folgendem Befund:

Bei Verbindungen des Methämoglobins mit Komplexpartnern der 6.Koordinationsstelle (von denen inzwischen eine große Zahl bekannt und untersucht ist) findet man alle Zwischenstufen zwischen Sextett- und Dublett-Fe^{3+}-Spinmagnetismus. Dieser magnetische Übergangscharakter befindet sich in Einklang mit der stufenweisen Änderung einer Anzahl anderer Eigenschaften (Spektrum, Stabilität, Bindungsenthalpie- und -entropie).

Der Multiplettabstand wird durch die Bindung beeinflußt, wodurch je nach der Bindungsfestigkeit bei Zimmertemperatur eine verschieden große Multiplettbesetzung auftritt. Im Vortrag wird vor allem auf die strukturchemischen und biochemischen Aspekte solcher Multiplettgleichgewichte eingegangen.

Aus dem experimentellen Material über Fe^{3+}-Komplexe ergibt sich bekanntlich die Einteilung in die beiden Gruppen "starkes" ($\Delta \approx 50000$ cm^{-1}) und "schwaches" ($\Delta \approx 10000$ cm^{-1}) Ligandenfeld. Die für die Spinverteilung wichtige Differenz zwischen dem d_γ-d_ϵ-Energieunterschied Δ und der Energie der Spinaustauschwechselwirkung π ist in beiden Fällen wesentlich größer als kT bei Zimmertemperatur (ca. 200 cm^{-1}).

Die Besonderheit des Porphyrin/Protein-Ligandenfeldes liegt neben der Symmetrieerniedrigung gegenüber der kubischen MX_6-Symmetrie ($O_h \longrightarrow D_{4h}$ bzw. C_{4v}) sowie dem durch den Porphyrin-π-Elektronenstrom bedingten sekundären magnetischen Moment im Zentrum des Ligandenfeldes vor allem darin, daß die Differenz Δ-π in den Bereich von ≈ 200 cm^{-1} fällt (4). Damit treten folgende experimentell überprüfbare Eigenschaften auf:

1. Geringe Variationen des Ligandenfeldes bedingen verschiedene Grundzustände $^{2S+1}\Gamma$ verschiedener Entartung und mit verschiedenem Spin (z.B. $^2T_2$, $^4T_1$, $^6A_1$).

2. Temperaturabhängigkeit der Multiplizität des Grundzustandes und damit des magnetischen Momentes bei solchen Liganden, bei denen $E(^{2S+1}\Gamma) - E(^{2S+1}\Gamma') \approx kT$.

3. Die relativ dicht beieinanderliegenden Zustände $^6A_1$, $^4T_1$ und $^2T_2$ bedingen große Werte der Konstanten D und a des tetragonalen bzw. kubischen Terms des Spin-Hamiltonoperators und damit die ungewöhnlich hohe Nullfeld-Aufspaltung (≈ 10 cm^{-1}) sowie die starke Anisotropie der g-Faktoren der high-spin-Fe/Porphyrin-Komplexe (5-7).

Alle unter 1. bis 3. aufgeführten Eigenschaften sind in den letzten Jahren experimentell nachgewiesen worden (1-6). Die relative Zuordnung der magnetisch wirksamen MO's entspricht bei allen Fe-Porphyrinkomplexen der Reihenfolge (geordnet nach zunehmender Energie) d_{xy}; d_{xz}, d_{yz}; d_{z^2}; $d_{x^2-y^2}$. Was sich jedoch durch kleine Variationen von Stärke und Symmetrie des Ligandenfeldes ändert, ist vor allem der Wert von Δ. Der Übergang vom Hämin zum Methämoglobin hat offenbar zwei Auswirkungen:

a) Durch die neue N-Bindung (Imidazol-N) wird Δ vergrößert und kommt in den Bereich, wo Δ-$\pi \approx kT$ und die Voraussetzung der oben genannten Eigenschaften 1. bis 3. gegeben ist.

b) Das Protein-Ligandenfeld ist sterisch so variabel, daß die oben unter 1. genannte Voraussetzung besonders gut erfüllt ist.

Offenbar sind die Voraussetzungen für 1. bis 3. bei den eiweißfreien Fe-Porphyrinen noch nicht vollständig vorhanden. So zeigen z.B. die Protohämin-Komplexe mit Cl$^-$, Br$^-$, J$^-$, SCN$^-$, COO$^-$ alle den theoretischen Spinmagnetismus der $d_\epsilon^3 d_\gamma^2$-Besetzung ohne Bahnanteil (8). Bei bestimmten Methämoglobinverbindungen (z.B. mit NO$_2^-$ oder OH$^-$) liegen diese Voraussetzungen jedoch vor. Damit läßt sich für die genannten Methämoglobinverbindungen Δ abschätzen. Da sich aus spektroskopischen Daten für $\pi \approx 30000$ cm^{-1} ergibt, beträgt also auch $\Delta \approx 30000$ cm^{-1}.

Bei den Methämoglobinverbindungen mit Temperaturgleichgewichten hat Δ einen Wert, für den $E(^6A_1)$ der Größenordnung nach um 10^1 cm^{-1} höher als $E(^2T_2)$ liegt. Daher nimmt das magnetische Moment dieser Verbindungen mit steigender Temperatur zu.

Die Beeinflussung von Δ läßt sich an Hand der von SPANJAARD und BERTHIER (9) sowie OHNO, TANABE und SASAKI (10) durchgeführten halb-empirischen MO-Rechnungen (HELMHOLZ-WOLFSBERG-Methode) noch präzisieren: Durch Liganden der 5. und 6. Koordinationsstelle sind neue Kombinationsmöglichkeiten von Me-d- und Liganden-Symmetrieorbitals gegeben, u.zw. (in Klammern die Ligandenorbitals, die zu gleichen irreduziblen Darstellungen von D_{4h} gehören): $d_{x^2-y^2}$ (σ B_{1g}), d_z^2 (σ A_{1g}), d_{yz}, d_{xz} (π E_{1g}). Dadurch verändert sich die Termlage der spinkritischen $d_{x^2-y^2}^*$ und d_z^{*2} (σ-Bindungen) sowie d_{yz} und d_{xz} (π-Bindungen)-orbitale.

Je nach dem Überlappungsgrad, insbesondere von B_{1g}- und vor allem A_{1g} (d_z^2)-Symmetrieorbitals wird $\Delta = E (d_{x^2-y^2}^*, d_z^{*2}) - E (d_{yz}, d_{xz}, d_{xy})$ erhöht oder erniedrigt. Dadurch können Oxydations-Reduktions-Potential (Abstand zwischen höchstem besetztem und niedrigstem unbesetztem Orbital) und Ladungsverteilung zwischen Metall- und Liganden-Orbitalen durch Proteinkonformationsänderungen (Änderung des Überlappungsgrades der Fe-Imidazol-N-Bindung) gesteuert werden.

Die einzigartig hohe biochemische Spezifität des Systems Me/Porphyrin beruht also nicht zuletzt auf der Größenordnung von Δ und dem Mechanismus der Δ-Regulation durch kleinste Konformationsänderungen der jeweiligen Proteinmatrix, in die das Hämsystem eingebettet ist. Lange bekannte Erscheinungen wie z.B. Bohreffekt und Häm-Häm-Wechselwirkung werden durch diesen Regulationsmechanismus erklärt.

Literatur

1. HABERDITZL und HAVEMANN, Chem.Technik, 8, 418 (1956).
   a) HABERDITZL und HAVEMANN, Z.phys.Chem., 209, 135 (1958).
2. GEORGE, BETTLESTONE und GRIFFITH, Haematin Enzymes, S.105, Oxford 1961.
3. SCHELER, BLANCK und GRAF, Naturwiss., 50, 500 (1963).
4. Vortrag zum Symposium "Physikalische Chemie biogener Makromoleküle", Jena, Sept.1963 (im Druck).
5. BENNETT, INGRAM, GEORGE und GRIFFITH, Nature, 176, 394 (1955).
6. GIBSON, INGRAM und SCHONLAND, Discuss.Faraday Soc., 26, 81 (1958).
7. GRIFFITH, Proc.Roy.Soc., A 235, 23 (1956).
8. HABERDITZL, MADER und HAVEMANN, Z.Phys.Chem., 218, 71 (1961).
9. SPANJAARD und BERTHIER, J.Chim.Phys., 1961, 169.
10. OHNO, TANABE und SASAKI, Theoret.Chim.Acta, 1, 378 (1963).

6 C 2

A COMPARISON BETWEEN THE REACTIVITY OF IRON IN FERRYMYOGLOBIN AND COBALT IN SOME VITAMIN B_{12} DERIVATIVES.

G.I.H. Hanania and D.H. Irvine

American University of Beirut, Beirut, Lebanon, and
University of Ibadan, Ibadan, Nigeria.

Ferrimyoglobin, a coordination compound of iron(III), and aquocobalamin, factor B, and factor V_{1A}, which are cobalt(III) coordination compounds related to vitamin B_{12}, are unsymmetrical octahedral metal chelates with markedly similar structures and reactivities (1).

In ferrimyoglobin, the iron(III) atom is firmly bonded to the polypeptide chain at an imidazole nitrogen of a histidine residue, to four pyrrolic nitrogens in the planar porphyrin ring, and to a water molecule which is easily replaced by ligands (2). It carries a net charge of +1, but its effective charge is influenced by other groups on the molecule and varies with pH (3).

In the vitamin B_{12} derivatives, the cobalt(III) atom is also loosely bonded to a water molecule, and to the four nitrogens of a corrin ring, the latter differing from a porphyrin ring essentially in having one less methine bridge (4). But, whereas aquocobalamin has the first coordination position occupied by a benzimidazole glyoxalinium nitrogen from the nucleotide, it is a cyanide group which occupies this position in factor B and in factor V_{1A} (4). Factor V_{1A} differs slightly from factor B in having a terminal carboxylic group, instead of an amino alcohol group, in one of the side chains of the corrin ring (5). Charges are also different. Thus in factor B, the metal carries a net charge of +1, which is also the net charge of the molecule as a whole; and in factor V_{1A} it is +1 for the metal and zero for the molecule.

In this paper, we report thermodynamic data which we have obtained for these compounds in three types of equilibria. The equilibria were studied spectrophotometrically in dilute aqueous solution, using sperm whale ferrymyoglobin, aquocobalamin, factor B and factor V_{1A}.

Ionization of metal-bound H_2O in ferrymyoglobin and vitamin B_{12} derivatives

pK values for the acid ionization of the metal-bound water were obtained from measurements made on buffered solutions covering a wide range of pH, ionic strength I, and temperature. Thermodynamic values of pK^O were obtained by linear extrapolation using

$$f(I) = A \cdot I^{\frac{1}{2}} \Big/ (1 + B.a.I^{\frac{1}{2}})$$

where A and B are the Debye-Hückel constants, and "a" the distance of closest approach between the compound and small ions in solution, is taken as 1,5 Å. This function is suggested by theoretical works (6-8). Enthalpies were obtained from the plots of pK against 1/T. The data at 25° are summarized in Table I.

TABLE I

IONIZATION OF THE IRON-BOUND H_2O IN FERRIMYOGLOBIN; AND OF THE COBALT-BOUND H_2O IN FACTOR B, AQUOCOBALAMIN, AND FACTOR V_{1A} AT 25°

	pK°	ΔG° kcal.mole^{-1}	ΔH° kcal.mole^{-1}	$-T\Delta S^\circ$ kcal.mole^{-1}	ΔS° (e.u.)	charge change
ferrimyoglobin	8,85	12,08	$7,10 \pm 0,5$	4,98	-17 ± 2	$+1 \rightarrow 0$
factor B	10,96	14,95	$9,25 \pm 0,5$	5,70	-19 ± 2	$+1 \rightarrow 0$
aquocobalamin	7,65	10,43	$4,60 \pm 0,5$	5,83	-19 ± 2	$+1 \rightarrow 0$
factor V_{1A}	11,04	15,06	$6,18 \pm 0,5$	8,88	-30 ± 2	$0 \rightarrow -1$

Thus for ferrimyoglobin, factor B, and aquocobalamin, ΔS° of ionization is approximately constant and equal to the entropy of ionization of a free water molecule; the effect of bonding to the metal atom on the ionization of the water molecule is almost entirely an enthalpy effect. The order of exothermicity is: aquocobalamin > ferrimyoglobin > factor B. It seems likely that between aquocobalamin and ferrimyoglobin the difference is primarily the result of different electrostatic interactions. In aquocobalamin the secondary phosphate of the nucleotide, which contributes to the net charge of the molecule, is close to but not directly bonded to the cobalt, so that the latter in effect carries a net charge greater than unity. On the other hand the metal atom has the same net positive charge of +1 in both ferrimyoglobin and factor B, so that electronic effects must contribute to the different enthalpies of ionization.

Factor B and factor V_{1A} provide an interesting comparison. The only difference between these two molecules is the replacement of an uncharged amino alcohol by a negatively charged carboxylate group. The difference in their thermodynamics of ionization thus reflects the electrostatic effect of a negative group situated at some distance from the reactive site. A comparison of the data shows that there is scarcely any change in ΔG°, but there are comparatively large changes in ΔH° and $T\Delta S^\circ$ which nearly compensate each other. A similar compensation of ΔH and $T\Delta S$ has been observed by BEETLESTONE and IRVINE (9) in their study of the ionization of a number of mammalian methaemoglobins, and they have discussed the implications of this in some detail (10).

Ionization of imino NH in the ligand imidazole of the imidazole complexes

In neutral solution, imidazole forms well defined complexes with factor B and aquocobalamin. At pH > 11 both complexes undergo reversible spectral changes, and this is attributed to the ionization of the free imino =NH group. These ionizations have been investigated by direct spectrophotometric measurements. In the case of ferrimyoglobin direct determination of this ionization is not possible as excess imidazole tends to cause denaturation of the protein, and so the data were obtained from an analysis of the pH variation of the formation constant for the imidazole complex. Table II summarises the results, and also includes data for the corresponding ionization in free imidazole. There is a marked change in the pK of ionization in going from

TABLE II

IONIZATION OF IMINO =NH IN IMIDAZOLE COMPLEXES AT 25°

	pK^O	ΔG^O kcal. mole^{-1}	ΔH^O kcal. mole^{-1}	$-T\Delta S^O$ kcal. mole^{-1}	ΔS^O (e.u.)	charge change
ferrimyoglobin	10,30	14,05	$10,5 \pm 1,0$	3,55	-12 ± 3	$+1 \longrightarrow 0$
factor B	11,38	15,50	$12,0 \pm 1,0$	3,50	-12 ± 3	$+1 \longrightarrow 0$
aquocobalamin	10,25	14,00	$10,4 \pm 1,0$	3,60	-12 ± 3	$+1 \longrightarrow 0$
imidazole	14,45	19,70	$17,8 \pm 1,8$	1,90	-6 ± 6	$0 \longrightarrow -1$

the free ligand to the ligand bound in the complex, and this is largely an enthalpy effect. This result is in accord with earlier observations on other systems (11, 12). Quantitative interpretation of the data is not possible, as there is some uncertainty in ΔH^O for the aquocobalamin system, which is based on two temperatures only. The increased exothermicity of the ferrimyoglobin system relative to factor B is however in accord with the greater conjugation of the porphyrin ring.

TABLE III

FORMATION OF IMIDAZOLE COMPLEXES AT 25°

	10^{-3} K	$-\Delta G^O$ kcal. mole^{-1}	ΔH^O kcal. mole^{-1}	$-T\Delta S^O$ kcal. mole^{-1}	ΔS^O (e.u.)
ferrimyoglobin	0,185	3,10	$-4,45 \pm 0,7$	1,35	$-4,5 \pm 2,5$
factor B	12,2	5,58	$-6,30 \pm 0,5$	0,72	$-2,4 \pm 1,7$
aquocobalamin	39,0	6,26	$-7,30 \pm 0,5$	1,04	$-3,5 \pm 1,7$

Formation of imidazole complexes

Data for the formation of the imidazole complex of ferrimyoglobin, factor B, and aquocobalamin, are recorded in Table III. The values of K are the pH independent formation constants, and they are computed from the general equation;

$$K = K_{obs} (K_{Im} + H)(K_M + H) \Big/ K_{Im} (K_{NH} + H)$$

where K_{obs} is the measured constant at a given pH, K_M is the ionization constant of the metal-bound H_2O, K_{NH} is the ionization constant of the free imino $=NH$ in the imidazole complex, and K_{Im} is the acid ionization constant of the imidazolinium $\geqslant NH^+$ in imidazole. The reactions are all exothermic and show about the same entropy change. However, the affinity of ferrimyoglobin for imidazole is much lower, and results from its more unfavourable ΔH of reaction. Several factors probably contribute to this net effect.

References

1. GEORGE, P., IRVINE, D.H., and GLAUSER, S.C., Annals of New York Academy of Sciences, 88, 393 (1960).
2. CULLIS, A.F., MUIRHEAD, H., PERUTZ, M.F., ROSSMAN, M.G., and NORTH, A.C.T., Proc.Roy.Soc., A 265, 161 (1962).
3. GEORGE, P., and HANANIA, G.I.H., Biochem.Journal, 65, 756 (1957).
4. SMITH, E., Lester, "Vitamin B$_{12}$" Methuen monograph, London, 1960.
5. FOLKERS, K. (private communication).
6. KIRKWOOD, J.G., J.Chem.Phys., 2, 351 (1934).
7. TANFORD, C., and KIRKWOOD, J.G., J.Am.Chem.Soc., 79, 5333 (1957).
8. BEETLESTONE, J.G., and IRVINE, D.H., Proc.Roy.Soc. (in press).
9. BEETLESTONE, J.G., and IRVINE, D.H., (unpublished results).
10. BEETLESTONE, J.G., and IRVINE, D.H., Proc.Roy.Soc. (in press).
11. HANANIA, G.I.H., and IRVINE, D.H., J.Chem.Soc. (London), 1962, 2745.
12. HANANIA, G.I.H., and IRVINE, D.H., J.Chem.Soc. (London), 1962, 2750.

6 C 3

STRUCTURES AND REACTIVITIES OF COMPLEXES BETWEEN BIVALENT CATIONS AND NUCLEOTIDES.

H. Brintzinger

Institute of Inorganic Chemistry, University of Basel, Switzerland.

Investigations on structural data of ternary enzyme-ATP-metal ion complexes have recently been published by COHN (1). A refinement of such structural data would be facilitated by a thorough investigation of the coordination chemistry of the respective binary nucleotide-metal ion complexes; i.e. of the question: How and to what extent do the different parts of the nucleotide ligands - N-heterocyclic and phosphate moieties - interact with a metal ion, if a binary complex is formed in aqueous solution?

Interaction of phosphate ligand groups with metal ions are especially apt to be studied by IR measurements in the 1300 - 900 cm^{-1} region in aqueous solution (2). The degenerate antisymmetric stretching vibrations of $-PO_3^{2-}$-groups split to a triplet if a metal ion is coordinated to one of the phosphate oxygens. IR spectra of Ca^{2+}-, Mg^{2+}-, Mn^{2+}-, Co^{2+}- and Ni^{2+}-complexes of model ligands such as methyl phosphate show that a $-PO_3^{2-}$-group has no great tendency to penetrate the inner coordination sphere of these metal ions; the overwhelming majority of these complex molecules exists as hydrated (outer sphere-) ion pairs, and thus resemble the respective sulfate complexes. With more highly charged phosphate ligands, such as ATP^{4-}, the hydration shell between these cations and the $-PO_3^{2-}$-group breaks down to a somewhat greater extent, so that hydrated and directly coordinated ion pairs exist in equilibrium in comparable amounts. In Zn^{2+}- and Cu^{2+}-complexes of all $-PO_3^{2-}$-ligands studied, species with direct coordination between metal ion and phosphate group predominate. With $>PO_2^-$-ligand groups similar features can be observed.

The question of metal ion-adenine interaction in ATP complexes has been studies by UV measurements in aqueous solution (3). Difference spectra of the ATP complexes versus free ligand, and comparison with adenosin complexes, indicate that equilibrium {1} is far to the left for Ca-ATP, Mg-ATP, Mn-ATP, Co-ATP, Ni-ATP and Zn-ATP, while in solution of Cu-ATP species II predominate, in which Cu^{2+} is bound to the adenine ring too.

$$\{1\}$$

Stability constants as they are usually employed - such as

$$K^M_{ML} = \frac{[ML]}{[M][L]} \quad \text{or} \quad K^M_{MLH} = \frac{[MLH]}{[M][LH]}$$

- in the case of these complexes contain concentration terms like [ML] or [MLH] which in fact are summations over the concentrations of complex isomers. These "summary equilibrium constants" are of limited value in elucidation of the ionic species present in solution. In some cases it is not even possible to ascertain meaningful numerical values for K^M_{MLH} - namely if not only hydration isomers but also different tautomeric forms are contained in [MLH] (4).

In the catalysis of the non-enzymatic hydrolysis of ATP by metal ions (5,6) the binding of the metal ion to the adenine ring seems prerequisite (7). Structural features of this reaction will be discussed on the basis of experimental data, e.g. the inversion of the sign of the rotatory dispersion spectrum during hydrolysis of Cu-ATP to Cu-ADP (7).

References

1. COHN, M., Biochemistry, 2, 623 (1963).
2. Preliminary communication BRINTZINGER, H., Biochim. Biophys. Acta, 77, 343 (1963).
3. SCHNEIDER, P.W., BRINTZINGER, H., and ERLENMEYER, H., Helv. Chim. Acta, in press.
4. Diss. SIGEL, H., Basel.
5. TETAS, M., and LOWENSTEIN, J.M., Biochemistry, 2, 350 (1963).
6. SCHNEIDER, P.W., BRINTZINGER, H., and ERLENMEYER, H., Experientia, 19, 623 (1963).
7. Diss. SCHNEIDER, P.W., Basel.

N,N-DIMETHYLACETAMIDE (DMA) COMPLEXES OF URANIUM (IV) MIXED HALIDES.

K.W. Bagnall

Chemistry Division, AERE Harwell, Didcot/Berks., U.K.

The uranium tetrachloride-DMA complex, $2 UCl_4 . 5 DMA$, and the corresponding neptunium (IV) and plutonium (IV) complexes have been known for some years (1) but, apart from the probable presence of one or more bridging DMA molecules in the compounds, no data on their constitution are available. Although they are essentially non-electrolytes in organic solvents such as DMA, methyl cyanide or nitromethane, the chloride ion can be completely replaced by nitrate using the silver salt (2), thiocyanate with the potassium salt (3) or perchlorate, again with the silver salt (4), the resulting compounds having the compositions $[U(NO_3)_4]_2(DMA)_5$, $U(SCN)_4(DMA)_4$ and $U(ClO_4)_4(DMA)_6$. The first two of these are non-electrolytes and the third is, as would be expected, a 4:1 electrolyte in nitromethane (5). The chloride ion can also be replaced by iodide on treatment of the chloride complex with a large excess of sodium iodide in non-aqueous solvents, but this does not go to completion in organic solvents at room temperature, unlike the other replacement reactions. With DMA as solvent, the product of the reaction is the pale green complex $UI_2Cl_2(DMA)_5$ even at 100^0 (5). This compound is precipitated from solution in DMA by ethyl acetate, in which the excess of sodium iodide and any uranyl halide formed by atmospheric oxidation are both soluble. The dichlorodiiodide complex is also obtained when the reaction is carried out in methyl cyanide at room temperature, the filtrate evaporated to dryness and the residue dissolved in DMA, from which the desired product is again precipitated by ethyl acetate. However, when the residue from the methyl cyanide solution is dissolved in nitromethane, more sodium chloride is precipitated and the final product is the yellowish-green complex $UI_3Cl(DMA)_5$. Both of these compounds have molar conductivities close to those expected for 2:1 electrolytes and potentiometric titration of the two compounds in methyl cyanide solution against silver perchlorate shows that half the halogen is outside the coordination sphere. Analysis of the precipitated silver halide confirmed that this was iodide only.

It seemed unlikely that uranium (IV) would exhibit a coordination number of 7 in these compounds ($[UCl_2(DMA)_5]^{2+}I_2$ and $[UICl(DMA)_5]^{2+}I_2$) in contrast to the more common coordination numbers of 6, as in the $[UCl_6]^{2-}$ ion, or 8, as in the $[U(NCS)_8]^{4-}$ ion (6) or in $U(SCN)_4(DMA)_4$. It was therefore probable that the complexes were dimeric as in the tetrachloride compound, and that the difficulty of replacing all of the chloride in the complexes might be due to the presence of a double chlorine bridge. The suggested structures of the complexes are as shown below

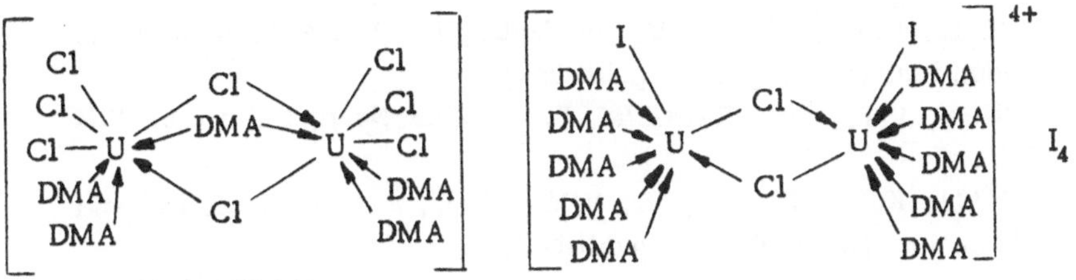

$(UCl_4)_2(DMA)_5$ $[UI_3Cl(DMA)_5]_2$

although the ligands are not necessarily arranged in the order as shown above since isomeric forms should be possible. The observed molar conductivities would also be consistent with these formulations.

Attempts have been made to establish the presence of chlorine bridges by far IR-spectroscopy, but this was inconclusive since DMA absorbs in the 260 cm^{-1} region where the metal-chlorine and chlorine bridge frequencies would be expected. Further work is in progress on the problem of establishing the presence of chlorine bridges.

Uranium and thorium tetraiodides both form complexes with DMA, of composition $UI_4(DMA)_4$ and $ThI_4(DMA)_6$, both of which are close to 2:1 electrolytes in nitromethane, but in very dilute solution ($< 10^{-4}$ M) the molar conductivity of both tends towards the value for a 4:1 electrolyte, presumably because of entry of the solvent into the coordination sphere and displacement of the iodide ion. The uranium compound can also be obtained in poor yield from the chloride complex and sodium iodide in methyl cyanide at 50° in an inert atmosphere, but the reproducibility of the preparation under these conditions is poor.

Tetraphenylarsonium salts of the hexaiodocomplexes (7) of thorium (IV) and uranium (IV), $[Ph_4As]_2MI_6$, have also been obtained by crystallisation of the tetraiodide/methyl cyanide complexes with tetraphenylarsonium iodide in methyl cyanide solution in the absence of free iodine which, when present, precipitates as Ph_4AsI_3.

References

1. BAGNALL, K.W., DEANE, A.M., MARKIN, T.L., ROBINSON, P.S., and STEWART, M.A.A., J.Chem.Soc. (London), 1961, 1611.

2. BAGNALL, K.W., ROBINSON, P.S., and STEWART, M.A.A., J.Chem.
Soc. (London), 1961, 4060.

3. BAGNALL, K.W., BROWN, D., and COLTON, R., J.Chem.Soc. (London),
(to be published).

4. BAGNALL, K.W., BROWN, D., and DEANE, A.M., J.Chem.Soc.
(London), 1962, 1655.

5. BAGNALL, K.W., BROWN, D., du PREEZ, J.G.H., and JONES, P.J.,
J.Chem.Soc. (London) (to be published).

6. MARKOV, V.P., and TRAGGEIM, E.N., Zhur.Neorg.Khim., 6, 2316
(1961).

7. BAGNALL, K.W., BROWN, D., JONES, P.J., and du PREEZ, J.G.H.,
J.Chem.Soc. (London) (to be published).

7 C 2

ON THE COMPLEX PHOSPHATES OF SOME RARE METALS.

I. Tananaev
Institute of Chemistry, Academy of Science, Moscow, UdSSR.

Metal phosphate complexes attracted considerable attention in connection with their application as protective thermostable covering materials for metals, mineral binding substances, inorganic ion exchangers and other materials. However, reactions which lead to their synthesis or thermal decomposition have been scarcely studied. In particular, there are few data about properties of acid and other complex or polymer phosphates.

Data will be presented on the process of formation, decomposition and properties of acid and mixed phosphates and in some cases pyrophosphates of lanthanum, cerium, gadolinium, yttrium, gallium, indium, zirconium, germanium, niobium and tantalum.

Some regularities are ascertained about solubilities in solutions of ortho- and pyrophosphoric acids and mutual ion exchange with alkali metals and hydrogen in the complex phosphates.

Considerations are made about structural changes by the thermal de- composition and the conversion into phosphate compounds according to data obtained from thermogravimetric measurements, X-ray analysis and IR spectrophotometry.

Recommandations are given about the utilisation of phosphates of the rare elements in particular for analytical chemistry.

KINETICS AND MECHANISM OF NUCLEOPHILIC SUBSTITUTION IN AN ACIDO-PENTAMMINE-COBALT(III) COMPLEX IN AQUEOUS SOLUTION.

D. Banerjea and T.P. Das Gupta
Department of Chemistry, University of Burdwan, Burdwan,
West Bengal, India, and Department of Chemistry,
Regional Engineering College, Durgapur-9, West Bengal, India.

A considerable amount of controversy appears to exist in the literature (1) regarding the mechanism of base hydrolysis of acido-pentammine-cobalt(III) complexes. Detailed investigations on the kinetics of base hydrolysis of several complexes of this type, $[Co(NH_3)_5X]^{2+}$, where $X = Cl^-$, Br^- or N_3^-, have been reported by earlier investigators (2-4). It is known (5) that in $[Co(NH_3)_5(S_2O_3)]^+$ the thiosulphate is so firmly bound that on treating with potassium cyanide all the ammonia are displaced leading to the formation of $[Co(CN)_5(S_2O_3)]^{4-}$.

It was found that OH^-, Cl^- and NH_3 lead to a slow displacement of $S_2O_3^{2-}$ from the complex. In the case of OH^- ion the rate has been found to obey the following relation:

$$Rate = k''_{OH^-} . [Complex][OH^-] \qquad \{1\}$$

while in the case of Cl^- and NH_3 the following rate law has been observed:

$$Rate = k'_{H_2O} . [Complex] + k''_R . [Complex][R] \qquad \{2\}$$

where $R = Cl^-$ or NH_3 and k'_{H_2O} is the pseudo first order rate constant for reaction with water (solvent) while k''_R is the usual second order rate constant for reaction with reagent, R.

The reaction with OH^- ion (base hydrolysis) has been investigated in potassium nitrate medium at constant ionic strength. For studying the reaction with ammonia, ammonium nitrate was used to suppress the dissociation of ammonia and also for maintaining ionic strength constant during the reaction. In case of chloride ion as the reagent mixtures of KCl, $BaCl_2$ and $LaCl_3$ in various requisite proportions were used in order to change the concentration of chloride ion without changing the ionic strength.

A plot of $k_{obs.}$ vs. $[OH^-]$ gave a straight line (slope of this being the value of k''_{OH^-}) passing through the origin ($k_{obs.}$ being the pseudo first order rate constant in a solution having excess of OH^- ion), showing that the rate in this case obeys the relation {1}. For reactions with Cl^- or NH_3 the plot of $k_{obs.}$ vs. [R] (where $R = Cl^-$ or NH_3) gave a straight line for each reagent. The two straight lines thus obtained have different slopes (values of k''_R) but exactly the same intercept (value of k'_{H_2O}). This suggests that the rate in these cases obeys the relation {2} given above. Values of the various rate constants determined at different temperatures are given in Table I along with the values of $\Delta H^{\neq}$ and $\Delta S^{\neq}$ evaluated as usual by making use of the Eyring equation. Effect of ionic strength on the rate of base hydrolysis has been investigated in detail and the results given in Table II show considerable retardation of the rate due to increase in ionic strength as is to be expected for a reaction of the S_N2 type between oppositely charged ions. The order of reactivity of the reagents of each class, namely $OH^- > Cl^-$ and $NH_3 > H_2O$ suggests that the rate decreases with decreasing nucleophilic character of the reagent. This also supports a mechanism of the S_N2 type. It may be mentioned that a similar mechanism has been suggested recently (6) for the base hydrolysis of the nitro-pentammine-cobalt(III) ion. A little earlier GREEN and TAUBE (7) have shown by an isotopic fractionation technique that the base hydrolysis of several labile cobalt (III) complexes of the type $[Co(NH_3)_5X]^{2+}$ (where $X = Cl^-$, Br^- or NO_3^-) proceeds by a common mechanism, possibly S_N1, while the relatively inert $[Co(NH_3)_5F]^{2+}$ reacts with OH^- by a different mechanism. It is thus likely that the inert complexes, including the one which we have investigated, react by a S_N2 process as was suggested originally by BROWN, INGOLD and NYHOLM (8), rather than by the S_N1 process suggested by BASOLO and co-workers (2).

TABLE I. RATE CONSTANTS AT DIFFERENT TEMPERATURES ($\mu = 1$).

Rate Constant	20°	25°	30°	35°	$\Delta H^{\neq}$ kcal/mole	$\Delta S^{\neq}$ e.u.
$k''_{OH^-} \cdot 10^5 [M^{-1} sec^{-1}]$	2,92	6,48	13,3	29,9	27,6	+14,4
	30°		35°	40°		
$k''_{Cl^-} \cdot 10^6 [M^{-1} sec^{-1}]$	1,12		1,62	2,4	14,7	-37,7
$k''_{NH_3} \cdot 10^7 [M^{-1} sec^{-1}]$	3,17		4,33	6,0	12	-48,5
$k'_{H_2O} \cdot 10^7 [sec^{-1}]$	2,83		4,5	7,83	18,9	-34,8

TABLE II. EFFECT OF IONIC STRENGTH ON THE RATE
OF BASE HYDROLYSIS.
(Complex 0,005 M; NaOH 0,1 M; 25°)

Ionic strength (μ)	0,1	0,2	0,35	0,55	1,0
$k_{obs} \times 10^6$ sec^{-1}	12,2	9,97	8,43	6,83	6,48

References

1. WILKINS, R.G., Quart.Revs., 16, 316 (1962), and references therein.
2. ADAMSON, A.W., and BASOLO, F., Acta Chem.Scand., 9, 1261 (1955);
   PEARSON, R.G., MEEKER, R.E., and BASOLO, F., J.Am.Chem.Soc.
   78, 709 (1956).
3. BRØNSTED, J.N., and LIVINGSTON, R., J.Am.Chem.Soc., 49, 435
   (1927).
4. LALOR, G.C., and MOELWYN-HUGHES, E.A., J.Chem.Soc. (London),
   1963, 1560.
5. RAY, P., J.Ind.Chem.Soc., 4, 325 (1927).
6. LALOR, G.C., and LONG, J., J.Chem.Soc. (London), 1963, 5620.
7. GREEN, M., and TAUBE,H., Inorg.Chem., 2, 948 (1963).
8. BROWN, D.D., INGOLD, C.K., and NYHOLM, R.S., J.Chem.Soc.
   (London), 1953, 2678.

17 C 4

UNUSUAL STABILITIES OF THE PYRUVATE-GLYCINATE MIXED COMPLEXES WITH SOME DIVALENT METAL IONS.

D.L. Leussing
Chemistry Department, Ohio State University, Columbus, Ohio, U.S.A.

Transamination between oxo- and amino-acids is one of the enzymatic reactions that can be duplicated in metal ion model systems (1-4). The reactive intermediates appear to be "mixed" complexes in which it is thought the ligands are condensed as Schiff bases. Recently, ways of applying pH-titration techniques to these systems and analyzing the data using high speed computers have been proposed (5). This earlier study (5) which concerned the Ni(II)-pyruvate-glycinate system has been extended to an examination of the Ca(II), Mn(II) and Zn(II) systems at 25°. An attempt was also made to obtain the heats and entropies of formation of the Mn(II), Ni(II) and Zn(II) complexes from additional titrimetric data at 10 and 40°.

The experimental procedures involved the measurement of pH during the titration with sodium glycinate of solutions of metal chlorides containing varying amounts of pyruvic acid and inert electrolyte. The reactions at 10 and 25° proceeded smoothly at moderate rates with the exception of Ni(II) at 10° which required 24 hours for equilibration. At 40° a slow secondary reaction was observed to occur. Evidence shows that this side reaction is the catalyzed dimerization of pyruvate (6).

TABLE I

THE LOGARITHMS OF THE FORMATION CONSTANTS

$$\beta_{MP_pG_g} = \frac{(MP_pG_g)}{(M^{++})(P^-)^p (G^-)^g}$$

Cumulative Constant	Ca 25°	Mn 10°	Mn 25°	Ni 10°	Ni 25°	Zn 10°	Zn 25°
β_{MP}	$0,8^a$	$1,2_8$	$1,2_6$	$1,4_8$	$1,1_8$	$0,9_8$	$1,2_8$
β_{MG}	$1,4^b$	2,66	2,60	5,73	5,66	4,96	4,88
β_{MG_2}	-	4,71	4,58	10,80	10,51	9,24	9,01
β_{MG_3}	-	6,0	5,7	14,4	14,0	11,9	11,0
β_{MPG}	3,9	5,51± 0,02	5,36± 0,02	8,19± 0,07	8,09± 0,02	7,61± 0,02	7,53± 0,02
β_{MPG_2}	-	7,7 ± 0,6	6,9 ± 0,3	13,57± 0,09	13,00± 0,04	11,54± 0,25	12,0 ± 0,1
$\beta_{MP_2G_2}$	-	10,25± 0,03	9,79± 0,03	15,76± 0,05	15,29± 0,02	14,54± 0,02	14,25± 0,02

a. SCHUBERT and LINDENBAUM quoted in "Stability Constants".

b. DAVIES, quoted in "Stability Constants".

These data clearly show the enhancement in the stabilities of the "mixed" species. For example at 25° the constant for the addition of pyruvate (or glycinate) to the glycinate (or pyruvate) complex is about 19-23 times greater than it is for the addition of the ligand to either the aquo Ni(II) or Zn(II) ion. Mn(II) is somewhat unusual since the prior coordination of one of the components of the mixed complex produces a 32 fold increase in the constant for the coordination of the other.

Calculation of the E_r Mn-Zn values (7) from the free energy data at 25° yields 50,7 and 53,7 kcals/mole for the MG^+ and MG_2^0 complexes and 50,6 and 53,7 kcal/mole for MPG and $MP_2G_2^{2-}$ (P^- = pyruvate, G^- = glycinate). The ligand field stabilization energies obtained using these results for E_r are 36,1, NiG^+; 38,2, NiG_2; 35,6, NiPG; and 37,6, $NiP_2G_2^{2-}$. However, optical data indicate the ligand fields in the "mixed" complexes are greater per nitrogen than those of the simple glycinate species. The maxima (in cm^{-1})

of the first Ni(II) d-d transitions lie at 9660, NiG^+; 9900, NiG_2; 9780, NiPG and 10200, $NiP_2G_2^{2-}$.

The reason for the low δH_{Ni} values using the formation constants can be seen in the heat and entropy data summarized in Table II.

TABLE II.

APPROXIMATE HEAT AND ENTROPY CONTRIBUTIONS TO COMPLEX
FORMATION AT 25^0 (kcal/mole)

Complex	Mn		Ni		Zn	
	ΔH	$T\Delta S$	ΔH	$T\Delta S$	ΔH	$T\Delta S$
MG^+	-1,4	2,2	-1,9	5,9	-2,1	4,6
MG_2	-3,3	2,9	-7,5	6,8	-5,9	6,4
MPG	-4,0	3,3	-2,6	8,5	-1,9	8,4
$MP_2G_2^{2-}$	-11,8	1,6	-11,9	9,0	-7,5	12,0

Although these figures have an uncertainty of no less than ± 1 kcal/mole owing to the limitations of the titrimetric data, semi-quantitative conclusions still can be drawn from them. In the Ni(II) and Zn(II) mixed complexes the entropy terms make important if not preponderant contributions to the free energies. The entropy contributions are much smaller with Mn(II). These discrepancies between Mn(II) and Zn(II) cause the free energy E_r values to be too large resulting in turn in low results for the δH_{Ni}. More realistic δH_{Ni} values are obtained using the heats. The slightly higher optical fields in the mixed complexes over those of the simple glycinate complexes is consistent with some degree of unsaturation at the coordinated nitrogen atoms in the former.

An inversion in the normal course of reaction heats is also observed in the mixed species, with those obtained for Mn(II) appearing to be greater than those found for the corresponding Zn(II) complexes. A possible explanation for this behavior is a contribution from π bonding between Mn(II) and the azomethine group. Further evidence for this is furnished by intense absorption bonds lying in the near ultra-violet and which overlap into the visible region imparting a yellow color to the Mn(II)-pyruvate-glycinate complexes. The Ca(II) and Zn(II) complexes on the other hand are colorless. A contribution from π bonding, also, accounts for the relatively greater enhancement noted above in the step-wise formation constants of the Mn(II) complexes.

References

1. NUNEZ, L.J., and EICHHORN, G.L., J.Am.Chem.Soc., **84**, 901 (1962).
2. MIX, H., Z.physiol.Chem., **315**, 1 (1959).
3. CHRISTENSON, H.N., J.Am.Chem.Soc., **80**, 2305 (1958).
4. DAVIS, L., RODDY, F., and METZLER, D.E., J.Am.Chem.Soc., **83**, 127 (1961).
5. LEUSSING, D.L., J.Am.Chem.Soc., **85**, 231 (1963); Talanta, **11**, 189 (1964).
6. WOLFF, L., Annalen der Chemie, **305**, 154 (1899).
7. GEORGE, P., and McCLURE, D.S., Progr.Inorg.Chem., **1**, 381 (1959).

7 C 5

TRANS-CONFIGURATION IN DIMETHYLTIN AND DIMETHYLLEAD BIS(ACETYLACETONATES).

Rokuro Okawara, Yoshikane Kawasaki and Toshio Tanaka
Department of Applied Chemistry, Osaka University,
Higashinoda Miyakojima, Osaka, Japan.

Recently, a series of diorganotin bis(8-hydroxyquinolinates) and diorganotin (2,2'-bipyridyl)dichlorides were characterized (1,2), and all of these compounds seemed to have chelating structure containing hexa-coordinated tin atom. It was not clear, however, whether they have cis- or trans-configuration concerning the methyl groups. Is is interesting to find out one of the isomers which is either cis- or trans-configuration among the similar compounds having hexa-coordinated tin atom.

As an attempt, we have prepared dimethyltin and dimethyllead bis(acetylacetonates). To methanol solution of dimethyltin dichloride and sodium methoxide (the mole ratio of 1 : 2), a slight excess of acetylacetone was added and refluxed for a few minutes. The precipitated sodium chloride was filtered, the filtrate was cooled, and dimethyltin bis(acetylacetonate) was gradually crystallized out. Recrystallization was carried out from methanol. M.p. 177-178°. Anal. Found: C 41,03; H 5,83; Sn 34,2. Calc. for $C_{12}H_{20}O_4Sn$: C 41,54; H 5,81; Sn 34,21%. To a suspension of dimethyllead dichloride and acetylaceton (the mole ratio of 1 : 2) in ethylacetate, diluted aqueous ammonia was added until the dimethyllead dichloride was dissolved. The organic layer was slowly evaporated and transparent crystalline dimethyllead bis(acetylacetonate) was obtained. This compound was also obtained by the same procedure with the case of dimethyltin bis(acetylacetonate). Recrystallization was made from methanol. M.p. 163-163.5°. Anal. Found: C 33.10; H 4.58; Pb 47.5. Calc. for $C_{12}H_{20}O_4Pb$: C 33.10; H 4.60; Pb 47.58%.

The infrared spectra of these compounds were measured as mulls in nujol and in hexachlorobutadiene, using Hitachi EPI-2G Grating Infrared Spectrophotometer for 2-25 μ and Leitz Infrared Spectrophotometer equipped with CsBr optics for 15-33 μ. The most interesting one in the spectra of dimethyltin bis(acetylacetonate) was that only strong asymmetric stretch of the Sn-C bond was observed at 556 cm^{-1} and the symmetric one disappeared, for there was no absorption down to near 430 cm^{-1}. In dimethyllead bis(acetylacetonate), similarly, only asymmetric stretch of the Pb-C bond was found at 541 cm^{-1}. Considering from the selection rule of the infrared spectra, two methyl groups in these compounds will be in the trans-position with the central tin or lead atom. The spectra of the ligand coordinated to the tin or lead atom were similar to that of copper acetylacetonate (3), in which four Cu-O bonds are considered to be equivalent. The four Sn-O or Pb-O bonds in these acetylacetonates seem, therefore, to be nearly equivalent. From these facts, the most probable structure of dimethyltin and dimethyllead bis(acetylacetonates) is considered as shown below.

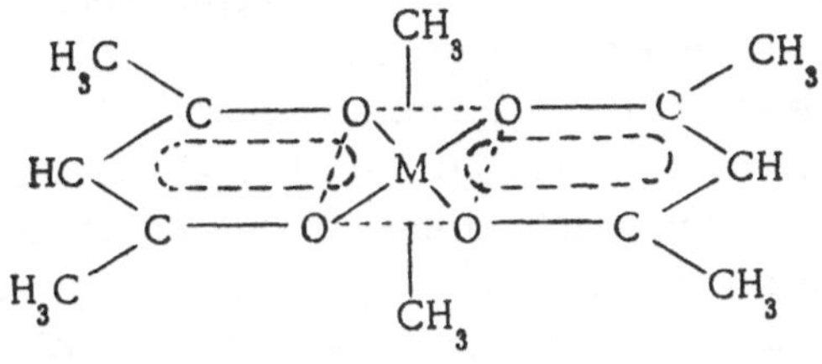

Trans-dimethylmetal bis(acetylacetonates)
M = Sn, Pb

Similary, bis(acetylacetonate) derivatives of diethyltin (m.p. 86.5-87.5°), methylchlorotin (m.p. 135-136°) and dichlorotin (m.p. 202-203°) were synthesized. Diethyltin bis(acetylacetonate) has trans-configuration, but that of others are now in study.

References

1. TANAKA, T., KOMURA, M., KAWASAKI, Y., and OKAWARA, R., J.Organometal.Chem., in press.
2. BLAKE, D., COATES, G.E., and TATE, J.M., J.Chem.Soc. (London), 1961, 756; ALLESTON, D.L., and DAVIES, A.G., J.Chem.Soc. (London), 1962, 2050.

3. NAKAMOTO, K., and MARTELL, A.E., J.Chem.Phys., 32, 588 (1960); NAKAMOTO, K., McCARTHY, P.J., and MARTELL, A.E., J.Am.Chem. Soc., 83, 1272 (1961); MIKAMI, M., NAKAGAWA, I., and SHIMANOUCHI, T., The 17th Annual Meeting of Japan Chem.Soc., 31X-21 (1964).

7 C 6

SOME RECENT RESULTS ON THE HYDROLYSIS EQUILIBRIA OF METAL IONS.

George Biedermann
Department of Inorganic Chemistry, Royal Institute of Technology, Stockholm, Sweden.

This survey summarizes the methods adopted in this laboratory for the study of the hydrolysis of the chloride complexes of the thallium(III), iron(III) and zinc(II) ions. The works discussed here represent a part of a general program initiated with the purpose to investigate the influence of chloride complex formation on hydrolysis. We hope that a systematic study of the chloride complexes will provide a basis for finding quantitative correlations between the hydrolysis mechanism and the ionic charge, size and structure.

Since many metals and metal oxides dissolve easily in hydrochloric acid, a great number of analytical and preparative methods utilize chloride solutions. Therefore, it is of considerable practical interest to ascertain the composition of the hydrolyzed species prevailing in chloride media. We intend to determine also the compositions and the solubility products of the basic salts which are the ultimate products of the hydrolysis process. The study of the solubilty equilibria may lead to the development of new methods for the separation of metal ions and it may prove to be useful to explain the analytical data obtained in sea water.

Experimental approach. As the first step the hydrolysis equilibria were studied in solutions where the $[Cl^-]_{total}$ was maintained at a high level by adding NaCl. By measuring the equilibrium concentration of hydrogen ions and when possible also that of the unhydrolyzed metal ions in such solutions of exactly known analytical composition, the number of OH groups, q, and metal ions, p, present in the hydrolysis products could be determined. No information can, however, be obtained on the number of chloride ions, r, bound to the hydrolyzed species because the $[Cl^-]$ was kept essentially constant.

To find also the prevailing values of r solutions containing a high concentration of an indifferent salt, e.g. $NaClO_4$, have to be investigated. In
order to avoid serious variations of the activity factors the total anion molarity
is to be kept constant and the sum of the equivalent concentrations of the
reacting species should not exceed 10% of the total anion molarity.

An effective approach for the ascertainment of the composition of the
hydrolysis products is to study, e.g. by measuring with emf methods the
concentration of the free metal ions, $[Me^{z+}]$, the chloride complex formation
at various levels of $[H^+]$. In each series of emf measurements $[H^+]$ and $[Me^{z+}]$
total are kept constant while the $[Cl^-]_{total}$ is successively increased. On the
basis of such measurements apparent formation constants, β_r', can be evaluated
for each species containing r chloride ions. These constants are defined by the
equation

$$\beta_r' = \beta_r (1 + \Sigma \gamma_{r,p} h^{-p}) \qquad \{1\}$$

where $h = [H^+]$, $\beta_r = [MeCl_r^{(z-r)+}][Me^{z+}]^{-1} [Cl^-]^{-r}$ and $\gamma_{r,p} = $
$[MeCl_r(OH)_p] h^p [MeCl_r]^{-1}$. Examination of the dependence on h of the β_r'
functions yields the prevailing values of p.

Results. All the equilibria were studied at $25,0\pm0,1^0$.

1. The hydrolysis of the iron(III) chloride complexes was studied (1) by
measuring h and the concentration of the unhydrolyzed iron(III) species $[Fe^{3+}]$,
in a series of iron(III) chloride solutions of varying acidity. The investigation
covers the [Fe(III)] range 10^{-3} to $2,5.10^{-2}$ M. The test solutions were made to
contain 0,5 M $[Cl^-]_{total}$ by adding NaCl. The value of h was determined by
means of a glass half-cell and the value of $[Fe^{3+}]$ by measuring the emf of
cells containing the Fe^{3+}-Fe^{2+} couple.

The hydrolysis proved to be a slow process leading to the formation of a
precipitate. After three weeks $\log [Fe^{3+}]$ and $\log h$ were found to attain values
which did not show further systematic change for several months. The $\log h$
and $\log [Fe^{3+}]$ data obtained in aged solutions could be explained by assuming
the equilibrium

$$Fe^{3+} + 2,7 H_2O + 0,3 Cl^- \rightleftharpoons Fe(OH)_{2,7}Cl_{0,3}(s) + 2,7 H^+$$
$$\{2\}$$
$$\log K = \log \{[Fe^{3+}]^{-1} h^{2,7}\} = -3,05 \pm 0,10$$

The [Fe(III)] of the aged solutions, from which the precipitate was
removed by centrifugation, was determined analytically (2), and in each

case it proved to agree to within $\pm 2\%$ with the value of $[Fe^{3+}]$. Thus the concentration of the soluble hydrolyzed species is very low. All the precipitates have shown identical X-ray diagrams which were found to be very similar to that ascribed by previous authors (3, 4) to "β-FeOOH".

2. The hydrolysis of the zinc chloride complexes was studied (5) by measuring with a glass half-cell h of zinc chloride solutions which were made to contain 5 M $[Cl^-]_{total}$ by adding NaCl. The $[Zn(II)]$ of the test solutions ranged from 0.01 to 0.2 M.

Treatment of the data according to the self-medium method (6) indicated that the main hydrolysis products contain one OH group. In order to find the number of zinc atoms present in the hydrolyzed species the dependence of $[Zn(II)]$ of the monoligandic constants, $K_1 = \sum\limits_{q} [Zn(II)]^q \beta_{1,q}$, was examined. It could be concluded by using this method that hydrolysis products containing one as well as two zinc atoms are present and the following formation constants were estimated

$$Zn^{2+} + H_2O \rightleftharpoons ZnOH^+ + H^+$$

$$\log \beta_{1,1} = \log \{[ZnOH^+] \, h \, [Zn^{2+}]^{-1}\} = -10.2 \pm 0.2$$

$$\{3\}$$

$$2\,Zn^{2+} + H_2O \rightleftharpoons Zn_2OH^{3+} + H^+$$

$$\log \beta_{1,2} = \log \{[Zn_2OH^{3+}] \, h \, [Zn^{2+}]^{-2}\} = -8.2 \pm 0.1$$

The emf of the cell

$$- Zn\text{-}Hg\,|\,B \text{ M } Zn(II),\ 0.001 \text{ M } H^+,\ (5.000 - 2B)\text{ M } Na^+,\ 5.000 \text{ m } Cl^-\,|\,RE +$$

where RE denotes a reference half-cell, was measured as a function of B in the range 0.01 to 0.25 M. The data could be interpreted by assuming the equilibrium

$$ZnCl_3^- + Cl^- \rightleftharpoons ZnCl_4^{2-} \qquad \log K_4 = -0.3 \pm 0.1 \quad \{4\}$$

Thus the symbol Zn^{2+} used in $\{3\}$ represents presumably the sum $[ZnCl_3^-] + [ZnCl_4^{2-}]$.

3. The hydrolysis equilibria of the thallium(III) chloride complexes was investigated (7) by measuring with the Tl^{3+} - Tl^+ half-cell the $[Tl^{3+}]$ of solutions of the general composition: B M Tl(III), H M H^+, (3.000 - 3B -H)

M Na$^+$, X M Cl$^-$, (3 - X) M ClO$_4^-$. The value of B ranged from 10^{-4} to 2.10^{-3} M. In each series of measurements B and H were kept constant while X was stepwise increased at least to 50 B. By applying the method described in the section "Experimental approach" we have found that the hydrolysis of the species TlCl^{2+}, TlCl$_3$ and TlCl$_4^-$ are negligible in the acidity range studied and that the hydrolysis of the TlCl^{2+} ion can be explained by assuming the equilibrium

$$TlCl^{2+} + H_2O \rightleftharpoons TlClOH^+ + H^+$$

$$\log K_a = \log \{h\,[TlClOH^+][TlCl^{2+}]^{-1}\} = -1,9 \pm 0,1 \quad \{5\}$$

References

1. BIEDERMANN, G., and CHOW, T.J., to be published 1964.
2. BERECKI-BIEDERMANN, C., to be published 1964.
3. WEISER, H.B., and MILLIGAN, W.O., J.Am.Chem.Soc., 57, 238 (1935).
4. MACKAY, A.L., Mineral.Mag., 32, 545 (1960).
5. BIEDERMANN, G., and HIETANEN, S., to be published 1964.
6. HIETANEN, S., and SILLÉN, L.G., Acta Chem.Scand., 13, 533 (1959).
7. BIEDERMANN, G., and SPIRO, T., to be published 1964.

17 C 7

CHLORO-COMPLEXES OF PENTAVALENT NIOBIUM, TANTALUM, PROTACTINIUM, TUNGSTEN AND URANIUM.

D. Brown
Chemistry Division, AERE Harwell, Didcot/Berks., U.K.

Although oxychloro-complexes of niobium (V) (1), tungsten (V) (2) and uranium (V) (3) have been known for several years only niobium (V) and tantalum (V) were known to form hexachloro-complexes (e.g. 4,5) when this study was started and even these had only been prepared by heating the appropriate chlorides together in a sealed system. As this work was being completed however ADAMS et al.(6) reported the preparation of tetraethylammonium hexachloro-complexes of niobium (V), tantalum (V) and tungsten (V) from solutions of the chlorides in thionyl chloride, one of the solvents in which we had conducted our own investigations. Using thionyl chloride, thionyl chloride-iodine mono-chloride mixtures or methyl cyanide as solvent we have prepared (7,8) hexa-chloro-complexes of the type M(I)M(V)Cl$_6$ (M(I) = Cs$^+$, NH$_4^+$, [NH$_2$(CH$_3$)$_2$]$^+$, [N(CH$_3$)$_4$]$^+$, [N(C$_2$H$_5$)$_4$]$^+$, [Ph$_4$As]$^+$; M(V) = Nb, Ta, Pa, W and U) and the

tetramethyloctachloro-complexes $[N(CH_3)_4]_3PaCl_8$ and $[N(CH_3)_4]_3UCl_8$. In
addition, evidence for the heptachlorouranate(V)anion, $[UCl_7]^{2-}$, was obtained
by conductimetric titration in thionyl chloride but no solid complex was isolated. The presence of a peak at 3275 cm^{-1} in the absorption spectra of thionyl
chloride solutions of uranium pentachloride and tetramethylammonium hexa-
chlorouranate(V), and its absence in the octachlorouranate(V) spectra, indicates
that the hexachlorouranate(V) anion is solvated, probably with the uranium again
becoming octacovalent. Attempts to prepare octachloro-complexes of
niobium (V), tantalum (V) and tungsten (V) and the tetraphenylarsonium octachlorouranate (V) were unsuccessful.

With a common cation all the elements give isostructural complexes of
low symmetry and the X-ray powder data have not yet been successfully analysed. Metal-chlorine vibrational frequencies (ν_3) in the far infra-red occur
between 336 and 302 cm^{-1} for the hexachloro-complexes and a shift to
290 cm^{-1} is observed, as may be expected, for the octachloroprotactinate (V).
The magnetic properties of the uranium (V) and tungsten (V) compounds have
been examined; Curie-Weiss dependence was observed for the hexachloro-
uranates (V) from 310° K to temperatures varying between 140 and 205°K;
below this temperature marked deviations occur. The moments at the higher
temperatures range from 1.62 to 2.14 B.M. but large values of Θ, the Weiss
constant, were observed in all instances. In contrast, the octachlorouranate (V)
complex showed temperature-independent paramagnetism ($\chi_m = 1100.10^{-6}$ c.g.s.
units) between 310 and 250°K and the hexachlorotungstates (V) were antiferro-
magnetic, behaviour similar to that observed (9) for the hexafluorotungstates (V).

A simple preparative procedure for the group 5B pentachlorides used for
the above work was also developed.

Niobium hydroxide reacts vigorously with thionyl chloride at room
temperature, dissolving almost completely within 24 hours; the yield of
niobium pentachloride, after vacuum evaporation of the solution and
sublimation of the resulting solid, is greater than 95%. In addition to the
obvious preparative advantages the pentachloride obtained in this manner is
completely free from niobium oxytrichloride, $NbOCl_3$, which is almost always
formed to some extent in other preparative procedures. Tantalum pentachloride
may be prepared in greater than 60% yield in a similar fashion. However, the
solid obtained from thionyl chloride solutions of protactinium (V) is the complex
$SO^{2+}(PaCl_6)_2^-$. The IR-spectrum of this complex has a single peak at 1406 cm^{-1}
probably due to the sulphur-oxygen vibration and consistent with the above
formulation. Vacuum decomposition above 150° gives some protactinium
pentachloride and an unidentified black residue.

References

1. WEINLAND and STORZ, Ber., **39**, 3057 (1906); Z. anorg. allg. Chem., **54**, 223 (1907).
2. COLLENBERG, Z. anorg. allg. Chem., **102**, 259 (1918); COLLENBERG and GUTHE, Z. anorg. allg. Chem., **134**, 317 (1924).
3. BRADLEY, CHAKRAVARTI and CHATTERJEE, J. Inorg. Nucl. Chem., **3**, 367 (1957).
4. MOROZOV and KORSHUNOV, Zhur. Neorg. Khim., **1**, 145 (1956).
5. MOROZOV, KORSHUNOV and SIMONICH, Zhur. Neorg. Khim., **1**, 1646 (1956).
6. ADAMS, CHATT, DAVIDSON and GERRATT, J. Chem. Soc. (London), **1963**, 2189.
7. BAGNALL and BROWN, J. Chem. Soc. (London), in press.
8. BAGNALL, BROWN and du PREEZ, J. Chem. Soc. (London), in press.
9. HARGREAVES and PEACOCK, J. Chem. Soc. (London), **1958**, 3776.

17 C 8

POSSIBLE KINETIC CONSEQUENCES OF ENTROPY CORRESPONDENCE.

James W. Cobble

Department of Chemistry, Purdue University, Lafayette, Indiana, U.S.A.

During the last few years, CRISS and COBBLE (1) have developed a principle of entropy correspondence for aqueous ions which predicts ionic heat capacities over wide ranges of temperature (0 - 300°). The entropy relations can be summarized as:

$$\bar{S}^{o\;abs.}_{(t_2)} = a_{(t_2)} + b_{(t_2)} \bar{S}^{o\;abs.}_{(25)} \qquad \{1\}$$

where $a_{(t_2)}$ and $b_{(t_2)}$ are constants depending only upon the temperature and the general class of ion (i.e., cation, anion, complex ion). The entropy symbols, $\bar{S}^{o\;abs.}$, refer to the standard partial molal entropy on the "absolute" scale. On this scale, the entropy of $H^+(aq)$ at 25° is -5.0 cal. mole^{-1}deg.$^{-1}$.

From a standard definition of heat capacity in terms of entropy, it follows that:

$$\overline{C}_p^{\,o}\Big]_{25}^{t_2} = \alpha_{(t_2)} + \beta_{(t_2)} \overline{S}^{\,o}_{(25)}^{\,abs.} \qquad \{2\}$$

where $\alpha_{(t_2)}$ and $\beta_{(t_2)}$ are constants calculated from $a_{(t_2)}$ and $b_{(t_2)}$ given in $\{1\}$. The symbol $\overline{C}_p^{\,o}\Big]_{25}^{t_2}$ represents the average value of the ionic heat capacity between 25° and the temperature t_2. Values of the constants α and β and the heat capacities of $H^+(aq)$ as a function of the temperature, t_2, are tabulated in Table I. Such constants are based upon all known available thermodynamic data for ionic solutions.

TABLE I

HEAT CAPACITY [a] CONSTANTS FOR $\{2\}$

Temp. (°C)	Cations		Anions		Oxy-anions		Acid Oxy-anions		$C_p\big]_{25}^t H^+(aq)$
	α	β	α	β	α	β	α	β	
0 [b]	98	-1,68	-103	0,07	-325	4,46	-283	6,05	63
60	35	-0,41	-46	-0,28	-127	1,96	-122	3,44	23
100	46	-0,55	-58	0,00	-145	2,24	-135	4,24	31
150	46	-0,59	-61	-0,03	-146	2,27	-144	4,43	33
200	50	-0,63	(-65)	-0,04	-159	2,53	-153	4,72	35

a) In cal. mole^{-1} deg.$^{-1}$.

b) The values at 0° have been evaluated by Dr. R. E. MITCHELL of these laboratories; see R. E. MITCHELL, Ph. D. thesis, Purdue University (1964).

It is of interest to question whether the entropy correspondence relationships can be applied to chemical kinetics in the estimating of heat capacities for ionic activated complexes. In general these species will differ from ordinary ions in their geometries, and possibly by the failure of the hydration spheres to properly orient around the newly formed species before their passage over the potential barrier. However, POWELL has already demonstrated (2) that the entropies of complex ionic activated complexes can be estimated reasonably well by the same methods used for ordinary stable complex ions. This is probably due to the fact that most of the entropy change involved in the association of ions is due to electrical Born effects.

It follows that ionic heat capacities may also be reasonably approximated by the heat capacities of similar stable species. The present analysis, based upon this assumption, would indicate that very large values can be expected for some heat capacities of activation. The very fact that large $\Delta S^{\ddagger}$ values have been observed for many such reactions in the past should have been reason enough to suspect that similarly large $\Delta C_p^{\ddagger}$ effects should exist. The effects of charge, size and dielectric constant affect both the ionic entropy and heat capacity, and large and negative values of $\overline{C}_{p_2}^{o}$ for ionic solutes are not uncommon (3). Further, a number of investigations have actually determined large values of $\Delta C_p^{\ddagger}$ for reactions involving polar or charge-separated activated complexes (4-6).

TABLE II

ESTIMATED VALUES OF $\Delta C_p^{\ddagger}$ FOR REPRESENTATIVE REACTIONS

Rate Determining Reaction	Reference	$\Delta S^{\ddagger}$ (cal. mole⁻¹ deg.⁻¹)	$S^{\ddagger\,a)}$ (abs. scale)	$\Delta C_p^{\ddagger}\,]_{25}^{100}$ (cal. mole⁻¹ deg.⁻¹)
$OCl^- + OCl^- \rightarrow [OClOCl]^{2-\ddagger}$	7	-20	7	98
$[S\cdot O_3]^{2-} + [S_2O_3]^{2-} \rightarrow [O_3S\cdot O_2S_2]^{4-\ddagger}$	8	-31	-3	75
$[Fe\cdot(CN)_6]^{3-} + [Fe(CN)_6]^{3-} \rightarrow [Fe\cdot(CN)_6Fe(CN)_6]^{7-\ddagger}$	9	-41	74	52
$HO_2^- + Co^{3+} \rightarrow [HO_2Co]^{2+\,\ddagger}$	9	55	-18	small, neg.
$[Pu\cdot(OH)]^{3+} + Pu^{3+} \rightarrow [Pu\cdot(OH)Pu]^{6+\,\ddagger}$	10	-32	-126	-29
$[Fe\cdot(OH)]^{2+} + Fe^{2+} \rightarrow [Fe\cdot(OH)Fe]^{3+\,\ddagger}$	12	-18	-88	-35
$Cr^{2+} + [(H_2O)Co(NH_3)_5]^{3+} \rightarrow [Cr(H_2O)Co(NH_3)_5]^{5+\,\ddagger}$	13	-48	-23	~-40
$H^+ + VO_2^+ + Fe^{2+} \rightarrow [HVO_2Fe]^{4+\,\ddagger}$	11	-37	-90	-53
$V^{3+} + VO_2^+ + H^+ \rightarrow [(VO_2)_2H]^{5+\,\ddagger}$	14	-24	-104	-70

a) Values recorded are at 25° and are based upon the "absolute" scale where the entropy of $H^+(aq)$ is -5.0 cal. mole⁻¹ deg⁻¹.

Table II contains a summary of estimated values of $\Delta C_p^{\ddagger}$ for a variety of solution reactions for which $\Delta S^{\ddagger}$ data are available. Values of the average heat capacities between 25 and 100° were calculated for the reactants and estimated for the activated complexes from {2} and the constants in Table I. Such calculations may be subject to large errors, but nevertheless should prove useful in a number of ways. The constants for coordinated anions were all approximated from those known for oxy-anions. It should be noted that the sign of $\Delta C_p^{\ddagger}$ is not necessarily the same as the sign of $\Delta S^{\ddagger}$.

In the usual statistical thermodynamic treatment of reaction rates for bimolecular reactions, the specific rate constant for the rate determining step is approximated by:

$$k^0 \cong \left(\frac{kT}{h}\right) e^{-\Delta H^{\ddagger}/RT} \cdot e^{-\Delta S^{\ddagger}/R} \qquad \{3\}$$

where $\left(\dfrac{kT}{h}\right)$ has a value of $6 \cdot 10^{12}$ sec^{-1} at 25^{0}. Most of the present values for ΔS solution reactions result from such an analysis of kinetic data over limited temperature ranges. If the functions $\Delta H^{\ddagger}$ and $\Delta S^{\ddagger}$ have any thermodynamic significance at all, then they must be temperature dependent. $\{3\}$ can be expanded, using ordinary thermodynamic methods, to give $\{4\}$ which predicts the extended temperature behavior of the rate constant:

$$\ln \frac{k_2^0}{k_1^0} = \ln \frac{T_2}{T_1} - \frac{\Delta H_1^{\ddagger}}{R} \left(\frac{1}{T_2} - \frac{1}{T_1}\right) - \frac{\Delta C_p^{\ddagger} \rbrack_1^2}{R} \left(\frac{\Delta T}{T_2} - \ln \frac{T_2}{T_1}\right)$$

$$\{4\}$$

The value $\Delta H_1^{\ddagger}$ will differ significantly from the average value $\overline{\Delta H}^{\ddagger}$ given in $\{3\}$, if $\Delta C_p^{\ddagger} \rbrack_1^2$ is significant between the temperatures T_1 and T_2. Previous authors have almost uniformly assumed that the product

$$\frac{\Delta C_p^{\ddagger} \rbrack_1^2}{R} \left(\frac{\Delta T}{T_2} - \ln \frac{T_2}{T_1}\right)$$

is small enough to be ignored. The numbers given in Table II indicate that this may not be a valid assumption, particularly for reactions having modest $\Delta H^{\ddagger}$ values. Further, the observation that a plot of $\log \dfrac{k^0}{T}$ vs. $\dfrac{1}{T}$ appears to be linear over a limited temperature range is not a sufficient criterion for assuming that the $\Delta C_p^{\ddagger}$ term is negligible. This method is perhaps the poorest one available for attempting to obtain information on $\Delta C_p^{\ddagger}$, even though slight changes in the slope of such a plot can cause substantial errors in $\Delta H^{\ddagger}$. A far better procedure is to estimate values of $\Delta C_p^{\ddagger}$ by the method previously outlined, and to use a rearranged form of $\{4\}$:

$$\ln k_2^o - \ln \frac{T_2}{T_1} + \frac{\Delta C_p^{\ddagger} \big]_1^2}{R} \left(\frac{\Delta T}{T_2} - \ln \frac{T_2}{T_1} \right) = \ln k_1^o - \frac{\Delta H_1^{\ddagger}}{R} \left(\frac{1}{T_2} - \frac{1}{T_1} \right)$$

{5}

A plot of the left hand side of {5} against $\frac{1}{T}$ will give much more accurate values of $\Delta H_1^{\ddagger}$ (or $\Delta H_t^{\ddagger}$ at any temperature) than will assuming the $\Delta C_p^{\ddagger}$ term to be zero.

Kinetic reactions can reasonably be divided into two classes: those having positive values of $\Delta C_p^{\ddagger}$, with the others having negative values at 25° (or any other convenient reference temperature). The latter class are particularly interesting. At a temperature such that

$$-\Delta H_1^{\ddagger} = \Delta C_p^{\ddagger} \big]_1^2 (t_2 - t_1) \qquad \{6\}$$

$\Delta H_2^{\ddagger}$ becomes zero. For temperatures above this inversion point, $\Delta H_2^{\ddagger}$ will become negative, and the rate determining step will proceed more slowly as the temperature is increased further. This retrograde temperature region depends upon a very literal interpretation of the statistical thermodynamic rate equation {3}. Whether the conditions and assumptions involved in the derivation of {3} can accomodate negative activation energies is a mute question. However, there are a number of reactions where this interesting point can be tested between 25 and 100°, particularly for those having already small activation energies at room temperature. (Unless $\Delta S^{\ddagger}$ is also rather negative, some reactions will proceed too rapidly to measure over the whole temperature range of interest.) It is perhaps reasonable to enquire whether examples of kinetic systems having negative activation energies at room temperature (for the rate determining step) have been mistakenly interpreted as being due to some more complex reason than is actually required.

This research was supported by the National Science Foundation.

References

1. CRISS, C.M., and COBBLE, J.W., to be published; see CRISS, C.M., Ph.D. thesis, Purdue University, Lafayette, Indiana, U.S.A. (1961).
2. POWELL, R.E., J.Phys.Chem., **58**, 528 (1954).

3. PITZER, K.S., and BREWER, L., Thermodynamics, 2nd ed., McGraw-Hill Book Co., New York, N.Y., p.400 (1961).

4. PINSENT, B.R.W., PEARSON, L., and ROUGHTON, F.J.W., Trans. Faraday Soc., 52, 1512 (1956).

5. APPELMAN, E., ANBAR, M., and TAUBE, H., J.Phys.Chem., 63, 126 (1959).

6. ROBERTSON, R.E., HEPPOLETTE, R.L., and SCOTT, J.M.W., Can.J. Chem., 37, 803 (1959).

7. FOERSTER-DRESDEN, F., and DOLCH, P., Z.Elektrochem., 23, 137 (1917).

8. AMES, D.P., and WILLARD, J.E., J.Am.Chem.Soc., 73, 164 (1951).

9. As summarized by BASOLO, F., and PEARSON, R.G., "Mechanisms of Inorganic Reactions", John Wiley and Sons, New York, N.Y., p.318 (1958).

10. KEENAN, T.K., J.Phys.Chem., 61, 1117 (1957).

11. DAUGHERTY, N.A., and NEWTON, T.W., J.Phys.Chem., 67, 1090 (1963).

12. SILVERMAN, J., and DODSON, R.W., J.Phys.Chem., 56, 846 (1952).

13. TAUBE, H., Can.J.Chem., 37, 129 (1959).

14. DAUGHERTY, N.A., and NEWTON, T.W., J.Phys.Chem., 68, 612 (1964).

AUTHOR INDEX